THE STUDY OF URBAN GEOGRAPHY

Fourth Edition

THE STUDY OF URBAN GEOGRAPHY

Fourth Edition

HAROLD CARTER

Formerly Gregynog Professor of Human Geography,
University of Wales, Aberystwyth

ARNOLD

A member of the Hodder Headline Group
LONDON • NEW YORK • SYDNEY • AUCKLAND

Fourth edition published in Great Britain 1995 by
Arnold, a member of the Hodder Headline Group
338 Euston Road, London NW1 3BH

Co-published in the United States of America by
Oxford University Press Inc., 198 Madison Avenue,
New York, NY 10016

British Library Cataloguing in Publication Data
A catalogue recored for this book is available from the British
Library

Library of Congress Cataloging-in-Publication Data
A catalog record for this book is available from the Library of
Congress

ISBN 0 7131 6589 8

3 4 5 6 7 8 9 10   98 99 00

Composition by Scribe Design, Gillingham, Kent
Printed and bound in Great Britain by The Bath Press, Bath

To Mari

CONTENTS

PREFACE TO THE CURRENT EDITION

In the preface to the third edition it can be seen that a choice was considered between a revision and a complete rewriting. A revision was chosen. That, however, was over a decade ago and for the fourth edition the choice was obviously a rewriting. That has been undertaken, although still preserving much of the basic material of the original book. As the first chapter sets out, both the nature of the city, the object studied, and the character of urban geography have changed. This new edition is an attempt to adapt to a new environment. But it is also written in the belief that however disparate the strands which make it up, there is still a common core, an urban geography, which represents a coherent view of the city as a feature in space and with its own internal structure.

Much of the original book was written while I was a visiting professor at the University of Cincinnati. I owe a great debt to the stimulus of that year and of subsequent contacts which I have always valued and which I am happy to acknowledge.

At Aberystwyth I have been greatly indebted to all my former colleagues and to generations of research students, but especially to Dr C. R. Lewis with whom I shared the teaching of urban geography. It is also proper that I should acknowledge the constant willing help I have had from the staff of the Hugh Owen Library of the University of Wales, Aberystwyth.

I always remember a phrase from a preface to a book by the late Dudley Stamp. He wrote, 'I have necessarily put my sickle into other men's corn'. That is equally true of this book, for inevitably a work which attempts to summarize a systematic area of study must draw heavily on others' research. There are specific acknowledgements in the text but a much wider recognition must be made here.

Finally, I would again pay tribute to the help and encouragement I have received from my wife, Mari, to whom the previous editions have been dedicated. Over the years this book has become more and more a joint effort and a recognition of that is only proper.

Harold Carter
Aberystwyth, Autumn 1994

PREFACE TO THE FIRST EDITION

This book is intended to provide an outline of urban geography for those undergraduates at universities or students in further education who wish to specialize in urban geography to an extent greater than is usually provided by general courses on human geography.

Changes in the concepts of geographers, and in academic work generally, make it clear that such a work cannot be concerned with discrete blocks of subject matter, for the general problem under review is urbanism and the approach geographical. But this opens up a vast field of enquiry and within that field any finite study of this nature must be selective and present a personal view of what an undergraduate can and should be expected to consider. This view is my own and I am solely responsible for it. What I hope this volume does is provide a continuation from the elementary and general texts in human geography and a lead towards the more specialized studies which the graduate student would wish to follow. Perhaps it is worth adding that in the title the emphasis should be on *study* and not on the definite article!

This view of the study of urban geography has been built up by experience in teaching at Aberystwyth and by contact with fellow urban geographers in Britain and America. It is impossible to acknowledge all those who have in some way helped but I would particularly wish to mension Professor Robert McNee and Howard Stafford (Jr) of the Department of Geography of the University of Cincinnati. Much of the material was discussed with them and with the graduate students of that Department in 1967–68. I would also like to put on record my appreciation of a Fellowship from the US National Science Foundation which enabled me to visit the United States and work for a year at Cincinnati. It was during that year that much of the basis of this book was established. I would also like to acknowledge the friendly and helpful comments of two other members of that Department, Dr K. B. Ryan and Dr Peter Halvorson, now of the Department of Geography at the University of Connecticut (Storrs).

I am also greatly indebted to my own graduate students who, from time to time, have critically examined much of the material and I would especially like to thank Dr W. K. D. Davies now at the University of Calgary and Dr C. R. Lewis of the Department at Aberystwyth. Finally Dr Ronald Jones of Queen Mary College read the manuscript and made many valuable suggestions both as to content and presentation.

The maps for the book were drawn by Mr Morlais Hughes, Mr M. Gelly Jones and Mr E. James, and I am grateful for their cartographic expertise. I would also like to acknowledge the ready help of Mrs Mair Jenkins in preparing the final version.

Lastly I would like to acknowledge the assistance of my wife in both a general and a particular sense. Not only has she accompanied me on many 'field excursions', and not always to the most attractive parts of towns and cities, but she also read and typed the original manuscript. It is appropriate that the volume be dedicated to her.

PREFACE TO THE SECOND EDITION

In this second edition an attempt has been made to extend two themes which were introduced but not developed in the first edition. These are the behavioural approach, both in relation to central place concepts and the choice of residential location, and the citizen's image of perception of city space. Necessarily much is omitted or dealt with summarily, for the constraints of time, length and costs have all to be considered. Even so the conclusion has been completely rewritten in an attempt to relate the book to some broader issues. In addition a number of minor changes have been made and the references and recommended reading lists have been brought up to date.

I have had the advantage of discussing much of this material with Professor K. Corey of the Department of Community Planning at the University of Cincinnati and Professor W. K. D. Davies of the Department of

Geography at Calgary, and I am grateful for their advice. My particular thanks are due to my colleague Dr C. R. Lewis, who shares with me the teaching of urban geography at Aberystwyth. He has helped both through constant discussion and by reading the additional matter. As usual, however, the final responsibility rests with the author.

PREFACE TO THE THIRD EDITION

The choice in the preparation of this edition lay between either a substantial revision and updating but preserving much of the original, or a complete rewriting of the whole book. The first of these alternatives was chosen since it seemed sensible to build on a successfully established foundation. There have been considerable modifications. Two chapters of the second edition have been deleted, although some of their content has been retained and included in other chapters. Three new chapters have been added, on the housing market, on spatial inequalities in the city, and on the city in the developing world.

Both in the Introduction and in the last chapter some comment is made on the proposition that the urban geography here presented is outmoded since it deals with some observed characteristics of urbanism whilst not establishing in the first instance a view of the controlling socio-political and economic system from which those characteristics are ultimately derived. In short, the view that the starting point should be the general operation of the advanced capitalist system is not adopted and the 'facts' of urbanism somewhat arbitrarily constitute the point of departure. The basis, therefore, is an attempt to present in some logical form and order the work that has been published by urban geographers; it is a responsive rather than an innovative basis. The justification is that for the beginning student such an approach is the most appropriate, it starts from the known environment rather than a 'first cause'.

All social science can be regarded as value orientated. The values on which this book is based are, it is hoped, made clear. Certainly no assumption of the presentation of an 'absolute' set of 'facts' is made.

Once again I must acknowledge the great help I have had by discussing much of the specific material, as well as general principles, with my colleagues in the Department of Geography of the University College of Wales at Aberystwyth. I must again especially note the advice of Dr C. R. Lewis. I am also greatly indebted to Miss Linda James who has been responsible for typing not only the new chapters but also the many minor insertions and changes which have been made for this edition.

ACKNOWLEDGEMENTS

The author and publisher wish to thank the following for permission to reprint or modify copyright material: Academic Press for a figure from 'Urban and rural settlement in the Frankish Kingdom' by K. Böhner in *European towns. Their archaeology and early history* edited by M.W. Barley, 1977 (Figure 19.5); Aldine Inc., Chicago, for two figures from 'How citizens view two great cities' by D. Francescato and W. Mebane in *Image and environment* edited by R.M. Downs and D. Stea, 1973 (Figure 16.1A & B); Allen and Unwin for two figures from *Outcast Cape Town* by J. Western, 1981 (Figures 14.10 & 14.11); Association of American geographers for a figure from 'Urban graffiti as territorial markers' by D. Ley and R. Cybriwsky, *Annals* **64(4)**, 1974 (Figure 16.3), three figures from 'The gentrification of Harlem' by R. Schaffer and N. Smith, *Annals* **76(3)**, 1986 (Figure 14.22); E.J. Brill, Leiden, for a figure from 'Land use in the urban core' by W. Hartenstein and G. Staak in *Urban core and inner city* edited by W.F. Heinemeijer, M. Van Hulten and Hans D. de Vries Reilingh, 1967 (Figure 10.2); E.J. Brill for a figure from 'The delimitation of the towncentre in the image of its citizens' by H.-J. Klein in *Urban core and inner city* edited by W.F. Heinemeijer *et al.*, 1967 (Figure 16.4); Cambridge University Press for a figure from The growth of London AD 1660–1800' by O.H.K. Spate in *An histori-cal geography of England before 1800* edited by H.C. Darby, 1948 (Figure 10.8), a figure from *Urban inequality under socialism* by D.M. Smith, 1989 (Figure 17.5); the author and David & Charles for a figure from *Urban geography* by D.T. Herbert, 1972 (Figure 12.8); Gustav Fischer Verlag, Stuttgart, for a figure from *The economics of location* by A. Lösch, 1939 (trans-lation by W.H. Woglom, 1954, Yale University Press) (Figure 3.9); David Fulton Publishers for a figure from *The European city. A western perpsective* edited by D. Burtenshaw, M. Bateman and G.J. Ashworth, 1990 (Figure 4.14), two figures from *The European city. A western perspective* edited by D. Burtenshaw, M. Bateman and G.J. Ashworth, 1991 (Figures 8.2B & 9.4); George Allen and Unwin for a figure from *The containment of urban England* by P. Hall, 1973 (Figure 4.7); Guilford Press for two figures from *The Geography of urban trans-port* edited by S. Hanson (Figures 9.2A & 9.3); Heinemann Educational Books for two figures from *Growth centres in the European urban system* by P. Hall and A. Hay, 1980 (Figure 3.10); Hutchinson and Co. for a figure from *City systems in advanced economies* by A. Pred, 1977 (Figure 6.1); Institute of British Geographers for a figure from *Alnwick, Northumberland: a study in town plan analysis* by M.R.G. Conzen, 1960 (Figure 19.7), a figure from 'Fringe belts: a neglected aspect of urban geography' by J.W.R. Whitehand, *Transactions* **41**, 1967 (Figure 19.9), a figure from 'Structural models of retail distribution' by R.L. Davies, *Transactions* **57**, 1972 (Figure 10.12), two figures from 'Religious residential segregation in Belfast in mid-1969; a multi-level analysis' by F.W. Boal and M.A. Poole in *Social patterns in cities* edited by B.D. Clark and M.G. Gleave, 1973 (Figures 14.3 & 14.4), a figure from 'The organization of Quito's urban markets: towards a reinterpretation of periodic central places' by R. Bromley, *Transactions* **62**, 1974 (Figure 5.2), a figure from 'Retail patterns in the Victorian city' by G. Shaw and M.T. Wild, *Transactions* New Series, 1979 (Figure 10.9), a figure from *The making of the urban landscape* by J.W.R. Whitehand, 1992 (Figure 19.11); for two figures from 'Distribution of schizophrenics in Nottingham' by J. Giggs, *Transactions* **59**, 1973 (Figures 14.17 & 14.18); one figure from 'The study of delinquency areas: a social geographical approach' by D.T. Herbert, *Transactions* New series **1(4)**, 1976 (Figure 14.20), and for a figure from 'Rural periodic day markets in part of Yorubaland' by B.W. Hodder, *Transactions* **29**, 1961 (Figure 18.1); the editors for a figure from 'The urban process under capitalism: a framework for analysis' by D. Harvey, *International Journal of Urban and Regional Research* **2(1)**, 1978 (Figure 14.15); La Documentation Française for a figure from *Les villes 'Europëenes'* by R. Brunet, 1989 (Figure 4.13); the author, Longman Southern Africa (Pty) Limited and the Human Sciences Research Council, Pretoria, for two figures from *Land use in central Cape Town: a study in urban georgraphy* by D.H. Davies, 1965 (Figures 10.4 & 10.5); Longman Group Limited for a figure from *People and housing in Third World cities* by D.J. Dwyer, 1975 (Figure 18.8), a figure from *The city's countryside* edited by C.R. Bryant, L.H. Russwurm and A.G. McLellan, 1982 (Figures 15.1 & 15.2); Los Angeles Department of City Planning for a figure from *The visual environment of Los Angeles*, 1971 (Figures 16.2A, B & C); North Staffordshire Polytechnic for a figure from 'The segregation of the New Commonwealth population in Wolverhampton Municipal Borough, 1961–1981' by D.J. Evans, Occasional Papers **3**, 1984 (Figure 14.7); Oxford University Press for two figures from *Urbanization and settlement systems* edited by L.S. Bourne, R. Sinclair and K. Dziewonski, 1984 (Figures 4.11 & 17.1); Praeger for a figure from *Quality of life indicators in US metropolitan areas* by Liu, 1976 (Figure 6.2); Prentice Hall Inc. for two figures from *Geography of market centres and retail distribution* by B.J.L. Berry, 1967 (Figures 5.1 & 5.4), a figure from *The growth of shopping centres in the USA, 1957–79* by P. Muller, 1981 (Figure 5.7); Princeton University Press for two figures from *The making of urban America. A history of urban planning in the United States* by J.W. Reps, 1965 (Figures 19.3 & 19.4), a figure from *Rabat: Urban apartheid in Morocco* by J.L. Abu-Lughod, 1980 (Figure 18.6) and for a figure from *Cairo: 1001 years of the city victorious* by J. Abu-Laghod, 1971 (Figure 18.3); the Department of Geography, Queen Mary College, University of London for a figure from 'Crime rates as territorial social indicators' by D.M. Smith, *Occasional papers* **N11**, 1974 (Figure 14.19); Routledge & Kegan Paul and the Humanities Press Inc. for a figure from *The west European city* by R.E. Dickinson, 1951 (Figure 11.2); the author for two figures from *The middle order towns of Wales* by Gwyn Rowley, University of Wales Ph.D. Thesis, 1967 (Figure 4.2A & B); Routledge and Kegan Paul for a figure from *housing and residential structure* by K. Basset and J. Short, 1980 (Figure 13.3); Rowman and Littlefield Publishers for a figure from *Atop the urban hierarchy* by R.A. Beauregard, 1989 (Figure 11.12); Schenkman Pub. Co. for a figure from *Fitzgerald; the geography of a revolution* by W. Bunge, 1971 (Figure 14.14); Toronto University Press for four figures from *The structure of urban systems* by J.U. Marshall, 1989 (Figures 4.3, 4.4, 4.5 & 4.6); University of Birmingham for *A glossary of urban form* edited by J. Larkham and A.N. Jones, 1991 (Figure 19.8); the Regents of the University of California Press for two figures from *The social areas of Los Angeles* by Eshref Shevky and Marilyn

Williams, 1949 (Figures 12.1 & 12.2); the University of Chicago Department of Geography and the author for a figure from *Private redevelopment of the central city* by L.S. Bourne, *Research Paper* **112**, 1967 (Figure 14.13); two figures from 'Cultural differences in consumer travel' by R.A. Murdie, *Economic Geography* **41**, 1965 (Figures 5.10 & 5.11), for a figure from 'Delimiting the CBD' by R.E. Murphy and J.E. Vance Jr., *Economic Geography* **30**, 1954 (Figure 10.1); for three figures from 'The industrial revolution and the emergence of Boston's CBD' by D. Ward, *Economic Geography* **42**, 1966 (Figure 10.6A, B & C); the editor for a figure from 'The spatial organization of residential areas in Accra, Ghana, with particular reference to aspects of modernization' by R.R. Brand, *Economic Geography* **48**, 1972 (Figure 18.4); the editor for a figure from 'The squatters of Lima: who they are and what they want' by F.M. Andrews and G.W. Phillips, *Journal of Developing Areas* **4**, 1970 (Figure 18.9), a figure from *The factorial ecology of Toronto* by R.A. Murdie, 1969 (Figure 12.6); The University of Hull Publications Committee for a figure from 'The segregation of immigrant communities in the city of Birmingham, 1961' by P.N. Jones, *University of Hull Occasional Papers in Geography* **7**, 1967 (Figure 14.6); the University of Kansas Press for a figure from *Distribution of land values in Topeka* by D. Knos, 1962 (Figure 7.6); the University of Illinois Press for a figure from *Urban land-use planning* by F.S. Chapin, 1965 (Figure 7.7); Eastern Michigan University for a figure from *Community interaction and racial integration in the Detroit area: an ecological analysis* by R.V. Smith, S.F. Flory, R.L. Bashshur and G.W. Shannon, 1967 (Figure 12.7); University of Novarra and the editors for five figures from *The changing geography of urban systems* edited by L.S. Bourne *et al.*, 1989 (Figures 4.10, 5.15A, B & C & 6.4) The University of Pennsylvania and the editor for a figure from 'The location of urban land uses' by L.K. Loewenstein, *Land Economics* **39**, 1963 (Figure 11.4); University of Toronto, the editors and authors for two figures from *Urban systems development policies: the Alberta experience* by W.K.D. Davies, 1991 (Figures 5.13 & 5.14); University of Wales Press for a figure from *The Roman frontier in Wales* by V.E. Nash-Williams, 1954 (Figure 19.6), three figures from *Urban social structure. A multivariate-structural analysis of Cardiff and its region* by W.K.D. Davies, 1983 (Figures 12.3, 12.4 & 12.5); the University of Washington Press for a figure from 'The spatial organization of business land uses' by B.J.L. Berry in *Studies in highway development and geographic change* edited by W.L. Garrison *et al.*, 1959 (Figure 7.4 which is the basis of Figure 7.5), a figure from *Studies of the CBD and urban freeway development* by E. Horwood and R. Boyce, 1959 (Figure 10.3); John Wiley & Sons for three figures from 'Spatial structure in East European cities' by F.E.I. Hamilton in *The socialist city* edited by R.A. French and F.E.I. Hamilton, 1979 (Figures 17.2, 17.3 & 17.4), a figure from *Functional regions for the population census of Great Britain* edited by M.G. Coombes *et al.*, 1982 (Figure 4.9). The publisher and author also acknowledge the use of non-copyright material from *The structure and growth of residential neighbourhoods in American cities* by H. Hoyt, U.S. Government Printing Office, 1939 (Figure 7.3).

1

THE STUDY OF URBAN GEOGRAPHY

1.1 INTRODUCTION

Urban geography became a distinctive systematic specialism taught in university geography departments of the English-speaking world in the period after the Second World War. It was certainly not included in courses as a separate topic in the same way as geomorphology or climatology, or indeed political and economic geography, prior to that period. This is easily understandable. Urban geography cannot claim to be a systematic study in the sense that it is concerned with those processes which, in the context of a culture, operate to create spatial patterns. These processes are economic, socio-cultural and political and their study rightly generates the standard systematic themes within human geography. Urban geography, in contrast, considers all these processes in relation to one phenomenon, the city. It tended therefore to concentrate on consequence, rather than process, and to have more in common with regional geography. But towns have always been of interest to the geographer and from the earliest times regional geographies have dealt with them. Thus Strabo in his *Geography* was well aware of the importance of location, 'the natural advantages [of a place] should always be mentioned, since they are permanent. Advantages which are adventitious are liable to change ... those which continue, come to be regarded by posterity, not as works of art but as the natural advantages of a place; these, therefore, it is evident we must notice' (Hamilton and Falconer, 1912: 287).

But in spite of the endeavour to place emphasis on advantages of location, the geographical study of towns became essentially descriptive. Strabo himself wrote of Lyons, 'Lugdunum, situated on a hill, at the conference of the Saone and the Rhone, belongs to the Romans. It is the most populous city after Narbonne. It carries on a great commerce, and the Roman prefects here coin both gold and silver money' (Hamilton and Falconer, 1912: 287). This is followed by a description of the temple dedicated to Caesar Augustus. This sort of 'digest' account long stood as the standard way of dealing with towns although occasionally location was seen as a controlling factor. In the great atlas of town plans *Civitates orbis terrarum* published in the late sixteenth century, the wealth of Lyons is clearly related to its position, 'Its wealth comes from the afore-mentioned rivers, for because they pass many towns and flow into the sea, and because the city stands in the centre of Europe and is counted the heart of France, such rivers are a good means of conveying all things out of and into all the chief countries of Europe' (Oehme, 1965: 73). But in general such relationships were rarely stressed. A typical example is William Frederick Martyn's *The geographical magazine: or new system of geography* published in 1793. He comments that 'it is impossible in a work of this nature to expatiate on everything beautiful or curious in the various cities and towns which present themselves in different countries; nor can we do justice to the numerous architectural beauties with which England abounds' (Martyn, 1793, **2**: 404). After dealing with London, Martyn continues, 'Bristol, reckoned the second city of England for its extent and population, is more remarkable for its commerce and opulence than for any curious or beautiful structures it contains: and indeed all the other towns and cities of England have little more to recommend them to our notice than their commerce and the conveniency of their situations' (Martyn, 1793, **2**: 405). The whole geography of towns was in this way briefly dismissed, although the growing influence of industrialism meant that the detailed recording of the nature of trade and commerce became more important.

Even so there was little formal method. John Pinkerton in his *Modern geography, a description of the empires, kingdoms, states and colonies ... in all parts of the world*, published in 1807 wrote, 'In giving a brief account of the chief cities and towns of England, a few of the most important shall be arranged according to dignity, opulence and population; and the others shall be stated without preference, in a kind of progress from the south-west to the north' (Pinkerton, 1807: 77). In his progress the author becomes uneasy and

excuses his omissions, 'In a chorography of England, Leicester and Shrewsbury might deserve description, but its geography can only embrace the most important topics' (p.89). From this it appears that geography took on the aspect of a descriptive gazeteer, although it was only concerned with the principal parts: 'It's something like learning Geography ... Principal Rivers ... Principal Mountains ... Principal Towns – Why what are those creatures making honey down there?' Thus Alice built up her geography of Looking Glass Land, although with an astute aside on the economic base, and it is not surprising that this approach resulted in a demand for an association, rather than a narration, of facts. This reaction was clearly apparent in the first issue in 1901 of the periodical now called *Geography*, which contained a paper on 'The position of towns'. 'The magnificence of a town's buildings', wrote the author, 'the greatness of its population and commerce are stated as though they were causes of the town's importance, instead of being the most convincing proofs of peculiar advantages of position' (Dickinson, 1901–2: 97). And he concluded, 'Let us once and for all give up the rote learning of towns, products and points of interest as separate facts in favour of a general but comprehensive grasp of distribution and the logical consequences of physical position' (Dickinson, 1901–2: 108).

The replacement of description by interpretation of location laid the foundations for urban geography to develop as a systematic study. The first decade of the present century saw the appearance of two major works. Karl Hassert's *Die Stadte geographisch betrachtet* published in Leipzig in 1907 was the first volume to present an outline of urban geography. Raoul Blanchard's *Grenoble, étude de geographie urbaine* published in 1911 was the first classical study of a single city. To a large extent, developments in urban geography mirrored those in geography as a whole. At this period the subject was finding a basis in working out the consequences of the physical environment, for in spite of the nebulous concept of 'man, the master of the possibilities' of the physical world, the great regional monographs were firmly based on lithology and relief. The association between the 'separate facts' lay in the causal effects of physical geography. It was inevitable that at the time the unifying basis of town study should be found in those factors of location which controlled urban development. Blanchard wrote in the foreword of his book, 'The basic concept of this study is to explain the origin and development of the town as a function of the physical conditions of its situation', and in the last lines he concluded, 'From its origin right down to its present extension, Grenoble is the town at the junction of types of terrain, at the confluence of rivers. In spite of human changes nature always asserts its rights, even on an organism as complex as a town' (Blanchard, 1911: 5).

There followed a whole series of studies which can conveniently be called the 'site and situation' variety, in which the main end was to demonstrate that the character of towns was to be derived from their physical locations. It is interesting to note that the only map, apart from one of general location in H. J. Fleure's paper 'Cities of the Po Basin, an introductory study' (1924), is one of January and July temperatures. However, the key word in such studies was 'nodality' – hence the vast range of diagrams attempting to demonstrate the nodal situation of settlements. In this sort of context there was little room for the further development of a true urban geography and little incentive to gather a multitude of single studies under the heading of the geography of towns.

The first general review of urban geography was made by Aurousseau in 1924. He comments that city geography embraces such a large section of human geography that it is hardly a specialization at all. Thus, at the outset, and in consequence of the problems involved in identifying urban geography as a systematic study, he is unsure about the nature of urban geography. An introductory section on method outlines Blanchard's approach and leads into the following statement: 'It is an astonishing fact that the greatest interest has centred upon the individual town. Geography is so deeply concerned with the distribution of things that an interest in town distribution seems to be an obvious considera-

tion. Little attention has been given to it' (Aurousseau, 1924: 445). Aurousseau then proceeds to note Fleure's contributions to the regional study of towns (Fleure, 1920, 1924) and discusses the beginnings of functional study together with the early work on town status. Examples of town studies from different parts of the world are then given, and in conclusion 'the lack of extended studies in the United States' is recorded! The work makes extremely interesting reading for it stands at a point when rapid changes were taking place, the methods of the past and the problems of the future are uneasily associated in a review of a nascent specialism.

The changes noted above were largely a result of the reaction against the restricted aim of many town studies and this reaction was derived from two sources. The first was a direct rejection of the stereotyped 'site and situation' formula as it was slowly realized that such a limited consideration could not be abstracted from what was a complex, functioning economic and social system. Crowe, writing on methodology in 1938, seized on the treatment of towns as indicative of the inability of geographers to penetrate beyond the superficial. He criticized the tendency 'to crystallize ... upon the distribution of inanimate objects and the morphology of static patterns'. He pointed out that the application of the 'site and situation' formula was meaningless 'where site had nothing but historical interest and situation was viewed in terms of routes and not currents of movement' (Crowe, 1938: 18). But already, in 1933, Walther Christaller's great work on the central places of south-west Germany had been published and, although its impact was not to be felt until the post-war years, the revolution Crowe had demanded was under way.

The second source of reaction came from the nature of town growth itself. The vast extension of urban areas under the stimulus of new modes of transport brought severe problems of interpretation. A site-situation approach became meaningless when the large urban agglomerations had to be considered. In 1915 Patrick Geddes had been forced to devise a term for these new growths and the word 'conurbation' (Geddes, 1949: 14–15) came into circulation. The simple growth plan, the main element of the morphological approach, was increasingly shown to be inadequate. In the 1920s the Chicago school of human ecologists was already considering the variety of economic and social forces which resulted in the segregation of urban land-uses. The attention of geographers was thus directed towards the complexity of the townscape and away from the apparent simplicity of growth and general plan.

By the end of the Second World War, therefore, the situation had been reached where the foundation for a rapid growth of urban geography had been laid. Many of the basic ideas had been propounded, although generally in isolation rather than as an integrated systematic study. In many ways the subsequent expansion in urban geography was mainly concerned with the exploitation of ideas already in existence in the 1930s.

From a purely practical point of view there was an urgent need, certainly in most European cities, to deal with the terrible conditions brought about by uncontrolled nineteenth-century development. Wartime bombing resulted in the need for reconstruction. Redevelopment and reconstruction both demanded planning. In one of the earliest post-war studies R. E. Dickinson wrote, 'This book is not about planning. It is concerned with certain aspects of the inherent spatial or geographical structure of society upon which planning must be based, and it insists that knowledge of the anatomy of society must precede the treatment of its defects' (Dickinson, 1947: xiii, 1964). The employment of many geographers in town planning and the interaction between academic urban geography and the practical and applied spheres of planning provided an active stimulus to development. Commercial concerns began to realize the need for rigorous analysis before developments were started and the study of store location and market survey (Applebaum, 1961) also impinged on methods of investigation in urban geography.

Finally there were changes within academic geography itself which made the development of systematic studies more acceptable. The 'physical basis' was deprived of its predominant

role; and, of the infinite variety of factors operative in the creation of the differentiation of the earth's surface, no single one was given an *a priori* importance. In economic geography this change slowly released a flood of pent-up energy seeking the universal rather than the multiplication of detail. This flood finally burst through in urban geography, especially in model building. At the same time, urban areas had become such important parts of the landscape in Western countries that the simplest geographical description had to come to terms with them: the need for systematic study was evident. The eastern states of America had to be interpreted in terms of 'Megalopolis' (Gottmann, 1961); the foundations of the contemporary geography of The Netherlands was seen to lie in Randstad Holland, not the sub-boreal peat.

All these conditions formed a great stimulus to the geography of towns and an enormous mass of literature accumulated. But urban geography was never at the outset conceived as a well-defined systematic study. Topsy-like it 'just growed', although to such an extent that there is no need to present any argument for its recognition. But it had been developed by a large number of specialist workers engaged on particular aspects and at this stage, therefore, it became necessary to show that it formed a coherent field of study based on geographical principles of investigation. This need dominated much of methodological writing on urban geography during the 1950s. Mayer, in a series of papers (1954), outlined the main points on which geographers had concentrated and demonstrated these as forming related parts of a coherent systematic study. The general content of urban geography as propounded at that time can be summarized as follows.

At the outset the major concern was conceived to be a geographical one: 'Geography', wrote Hartshorne, 'is concerned to provide accurate orderly and rational description and interpretation of the variable character of the earth's surface' (Hartshorne, 1959: 21). Since it was hoped, probably in vain, to be rarely inaccurate, seldom disorderly and never irrational, it was possible to omit the adjectives and argue that the geographer was concerned with the analysis of the variable character of the earth's surface. On that surface, the populations and the buildings agglomerated together to make up towns constituting the special interest of the urban geographer; they were to be abstracted from the totality for separate, topical study. Since the bulk of the population of the Western world lived in towns, and the problems of the urban environment were paramount, the importance of the study in academic geography and its relevance to applied geography needed no further stress.

The argument continued: the town as a unit feature of the earth's surface has, like all other features, two associated aspects. The first is location or position, the second is form or internal structure. In simplest fashion these two aspects emerge clearly when differences of scale are considered. On the atlas map towns are represented by conventional symbols and the main geographical implication is a concern with location, or with the town as a distributed feature. On maps of the scale of 1:50,000 or 1:10,000, the town is no longer shown by conventional means but is seen to have individual form or shape as well as internal structure, although this is, as yet, displayed only in a generalized way. On the scale of 1:2500 or 1:1250 the internal structure becomes apparent in detail, in the form of streets, blocks and individual buildings, although still in an unreal, two-dimensional way. Finally on the ground, the real town is seen to be three-dimensional, for each building has height and this must be added to the geographer's appreciation. Indeed, the geographer, confronted with the town fabric, is also made to realize that the fourth dimension of time is involved, for many of the structures and much of the form are inherited from past periods. But this progress through the scales emphasizes the two main themes of study introduced at the outset: the town as a distributed feature and the town as a feature with internal structure, or in other words, the town *in* area and the town *as* area.

It is axiomatic that location can only be understood through function; what a town does, or did in the past, determines its location and controls its growth. It is possible to derive two associated concepts from the consideration of these functions:

1. **The nature of urban functions.** What are the sorts of activities which dominate in any one town out of the whole range which towns perform? In particular, to what extent are specialized functions carried out and what are they?

2. **The stature (status) of urban services.** To what extent does the town take part in central place activities (excluding specialized functions)?

It is dangerous to confuse these two although they are so clearly bound together: what is needed is 'a separate study of resource oriented functions and other activities whose location central place theory cannot explain and the subsequent superimposition of such functions on to the areal pattern of central places' (Isard, 1960: 227) Here therefore were two critical lines of investigation. The link between them is associated with accessibility and the transport role has to be isolated as a distinctive feature to be considered apart.

The morphology of towns, or the study of internal structure, is related to three variables. These are plan or layout, which includes the contribution of both streets and plots or lots; land-use or the function of buildings; and the architectural style of buildings. These three vary independently and produce an infinite variety of urban scene. Each has to be considered separately and then the relationship between them demonstrated in a study of the whole townscape.

Into this situation of the slow crystallization of urban geography into a clearly structured systematic study, came the 'quantitative revolution'. The use of the description 'quantitative revolution' is misconceived. Essentially what occurred was the impact of logical positivism, transferred from other social sciences to geography, with the vision that a theoretical geography could be developed with the intellectual cachet of scientific method. Out of theory came model building, and the models, as well as theory, had to be tested in the real world. That testing involved statistical analyses of data, hence the quantitive revolution. The specific point at which this occurred, in human geography as much as in urban geography, was that where theory already existed for, as has been noted, Christaller's central place theory dated from 1933. But associated with it and growing out of it was the conviction that all urban geography could be developed in that way since social patterns were to be the matter of a locational science. Thus in intra-urban terms the residential areas of the city were to be disentangled into a clarity, albeit a complex one, by factorial ecology. It was an exciting and stimulating period as geographers who lived through it will bear witness. It seemed that the uncertainties of an academic geography which had been condemned as descriptive could be set aside, and the sterility of arguments over determinism and possibilism forgotten as irrelevances. A 'New Geography' had emerged and urban geography could lay claim to being the leading edge of innovation.

There was apparent certainty and order and in consequence there was a brief period of what can justifiably be termed a smug and arrogant satisfaction in urban geography. Quantification had transformed vague descriptive statement into the precise analysis of models derived from theory based on deductive process. The complexities of the urban world lay as open to scientific investigation as those of the physical universe. It was simply a question of adequate mathematics. But such a situation was not to last long. The collapse came in a number of ways but possibly initially where statistical analyses had been effectively applied, in social area analysis and factorial ecology, for they revealed that simple class-based contrasts were inadequate reflections of the manifest internal variations within residential areas. Moreover, areas having been derived from factorial ecologies, there seemed but little purpose other than to dispute whether they constituted zones, sectors or a set of separate nuclei. Attention, once switched to social class, life cycle and ethnicity as bases of

variation, inevitably continued to probe those factors. As that took place, it became evident that explanation conceived at an aggregate level was inadequate and that individual perception, behaviour and choice were not only critical but often divergent from both the assumptions and the predictions of theoretical models. Again, in relation to central places, theory moved further and further into abstract mathematical formulation on the one hand, a move which seemed to have little practical relevance, while on the other the replication of empirical studies added nothing but case examples, which by their nature could do nothing to confirm or reject a theoretical formulation.

The way forward became associated with process or behavioural studies, where, it was conceived, deeper rooted and more satisfactory explanations would be found. Central places, or shopping centres, obtained their statuses by the preferred behaviour of consumers and most certainly it could be demonstrated that such behaviour was not that predicted by central place theory. Likewise residential character was created through the locational choices of individual families, and in their perception of the qualities of areas and in their behaviour in choosing, more convincing insights into the creation of social areas were to be found, as opposed to their identification and description. Urban geography moved into a predominantly behavioural phase when the literature was dominated by studies of consumer behaviour and residential choice, closely associated with investigations into people's perception of the townscape and the opportunities it offered.

In turn, dissatisfaction with behavioural studies has grown rapidly, largely because of the basic assumption of the unconstrained nature of choice. It was apparent that those with few resources to deploy had no choices to make, but were at the mercy of the 'system'. This awareness of deprivation was heightened by the actual physical process of inner city decay which characterized the Western world, and also by the increasing impact of studies of the spontaneous settlement about the cities of the developing countries.

Urban geography in this way became dominated by what can be called 'social conscience' studies, which sought to identify the roots of spatial inequalities in urban terms. These ranged from the relative deprivation of immobile residents of small country towns who are effectively isolated from standard urban services, to the limitations of life opportunities of those resident in the inner city. These sorts of people, it was contended, exercise little choice, for their lifestyles are dependent on the major public and private institutions which manipulate the space economy. Hence, description of the patterns of deprivation has become associated with the analysis of the operation of those institutions which control the city. These radical reactions were to introduce another element into play, that of involvement. Urban geographers had, since the Second World War, acted as consultants in planning, but this idea of involvement was very different, for it demanded that academics should not set themselves apart as objective observers and analysts of the city but should become propagandists and activists promoting change. This was epitomized in what was the most celebrated work in the genre, William Bunge's *Fitzgerald: the geography of a revolution* (1971). It was to be followed by a spate of books all concerned with the social evils necessarily generated by the capitalist city. To some extent these concerns carried a methodological implication for patterns became a relatively trivial end product; what mattered was the organization of the economic and social system which produced them.

R. J. Johnston effectively summarized these changes in relation to the contents of human geography as they appeared in the late 1970s:

> Three branches are identifiable in the current literature of urban geography: the first is a quantitative approach to description, based on a nomothetic philosophy in which the geographer's role is to document the spatial organization of society.... Second is the so-called behavioural approach ... which proposed ... the study of individual activities within their perceived worlds. Finally, there is the approach variously termed 'radical' and 'structuralist' which stresses the constraints that society as a whole, and particularly certain groups within it, impose on the behaviour of individuals. (Johnston, 1977).

But the notion of three paradigms existing together in some form of a three-pronged trident diverging from the unity of a common handle was illusory. The emergence of contrasted methodologies was the indication of the onset of post-modernism in urban geography, the replacement of a well-structured consensus-based unity by an incoherent diversity that has been made even more acute by the increasingly diminished emphasis placed on the fundamental spatial and locational themes of geography. In a sentence which could only have appeared in a social science, Knox has written, 'whereas modernism is paradigmatic, universalistic, purposive, hierarchical, synthetic, selective and concerned with master codes and metanarratives, postmodernism is syntagmatic, playful, anarchical, antithetical, combinatorial, idealectical, localistic and anti-narrative' (Knox, 1991). Perhaps it can be summarized as 'anything goes' which casts some light on the city from whatever the angle it is generated: an attitude reinforced by the widespread and often thoughtless commendation of inter- and multidisciplinary approaches as of necessity good in themselves. No better has this been illustrated than in the third of Johnston's paradigms, that of radical geography. Its central tenet was a critique of the operation of *laissez-faire* market forces and the inequities and iniquities which were generated by them. But any critique which is radical must believe in replacement root and branch, that is, it must have an alternative system to offer. That was the socialist city, the city of planning controls and central direction ensuring egalitarianism as well as plenty for all. But recent events have revealed that as no more than an Utopian dream which, when translated into reality, brings conditions even worse than those of market forces. The collapse of socialist regimes around the world in the 1990s, and the clear indication that they were only kept in operation by rigid totalitarian control, which bred massive bureaucracies and inefficiency without ever achieving the ideals, has greatly tarnished the image. The critique generated by a mass of 1980s radical literature now looks outdated and outmoded. The problems certainly remain but the solutions are to be sought elsewhere.

Deprived of the force of its radical political bases – social democrats now seem to claim that they can run the capitalist city more efficiently – much of the *avant-garde* analysis of the city has perforce sought other viewpoints. Thus the central and traditional triad of social class, life cycle and ethnicity as the components of both perception and of spatial character, have been joined by a series of other viewpoints linked to lifestyle, including especially that of women via a feminist geography and of homosexuals via a gay and lesbian geography. It is perhaps surprising that a 'green' geography of the city has yet to be written, although it makes sporadic appearances in condemnation of private transport, road building and extension of suburbia on to green belt land.

When the present author wrote the first edition of this book in the late 1960s – published in 1972 – it was possible to give clear shape and form to a nascent urban geography. That first edition in modernist mode carried a diagram confidently setting out the concerns of the systematic specialism and demonstrating its holistic character by showing the links between them. Possibly books such as that set the agenda of what was to be the urban geography of the 1970s. But later editions recorded changes, and now no such harmony exists. Both in content and methodology there is an eclecticism and uncertainty which seeks justification under the umbrella of post-modernism.

The review of recent development must necessarily impact upon the preparation of a book which claims to encompass the study of urban geography. There is some justification in the argument that no such systematic discipline is now tenable. Many of the areas of study which urban geography introduced have grown into distinctive specialisms of their own in a sort of family tree of succession. An obvious example is retail geography which has absorbed much of central place theory and consumer behaviour adding to it a great deal which is aspatial. This is a textbook and in any review of courses as they are taught in British universities in a three year degree scheme, it will be apparent that urban geography has slipped down the years of study. Whereas in the 1960s and 1970s it was a third year specialism, now it is little more than a first year course, often no more than part of one, and constituting an introduction to the greater specialisms to follow, ranging from town-plan analysis to housing studies, even public finance or the impact of public policy on living standards and quality of life.

This growing disjointedness of content developed in parallel to transformations of the object of study itself. Early urban geography arose in relation to discrete, free-standing cities. Conurbation although apparent made little impact and Los Angeles, the classic set of suburbs in search of a city, was no more than an exception, more of a lesson in what planning could avoid than an indicator of things universally to come. But come they did and metropolitan and megalopolitan developments totally transformed the free-standing city and accordingly newer outlooks and different methods were demanded.

It follows that in this Introduction to the fourth edition, unlike in the late 1960s, it is impossible to present a clear consensus view of content. Urban geography is now more of an amalgam of different approaches and considerations of the city, nearer perhaps to a multidisciplinary analysis of its object rather than a neatly structured systematic geographical study. It is true that there are three basic components, urban populations, urban functions and the internal structure of cities which necessarily provide the basic outline, but within those sections and beyond them the contents are far less secure.

This book however retains the basic structure of its predecessor. The first imperative is to review the growth of urban populations across the world and to consider major changes which are taking place. The second section must deal with the functions of towns, for what they do and how they do it determines location and growth. The third part then has to consider internal structure both in terms of conventional land-uses and also in the way in which people work and live. It is here that the greatest transformation has taken place in urban geography over recent times. The town can no longer be viewed simply as an object of study, much in the way a geologist would appraise a piece of rock, however complex and far reaching an explanation may have to be. For a town is much more than that; it is the environment where the greater proportion of the world's people spend their lives and the nature and quality of those lives is a central theme which humanistic geography has placed on the agenda. This is not an urban geography which is in any way prescriptive but, even so, it cannot ignore the fact that aspirations for the good life have to be met in the city and the degree to which they are met, in so far as it is spatially related, is a valid area of study. The viewpoint broadens, therefore, from the attempt to understand mechanisms which control

patterns and create differentiations within a highly specific part of the earth's surface, to an estimate of the efficacies of the product in the light of the welfare of its inhabitants.

As urban geography has become less unified so it becomes more difficult, even less academically viable, to present a neat structure. Each urban geographer will have his or her priorities so that a book such as this, which some twenty years ago could claim to be a synthesis based on consensus, is now a good deal less than that, for it is rather one person's view of what is most pertinent. Omissions may be as revealing as inclusions. But, nevertheless, it is intended to be synthetic in that it attempts to bring together the critical fields and diverse strands and demonstrate their interaction. It is firmly and uncompromisingly based on the belief that the geographer's fundamental concern is spatial and that the description and explanation of pattern is fundamental and has as important a role to play in urban studies as any other approach. Unless all this is believed and attempted with conviction then such an urban geography should surely not be written.

THE PROCESS OF URBANIZATION

Introduction

The existence of a branch of systematic geography which is called 'urban geography' implies that there is a discrete area of study which can be legitimately set apart from other systematic areas, including, specifically, a rural geography. The first and most immediate problem that arises, therefore, is that of the definition of what is to be regarded as urban. What does the term 'town' or 'city' imply as contrasted with 'village' or 'hamlet', and where lies the real distinction between that part of the total settlement pattern called 'urban' and that called 'rural'? It is at once apparent that the differences between town and country which are so easily made in everyday life become very difficult to transpose into precise and scholarly terms. A discussion of these difficulties is best undertaken against the background of the varied definitions used throughout the world. What follows is an arbitrary selection, but one made to encompass and illustrate all the varied bases which are used

to define the urban population. They are all taken from the *United Nations demographic yearbook 1988* (United Nations, 1990).

Definitions of the urban population: a sample

A wide range of examples is included covering all the continents:

1. **Botswana.** Agglomerations of 5000 or more inhabitants where 75 per cent of the economic activity is of the non-agricultural type.
2. **Egypt.** Governorates of Cairo, Alexandria, Port Said, Ismailia, Suez, frontier governorates and capitals of other governorates as well as district capitals (Markaz).
3. **Ethiopia.** Localities of 2000 or more inhabitants.
4. **Gambia.** Banjul only.
5. **Argentina.** Populated centres with 2000 or more inhabitants.
6. **Peru.** Populated centres with 100 or more dwellings.
7. **Canada.**
    (a) 1976. Incorporated cities, towns and villages of 1000 or more inhabitants and

their urbanized fringes; unincorporated places of 1000 or more inhabitants having a population density of at least 1000 per square mile or 390 per square kilometre and their urbanized fringes.

(b) 1981. Places of 1000 or more inhabitants having a population density of 400 or more per square kilometre.

8. **USA** Places of 2500 or more inhabitants and urbanized areas.

9. **India.** Towns (places with municipal corporation, municipal area committee or cantonment board); also all places having 5000 or more inhabitants, a density of not less than 100 persons per square mile or 390 per square kilometre, pronounced urban characteristics and at least three-quarters of the adult male population employed in pursuits other than agriculture.

10. **Israel.** All settlements of more than 2000 inhabitants except those where at at least one-third of households participating in the labour force earn their living from agriculture.

11. **Japan.** Cities (Shi) having 50,000 or more inhabitants with 60 per cent or more of the houses located in the main built-up areas and 60 per cent or more of the population (including their dependants) engaged in manufacturing, trade or other urban types of business. Alternatively, a shi having urban facilities and conditions as defined by the prefectural order is considered as urban.

12. **Czechoslovakia.** Large towns, usually of 5000 or more inhabitants, having a density of more than 100 persons per hectare of built-up area, three or more living quarters in at least 15 per cent of the houses, piped water and a sewage system for the major part of the town, at least five physicians and a pharmacy, a nine year secondary school, a hotel with at least twenty beds, a network of trade and distributive services which serve more than one town, job opportunities for the population of the surrounding area, the terminal of a system of bus lines and not more than 10 per cent of the total population active in agriculture; small towns,

usually of 2000 or more inhabitants, having a density of more than seventy-five persons per hectare of built-up area, three or more living quarters in at least 10 per cent of the houses, piped water and a sewage system for at least part of the town, at least two physicians and a pharmacy, other urban characteristics to a lesser degree and not more than 15 per cent of the total population active in agriculture; agglomerated communities which have the characteristics of small towns in regard to size, population density, housing, water supply and sewage, and percentage of population active in agriculture, but which lack such town characteristics as educational facilities, cultural institutions, health services and trade and distributive services, because these facilities are supplied by a town in the vicinity.

13. **France.** Communes containing an agglomeration of more than 2000 inhabitants living in contiguous houses or with not more than 200 metres between houses, also communes of which the major portion of the population is part of a multi-communal agglomeration of this nature.

14. **Iceland.** Localities of 200 or more inhabitants.

15. **Netherlands.** Urban: municipalities with a population of 2000 or more inhabitants. Semi-urban: municipalities with a population of less than 2000 but with not more than 20 per cent of their economically active male population engaged in agriculture and specific residential municipalities of commuters.

16. **Norway.** Localities of 200 or more inhabitants.

17. **Portugal.** Agglomerations of 10,000 or more inhabitants.

18. **The former USSR.** Cities and urban-type localities officially designated by each of the constituent republics, usually according to the criteria of numbers of inhabitants and predominance of agricultural, or numbers of non-agricultual workers and their families. Since the constituent republics are the basis

this presumably still applies for Russia and the autonomous republics.

19. **Australia.** Population clusters of 1000 or more inhabitants and some areas of lower population (e.g. holiday areas), if they contain 250 or more dwellings of which at least 100 are occupied.

A review of these nineteen representative definitions reveals at least seven bases which are used either singly or in combination to identify the urban function of the population:

1. Specifically named settlements.
2. Settlements designated urban by administrative status.
3. A minimum population.
4. A minimum population density.
5. A contiguity constraint which is either inclusive in order to take in surburban and commuter areas or exclusive in order to set aside areas of loosely scattered settlement.
6. The proportion engaged in non-agricultural occupations according to the varying bases of adult males, households or total population.
7. Functional character.

Against the background of the definitions and the seven bases it is now possible to review the sources from which the difficulties of identifying the urban population are derived.

Problems of definition

The *United Nations demographic yearbooks* concentrate in each issue on a specific problem. As long ago as 1952 the yearbook was devoted to the problem of providing adequate data on the world's urban population. It concluded that, 'There is no point in the continuum from large agglomerations to small clusters or scattered dwellings where urbanity disappears and rurality begins; the division between urban and rural populations is necessarily arbitrary' (United Nations, 1955). Thus as one goes down the scale from the largest metropolitan area to the isolated farm it is impossible to identify satis-

factorily a dividing line which is conceptually meaningful. Seemingly the obvious way out of this dilemma would be the identification of an accepted minimum population and from time to time figures of 2000 or 2500 have been proposed. But apart from the fact that the figures derived will necessarily depend on arbitrary and, in some instances, anachronous boundaries (see p. 13) the range of minima employed in national censuses is, as the definitions included demonstrated, ample testimony to fundamental problems.

These problems are derived from differing economic and cultural situations. In Iceland, for example, a settlement of 200 or more people is normally concerned with urban functions, for the nature of the rural environment precludes large agglomerations solely concerned with agriculture. In contrast, in southern Spain or Italy rural settlement is basically agglomerated and 'villages' may reach totals of 8000 to 10,000 population. This is due partly to historical reasons, such as the insecurity of individual holdings in face of banditry, and partly to the whole nature of the economic organization of land holding which has never suffered 'revolution' as in most of north-west Europe. Portugal, significantly, adopts a figure of 10,000 as its minimum population for urban status, fifty times greater than that used by Iceland and Norway. This situation upsets traditional, simplistic views that towns are bigger than villages. It is also relevant that high population density is not directly associated with urbanization, which is another general misconception. Australia is one of the world's most highly urbanized countries, 85.7 per cent lived in cities at the 1981 census, but it has a very low national population density of 2 per square kilometre: India is not highly urbanized, 23.3 per cent were city dwellers in 1981, and 25.7 in 1991, but the country has a high population density of 242 per square kilometre. It is an occidental thesis that a high population density and urbanization are necessarily correlated. It is because of the meaninglessness of size as a simple discriminant that other criteria, such as employment proportions in agriculture or the presence of specific urban functions, have been brought into the definitions set out earlier.

This problem of definition has been made all the more difficult by the fact that the reality, and consequently the concept, of what is urban is not static but is subject to change. In early times the town meant pre-eminently 'market town'; the legal possession of a market by charter or grant virtually defined a town. Often the town was sharply distinguished from the surrounding countryside and in many cases town walls were tangible divides between the urban and the rural, although even then suburbs compromised the sharpness of the break. But changes during the eighteenth and nineteenth centuries produced even greater difficulties. Industrialization brought into being a large number of settlements which were certainly not villages, in the sense of being nucleated settlements of agricultural populations. But neither were they market towns serving the surrounding countryside. They were large tracts of bricks and mortar which broke across the old division. In Britain their status was significantly described under the Public Health Act of 1872 when they became 'Urban Sanitary Districts', the crucial and necessary urban function embodied in their description. Later, as functions broadened, they were called 'Urban Districts'. Ironically, but significantly in the context of this discussion, they were often conceived as being made up of a number of industrial villages. But there is a wealth of implication in the contrast between a 'town' and an 'urban district'. The name did not disappear in Britain until local government reorganization in 1974.

By the end of the twentieth century new developments have added still further problems of definition. The growth of massive urban aggregations, reflected in the successive introduction into the literature of urbanism of the terms conurbation, metropolis and megalopolis, has meant that few parts of highly developed countries are out of the range of urban influences, and most parts are saturated by them. Long distance commuting, made possible by urban freeways and mass transit systems, has given rise to a significant outflow of population,

even beyond the conventional surburban housing estates. Another name has been coined, the 'metropolitan villlage' which is made up of city workers who have invaded the countryside, which has itself long been shedding labour as agriculture has become more capital intensive. Even further, the acquisition of second homes by the affluent urban dwellers has produced what is termed 'seasonal suburbanization'. What was once the prerogative of the aristocracy and the gentry, a town house and a country estate, has extended down the social scale, changing somewhat in nature but fundamentally the same, so that urban influences are extended still further. In brief, in many parts of the world urban and rural are so thoroughly integrated that any distinction must be arbitrary and, indeed, of doubtful validity and purpose.

In spite of these difficulties the demand to identify an urban fraction remains and the way out of the situation, both in an academic context and in national censuses, has been twofold. The first has been to reinterpret the concept of 'market town' by an examination of the functions which the town performs for the surrounding countryside for, it is contended, a town is distinguished by its role as a central place for a tributary area. The rural countryside needs a focal point to which its produce can be sent for export and from which its needs can be distributed. This general role is represented in the town by shops, banks and offices (consider again the Czechoslovak definition) and the identification of these will indicate urban character. But such a procedure offers no simple solution for even a small village has an area-serving role represented by such functional elements as a small shop and/or post office, a petrol station/garage and, in Britain, a public house. So there is no question of the simple presence or absence of clearly definable 'urban' functions. Accordingly much more sophisticated assessments have to be made; but any boundaries set are arbitrary, and definitions become much more culturally dependent than the apparent, though illusory, universality represented by population. Thus in some countries, and at one time in the West, markets and fairs are held in open country and are not associated with settlements. This is still true of 'Sunday markets' in developed countries which are often held outside towns.

The second 'way out' has been to employ the proportion engaged in agriculture which purports critically to distinguish rurality. It is, perhaps, the most effective of measures. However, no agreed proportion exists, those in the selected definitions range between 25 and 33 per cent, and no standard basis is adopted, for total population, households and adult male population are all represented. Moreover the relative dispersal of rural population varies considerably and will have an effect on the figures produced. But perhaps, most fundamentally, the high capitalization of agriculture and the rural depopulation it engendered, together with the move into country villages of commuting urban workers, already indicated in the term 'metropolitan village', means that even proportions engaged in agriculture have diminished relevance.

The conclusion from this consideration of the changing nature of urbanism and the functional overlap between what is commonly regarded as town and likewise village has already been set out. Over much of the world the distinction between what is urban and what is rural has lost any real meaning and hence attempts at definition are themselves flawed from the outset.

Problems of reliability

Even if there were a standard and universal definition there would still remain what is perhaps an even more basic problem – that of reliability. The United Nations produces a four category quality code for its demographic data. These are, respectively, complete census of individuals, sample surveys, partial survey or partial registration and, finally, conjecture. Over much of the developing world figures are seldom derived from the first category, often from the fourth. The *UN demographic yearbook* notes that in regard to the tabulation of urban population, 'estimates ... have been limited to countries or areas for which estimates have been based on the results of a sample survey or have been constructed by the component method from the results of a population census or survey' (United Nations, 1990: 53). As a result the number of countries included is greatly limited and data are only provided for 112 countries or areas out of a total for the general tabulation of population of 218. It must be added that most of those missing are very small countries or 'areas'. The total population of the countries listed amounts to 4254 million. The data are not derived from a single year but range over the whole of the 1980s, so the best comparable figure for a total world population is the mid-decadal 4854 million. Even so, there are some strange omissions from the urban table. England and Wales are not included (although Scotland is), neither is Belgium, Italy, nor Norway and Sweden. Presumably, the exclusion of England and Wales is related to problems of definition after the local government changes of 1974 which did away with boroughs and urban districts. This introduces the rather different problem of the actual boundaries of towns and cities, but before moving on to consider that it is imperative to note that both the numbers and the percentages of population, the latter often given to decimal places, are extremely unreliable as well as being based on varying definitions. The exactitude of figures should not cloak the nature of their determination.

Introduction

The problems which are inherent in basic definition and reliability are compounded by further difficulties over the identification of the areas which are to be taken as limiting a settlement. The situation is well illustrated by an historical example. Meuriot (1911) in the early part of the century demonstrated how two totally different interpretations of urban growth could be derived from a set of figures relating to population change (Table 2.1). The conclusion from Table 2.1 is clear; Vienna was growing more

Table 2.1 *A comparison of the population of Vienna and Berlin according to administratively defined areas, 1900–10*

	Population 1900	Population 1910	Increase 1900–10
Vienna	1,674,000	2,030,000	356,000
Berlin	1,888,000	2,070,000	182,000

quickly than Berlin. But that is the opposite of the anticipated contrast between the two cities in the light of the contemporary political and economic trends. Table 2.2 presents a closer scrutiny of the position based not on administratively bounded areas but on the actual extent of settlement. When that is done the relation between the growth of the cities between 1900 and 1910 is reversed. Two diametrically opposed interpretations can apparently be dependent on the areas used.

Underbound and overbound cities

This problem, which is a universal one critical to all calculations of the urban fraction, is usually stated in terms of 'under' or 'overbound' cities. In the underbound city the administratively defined area is smaller than the physical extent of settlement. In the overbound city the administrative area is greater than the physical extent. The 'truebound' city is one where the administrative bound is nearly coincidental with the physical extent. It is evident that whatever definition of 'urban' is used, be it population

numbers, employment characteristics or functional character, marked variations can occur dependent on underbound or overbound character.

Rules for the identification of urban areas

The standard way, at least within a country, to solve the above problem is to draw up a series of rules for the identification of urban areas. Unfortunately, as every eventuality is considered, these tend to become more and more complex. One example can be presented, necessarily from a developed country, since only such a country possesses the detail of information to devise sophisticated formulae.

The British census of 1981 responded to a request from the statistical office of the European Commission for the definition of larger urban agglomerations of over 100,000. It set out to identify what were called 'urban areas'. The starting point was the definition of urban land (OPCS Census, 1981; *Key statistics for urban areas*, 1984), which was deemed to be made up of:

1. permanent structures and the land on which they are situated;
2. transportation corridors (roads, railways and canals) which have built-up sites on one or both sides or which link up built-up sites which are less than 50 metres apart;
3. transportation features such as railway yards, motorway service areas and car parks (operational airfields and airports are also included);
4. mineral workings and quarries;

Table 2.2 *A comparison of the population of Vienna and Berlin in similarly defined built-up areas, 1900–10*

	Urban areas in hectares	Population in 1900 of comparable areas	Population in 1910 of comparable areas	Increase 1900–10
Vienna	17,000	1,674,000	2,030,000	356,000
Berlin	6,000	2,460,000	3,315,000	855,000

Table 2.3 *Urban areas by size and population. Great Britain totals are in 000s*

Size (persons present 1981)	Number of urban areas	Persons present 1981	Cumulative person	Cumulative per cent
1,000 plus	4	13,850	13,850	25.5
500,999	6	4,066	17,916	33.0
200,499	22	6,669	24,585	45.3
100,199	28	3,862	28,447	52.4
50,99	75	5,265	33,712	62.1
20,49	167	5,220	38,932	71.7
10,19	260	3,644	42,577	78.4
5,99	379	2,669	45,245	83.3
2,49	881	2,761	48,007	88.4
Under2	409	649	48,655	89.6
All urban	2,231	48,655	48,655	89.6
Rural remainder	–	5,630	54,285	100

Source: After Denham, 1984.

5. any area completely surrounded by built-up sites.

It is added that areas such as playing fields and golf courses are excluded unless they are completely surrounded by built-up sites as in point 5 above. This, of course, only characterizes the area and in order to qualify as urban certain minima are needed. These were set out first as a minimum continuous extent of 20 hectares with separate areas being linked if they were less than 50 metres apart, and, second, a minimum population of 1000. But here a problem arose for population figures were not available on the land-use basis. Hence, a switch was made to the data gathering basis of the census, the Enumeration District or ED. A proxy threshold was set at a minimum of four EDs. It is admitted that 'this resulted in the exclusion of some areas of urban land with more than 1000 persons, but very few of those above 2000 persons' (OPCS, 1984: 9). Thus at the critical lowest level of distinction, the point at which a crucial break is made between urban and rural, the division was essentially arbitrary and

variable in spite of all the rules devised.

The procedure by which the areas were defined takes another twelve paragraphs. In them such problems as the treatment of EDs, the boundaries of which did not coincide with the urban land as defined, are treated. Thus a point is reached at which

any areas of urban land each of 20 hectares or more, which were less than 200 metres apart and which had in aggregate four or more EDs were linked together to form continuous urban areas.... All other areas of urban land of 20 hectares or more, and lying more than 200 metres from any other built-up areas, were deemed non-urban and population in them was counted as part of the rural population.

(OPCS, 1984: 10).

Elaborate as this procedure is, there is one flaw and one paradox. The flaw, of course, is the arbitrary inclusion/exclusion based on the four EDs at the lower end. In terms of the total urban population and of the percentage urban it is of no great significance, but it highlights the problem of establishing a firm and easy break between town (urban) and village (rural). The paradox is that at the final stage the situation arises in which 'urban land' is deemed to be 'non-urban'! Again it is more than a question of semantics for the term 'built-up land' or even the older term 'developed land' could have been chosen to avoid the implication that a village consists of urban land which is only limited in its extent, and an arbitrary extent at that.

At the 1991 census effective definition by modern methods including satellite imagery, and an effective population base from a geographical information system, has provided a very fine meshed definition of *de facto* urban areas, but the basic problems will remain.

From these elaborate procedures what emerges is the most valuable of general analyses – that which tabulates urban settlements by number, size and the contribution of each size group to the total (Table 2.3) (Denham, 1984).

2.3 THE FACTS OF URBANIZATION

The basic data in relation to urban population for the middle of the present century were set out in the early 1970s by Kingsley Davies in his book *World urbanization 1950–1970* (Kingsley Davies, 1972). A near-contemporary United Nations publication, *Growth of the world's urban and rural population 1920–2000* (United Nations, 1969), provided a further source. Since then the main source has been the *United Nations demographic yearbooks*.

Perhaps the most significant feature of world urbanization is the recency of its dominance. It has been estimated that before the start of the nineteenth century only some 3 per cent of the world's population lived in towns of over 5000. At the present time the figure is probably about 40 per cent. Expressed in a different way, the world's population increased threefold between 1800 and 1860 but the world's *urban* population increased thirtyfold. Between 1950 and 1970 the world's urban population rose from 28.2 per cent to 38.6 per cent. The actual proportion for those countries included in the 1988 *United Nations demographic yearbook* is 40 per cent. The countries omitted would probably alter the percentage figure only marginally (though clearly not the absolute total) and something in the range of 42–46 per cent can be suggested for 1990. Perhaps the vagueness of a 'suggested' range might appear unacceptable, but given all the problems that have been discussed it is greatly to be preferred to a precise figure which would have no validity. Even so, the situation is certainly being approached where half the world's population can be considered as urban. That will surely have been reached by the year 2000.

This latter fact emphasizes that 'the most conspicuous feature of today's accelerated world population growth is its even greater rapidity of urbanization. In many periods of history, populations have grown, but the tempo and dimensions of recent years have never been equalled' (United Nations, 1969). It follows that urbanization is the predominant process in the spatial organization of the world's population and it is this which makes its geographical study an imperative and perhaps puts the niceties of definition into proper perspective. Ethnocentric Westerners, aware of the great changes brought about by industrialization in the nineteenth century, often regard that period as marking the peak of urban growth. That is not so, for whereas urban growth reached about 23 per cent per decade between 1850 and 1900, it achieved over 40 per cent between 1960 and 1970. But again such global figures have a questionable validity.

Table 2.4 sets out the percentage increases in the urban populations of a range of countries for a period which in most cases is *less* than a decade. The massive urban increase is unmistakably apparent.

A rather different measure is that of cities of over 1 million inhabitants, although here again problems of definition arise. Even so it has been indicated that the number was 24 in 1921, 41 in 1941 and 113 in 1961 (Linton, 1958). By 1988 there were 240 (*United Nations demographic yearbook*, 1990, Table 8), although once again definition is critical. The major trend between

Table 2.4 *Percentage urban growth in selected countries in the 1980s*

Country	Period	Percentage urban increase
Egypt	1979–88	30.0
Malawi	1979–88	97.3
Tanzania	1980–88	69.9
Zaire	1979–85	45.1
Brazil	1979–88	35.1
Bolivia	1979–88	45.5
Bangladesh	1979–82	39.7
Iran	1979–86	49.7
Syria	1979–88	45.6

Source: Calculated from *United Nations demographic yearbook,* 1990.

1921 and 1961 was the progressively southern location of the mean latitude, it having shifted from 44°30′ north to 35°44′ north. But the large growth of urban population in countries such as China has probably checked that progress. Indeed, the southern shift itself is largely irrelevant since what is critical is the swing from developed to developing countries.

A final point in the context of the facts of urbanization relates to scale size. It is apparent from the difficulties set out in defining and measuring the city and hence its size, that any comment will again rest to a great degree on definition. Thus, in 1983 in the USA when the definition of the Standard Metropolitan Statistic Area (SMSA) was revised, new concepts were introduced.

The 1980 standards provide that within metropolitan complexes of 1 million or more population separate component areas are defined if specific criteria are met. Such areas are designated Primary Metropolitan Statistical Areas (PMSA's (*sic*)); and any area containing PMSA'S is designated a Consolidated Metropolitan Statistical Area (CMSA) .... Of the 21 CMSA'S, most were previously recognised as standard consolidated statistical areas (SCSA'S), but 5 were formerly SMSA'S within which PMSA'S were newly qualified in 1983.

(US Dept. of Commerce, Bureau of Census, 1990)

In all this complexity it is difficult to identify what is the population figure properly to be taken. Thus, in 1988 the New York PMSA returned a population of 8,567,000 and Newark one of 1,886,000. But they are both part of the New York–Northern New Jersey–Long Island (NY–NJ–CT) CMSA with a population of 18,120,000. The same problem appears in the tabulation in the *United Nations demographic yearbook* of the population of capital areas and cities of over 100,000. Thus, in 1982 the city proper of Shanghai is shown as 6,292,960 but the urban agglomeration as 11,185,100.

Given all these difficulties it is still possible to make some comment on size scale. The first is that the size of some of the largest agglomerations is still increasing so that there seems no limitation to urban size, a point at which any increase becomes impossible. In 1979 the population of Mexico City was recorded as 14,750,182 but its immense growth in the last decade has greatly increased that figure and Dr Gustaus Garza, Director of the Centre for Demographic and Urban Development Studies at the College of Mexico, is quoted as saying, 'I predict that by the year 2010 between 30 million and 40 million people will be living in the Mexico City area' (Thompson, 1990). But, once again, estimates of size depend on definition for, note the phrase in the sentence just quoted – 'the Mexico City *area*' [my italics]. As noted earlier, central city has long given way to conurbation, a word introduced in 1915 (Geddes, 1949) and conurbation to metropolis, and metropolis to megalopolis, a word used first in 1961 (Gottmann, 1961). The last is the type of regional structure which makes the notion of the free-standing town and of an urban–rural divide anachronistic.

Certainly patterns of growth are changing. Especially in the West the old industrial metropolises are losing population from their central cities so that, in USA terminology, many of the PMSAs which developed in the nineteenth century are declining. Thus the Chicago–Gary–Lake County CMSA lost 290,000 people by out-migration between 1980 and 1988 and Detroit–Ann Arbor CMSA, 364,000, while Pittsburg PMSA lost 168,000. According to the British census (Census 1981, Preliminary report for towns, 1981) 'almost all the major cities and other large towns lost population between 1971 and 1981, while there was a growth in many smaller towns away from main centres of population and in the New Towns' (p.5). The Greater London Boroughs showed a total loss of 756,000 or 10.1 per cent. But all this is a process of internal structural change and does not imply that there is some maximum beyond which a megalopolis cannot reach. In essence a megalopolis eventually extends economically, culturally and physically so that the whole national population becomes part of its structure. This is implicit in Octavio Paz's conclusion on Mexico City in his book *The labyrinth of solitude* (1967); 'Either the developed Mexico will absorb and integrate the other, or the undeveloped Mexico, by sheer dead weight of demographic increase, will end up by strangling the developed Mexico.'

2.4 THE CAUSES OF URBANIZATION

Introduction

Given the overwhelming trend to urban growth across the world, it is essential to undertake a review of the underlying reasons for that trend. Those reasons are usually interpreted as being economic, again largely from Western experience, but there is also a range of cultural, social and political bases for urbanization.

Economic development

It is a truism that economic development and urbanization are intimately associated; the development process necessarily involves urban increase. Figure 2.1 is a typical form in which the

linkage is demonstrated. The proportion of the population living in towns is related to economic development measured by gross national product (GNP) per capita. At the heart of the association is the specialization of economic activity, which, indeed, is also seen as the key to the origins of urbanization. Childe (1950), when discussing this last subject in his paper on the urban revolution, postulated a number of features which distinguished towns from the older settlements and foremost among them was the specialization of occupations over and above those dictated by sex and age. In the new towns craftsmen no longer had to remain itinerant and by virtue of their particular skills, detach themselves from the group. By the use of the surplus product they could be translated into a specialized resident section of the new urban society within which therefore the classification of population by occupation became relevant. The emergence of an administrative class (made up of king and priests), the keeping of records, the development of the arts, as well as the extension of trade and the localization of special skills, are all integral parts of the same urban process. Urbanization is, in this way, seen as a product of increasing economic development and of advancing technology. The only way to advance from a subsistence economy is by the specialization of economic activities. But the specialists cannot be isolated for the linkages between them necessitate propinquity. That in turn leads to the accumulation of people which is the process of urbanization. This is reflected in Lampard's dictum; 'city growth is simply the concentration of differentiated but functionally integrated specialisms in rational locales. The modern city is a mode of social organization which furthers efficiency in economic activity' (Lampard, 1955: 92).

This process reached its apotheosis in the West during the nineteenth century with the coming of the factory. There production, and hence jobs, could be broken down into the smallest and most limited of tasks. This, too, was the essence of Fordism, for by the introduction of routine conveyer-belt procedures, even the need for skills, initially the impulse

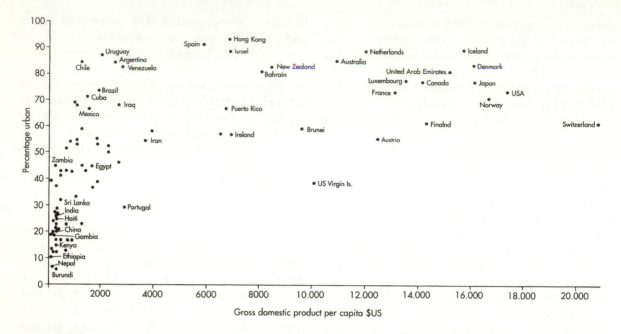

FIGURE 2.1   *Urbanization in relation to economic development. Sources: United Nations Demographic Yearbook 1988, Table 6. United Nations National Accounts Statistics Analysis of Main Aggregates, 1986, Table 1.*

towards specialization, could be dispensed with. But a demand for vast labour forces, necessarily concentrated in space, was generated and hence the creation of entirely new urban conglomerations in great contrast to the earlier discrete, free-standing towns.

Agglomeration economies

To an extent the operation of economies of scale is a restatement of the impact of economic development, but it stresses its cumulative and cyclical nature. Once a large city is created, then the attraction it offers, in terms of supplies of labour and capital, as well as the build-up of infrastructures, will of itself promote growth and initiate a rising spiral of development. This is partly why the largest cities tend to grow fastest. This will be considered in more detail in Chapter 6, but some further comments must be added here.

The first is the need to stress the attraction of greatly enhanced infrastructures in developing countries where industrialization is not the attractive force. In the largest cities the greatest opportunities are perceived in education and training, in health care and in the general improvement of living standards. That the streets of great cities are supposedly paved with gold is no more than a popular representation of the advantages which accrue to scale.

But it is also apparent that the reverse can happen and diseconomies be perceived. Hence the process of counterurbanization and the decline of the older, larger metropolitan centres already noted. The changes taking place do not all move in the same direction.

Political, cultural and social influences

Although it is viable to argue that massive Third World urbanization is, at least to a degree, a consequence of economies of scale, it is less convincing to associate it with development. Indeed it is possible to regard it as a function of underdevelopment. Such an argument must begin with the basic demographic condition of rapidly increasing populations due to declining death rates, but still high birth rates, a distinctive stage in the demographic transition. Kenya can be taken as an example. In 1948 its population was 5.4 million. At the last census in 1979 it was 15,327,203. This is an impossible increase for rural areas to sustain so that an urbanward movement ensues. Although the present urban percentage is only 15.5 it is increasing rapidly. Nairobi doubled its population between 1970 and 1980 and the city proper is estimated at 1,162,189 in 1985 with a further 442,368 in Mombasa.

These greatly increasing populations have been made footloose partly by political change. Most, though not all, of the Third World countries have been until recent times under direct control of colonial powers. That control by its very nature was exploitative and extractive. There were few attempts to develop industry, other than those processes essential before the transportation of raw materials could take place. There followed two associated characteristics. The first was that such exploitation was most easily and profitably accomplished in a rigid political and social system. The second was a cultural division of labour so that the higher order jobs were held in the hands of the colonial power and hence social mobility inhibited, just as a cultural divide was maintained. The break up of all these constraints in a period of decolonization, together with the stimulus of new national feelings, have brought about less stable and far more fluid conditions. This relaxation of constraints has triggered off urbanward migration for, as already demonstrated in the context of agglomeration economies, it is in the cities that new expectations of a post-colonial era are most likely to be met or are perceived as likely to be met. This is likely to be the central problem of post-apartheid South Africa.

But if colonial control has departed, neo-colonial exploitation remains, for these countries are seen as prime targets for manufacturers from developed countries. New markets are urgently sought by Western countries and the goods they export undermine the indigenous crafts of the villages, just as craft industry was itself destroyed in the nineteenth century in the West. The 'bazaar' economy is undermined and traditional craftsmen displaced. The consequence is a drift to the towns. Still further, much new rural development is based on high technology. That necessitates high capital and low labour inputs, so again the country population is displaced. Moreover, as Third World nations try to meet their burden of debt to Western bankers incurred to promote development, or even more fundamentally to offset disaster, they have to concentrate on the production of cash crops for sale to world markets in order to generate hard currency. To compete, such crops have to be produced on a large scale and as cheaply as possible. Again labour is displaced and gravitates to the towns.

That 'gravitation' is greatly encouraged by the nature of modern communications and an increased awareness, even if a mistaken one, of urban living standards. Even in the poorest of slums, television is to be found. But because of the costs of indigenous production of programmes, schedules are filled out with Western-made programmes portraying all the attractions of the consumer society and its affluence. Television, along with the other mass media, breaks down localisms, widens horizons and generates discontent with the rural environment and what it has to offer. New aspirations are aroused which can be met only in the city. Once one migrant has made the move others are encouraged to follow so that migrational chains between country and town are set up. As all this takes place the constraints of cultural and social togetherness are shattered and a footloose population crowds into cities. But there is no concomitant industrialization, no wide range of jobs available. And, also, there is no housing. Hence the massive squatter settlements which will be examined later (Chapter 18). At this point the major feature to be noted is the production of an urbanization without industrialization and an urban growth the cause of which, if it is ultimately economic, is the consequence of a nexus of political and social forces.

2.5 CONCLUSION: NATURE OF THE URBANIZATION PROCESS

Discussion of the divide between what is urban and what rural, revealed more and more as a somewhat barren exercise, has led in this chapter to a review of what can be called the urbanization process. It has become apparent that that process is highly complex and far removed from the simplistic notion of a set of discrete, free-standing towns adding to their populations by inward migration. Indeed, it is possible to identify at the crudest level two totally contrasted processes operating in the contemporary world.

The first of these is counterurbanization. It was defined in 1976 by Berry as 'a process of population deconcentration: it implies a movement from a state of more concentration to a state of less concentration' (Berry, 1976: 17). That movement was put forward as a 'clean break', explicitly reversing the trends which had dominated the nineteenth century and most of the twentieth century. But the concept of a clean break has been contested on a number of grounds. Champion sets these out as four progressive restrictions on the breadth of Berry's definition (Champion, 1989). The first is that counterurbanization cannot subsume the long-established process of suburbanization. Continued metropolitan growth and spillover beyond administrative boundaries might give the impression of a new trend, but it is no more than the continuation of one long established; it is no clean break. The second restricting interpretation is that what is taking place is no more than the long-established process of growth and decline within the urban system, and not the emergence of a new settlement pattern. The third restriction is that counterurbanization must involve measurable deconcentration as witnessed by a decline in the urban proportion. But as this chapter has shown, measurement is never easy and is a problem in itself. Finally, it has been argued that counter-urbanization cannot include those who move but retain an urban way of life, by long distance commuting for example: rural lifestyles must be adopted to justify the counterurbanizing description – or rural retreating as an alternative defining its true nature. Taking all these reservations into account, Champion provides his definition. 'Counterurbanization is deemed to be the prevailing tendency when the distribution of population is shifting from larger to smaller places, where "places" are defined in terms of relatively self-contained areas comprising an urban centre and its commuting and servicing catchment' (Champion, 1989: 32). This does not require the adoption of rural lifestyles, nor the shift of population from town to village. It does however exclude suburbanization and metropolitan expansion.

Champion's book entitled *Counterurbanization* contains nine studies, including the USA and Japan in addition to West European countries. The evidence suggests that such a process as defined by Champion appeared early in the second half of the twentieth century and reached a peak in the 1970s. It has since continued although not at the same level of intensity. From the studies, as many as seventeen reasons for counterurbanization are abstracted (Champion, 1989):

1. The expansion of commuting fields around employment centres.
2. The emergence of scale diseconomies and social problems in large cities.
3. The concentration of rural population into local urban centres.
4. The reduction in the stock of potential out-migrants living in rural areas.
5. The availability of government subsidies for rural activities.
6. The growth of employment in particular localized industries like defence and tourism.
7. The restructuring of manufacturing industry and the associated growth of branch plants.
8. Improvements in transport and communications technology
9. The improvement of education, health and other infrastructures in rural areas.
10. The growth of employment in the public sector and personal services.

11. The success of explicitly spatial government policies.
12. The growth of State welfare payments, private pensions and other benefits.
13. The acceleration of retirement migration.
14. The change in residential preference of working-age people and entrepreneurs.
15. Changes in age structure and household size and composition.
16. The effect of economic recession on rural–urban and return migration.
17. The first round in a new cyclic pattern of capital investment in property and business.

This is clearly a *mélange* of suggestions some of which, like the first, would be ruled out according to Champion's own restricted definition. Others are more concerned with regional shift, although that is part of the process of change. Perhaps the most significant of all is the eighth. In this chapter the prime economic basis for urbanization was seen to be specialization but with the critical need for the many specialists to be contiguous. That contiguity constraint was first greatly modified by the motor car and the urban expressway, and then subsequently broken by the development of electronic communications. Obviously goods have still to be moved even in an age when a premium is on information and its communication, but all the necessities which meant that distribution points had to be associated and located at a city centre have been undermined. Hence the growth of megalopolitan structures, great regional frameworks, and the restructuring of population distribution which has seen the decline of central cities and inner metropolitan areas. There is no doubt, too, that the 'ungovernable' great city with its high crime rate and physical menace has encouraged out-migration to the security of the small, homogeneous town. But even here two distinct changes are in operation. The one is the restructuring of the great metropolitan areas. Peter Hall identifies megalopolis England as:

a giant urban area only in the sense that there is a large tract of the earth's surface where the great majority of people depend on urban jobs and services, and where the impact of these jobs and services, in terms of measurements like commuter zones, service areas and the exchange of goods and information, expands to involve each part of the area in a complex series of interactions with the other parts.

(Hall, 1973: 320)

But even in that part outside megalopolis which Hall identifies as 'areas of standard urban hierarchies' actual behaviour is parallel to that in metropolitan areas. Mobility means that people can travel widely to satisfy different needs, and do so. The second change is the rural retreating, the shift of population from the metropolises to the remote country.

All this is, however, a very long way from the 'urban civilization without cities' (Kristol, 1972) which is seen as an ultimate end of counterurbanization. Certainly the city system is being restructured in a variety of ways and the urban hierarchies of the past are becoming less meaningful. Again remote rural areas once characterized by depopulation are being repopulated. But there is no sign of total structural change in the nature of settlement.

The second and contrasted process to the deconcentration in developed countries, is the concentration which dominates in the developing world, and on a quite massive scale as the example of Mexico City, already noted, illustrates. Daniel R. Vining (1985), in a paper called The growth of core regions in the Third World, has effectively demonstrated the way in which core regions, usually concentrated on capital cities, have dominated population growth in non-Western economies. He points out that:

in 1952 the regions of Ecuador that include Quito, the capital, and Guayaquil, the port, held 30 per cent of the national population. In 1982 the fraction was 42 per cent. If the current rate of redistribution persists, by the year 2000 more than half of the population will live in or near the two cities. Between 1950 and 1980 the fraction of Panama's population living in and around the capital increased from 31 to 46 per cent

(Vining, 1985: 27)

Vining also refers to Jabotabek, the sprawling metropolis centred in Jakata, which grew in population from 6.7 million in 1961 to 13 milllion in 1981, in order to emphasize the monopoly on national resources which such city regions exert. He notes that the Indonesian Ministry of Public Works estimated that $1.2 billion would be needed over 10 years to build a public transport system. But that was seven times the amount included in the whole national budget for all forms of public transport (Vining, 1985: 27).

These great agglomerations of anything up to 30 or 40 million people are very different from the earlier industrial conurbations of the West and it would be mistaken to try to derive some overarching urban process of concentration–deconcentration similar, or parallel, to the demographic transition. The prime aim of the succeeding chapters, therefore, must be to demonstrate how the systematic examination of geographic patterns and processes provides an understanding and an elaboration of what this chapter has sketched.

URBAN PLACES – CENTRAL, PERIODIC AND DISPERSED

3.1 INTRODUCTION

In the last chapter the general issue of urbanization in its world-wide distribution was discussed. The next stage is to change the scale and to examine more intimately those forces which bring about the actual distribution of towns in area. But the starting point is properly the same as for Chapter 2 where the key to general urbanization was identified as the specialization of occupations and the need for their association in area. In a more general way, any advance which is made from a self-contained subsistence economy can only be effected by specialization. It follows that the town is a point of specialized activities, a point carrying out tasks which are best performed either at central, accessible places, or where a high degree of population concentration is necessary. These latter two conditions are, of course, never separate and one can engender the other. A third activity can be added to them, that of the organization of the intermediary, long distance transport, so that the following

can be regarded as characteristically urban functions.

1. Central place functions, or general services, which are carried out for a more or less extensive but contiguous area.
2. Special functions, which are carried out for non-local, non-contiguous areas. These could include extractive and manufacturing industries with world-wide markets or minor industries whose distributive areas are smaller than the general service area.
3. Transport functions, which are carried out at break of bulk points along lines of communication.

To a degree transport functions can be associated with central place functions, since by their nature general services for a hinterland demand local distributive networks. The critical distinction, therefore, is between general and special functions for, as the terms imply, the former are of necessity characteristic of all towns, whereas the latter are not. In more direct and elementary terms, all towns have shops and a retail role is universal, but not all towns manufacture cars or mine coal.

It is for the above reasons that it is usually asserted that the essence of urban character, the universal attribute, is service for a tributary area and hence the study of central place functions became a highly developed field within urban geography. At this point a distinction must be proposed between economic theories of town origin and growth and more pragmatic evaluations. Few towns originated simply as service centres, at least in the economic sense. Most came into being for highly particular reasons, such as elements of effective military or colonial occupation or the exploitation and processing of point resources. Growth, as opposed to origin, might it is true show a greater dependence on local and regional services, but neither exclusively nor even predominantly so. It is worthy of note that when in 1840 in Britain the Select Committee on the Health of Towns found it necessary to introduce some orderly arrangement into their proceedings, it proposed a five-fold classification:

1. The metropolis
2. Manufacturing towns
3. Populous seaport towns
4. Great watering places
5. County and other considerable inland towns not being the seats of particular manufacturers.

If the metropolis (London) is set aside, then three of the remaining four groups are distinguished by special functions. Those with exclusively general functions are relegated to the fifth and final group.

It is now evident that there are two aspects of town function, closely interrelated but nevertheless conceptually quite different. Together they are the determinants of growth though not necessarily of initial location which again, it must be stressed, is a very different issue. The product of that growth is the population of the town, and crude size is most often taken as the measure of significance. But it is worth recalling that Christaller, the father of central place theory, carefully distinguished between a town's importance, which was reflected in its population, and its centrality, or the degree to which it served a tributary area and which could not be measured by population size alone.

In the broader historical sense it can be argued that the emphasis on special functions in urban growth is a consequence of industrialization in the West, a period of some 200 years which created highly specialized towns but which in the broadest purview of urbanization can be regarded as an aberrant phase. Indeed, post-industrialization is largely a reversion to earlier conditions. Deindustrialization has greatly diminished the coalfield-based, smoke-stack industries. Coal itself, once a great locational influence as a power source, is now largely irrelevant in that capacity. Settlements, like individual companies, and the association is not without its significance, have sought diversification. The result has been that the very high degree of specialization, which once so widely characterized the towns of the developed world, no longer exists. It has not disappeared. Detroit still specializes in car making, but that activity has been reduced in its dominance in Detroit and is more widely spread both in the USA and across the world. Effective communications have diminished the significance of locational constraints on forward and backward linkages. It has, therefore, become far less needful to classify towns by their specialisms in order to approach the problem of size and distribution. It follows that the analysis of general functions, through the theory of central places, becomes the first step in the comprehension of the detailed pattern of urbanization. Even so, there are difficulties. It was evident from Chapter 2 that in the West the concept of a hierarchically ordered system of cities within the national territory has been greatly modified both by agglomeration and deconcentration. Under such circumstances, and within megalopolitan structures, central place theory can no longer be used in its classical form as an explanatory basis for urban size and situation. At this stage three points need to be made.

1. It has been contended, most powerfully by de Vries in relation to Europe (de Vries, 1984) and Pred in relation to the USA (Pred, 1977) that industrialization saw no revolutionary overturning of the urban system but rather the confirmation of the mercantile structure of pre-industrial times. If this were so, then the significance of the traditional central place is emphasized.

2. Industrialization affected only a small part of the earth's surface so that there are still large areas where the discrete, free-standing city is the dominant settlement form and where it has not been subsumed into metropolitan regional systems.

3. The declining attention paid to central place theory in contemporary urban geography is part of the post-modern disintegration of a unified, structured, systematic study. The swing to the pursuit of socio-economic 'causes' has engendered a lack of interest in the location and distribution of cities. That 'lack of interest' is partly the consequence of the nature of present-day urbanization as has been indicated, but it is also related to the predominant interest in the geography of welfare. But here one comes back to the point of departure for the spatial aspect of welfare is the effective delivery of services, and central place theory was conceived about the way in which goods and services were distributed from central places. Whatever one's interpretation of the content of urban geography, therefore, a consideration of central place theory is the spatial starting point.

3.2 CENTRAL PLACE THEORY

No consideration of central place theory can proceed far without introducing the name of Walther Christaller whose work *Die zentralen Orte in Suddeutschland* (Central places in southern Germany) was published in 1933 and it is as well to begin an analysis of central place theory with Christaller's ideas (Baskin, 1966). He divided his book into three parts: the first, called the Theoretical Part, was concerned with the setting up of the theory; the second, the Connecting Part, considered practical methods whereby the theory could be tested in the real world; in the last part, the Regional Part, southern Germany was examined, and the methods devised in the second part were employed to substantiate the theory. The practical methods devised were not very successful and have not been used subsequently and it follows that the regional application is also of limited value. It is the theoretical part which is of greatest interest and worthy of close attention.

The introduction to the theoretical part of Christaller's book is entitled 'Are there laws which determine the number, distribution and size of towns?' and the major theme of the first part is contained in that title, for the aim of the author is to establish a deductive theory which reveals the 'Ordering principle' in the distribution of towns. Christaller justifiably set his ideas alongside those of Von Thunen in relation to agricultural productions and Weber in relation to industrial location. Foremost among them is that towns act as central places for the countryside, that they come into being to carry out at a central, accessible place the tasks which the life of the countryside creates. Christaller acknowledges that this basic assumption is derived from previous workers: as early as 1916 Gradman had contended that the distinctive role of a town was 'to be the centre of its rural surroundings and mediator of local commerce with the outside world' (Gradman, 1916), collecting and exporting the local products, importing and distributing the necessary goods and services which the countryside demands.

The significance of this role cannot be measured by the population of a town for whereas size might be a measure of 'importance', it is not a measure of centrality. Indeed it includes in an indivisible total the population due to those special functions which have already been noted (p. 25). Centrality, the degree to which a town serves its surrounding area, can only be measured in terms of the goods and services offered. There are variations in quantity as well as in quality in that offering; there are different orders of goods and services, for some are costly and purchased or needed infrequently and will need large populations to sustain them; others are everyday needs and will require small populations. From this two of the fundamental concepts emerge:

1. **Threshold population.** Christaller did not use this term but the concept is implied in his construct. The threshold is defined as the minimum population which is required to bring about the offering of a certain good for sale or to sustain any service; in economic terms this means the minimum demand to make such an offering viable. Assuming uniformity of income, consumption and taste it can be measured in terms of population numbers. In a concrete way this concept is easily demonstrated by the minimum varying population required to maintain successively a district nurse, a doctor, a specialist doctor such as a paediatrician, a general hospital and a specialized eye hospital. To evaluate these minima is much more difficult.

2. **Range of good or service.** This is the maximum distance over which people will travel to purchase a good or derive a service offered at a central place. At some range from the centre the inconvenience of travel measured in time, cost and trouble will outweigh the value or need of the good, or an alternative nearer centre becomes available. Again this can be visualized in the length of journey to buy bread, which is likely to be very small and hence there will be very frequent trips as against a journey to buy a coat, where in relation to the value of the article and the infrequency of demand a much longer trip would be tolerated. If a sick and dying person can only obtain remedial treatment in one place in the world, then the range of the service could be extended to the absolute maximum measured in miles. Again there are severe practical problems in that most journeys have multiple purposes, one can buy the bread and the coat on the same trip, but

Table 3.1 *The order and arrangement of centres in an urban hierarchy*

Type of centre i.e. rank or order	Number of places	Number of complementary regions	Range of region in kilometres
M (Marktort)*	486	729	4.0
A (Amstort)	162	243	6.9
K (Kreisstadt)	54	81	12.0
B (Bezirksstadt)	18	27	20.7
G (Gaustadt)	6	9	36.0
P (Provinzstadt)	2	3	62.1
L (Landstadt)	1	1	108.0

(*Source:* C.W. Baskin (1957): *A critique and translation of Walther Christaller's 'Die zentralen Orte in Suddeutschland'.* University of Virginia, unpublished PhD dissertation, 345–52).

* the terms used in this table are the ones most frequently employed in an English context, the first part referring to administrative areas such as Gau or Kreis, and the 'stadt' meaning city. The longer terms used by Christaller were Marktflecken (Market locations), Amsstadchen (Office towns), Kreisstadtchen (County seats in the American meaning of the term), Bezirkshauptorte (main district or regional centres), Gaubezirkshauptorte (Gau centres), Provinzialhauptorte (provincial capitals), Landeszentralen (major regional centres). Christaller also added Reichsteilstadte, or RT, for those cities which were more than major regional centres, but not national capitals, with populations of about 1 million and Reichshaupstadte or R (world cities or national capitals with a population of over 2 million.

these are excluded from the present theoretical consideration.

It is possible from the above to isolate two limits in relation to each good or service; one can be called a lower and the other an upper limit (Fig. 3.1). The lower limit is determined by the minimum demand necessary to ensure a commodity or service is offered, that is, the threshold; the upper limit is that beyond which a good will no longer be obtained from a centre, the range.

If these principles are now applied as controls to the development that would take place on an

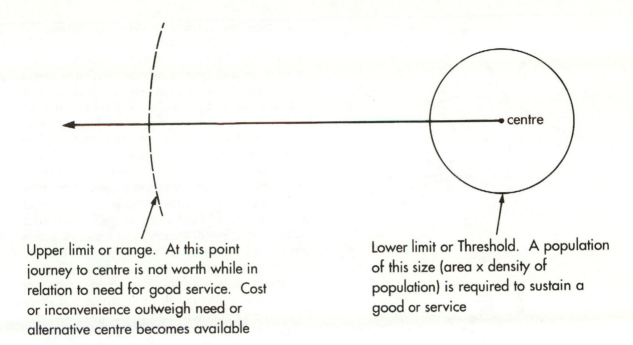

Upper limit or range. At this point journey to centre is not worth while in relation to need for good service. Cost or inconvenience outweigh need or alternative centre becomes available

Lower limit or Threshold. A population of this size (area x density of population) is required to sustain a good or service

FIGURE 3.1   *The lower limit (threshold) and the upper limit (range) of a good from a central place.*

isotropic surface, that is, a flat uniform plain of equal population density and with no variation in wealth or income, then it is possible to derive a model of town distribution.

At this point a word on nomenclature must be interposd. Christaller designated the various ranks of settlements by using the initial letter of the different orders of settlements as they could be identified in southern Germany. They are listed in Table 3.1 and used in this discussion. To continue the argument, a settlement given the rank 'B' is postulated. This serves the surrounding area and if one of the goods which it offers, number 21, be given the upper limit, or range of 21 kilometres, and if the lower limit or threshold is such that it can only be offered at B, then it will be supplied over an area of 21 kilometres radius about B. Now if the next central good, number 20, has a range of 20 kilometres, then there will emerge a ring 1 kilometre wide which cannot be served from B with that good. Further centres are therefore presupposed and if the lower limit, or threshold, is sufficiently elastic, then these must be of

the same B order. But they must be equidistant from B in terms of all the presupposed conditions and if the most closely packed equidistant distribution of settlement points is adopted, there will be six of these on a ring about B (Fig. 3.2). Christaller gives the distance between centres as 36 kilometres although the justification of this is provided later and such arithmetical distances should not properly be part of a deductive structure. For still lower order goods the next locations will be those at the

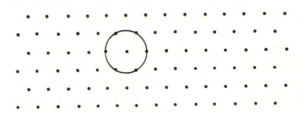

FIGURE 3.2   *The most closely packed equidistant distribution of points (settlements) and the smallest association of centres.*

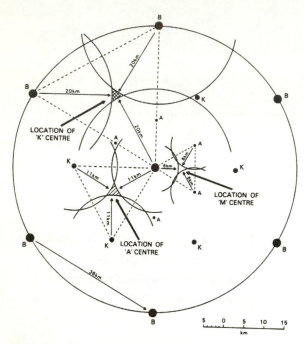

FIGURE 3.3 *The derivation of the urban hierarchy. This diagram is derived from C. R. Lewis, 1977: Central place analysis. Unit 10, Fundamentals of Human Geography, Open University.*

centres of equilateral triangles joining the B centres (Fig. 3.3); at these points K centres will emerge.

Now goods numbers 19, 18, 17, 16, 15, 14, 13 and 12 with ranges of 19, 18, 17, 16, 15 14, 13 and 12 kilometres can be offered effectively at the B and K centres but good number 11 cannot, for once more the unserviced 1 kilometre ring will emerge; a further series of lower order service centres has to be introduced. In this way, a whole hierarchy of central places emerges with towns of equal rank equidistant from each other. Christaller called the area which a town served the 'complementary region' and under the conditions above, these regions were circular. However, to avoid overlap and to match the densest distribution of settlement points, the circular regions were transformed into hexagons.

If there are further goods for which the threshold population required is greater than that provided by the B system, then one place alone from that system may be sufficient and it accordingly will acquire higher value as a central place and given the designation of a 'G' centre. It will have a range of 36 kilometres as suggested above, for it is now apparent that it is at a value three times that of the next lower range limit that transition from one rank of central place to another becomes necessary. New and higher order centres will therefore be found with ranges of 62 kilometres and 108 kilometres. At each of these distances new types of goods can be offered because the range, and thereby the threshold population, is increased. Each central place is able to offer all the goods of lower order centres and, in addition, a distinctive range of goods related to the increased size of its hinterland. On this basis a distinctive series of ranks emerges which is referred to as the urban hierarchy. The pattern finally produced is shown on Fig. 3.4 and in Table 3.1.

It should be apparent, however, that the threshold and range of any one good or service will be an arbitrary figure and consequently it is possible that each one will demand a different hierarchical structure. Christaller understood this to some extent and was aware that throughout the theoretical part he was describing a special case. In the case demonstrated above, there is a strict ordering whereby each settlement serves its own hinterland and an area/population equivalent to the hinterlands of two other settlements in addition, hence it has been called 'the rule of three'. or, using the constant k to express this, then $k = 3$ (Fig. 3.4). But still retaining the hexagonal structure of basic settlements Christaller realized that two other situations were possible where $k = 4$ and $k = 7$ as shown on Fig. 3.5. In order to resolve this problem Christaller associated a controlling principle with each of the three arrangements (Fig. 3.6).

1. The marketing principle: $k = 3$. All areas are served from a minimum set of central places.

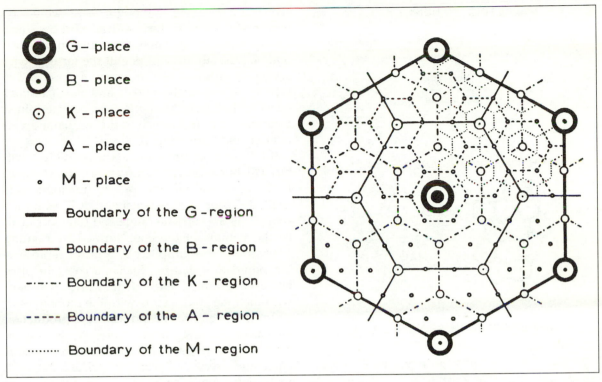

G – place

B – place

K – place

A – place

M – place

━━━ Boundary of the **G** - region

──── Boundary of the **B** - region

─·─·─ Boundary of the **K** - region

----- Boundary of the **A** - region

········ Boundary of the **M** - region

FIGURE 3.4 *The central place system after Christaller. In this construct* k=3.

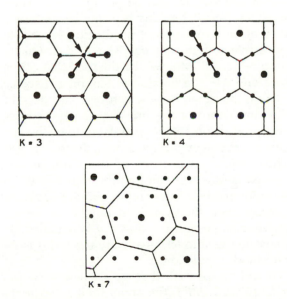

K = 3

K = 4

K = 7

FIGURE 3.5 *The central place system: the three smallest systems: k=3, k=4 and k=7 (after P. Haggett, 1965)*

2. The transport principle: $k = 4$. Here, the distribution is such that as many places as possible lie on main transport routes connecting the higher order centres.

3. The administrative principle: $k = 7$. Efficient administration is the control in this case and this will demand a clear separation of all complementary regions for they cannot be shared administratively.

However, these three principles of arrangement do little to offset the major problem. Christaller presents no argument to justify giving all goods the same thresholds and ranges, or a multiple of these basic measures, so that although his theoretical framework is sustained by these means its relation to reality is made more remote.

Much detail has been omitted, yet the preceding paragraphs give the core of Christaller's ideas. In a situation, not without precedent,

MARKETING PRINCIPLE K = 3

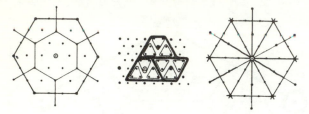

ADMINISTRATIVE PRINCIPLE K = 7

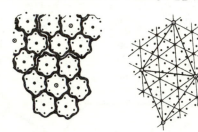

TRANSPORTATION PRINCIPLE K = 4

FIGURE 3.6 *The three controlling principles as identified by Christaller, corresponding to the three smallest systems in Fig. 3.5.*

another worker was arriving at similar conclusions virtually at the same time. This was August Lösch (1954) whose *Economics of location* was first published in 1939. Lösch was concerned with the central problem of the location of economic regions. His emphasis was, therefore, like Christaller's on economic factors as against what he termed the 'natural' or 'political'; raw materials were to be equally dispersed over a flat plain characterized by a pattern of self-sufficient farms equally spaced in the densest possible manner. Spatial differences would emerge from this postulated initial situation owing to the forces of concentration brought to play by the possibility of specialization and of the operation of

economies of scale, although there would be limiting forces in the form of transport costs and the benefits of diversified production. Lösch outlined the situation of one of the farmers who set out to produce beer, a secondary productive activity, it should be noted, and not a tertiary service. The size of his market can be determined from the relations implicit in the normal individual demand curve (Fig. 3.7).

If *OP* is the brewery price, the individual will buy *PQ*. But away from the centre of production, *P*, the price will increase due to transport costs until at *F* beer is so expensive that it cannot be sold. *PF* is the extreme range and total sales will be the volume of the cone formed by rotating *PFQ* around *P*, multiplied by a factor representing population density. So far a constant price has been assumed, *OP*, but that price will vary with the total sold, that is with the economies of scale that can be introduced as production increases. A new curve, therefore, is drawn, based on the volume of the demand cone calculated for a series of arbitrary prices, that is, it represents total demand as a function of brewery price (Δ). On this is superimposed a planning curve, the smallest average cost at which any amount can be produced (π). These must intersect before any beer can be sold, for if they do not, the costs of production exceed the price which sufficient people will pay. On the graph (Fig. 3.8) *MN* will be the total that can be sold and *MF* the maximum shipping distance, that is, the range over which beer can be sold. But the output of the brewery can be reduced and production still be profitable until the two curves are tangential (see broken line of Fig. 3.8). Thus *MiF* is a measure of the minimal area needed before beer can be produced and if this is multiplied by a factor representing population density, then the minimum number of people required to sustain the brewery is obtained. To translate this into terms already used, this is the threshold population.

It will be apparent that we are here very near to Christaller. *MF* is the upper limit and *MiF* the lower as defined above. The two authors have similar fundamental ideas but they have couched them in different arguments.

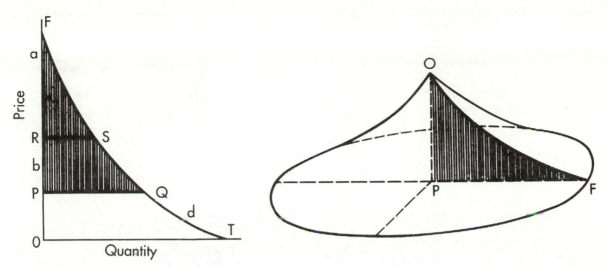

FIGURE 3.7 *Lösch's derivation of the market area and demand cone from the demand curve for a product as a function of distance. For explanation see text.*

Lösch proceeds to demonstrate that his analysis will result in a series of circular market areas but in order to cover the territory most econom-

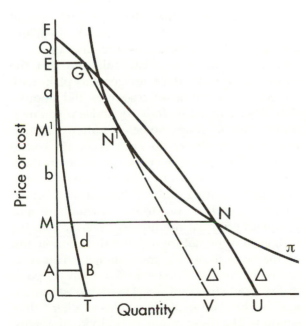

FIGURE 3.8 *Demand curves (Δ), that is quantity demanded as a function of price – and planning curve (Δ), that is the smallest average cost related to quantity produced.*

ically these are converted into hexagons and a proof provided that this is justifiable. The hexagonal structure is dependent on the number of farms required to institute the production of a commodity and this number will vary considerably from commodity to commodity. Lösch, unlike Christaller, allows for this fact and adapts it into his structure. Given the closest packed distribution of farms and hexagonal market areas the smallest number of farms which can be served is three, as in Fig. 3.5. This is the minimum threshold and thereafter the succession continues through four and seven (Fig. 3.5). This is in accord with the argument developed by Christaller, but Lösch continues the logic of this procedure for there is a whole series of succeeding arrangements out of which Christaller only isolated the three smallest cases. The whole series continues 3,4,7,9,12,13,16, etc. Lösch proceeds to consider the *ten* smallest areas and tabulates the relations between them (Fig. 3.9 and Table 3.2).

As intimated earlier, every good sold, every service offered will have a different lower and upper limit, a different threshold and range and indeed could be offered at a variety of different points. There is no reason therefore why a chaos of different meshes should not occur thrown

Table 3.2 *The ten smallest possible market areas*

Area no.	No of settlements completely supplied	Distances between centres	Range
1	3	$a \sqrt{3}$	a
2	4	$a \sqrt{4}$	a
3	7	$a \sqrt{7}$	a
4	9	$a \sqrt{9}$	$a \sqrt{3}$
5	12	$a \sqrt{12}$	$2a$
6	13	$a \sqrt{13}$	$a \sqrt{3}$
7	16	$a \sqrt{16}$	$2a$
8	19	$a \sqrt{19}$	$2a$
9	21	$a \sqrt{21}$	$a \sqrt{7}$
10	25	$a \sqrt{25}$	$a \sqrt{7}$

Source: After Lösch, 1954.

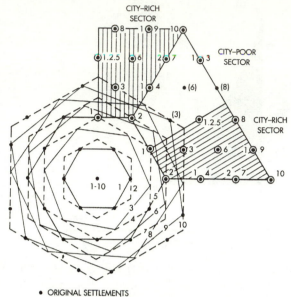

- ORIGINAL SETTLEMENTS
- CENTRES OF MARKET AREAS OF SIZES INDICATED BY FIGURES
  ALTERNATIVE REGIONAL CENTRES ARE IN PARANTHESES

FIGURE 3.9 *The ten smallest market areas (after Lösch, 1954).*

over the supposed uniform plain. Some order can be introduced by arbitrarily centring all the meshes on one point, which is *ipso facto* made the metropolis. Further, by rotating the various nets about this point city-rich and city-poor areas can be produced with a maximum degree of coincidence. But it must be emphasized that only by excessive and unreal simplification, so that a uniform structure is presupposed with a fixed *k*, can Christaller's model be derived. Under Lösch's scheme a hierarchy in the strictest sense of the term, that is with an equal and regular addition of the number of subsidiary places served, does not emerge. But distinctive groupings of the subsidiary places can be found and hence distinctive ranks identified. The confusion here is possibly semantic but clear comprehension is essential if empirical studies are to be properly conducted and interpreted. To conclude, therefore, it must be admitted that while the functional array of services does fall into distinct groupings in all cases, it is only when a fixed *k* is assumed that a strict hierarchy in the Christaller sense is obtained.

At this stage some brief review of the argument is needed. Models of the distribution of central places have been deduced and these hold good under the condition that special functions are excluded, or severely limited, and that service for a contiguous surrounding area is assumed as the sole urban role and that the earth's surface is a flat homogeneous plain with an even distribution of resources and population. That empirical study, or indeed 'common sense', might show no such pattern as that envisaged in theory to exist is of no consequence. The point at issue is whether these models are logical constructions which, although isolating only a limited number of factors operative on town distribution and postulating unreal conditions, nevertheless give insight into the nature of town distribution and of the way in which the national territory is served by towns. Moreover these concepts point in the right direction for they demonstrate the search for unified principles rather than continue the description of individual towns. They connect what had hitherto been isolated fact and they lead to further experimental observation by indicating the most pertinent questions we can ask.

Even so some basic difficulties arise in considering the central place model. Christaller's presentation clearly involved excessive simplification and the concept of market, transport and administrative systems appears as an unconvincing attempt to sidestep the difficulties which accumulate about this simple model. The adoption of Lösch's approach seriously modifies the notion of a clear-cut unequivocal hierarchy. Even more pertinent however, is that at least two other ways of ordering towns in a system have been propounded from an inductive, empirical basis and there appears a marked unconformity between these observances and theoretical models, These are respectively the Rank–size distribution and the Primate city distribution

3.3 THE RANK–SIZE DISTRIBUTION OF CITIES

The Rank–Size Rule for Cities was originally put forward by Felix Auerbach in 1913, but its development and popularization is associated with George K. Zipf in a volume published in 1941 and entitled *National unity and disunity*. In loose terms the 'rule' states that if the population of a town is multiplied by its rank, then this will equal the population of the largest and highest ranked city. It is often stated as

$$R^n SR = M$$

where R = the rank of a city, SR = the population of a city of rank R, and M and n are constants. But where R is 1, whatever the value of n, S equals M so that the constant M is the population of the largest city. It is also less reliably argued that in Western industrial countries the constant n is equal to unity so that the descriptive version given above holds good and a log-size plot against rank will give a straight line graph. Characteristic plots showing this for the period 1950–75 are displayed in Fig. 3.10(a).

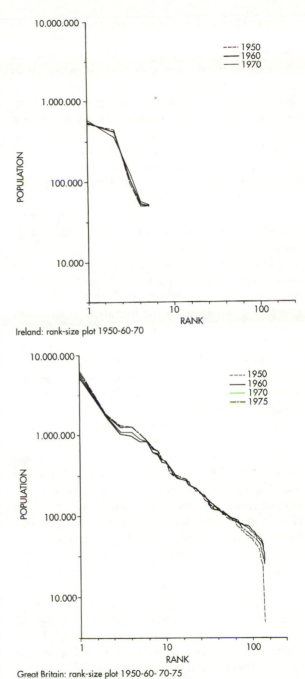

Ireland: rank-size plot 1950-60-70

Great Britain: rank-size plot 1950-60- 70-75

FIGURE 3.10 *Rank–size plots for Great Britain 1950, 1960, 1970, 1975 and for Ireland 1950, 1960, 1970 (after P. Hall and A. Hay, 1980). Source: P. Hall and A. Hay 1980: Growth centres in the European urban system. Heinemann Educational Books, Figures 4.18a and 4.18b, pp. 166–7.*

Superficially the rank–size rule seems to contrast with the conclusions of Christaller and Lösch in nearly every aspect. It is supposedly an observed fact and not a deductive model. It is true that Zipf saw it as part of a broader theory in which any economy was subject to two forces, the one moved towards diversification where a large number of communities benefited from being located near to raw material sources and minimized transport costs; the other force was one of unification where a small number of large communities minimized movement of finished goods to the consumer. From these opposing conditions a balance emerged which was reflected in the rank–size rule.

It is apparent at a superficial level that if the rank–size relationship implied in the rule is accepted as an accurate finding, then it completely contradicts the hierarchical situation derived from deductive argument. Christaller postulated a stepped and ranked distribution of size classes; Zipf's finding shows a smooth relation where no distinctive classes can be identified. But the conflict has been considerably exaggerated and at this stage there is no need to accept the view that the rank–size rule undermines the idea of a hierarchy. Thus, attempts were made to show that the two concepts are not incompatible even when applied to the same data (Berry and Garrison, 1958a) and indeed the very meaning of the word 'hierarchy' needs a clear and precise definition in this context. But there are more cogent and simpler bases for no undue disturbance being generated by the introduction of Zipf's work. The factual truth of the rule has been challenged and certainly, it has been shown that it applies only to large areas (Stewart, 1958). It measures size by population totals and not centrality as envisaged by Christaller and in consequence it embraces not only central place functions but special functions as well – those special functions which need to be examined separately from central place functions as indicated at the beginning of this chapter. When *all* urban functions are massed together then it is perfectly possible that the wide range of other functions, over and above central place functions, can transform a ranked hierarchical array into a continuous rank–size relation. The empirically observed rank–size rule need not therefore prevent the acceptance of the theoretical constructs of Christaller and Lösch.

3.4 PRIMATE CITY DISTRIBUTION

The rank–size graph for Ireland, shown in Fig. 3.10, is manifestly not a straight line. It rather suggests dominance by one (Dublin) or two (Dublin and Cork) cities. A proposal that such was a frequent situation was formulated by Mark Jefferson as long ago as 1939 in a paper entitled 'The law of the primate city'. Jefferson propounded the view that 'the largest city shall be super-eminent and not merely in size, but in national influence'. On an intuitive level this type of dominance of the whole economic, cultural and political scene by one city is associated with what is sometimes called 'The Great Tradition' or with a 'primary orthogenetic city' where all these definitions refer to a situation where the peoples who made up the pre-civilized folk more or less shared a common culture which was the matrix of the urban culture which developed from it.

This situation has clearly perplexed research workers on central place theory. Not only was there the problem of the way in which such a city system as envisaged by central place theory had emerged, if the concept has any meaning other than as a purely static and unreal postulate, but further, an observed and contrasted situation was proposed in the dominance of the primate city. The general response was that postulated by Hoselitz, who maintained that a 'comparative analysis of central place might show... the degree to which a rationalized western system of economic organization and activity had penetrated a given country and might be a fairly good measure of the breadth on which the economic development of a country has taken place' (Hoselitz, 1955: 278). That is, a hierarchical structuring is indicative of advanced economic development. This point is taken up again in Chapter 5 (86 *et seq.*) and Chapter 18 (p. 340).

3.5 CONCLUSION

At this point some conclusions can be drawn from the general consideration of central place theory and other interpretations of the relationship between settlements in national or regional systems. At the outset the notion of a 'system' must be retained for all interpretations necessarily assume that towns are not independent entities but interact with each other and that interaction extends from the smallest town to the largest city. What is less clear is the extent to which regions can be abstracted from national systems, for ideally the city system must be closed, a situation never attained in reality but maximized within national frontiers.

The conclusion from this chapter can be set out as follows:

1. **The evolution of the city system.** If different conditions can be discerned, such as hierarchical or primary arrays, then it is possible that they are not competing interpretations of contemporary structures but rather manifestations of structures which can be linked in some historical or evolutionary sequence. This has been discussed elsewhere (Carter, 1983) but will also be briefly reviewed in Chapter 5 (p. 85).

2. **Empirical testing.** To a large, even an exclusive extent, early work on central place theory entered a blind and unproductive alley. There were innumerable attempts (Berry and Pred, 1961; Andrews, 1970) empirically to establish urban hierarchies and to examine whether the levels or ranks identified had some universal validity. But a theory or model in the social sciences is not intended to portray a direct representation of reality, but rather to offer means of understanding a reality which is highly complex. It is an exercise in parsimony.

Much of the empirical work which dominated urban geography in the 1960s is now seldom cited and presumably considered irrelevant. What is still crucial, however, is the way in which a region or nation is served by the system of cities and what light that system throws on regional organization and structure.

3. **The distribution of service centres.** Central place theory by its isolation of the twin concepts of threshold and range and the study of the nature of their interrelationships, identified the crucial influences upon the emplacement of the offering of goods and services, both public and private. At a contrasted scale, therefore, it spawned extensive consideration of locational constraints and decision making via a whole range of offshoots such as the optimum location for public services, such as hospitals, or for private ventures, such as retail stores. But however far such studies, for instance retail geography, have moved to become highly sophisticated disciplines in their own rights, they owe their origins and find their roots in central place studies.

4. **Variations on the hierarchy.** There are two variations which can occur on the basic theme of the hierarchy of goods and services creating a hierarchy of towns. The first has already been suggested. It is possible that with modern mobility there is no overriding necessity for all the goods and services to be offered at one and the same place. That was a basic assumption made by both Christaller and Lösch. But it can now be rejected in that people can move easily around a succession of 'centres', each one possibly offering specialized shopping or services. It is possible to envisage a dispersed city. The second variation is that where a necessary threshold cannot be reached, then a solution can be offered by an adjustment of 'time', that is the goods and services can be offered periodically, at certain set times, so that the overheads of permanency are reduced. A further version of such an adaptation to low

thresholds is the mobile shop or library van which visits set locations at fixed intervals. At the lower end of the scale, therefore, modifications can occur so that goods and services can be provided even where the threshold at a single central place is well below that required for viability. Both these conditions will be taken up later.

The four items of this conclusion provide the basis for the continuation and development of the consideration of urbanization.

4

HIERARCHIES, SPHERES OF INFLUENCE AND URBAN SYSTEMS

4.1 INTRODUCTION

As the last chapter concluded, the consequence of the development of central place theory as the main prop of an emerging urban geography was the generation of a large range of studies which sought:

1. to test whether hierarchical structures could be identified in the real world; and
2. assuming that hierarchies did exist, to set out the system for particular regions or countries.

4.2 METHODS OF ESTABLISHING HIERARCHIES

The plethora of measurements of the urban hierarchy were based on the core notion of a count of centrally located private and public services. There were, however, three major problems.

1. There was no common consensus on what to count, especially in retailing. At the most elementary level there is a distinction which can be made between establishments, functions and functional units where the use of

   (a) establishments means that one shop whatever it sells is taken at its crudest value and considered as one unit. This does tend to be the form in which census data occur;
   (b) functions means that each type of goods outlet is counted regardless of the association in establishments.
   (c) functional units means a count of the total number of outlets regardless of type.

   An example will clarify these possibilities. There are three shops A, B and C. A is a grocer and a greengrocer with an off-licence for the sale of wines and spirits; B is a greengrocer and fishmonger; C is a grocer and butcher. In Britain all these combinations occur quite commonly. A count would reveal three establishments (A, B and C), five functions (grocer, greengrocer, wine and spirits, fishmonger, butcher) and seven functional units (two groceries, two green-groceries and one each wine and spirits, fishmonger, butcher). Clearly, surveys carried out on different bases, each of which can be justified in central place terms, can produce widely varying results. Moreover, such counts were common in the days before supermarkets and hypermarkets made assessment even more questionable. Floor space is, of course, a universal measure but seldom is it easily available.

2. There is a possibility of considerable variation dependent on the areal unit of study adopted. Thus data can be assembled for settlements including all the suburban shopping areas, or just for the central core. Again, the growth of large metropolitan areas and the development of out-of-town shopping centres have greatly changed a situation in which Christaller envisaged discrete, free-standing towns with single central business districts.

3. The third problem is less theoretical and more a practical and operational one. Any of the traditional approaches involves a count of establishments, functions or functional units. Such data are seldom available from censuses so that field work of a very time-consuming nature becomes necessary. Whereas that is feasible for limited areas where most of the early work seeking to identify hierarchies was done (Berry and Garrison, 1958b; Davies, 1967), it was impossible for large regions or nations. The way out was quite arbitrarily to select what were regarded as diagnostic criteria which could be assembled from easily available sources such as telephone or commercial directories. This was the form taken by the pioneering work of A. E. Smailes in Britain (1944, 1946) and of J. E. Brush in the United States (1953). Thus Smailes was intuitively aware of what he called 'the fully fledged town' which was identified as a distinctive stratum in the British urban hierarchy. He then argued that the level was characterized by the occurrence together of certain key establishments which made up a 'trait complex'. This was made up of:

A   Branches of three out of what were then called the five major banks together with a Woolworths store *diminishing to*
A^1 three branch banks only *and*
A^{11} two branch banks
B   Grammar School and hospital
B^1 only one of the above
C   Cinemas *diminishing to*
C^1 only one cinema
D   Publication of a local newspaper

The complete trait complex, *ABCD* was reduced to $A^1B^1C^1D$ or A^1BC for qualification as a 'fully fledged town'. Sub-towns which did not meet the minimum qualifications were identified and, by extension and the use of additional criteria, 'major towns', 'cities' and 'major cities' were introduced to produce a complete ranking of the urban settlements of England and Wales (Smailes, 1946).

Smailes's work is now mainly of historical interest. The merging of banks, and the proliferation of building societies, the attempted abolition of grammar schools, the closures and developments in the National Health Service and changes in the demand for public entertainment with the rise of television, all have made the criteria outlined above outdated. But it is a good example of the sort of work which characterized the initial introduction of central place ideas. Subsequently the impact of the so-called quantitative revolution in geography engendered more and more sophisticated procedures, though often based on sketchy data.

It is now possible to argue that most schemes still extant are variations on that introduced by W. K. D. Davies (1967) which is reasonably simple and effective. A location coefficient of a single outlet of any functional type is determined by the formula

$$C = \frac{t}{T}\,100$$

where C is the location coefficient of function t, t is one outlet of function t and T is the total number of outlets of t in the whole system. 'Multiplication of the relevant location coefficient by the number of outlets of each functional type present in a settlement gives the degree of centrality (centrality value) imparted to each settlement for every type of function. A functional index, the measure of centrality in Christaller's meaning, is derived by the addition of all the centrality values attained by any settlement' (Davies, 1967: 63). If there are 200 grocers in the area examined then the location quotient for that function is:

$$C = \frac{1}{200} \times \frac{100}{1} = 0.5$$

and if there are 23 grocers in settlement A the centrality value imparted to that settlement by grocers is $0.5 \times 23 = 11.5$. If there are only two jewellers' stores in the area, one of which is in A, then C is 50 and this is added to the 11.5 from the grocers, so that the Functional Index is $11.5 + 50 + n$, where n represents all the other functions identified. Essentially what the procedure is doing is dividing a theoretical total centrality in an area proportionally between the various settlements.

It is apparent that two of the problems discussed still remain and a third has been added:

1. It is still difficult to resolve what constitutes a settlement.
2. All the functional types of an area have to be enumerated constituting a major piece of data collection. If more sophistication is required some form of weighting can be introduced, ideally by turnover but that is never available so that some surrogate such as floor space can be adopted. Data assemblage then becomes even more difficult.
3. The added difficulty is that the method assumes a closed and bounded system with no external leakages. That is totally unreal.

The response to these problems which arise over the constructing of hierarchies has been to fall back on population figures, although it will be recalled from Chapter 2 that these suffer from the same problems over the definition of area. However, inadequate as they are, they are easily accessible and the standard means of scrutinizing hierarchies has become the graph of population against rank. Inevitably all the adventitious increments of population must blur the appearance of clear, discrete ranks of towns. It may well be that the move away from the recognition of urban ranks has been accelerated by the move away from the use of functional arrays and the adoption of population totals. It is also a response to the shifting emphases within urban geography itself. Initially concerns were primarily scholarly in the sense that basic questions were asked as to whether towns within nations or regions were grouped into distinctive ranks or levels in terms of the services they offered. That, in turn, was seen as a contribution to the comprehension of the factors which determined the spacing and size of towns, part of an answer to the basic geographical question of location. But the greater stress on relevance, perhaps even the nature of research funds, has seen that question set aside as against more pragmatic questions as to the role of the hierarchical structuring of services in planning both private and public. Large questions as to the nature of national hierarchies, or of international comparisons between systems in developed countries, have become less fashionable as against the basic analyses relevant to the planning of specific operations, such as retailing or public services.

4.3 METHODS OF ESTABLISHING TRIBUTARY OR MARKET AREAS

The investigation of the areas tributary to towns, variously given the names hinterland, urban field or urban sphere of influence, has perhaps remained more a part of geographical study, although also more directed towards specific planning ends than to the evaluation of regional or national structures. The theoretical establishment of urban hinterlands, or complementary regions as he called them, was not part of Christaller's original schema, at least they were not marked on his diagrams, although they were basic to the derivation of his ranking system. Theoretically every rank or level in the hierarchy would have a particular sphere of influence. Thus the town at the head of a five level hierarchy, as set out by Christaller, would have five distinctive spheres, the largest corresponding to the whole region and the others to each of the lower ranks. The existence and the definition of these various spheres depend on the assumption that people will travel to the nearest place at which a good or service is available. That assumption was always modified by multi-purpose trips, but increased mobility has brought the possibility of much wider choice in consumer behaviour where distance (cost and time) is not necessarily a constraint. Spheres of influence must remain no more than generalizations of movement and are not neat, discrete boundaries.

There have been two approaches to the identification of urban spheres of influence. The first looked outward from the town in order to identify the various areas served by it, though many in the public sector were statutorily determined. The second has looked inward from the countryside and has been more concerned with consumer behaviour and the ways in which people use the various centres. Given the general trend of interest towards consumer behaviour the second has become dominant. Indeed, the first is now seldom used although it is still a valid procedure, even if again mobility has done away with the neat compartmentalization of service areas which once covered countries and looser and more fluid structures are now in place.

Early attempts at defining urban spheres of influence were closely associated with Smailes's 'trait complex' for if the elements of that complex defined a 'fully fledged' (i.e. a particular rank or grade of) town, then the areal extent which those elements dominated, delimited the urban sphere of influence. Moreover, if they were associated within the town, then the areas over which they extended their influence should also coincide. 'It is usually found that the fields shown by various indices group themselves sufficiently to allow broad recognition of composite fields at a series of functional levels which correspond to the more clearly defined ranks of the urban hierarchy' (Smailes, 1947: 151).

The procedure, therefore, was to plot the areas which the selected criteria covered, for example the area from which a bank drew its customers, a hospital its patients, a cinema its clients, a school its pupils and a local newspaper its readers. An example of this method is given in Fig. 4.1 where the boundaries of seven selected functions exercised from Aberystwyth were indicated. It would be meaningless to resolve these lines into *one* sphere by any such method as identifying a median between the maximum and minimum areas, for at least two spheres are identifiable, an inner intensively dominated one and an outer extensive sphere with a possible intermediary area between, indicating the town operating at three hierarchical levels.

This method was and is a quick and useful one but it is open to the same sorts of criticism as the selection of the indices for a 'trait complex'; the choice is arbitrary and often an ill-considered rag-bag of areas is assembled. It would be no easy task to justify the seven criteria chosen for the map of Aberystwyth and a selection of seven others might possibly present a very different picture. For this reason, as well as the association of many areas with administrative boundaries, the method is now rarely found.

Since the sphere of influence is established about the town, one might well argue that the approach to its definition must be the reverse of that of ranking the town and be directed to the hinterland. The second group of methods, therefore, has sought to identify areas within which movement to centres takes place. Here the earliest methods, at a time before car ownership was common, were by examining the physical movement of vehicles, particularly buses. A flow diagram of bus services produced

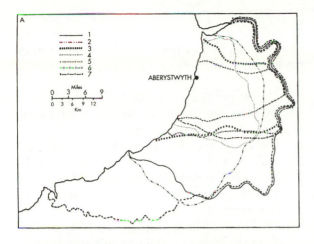

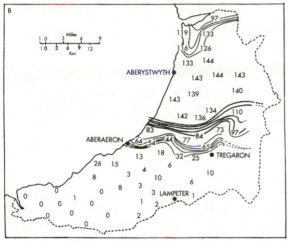

FIGURE 4.1 *The urban sphere of Aberystwyth in 1960 as demarcated by: A: an arbitrary array of criteria; B: a questionnaire survey. In Figure 4.1A the criteria used were – 1 An insurance company's area; 2 Baker's delivery area; 3 Agricultural Cooperative delivery area; 4 Bank area; 5 School catchment area; 6 Postal district; 7 Veterinary surgeon's area. On the lower map, the area of maximum purchase for twelve goods and services were asked of twelve people randomly selected for each kilometre square of the national grid. The replies nominating Aberystwyth were summed for each square and the totals used for drawing isopleths. Note the correspondence between the maps although the bases were quite different.*

a visual pattern of movement into (and out of) a centre and, by inspection, a sphere of influence could be delimited (Green, 1950). Apart from the fact that the private car now carries a large share of this movement, the crudity of the method meant that while quick and easy, it was not acceptable at a rigorous level of investigation. Even so, commuter movements have now become one of the major criteria, echoing these early methods as will be indicated later in the chapter (p. 51–2).

Most contemporary work uses a method which is a variation on that introduced by Bracey as long ago as 1953 (Bracey, 1953). A questionnaire is drawn up which aims to elicit the usual place at which a standard array of goods and services is obtained. Three difficulties immediately arise, all of which are implicit in the last sentence:

1. How is the sample of population to answer the questionnaire chosen? There is little doubt that consumer behaviour will vary with such characteristics as income, social class, possession of a private car and place of work. Properly, the sample should be stratified according to characteristics such as these, but since basic data are lacking it is seldom possible to do this in a systematic way, although questions as to occupation of household head and possession of a car and so on can be asked and used later as variables. The selection of respondents is, therefore, quite arbitrary, but with an attempt to obtain a fairly uniform ground coverage. Rowley (1967), in a study in Wales, obtained five completed questionnaires from each quarter kilometre grid square on the British Ordnance Survey map. If an adjustment to population is made, so that a given percentage sample is obtained, then the task is very great indeed and also very monotonous. There are inherent problems in selecting respondents even from electoral registers or other lists and these are seldom overcome, except in very small-scale studies where the behaviour of specific groups is the prime purpose.

2. How is the standard array of goods and services determined? C. R. Lewis (1970) in a study of only the small towns of mid-Wales identified 157 different functions. To put all these on a questionnaire, which is to be completed by large numbers of people, is not practicable and some selection *has* to be made. Rowley, in the study noted above, used 20 goods and services, and attempted to range these over high order and low order categories. But whatever is done the selection remains subjective and arbitrary.
3. What is meant by 'usual place of purchase'? Attempts can be made to refine this by defining it as the place at which the largest expenditure was incurred over the period of a week or longer, or just the last place of purchase can be recorded. Again, a measure of vagueness can creep in where the respondents are faced with providing an instant answer from recollected actions.

There is a solution to the last two difficulties and this is to have selected respondents keeping a complete account of the place and amount of all purchases over a given period – a complete diary of expenditure. This solution increases the difficulties of selecting the sample to an enormous extent and can only be used when answers to specific hypotheses as to consumer behaviour are required, rather than a general interpretation of urban spheres (see Chapter 5, pp. 79–83).

Once the questionnaires have been completed the procedure is straightforward. For each of the central places a point is awarded each time it is nominated as the place of purchase of a good or service. Each questionnaire cannot be treated as an isolated spot score since individual eccentricities have to be eliminated. A number of questionnaires are, therefore, aggregated for a given grid square on a map, for an administrative area or for an arbitrarily defined settlement. In Rowley's study five questionnaires for every quarter kilometre meant that, with twenty goods and services on the sheet, a total of one hundred answers was obtained, each nominating a centre used. In this case these could be treated directly as percentages and these scores can be used to denote the patterns of affiliation. Isopleths can be interpolated and the following points can be made:

1. The 50 per cent isopleth for any centre will indicate the point at which a settlement loses dominance.
2. The bunching of isopleths at any point will indicate a sharp change of gradient and the limit of a sphere of influence related to a bundle of functions at a given rank in the hierarchy.
3. The 1 per cent isopleth will mark the absolute limit of a settlement's influence.

An example of this sort of exercise is given in Fig. 4.2.

4.4 GENERAL CENTRAL PLACE ANALYSIS AND ITS APPLICATION

The construction of urban hierarchies for areas large and small, together with associated spheres of influence in some but by no means all cases, was a central theme of urban geography during the 1960s and continued into the 1970s. But such work fell away as the topic became worked out. That might seem a damning phrase but expectations had been raised that here was a truly theoretical geography but which real world investigation revealed as more and more flawed. To regard it as no more than an aid to interpretation of a much more complex problem was a significant retreat from the view that 'the initial and growing beauty of central place theory is geography's finest intellectual product' (Bunge, 1962: 129). During the decade of the 1980s only two significant general works have appeared on the topic. Leslie J. King's *Central place theory* (1984) was essentially a general statement of earlier work; in the art world it would have been called a 'retrospective'. Even so, it is a good, clear, basic statement. John U. Marshall's *The structure of urban systems* (1989) is virtually the only contemporary restatement. It is, therefore, both

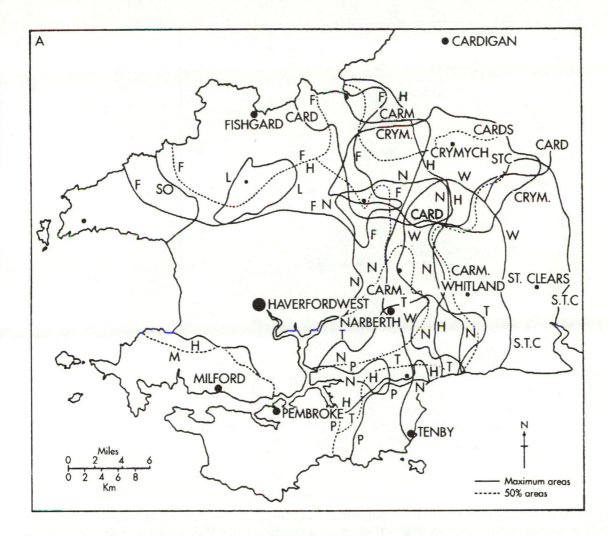

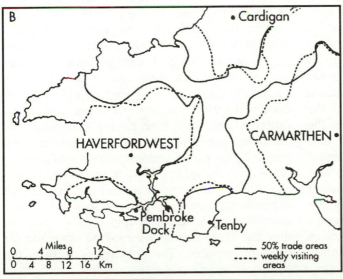

FIGURE 4.2 *South-west Wales: urban spheres of influence in 1967 (after G. Rowley). A: maximum and 50 per cent areas. The letters on the lines refer to the named centres. B: the 50 per cent areas and the areas delimited by weekly shopping visits.*

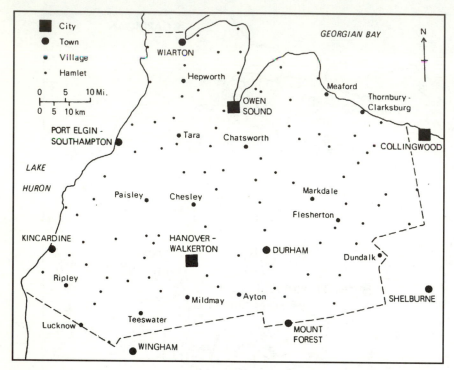

FIGURE 4.3 *The hierarchy of central places in Bruce and Grey Counties, Ontario. The method of classification is explained in the text (after J. U. Marshall, 1989). Source: J. U. Marshall 1989: The Structure of Urban Systems. Toronto: Toronto University Press. Fig. 7.4, p. 229.*

appropriate and revealing to use Marshall's work in demonstration of conventional central place analysis.

Marshall uses a form of Davies's method (p. 41) in the establishment of an urban hierarchy for an area in southern Ontario, made up of Grey and Bruce counties. The area, together with the classification of central places is shown in Fig. 4.3. It should be noted that three of the places are twin settlements. Table 4.1 lists the centres but omitting the 79 smallest; it also shows the populations, the numbers of central functions, the numbers of functional units and measures of nodality and centrality. Marshall relabels Davies's centrality index 'nodality' on the grounds that it incorporates the settlement's own contribution to its array of central functions – what he calls self-orientated nodality.

Centrality is derived by subtracting from the measure of nodality a sum which is calculated by finding the ratio of nodality points to population for the whole area and applying it to the nodality total for each centre to find the sum to be subtracted. The basic data were derived from field work and directories for the period 1972–4, and those dates are a significant indication of the time of the actual study as against its publication in 1989. However, having established the centrality measure, the critical problem still remains – do these demonstrate a set of discrete ranks or a continuum? Here the analysis loses its quantitative base and is related to a scrutiny of dependence derived from knowledge of the area; that is, how do people behave, how do they use these centres by moving up-grade for superior services. The conclusion is that 'the

Table 4.1 *Population and functional data for the twenty-five leading central places in Bruce and Grey counties, Ontario*

Central place[a]	Population[b]	Number of central functions	Number of functional units	Nodality[c]	Centrality[d]
Owen Sound	18,469	76	454	2,318	969
Hanover-Walkerton	9,542	69	309	1,263	566
Wiarton	2,222	50	105	412	250
Durham	2,448	56	103	397	219
Kincardine	3,239	56	138	452	215
Port Elgin-Southampton	4,891	56	152	559	202
Lucknow	1,047	42	63	212	136
Dundalk	1,022	41	66	191	117
Chesley	1,693	49	72	229	105
Markdale	1,236	43	69	192	102
Ripley	448	29	40	121	89
Flesherton	524	31	48	115	77
Thornbury-Clarksburg	1,609	41	79	182	65
Tara	643	29	42	107	60
Teeswater	983	35	54	116	45
Chatsworth	399	25	36	74	45
Meaford	4,045	51	107	339	44
Paisley	793	32	45	94	36
Ayton	423	25	30	63	32
Hepworth	372	17	28	49	22
Mildmay	963	30	42	90	20
Cargill	233	15	17	29	12
Holstein	176	10	12	24	11
Ceylon	66	6	6	16	11
Feversham	161	12	14	21	9

Source: After Marshall, 1989

[a]Places are ranked according to their centrality scores, which appear in the extreme right-hand column. Of the 79 centres not shown, none has a centrality score greater than 5.

[b]For 1971, the census year closest to the field seasons when the functional data were compiled.

[c]The 25 places listed in the table account for 95 per cent of the total nodality within the study area. The remaining 5 per cent is accounted for by the 79 small centres not listed.

[d]Values for both centrality and nodality are rounded to the nearest integer.

procedure described above results in the identification of four hierarchical orders. For convenience, we will refer to these orders, in ascending sequence, as hamlets, villages, towns and cities. [These labels, as used here, carry no implications with regard to the legal rank of incorporated centres.]' (Marshall, 1989: 288). The crucial difficulty is that in the end a statistical

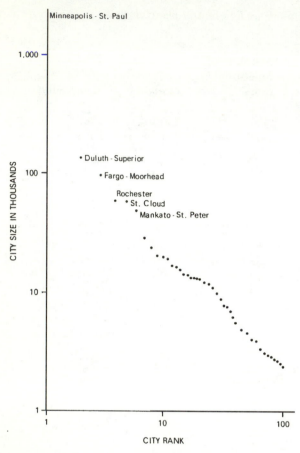

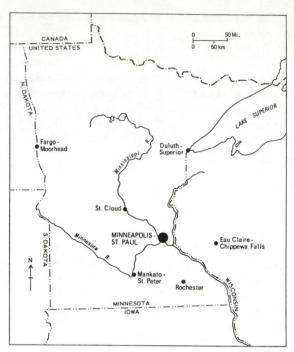

FIGURE 4.5 *Large central places near Minneapolis-St Paul, Minnesota (after J. U. Marshall, 1989). Source: As for Fig. 4.3. Fig. 7.2, p. 220.*

FIGURE 4.4 *Rank–size continuum of cities in southern and central Minnesota, 1980 (after J. U. Marshall, 1989). Source: As for Fig. 4.3. Fig. 7.1, p. 218.*

evaluation of centrality is followed by a subjective ordering into ranks. To a large degree purely statistical attempts to prove discrete hierarchical levels were chimerical and Marshall's 'eyeballing' is acceptable although indicative of the nature of central place resolution.

Marshall also presents an analysis of a larger area, though in a different location, the region tributary to Minneapolis–St Paul. He begins with a rank–size diagram (Fig. 4.4) which, apart from the major break between Minneapolis–St Paul and Duluth–Superior, seems to demonstrate that 'further evidence of hierarchical structuring appears to be absent' (Marshall, 1989: 219). But he continues that further inspection of the graph, together with analysis of figures for population sizes and distances apart, suggests that there are five medium sized towns within Minneapolis's orbit – Eau–Claire–Chippenwa Falls, Rochester, Mankato/St Peter, St Cloud and Duluth–Superior (Fig. 4.5). These, he argues, are similar in size, although Duluth–Superior has a larger population derived from its special function as ore exporter and steel town, are located at similar distances from Minneapolis, are evenly spaced about that city and each of them is the largest centre in its own area. Here, then, is a second rank in the hierarchy and the analysis is continued by shifting down a level and Mankato/St Peter is taken up

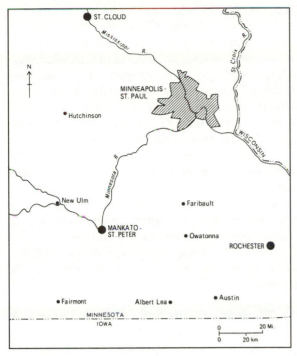

FIGURE 4.6 *Intermediate-level central places near Mankato-St Peter, Minnesota (after J. U. Marshall, 1989). Source: As for Fig. 4.3. Fig. 7.3, p. 224.*

(Fig. 4.6). Here again six smaller towns are found in its orbit having the same distinguishing characteristics, that is similarity of size and distance from Mankato, even spacing and dominance of their own areas, although two exceptions have to be observed. Again one of the towns in Mankato's sphere, New Ulm, is taken and the procedure repeated with seven smaller settlements identified in its orbit. Subsequently, a suggestion is made, which is not followed up, that two additional lower orders could be identified.

This, again, is based on no quantitative evaluation but Marshall argues that the critical issue, as in his detailed scale study, is the progressive dependency of each lower rank on the higher, which he again sets out in perceived terms. These sorts of relationships, he contends, form 'a town chain' (Brookfield and Hart, 1971) in

which each link in the chain is a hierarchical rank.

At this point a further question arises, one which Marshall does not answer, and that concerns the relationship between the Minneapolis–St Paul system in the United States and that based on Owen Sound in southern Ontario in Canada: that is, does the Owen Sound system have any relationship with that about Minneapolis or are they separate and unique? On examination the size of Owen Sound would appear comparable to New Ulm and, if an equivalence is accepted, the two can be merged into a hierarchy of six orders with a suggestion, if no more, of generality in a Western, developed economy. This was one of the assumptions of early empirical work on central places. As long ago as 1957 A. K. Philbrick, in a paper, the title of which indicates the presumptions of the time – Principles of areal functional organization in regional human geography – had set out to demonstrate a seven stage hierarchy of nested functions using Chicago as his exemplifying area, although New York stood out as constituting the highest rank. It is interesting that given the similar supremacy of New York, Marshall's scheme is also made up of seven ranks. But the general looseness of analysis and the complexity of the real world, especially in highly industrialized areas, has meant that such universal schemes have not been given much attention. One other point needs to be observed at this juncture. Although Marshall was able to use the number of urban functions collected from field work for his more local study, once he moved on to the metropolitan scale, he reverted to the use of population totals; measurement by the aggregate of functions becomes an immensely difficult task at the largest scale and unless arbitrary selection is used then reversion to size by population becomes a standard procedure.

The attempts, such as those by Marshall, to set up hierarchical systems for areas ranging from small localities to whole nations, were characteristic of urban geography during the 1960s and 1970s. It is true that the theoretical basis became stated in more sophisticated terms (Beavon,

1977) and that advanced statistical techniques of taxonomy were brought to bear in real-world analysis, but little fundamental progress was achieved; to repeat the analogy, it looked as if a rich seam had been worked out. But in spite of that disenchantment, which meant that research in central place systems became limited, there remained a demand for the knowledge of the basic urban structure of regions and countries since it was critically relevant to so many aspects of public and private planning. Two examples can be suggested on different scales.

Any urban authority concerned with its own economic standing and welfare must demand to know how much of the total expenditure on goods and services leaks out to other centres, presumably, but not necessarily, of a higher order. Any such study must involve a questionnaire survey of a sample population as to where goods are bought and services obtained. But it is unequivocably an exercise in central place analysis, however more or less sophisticated its statistical basis.

On a much larger scale, the whole issue of the revision of local government in Britain, which was to result in a new structure after 1974 was, once the divorce of town and country in the old system had been rejected, also an exercise in central place analysis. That it was not carried out with the logic and rigour that some demanded (Senior, 1969) is no denial that the basic principle and the central problem was the reconciliation of threshold and range, that is, of the numbers of people deemed economically necessary to support services, and the distances demanded of people and their representatives to travel to administrative centres. The agitation which has characterized the last decade of the century suggests that the revision was, and is, disliked since it sacrificed range, at the dominant county level, for the building of high thresholds. As Senior wrote, 'my colleagues express the view that ... an authority responsible for the personal services (among which they include education and home-building) ought to have a population of at least "around 250,000" ' (Senior, 1969: 3). Even so, much of the complexity and uncertainty is due to the fact that reality

is no longer made up of discrete settlements and well-defined spheres of influence. The spread of urban activities to city margins, the pervasive influence of residential extension into what were country villages, the growth of second and holiday homes and easy mobility have undermined the distinction between town and country on which central place systems depended, that is, a central place and tributary countryside. Indeed, megalopolitan growth, the massive extension of cities which Chapter 2 noted, has brought about concern for much broader contrasts, especially those between core and periphery; the terms may well appear similar but the scale is totally different.

Against the foregoing background a consideration of central place studies in the 1980s would suggest the following:

1. There is still the basic need to know how urban services are delivered via their aggregated location and the delivery range.

2. Any simple system of a Christaller type is most likely to be found in the period prior to 1950 and near the time when the theory was first published. It is now anachronistic to attempt to derive such structures except in so far as time and space, in a rather different way from the usual, have not converged. Some parts of even developed countries least affected by metropolitan structures will still preserve the older systems, but only to a degree. To descend to the anecdotal, although it is no more reprehensible than 'eye-balling', the present author, who lives in a town of some 10,000 (nearer 15,000 if suburbs are included), has within the last year bought clothes in a town some 25 miles away, furniture in another some 30 miles distant and what can be called arts/crafts in yet another at some 16 miles and these are repeated not one-off visits. The critical point is that all these centres visited are *smaller* than the town of residence; that is, it is not a process of shifting up the hierarchy for higher order goods. It is nothing more than treating the local network of centres much as a metropolitan inhabitant would treat out of town and subur-

ban centres. The key is easy mobility which consequently greatly modifies the traditional hierarchy of centres allowing specialization unrelated to rank order.

3. The earlier labour-intensive methods of establishing the ranking of towns have become outmoded, as also have the similar methods of defining spheres of influence. Again the reason is partly to be found in the nature of settlement growth itself, but is probably mainly due to the direct difficulty of data acquisition. It might be contentious, but not without foundation, to argue that that the development of theory and concern with the grand design created an impatience with the excessive demands which meticulous empirical study demanded.

4.5 RECENT STUDIES OF URBAN STRUCTURE; NATIONAL AND INTERNATIONAL

The consequence of the above problems has been a virtually universal fall-back to population figures as the basic measure of rank. The necessity of defining tributary fields, however, remained. But here, too, census data were to provide a way forward. Virtually all modern censuses collect information on the usual place of residence of respondents and the place of work, and, in most cases the means of transport, the so-called modal split. These data echo, as has been noted (p. 43), in a more complex way the early bus route analyses of Green. Thus the metropolitan statistical areas of the USA noted in Chapter 2 (p. 18) are defined on the basis of journey to work which has been gradually extended as the basic means of defining the structures of the urban systems of countries. Thus the Centre for Urban and Regional Development Studies (CURDS) justified its approach to defining functional regions in mainland Britain by arguing:

The functional approach deals with these patterns by defining areas on the basis of patterns of commuting linkage to a set of centres which have been identified from employment and retailing statistics.... In less heavily populated areas, therefore, separate employment centres can be distinguished and existing Districts can be split up to provide them with their tributary, or dependent, hinterlands. Conversely, in the more urbanised context, consistently defined city-region units can be produced by grouping areas together.

(CURDS, 1983: 1)

This type of approach was first applied to England and Wales by Peter Hall and set out in his book, *The containment of urban England* (1973). It is instructive to compare the two systems: that of Hall and that established by CURDS.

Peter Hall's urban structure of England and Wales is based on two building blocks. The first is the Standard Metropolitan Labour Area (SMLA) which is made up of a core containing administrative areas with a density of 5.0 workers per acre (12.3 per hectare) or a simple administrative area with over 20,000 workers, and a ring comprising contiguous areas which send over 15 per cent of their resident employed workers to the core. To qualify, the total population must be over 70,000. The second element is the Metropolitan Economic Labour Area (MELA) which is composed of the SMLA plus contiguous administrative areas which send more of their commuting workers to the SMLA core than to any other core. When the SMLAs are arranged in rank order, break points are identified by inspection (note the parallel with what Marshall did) and rank orders are proposed as in Table 4.2. Here then is a proposed four level ranking of the highest order centres in England and Wales, which is akin to that suggested by Marshall and, much earlier, by Philbrick. But Hall takes his analysis further (Fig. 4.7). Virtually the whole of the country, he contends, comes within the aegis of MELAs and only parts of the South-West and Wales really lie beyond their influence. From his building blocks he constructs Megalopolis England, a set of contiguous areas which gives the maximum concentration of people in relation to area. This description is of:

Table 4.2 *The ranking of Standard Metropolitan Labour Areas in England and Wales*

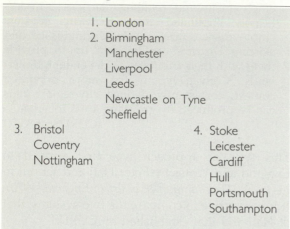

1. London	
2. Birmingham	
Manchester	
Liverpool	
Leeds	
Newcastle on Tyne	
Sheffield	
3. Bristol	4. Stoke
Coventry	Leicester
Nottingham	Cardiff
	Hull
	Portsmouth
	Southampton

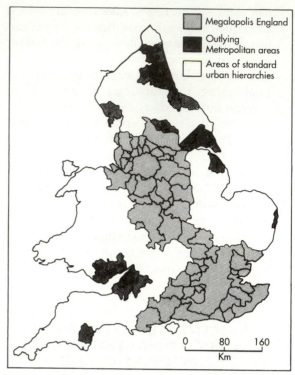

Megalopolis England

Outlying Metropolitan areas

Areas of standard urban hierarchies

0   80   160
Km

FIGURE 4.7 *The urban structure of England and Wales (after P. Hall, 1973). Source: P. Hall, 1973 The containment of urban England. London: George Allen and Unwin.*

a giant urban area only in the sense that here is a large tract of the earth's surface where the great majority of people depend on urban jobs and services, and where the impact of these jobs and services, in terms of measurements like commuter zones, service areas and the exchange of goods and information, expands to involve each part of the area in a complex series of inter-actions with the other parts.

(Hall, 1973: 320)

The implication is that within the megalopolis extensive commuting means that the town–country dichotomy is meaningless, while ease of movement to service centres means that traditional hierarchies lose meaning. Only in the marginal and peripheral areas do they survive. This is in keeping with the situation already noted (p. 50), with the proviso that modification is characteristic even in those areas of traditional hierarchies due to contemporary mobility.

An analysis of the United Kingdom carried out by the Centre for Urban and Regional Development Studies (CURDS, 1983) was in-tended to derive a set of functional regions for the purpose of the meaningful presentation and analysis of census data, and not primarily to establish an urban hierarchy. However, the outcome is closely related. The first necessity was the definition of the centres (cities) which

were to form the central places and they were established by the use of 'a quotient which compares each area with the national average for retail activities/resident population (which identifies Shop Centres) and for jobs by work-place/residents in employment (Work Centres); the threshold for Centre studies is set at 66 per cent of the national average' (Coombes *et al.*, 1982: 76). Qualification as shop and work centre is needed for eventual recognition. The subse-quent stages are complex in detail but are directly based on journey-to-work flows and the daily urban systems so generated. Unlike Hall's analysis, however, which was concerned with the definition of Megalopolis England the CURDS study is designed 'to exhaust the national territory', so that lower level areas are defined, those outside metropolitan dominance

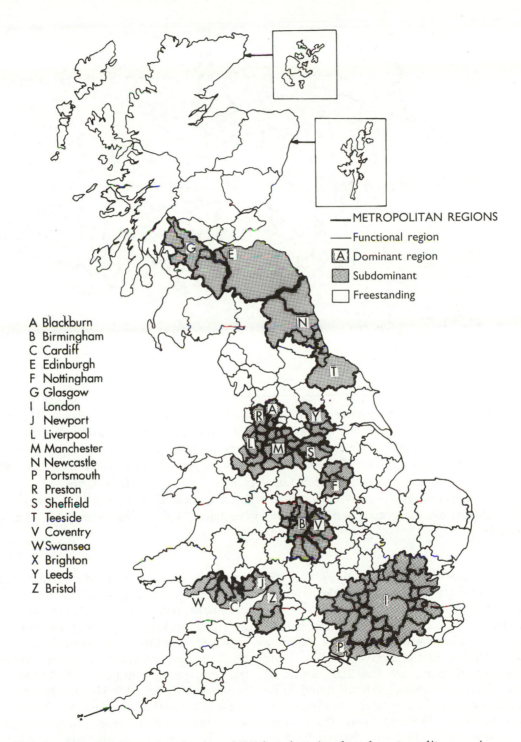

A Blackburn
B Birmingham
C Cardiff
E Edinburgh
F Nottingham
G Glasgow
I London
J Newport
L Liverpool
M Manchester
N Newcastle
P Portsmouth
R Preston
S Sheffield
T Teeside
V Coventry
W Swansea
X Brighton
Y Leeds
Z Bristol

METROPOLITAN REGIONS
Functional region
A Dominant region
Subdominant
Freestanding

FIGURE 4.8 *England, Scotland and Wales: functional and metropolitan regions (after CURDS). Source: University of Newcastle upon Tyne, Centre for Urban and Regional Studies. Fact Sheet 1, Dec. 1983.*

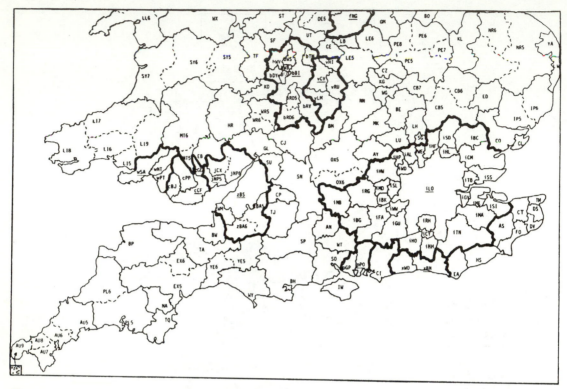

FIGURE 4.9 *Southern England and Wales: lower-tier regions (after CURDS). The heavy boundary denotes metropolitan regions, the continuous light boundary denotes functional regions and the broken boundary denotes rural areas. Underlined centres (e.g. cCF/Cardiff) are dominant functional regions. Those not underlined are subdominant. Those with two letters (e.g. EX/Exeter) are free-standing functional regions. Source: M. G. Coombes et al. 1982: Functional regions for the population census of Great Britain. Chapter 3 in D. T. Herbert and R. J. Johnson eds, Geography and the urban environment, Vol. 5, Chichester: John Wiley and Sons.*

being called 'free-standing' and those at a still lower level with populations below 50,000 constitute residual rural areas.

The structure which results has elements both in common and in contrast with that of Hall. The critical similarity is the derivation of a basic twofold organization. Of the 228 functional regions which are defined, 115 are found to be self-contained and designated Free-standing Regions. There are 53 Rural Areas. These correspond to the areas which are characterized by Hall as having standard urban hierarchies. There are differences in detail since Hall's attempt to define a continuous megalopolis

produces a less fragmented area than that of the CURDS analysis (Fig. 4.8). The 20 metropolitan dominants recognized by CURDS correspond to the four ranks set out by Hall, but with the expected addition of Edinburgh and Glasgow (Hall only considered England and Wales), and the unexpected addition of Blackburn and Newport. Figure 4.9 shows the clasification applied to southern England and Wales. The full key is not given but as an example, in the south-west Ex 5 refers to the free-standing Exeter Urban Region and Ex 6 to the tributary Tiverton Rural Area. The system is certainly not above criticism but it is effectively illustrative of the

Table 4.3   *Hierarchy of urban settlements in Switzerland*

Rank	Description	Population	Number	Cumulative percentage of population
1	Largest	over 200,000	5	33.3
2	Higher medium	80–200,000	5	42.1
3	Medium	50–80.000	9	50.9
4	Lower medium	30–50.000	7	55.2
5	Small	20–30.000	7	57.7
6	Isolated towns	10–20.000	15	61.5

Source: After Racine and Cunha,1989.

way in which the attempt to establish national hierarchies of discrete, free-standing towns with nested spheres of influence has given way to the setting up of metropolitan structures, based on daily commuting as more effective and relevant representations of national urban systems. This situation can be illustrated by considering, even though very briefly, two further greatly contrasted examples.

The first, the Swiss Urban System, has been the subject of a study by Racine and Cunha (1989). They note that the *Office fédéral de la statistique* identifies two levels within the urban network: large agglomerations, which together define the major urban zones of the country; and isolated towns. The members of the agglomerations are classified, by population totals be it noted in relation to earlier comment (p. 41) into five levels (isolated towns excepted) (Table 4.3).

The distribution of the five ranks is shown on Fig. 4.10 where the main axes of development and areas of concentration are shown. The latter are made up of two so-called golden triangles, the first having apices at Basle, Zoug and St Gall and being focused on Zurich, with a second in the Basle–Berne–Lucerne area. A third – the 'Triangle d'Azur' – has its apices at Geneva, Lausanne and Neuchatel. As Racine and Cunha note, the first of the three contained 64 of the 100 main industrial, commercial and service enterprises, 68 per cent of the turnover of the 100 main industrial enterprises and 21 out of the 25 enterprises the turnover of which was over 2 billion Swiss francs. Within this clear metropolitan dominance however, the axis from Geneva to St Gall emerges as a Swiss megalopolis which, even in 1975, had been envisaged as 'one enormous conurbation, a *Bandstadt* stretching across Switzerland from the Lake of Geneva to the Lake of Constance, a sort of Swiss Megalopolis' (LeBeau, 1975). Racine and Cunha comment:

A multi-centred city perhaps? Although these centres are clearly distinguishable on a density chart, the outline of the central urban blocks will become more and more irregular until the centres look like so many spiders in the middle of their webs, and, just like Zurich along the shores of the lake, they will have grown long, urban tentacles.

(Racine and Cunha, 1989: 159)

Here then is a prime example of a post-industrial urban regional structure, totally different from anything envisaged in classical central place theory. Indeed, it is possible to envisage that the contrasted sections of Switzerland as of Britain represent on the one hand a pre-industrial situation where the conditions which gave rise to central place theory still appertain in marginal and peripheral contexts, and a post-industrial situation where major metropolitan agglomerations merge into megalopolitan structures.

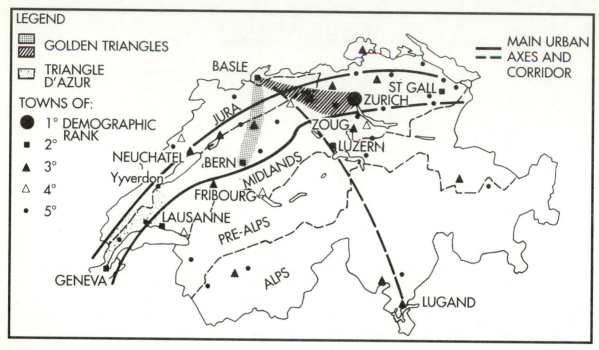

FIGURE 4.10 *The Swiss urban system (after J-B. Racine and A. Cunha, 1989). Source : L. S. Bourne et al. ed., 1989: The changing geography of urban systems. Fig. 1, p. 149. Department of Human Geography, University of Navarra. Copyright the editors.*

The second example is Spain. A ranking which is based on the number, class and size of employment in tertiary functions has been derived by Ferrer and Precedo (1984) and is set out in Table 4.4 and Fig. 4.11. This gives, it is claimed, six discrete levels, as opposed to a rank–size continuum on a population basis. The suggested equivalence to the $k = 3$ system proposed by Christaller is indicated in Table 4.4. The basic structure is bipolar, based on the national metropoles of Madrid and Barcelona. Below this are the seven regional metropolises, three of which, La Coruna, Valladolid and Palma, are ranked much higher by their service content than their population. Below this again is a varied ensemble of medium cities that function as tertiary sub-regional or provincial centres' (Precedo Ledo, 1989: 47). And again, below that, there is a remainder of the 34 cities usually serving less developed rural areas or less populated provinces where thresholds are

lowered. The system can be extended into an array of smaller towns and 'cabeceras comocales' as the lowest rank. Here then would seem to be a classic representation of a hierarchical system in a country removed from the industrial heart of Europe. But interpretation soon modifies that simplicity. The system can be divided into two elements (Fig. 4.12). The first is the core which is made up of Madrid together with the Cantabrian–Basque fringe and the eastern Mediterranean fringe, while the second, the economic, though not the geographic periphery, constitutes the remainder. Galicia and Atlantic Andalusia show tendencies to develop as the core. Ferrer *et al.* (1989) note the trend for Madrid to grow outward.

This spread growth above and beyond provisional limits has affected nearby cities such as Guadalajara, Toledo, Avila and Segovia through the emergence of three distinct development

Table 4.4 *The urban hierarchy in Spain*

Levels	Categories	No. of central places Observed a	b	Theoretical
I	National metropolises	2	2	2
II	Regional metropolises	5	7	4
III	Middle level metropolitan areas	13	10	12
IV	Middle level cities	43	43	36
V	Small cities	89	90	108
VI	*Cabeceras comarcales*	320	320	324

Notes
a According to method of Ranks
b According to typical deviation   472   472   486

The centrality of each city in the first three levels according to method of Ranks is as follows:
Level I   Madrid (197.3), Barcelona (131.0);
Level II   Valencia (32.0), Sevilla (26.5), Bilbao (23.8), Zaragoza (19.2), Málaga (16.6);
Level III Alicante (14.2), La Coruña (13.1), Oviedo (12.5), San Sebastián (12.2), Granada
           (10.5), Córdoba (10.2), Vigo (10.1), Murcia (9.9), Santander (9.7), Valladolid (9.6),
           Pamplona (8.3), Cádiz (8.6) and Salamanca (7.9).

axes. This new structure, which includes a central metropolitan municipality now experiencing a population loss, a strengthened outlying sub-metropolitan area and three axes leading towards the above mentioned peripheral provincial capitals, reveals a transition away from what was a great metropolis towards what may become a complex metropolitan zone.

(Ferrer *et al.*, 1989: 33)

Similar processes can be envisaged in relation to the other metropolitan centres, while the isolated urban nuclei of the periphery retain their traditional roles. Even in a country where the tenets of central place theory seem to provide the bases of explanation for the sizes, if not the locations, of settlements, strong forces towards metropolitan growth are apparent and become critical to the interpretation of the urban system.

This type of interpretation, moreover, can be continued from a national to a European scale.

Figures 4.13 and 4.14 reproduce two interpretations of the urban system of Europe, both made before the unification of Germany. The one after Brunet sets out a typology of cities (Brunet, 1989) while the second map after Burtenshaw *et al.* translates it into an interpretative scheme (Burtenshaw *et al.*, 1991). Presumably the closer economic and political integration of Europe, whatever its precise form, will strengthen those processes which are creating an interacting European urban system. The dominant element in that system is what Burtenshaw *et al.* (1991) call 'the Metropolitan Spine' which extends from Birmingham to include the south-east of England (note that it is less than Megalopolitan England) through the Low Countries and the Rhineland to Switzerland and northern Italy. Perhaps one might have expected it also to take in Paris, which is surely more than a fringe of that central spine. Outside this there are two sub-systems. The first is the 'North of the South', the area of most dynamic growth along the

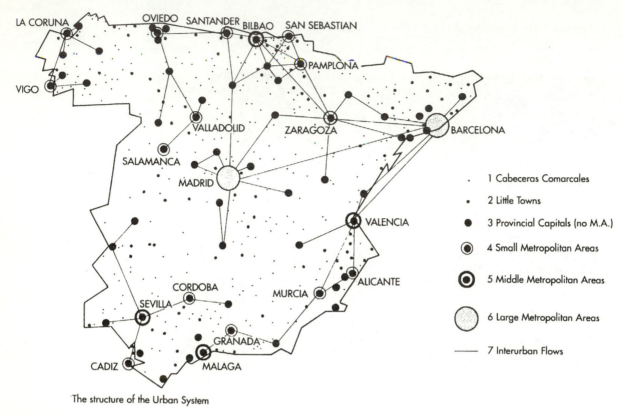

The structure of the Urban System

FIGURE 4.11 *The structure of the Spanish urban system (after M. Ferrer and A. Precedo, 1984). Source: L. S. Bourne, R. Sinclair and K. Dziewonski eds. 1984: Urbanization and settlement systems. International perspectives. Oxford: Oxford University Press. Fig. 14.3, p. 314.*

Mediterranean from Madrid and Barcelona, the constituents of the bipolar system in Spain, to Lyons, Geneva and Venice, crossing and reinforcing the metropolitan spine and emphasizing the significance of the Swiss Bandstradt. The second sub-system is much less coordinated, including what Burtenshaw *et al.* call 'the cities of the North and East', a set of independent metropolises of which Edinburgh–Glasgow and Vienna are contrasted examples. The peripheries to these strong organizing systems are made up of the western fringes, called the 'Finisterres', and the southernmost fringes of the Mediterranean, called simply 'the South'. That Ireland should be part of 'the South' rather than of the western periphery seems strange. This brief consideration of Europe suggests that the

universal process at work is the concentration of economic advancement and power in the core areas and that the same process operates at every scale. Thus, although Ireland may be part of the European periphery, within the country the concentration on metropolitan Dublin (see Fig. 3.10, p. 35) mirrors the larger scale contrasts, as does the significance of Cardiff within Wales.

4.6 CONCLUSION

This chapter began with a discussion of the attempts to derive empirical structures from the central place theory as developed by Walther

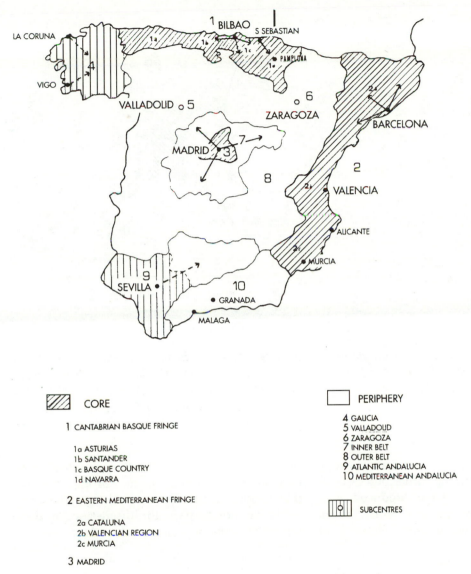

FIGURE 4.12 *The core–periphery model applied to the Spanish urban system (after M. Ferrer et al., 1989). Source: M. Ferrer et al., 1989: Recent evolution of the Spanish urban system. In L. S. Bourne ed. as Fig. 4.10, Fig. 2, p. 32.*

Christaller and it has concluded with the discussion of the emergence of megalopolises on a national and international scale. The two indicate the major themes in the interpretation of urban systems. In many parts of the world discrete urban settlements serve surrounding and tributary areas and a hierarchy of centres with nested spheres of influence have been created by the needs of the countryside for central and accessible services. But the excessive urbanization discussed in Chapter 2, together with the processes of central area population

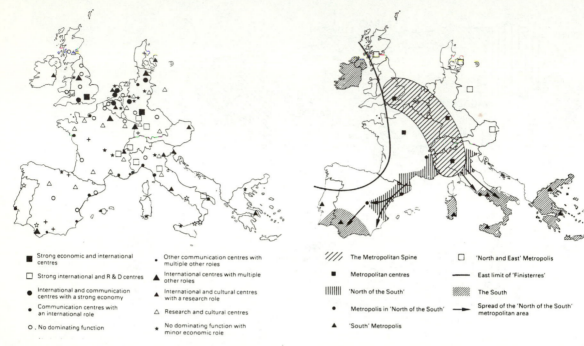

■	Strong economic and international centres	•	Other communication centres with multiple other roles
□	Strong international and R & D centres	▲	International centres with multiple other roles
●	International and communication centres with a strong economy	▲	International and cultural centres with a research role
•	Communication centres with an international role	△	Research and cultural centres
O	No dominating function	★	No dominating function with minor economic role

/////	The Metropolitan Spine	□	'North and East' Metropolis
■	Metropolitan centres	——	East limit of 'Finisterres'
||||||	'North of the South'	/////	The South
•	Metropolis in 'North of the South'	➝	Spread of the 'North of the South' metropolitan area
▲	'South' Metropolis		

FIGURE 4.13 *The European urban profile (after R. Brunet, 1989). Source: R. Brunet 1989: Les villes 'Européenes'. Paris, La Documentation Française.*

FIGURE 4.14 *Urbanised Europe 1990 (after Burtenshaw et al., 1991. Source: D. Burtenshaw, M. Bateman and G.J. Ashworth 1991: The European City. A Western Perspective. Fig. 13.2, p. 303. London: David Fulton Publishers.*

decline and surburban and exurban growth, themselves derived from the earlier processes of industrial concentration, have produced new forms of metropolitan growth which break across and transform the simplistic views of an urban hierarchy. Moreover, the growth of international communities, as in Europe, suggests that national territories are no longer the effectively bounded space for discussion of these largest systems.

CENTRAL PLACE THEORY: PROBLEMS AND DEVELOPMENT

5.1 EARLY PROBLEMS DERIVED FROM EMPIRICAL STUDIES

At this point it is useful to look back upon the early difficulties which arose in relation to central place theory before the development of large metropolitan structures themselves introduced new complications.

Christaller's study of central places in southwest Germany was aimed at answering the question whether there were general laws which governed the size and spacing of towns. His model was designed to provide an affirmative answer by revealing the underlying regularity in town distribution. Early attempts at empirical investigation accepted Christaller's thesis as proven, and questioning neither his assumptions nor the logic of his argument, they set out to show its applicability to the real world. But these somewhat uncritical studies were soon paralleled by others which found little of the regularity which Christaller's model predicted.

The first challenge was directed at the point where central place theory seemed clearly inapposite. A cursory inspection of the distribution of towns in most areas of the world reveals no sign of a hexagonal lattice. If, obeying the demands of threshold and range, particular grades of towns were marked by characteristic assemblages or bundles of functions, they seemed most certainly not to display any regularity in distribution. The time lag before any serious work on this aspect of central place theory was undertaken was due partly to the fact that many investigators assumed it was self-evident that no such spatial arrangement appertained, and partly to the lack of adequate techniques for examining the distribution patterns.

These were eventually derived from the investigations of plant ecologists, and a series of papers by Dacey applied nearest neighbour analysis to town (central place) distribution (Dacey, 1960, 1962, 1964, 1967). The central concept of nearest neighbour analysis is randomness. When there is complete absence of a systematic pattern of points in a specified region, the distribution of points is called random. A pattern that is not random is either more clustered than random or more uniform than random (Dacey, 1962). Christaller's thesis unequivocally implies that the distribution of central places is uniform and that a situation 'more uniform than random' should be shown regardless of rank if all the central places of an area are analysed. Dacey (1962) demonstrated that, using Brush's data for south-west Wisconsin, the system most closely approximated a random condition. From this three implications were drawn:

1. There *was* a hierarchy of central places in the area but spatial distribution did not conform to central place theory.
2. A hierarchy did exist but had not been properly defined.
3. Central place theory did not apply to market towns in south-west Wisconsin.

Taking into account the inadequate diagnostic basis of Bush's ranking, the third implication was at least possible. Later work by L. J. King (1961, 1962) was no more successful in identifying uniform lattice patterns. Using the nearest neighbour statistic (Rn) with $Rn = 0$ indicating a clustered situation, $Rn = 1$ a random situation and $Rn = 2.15$ a uniform lattice, King examined twenty sample areas and found that his results showed a range of $Rn = 0.7$ for an area in Utah to $Rn = 1.38$ for part of Missouri. Although a claim was made that some of the samples, such as the Missouri one quoted above, could be classified as 'approaching uniformity', the general conclusion must be that these patterns approximate to a 'random condition'.

There have been later attempts to rescue an element of regularity from these apparently random distributions such as that of Medvedkov using the concept of entropy. Entropy is a notion borrowed from the statistical theory of information. Entropy helps to measure disorder in settlement patterns, so that the random and uniform components are effectively separated and measured (Medvedkov, 1967a and b; Semple and Golledge, 1970). This means that any settlement pattern is assumed to have a regular and a random component and the calculation of the entropy value is designed

to disentangle these two parts indicating the degree of disorder. This seems of comparatively little value, and of questionable validity, in relation to a static distribution, but may possibly be useful in assessing whether a distribution is becoming more uniform with time. But it is possible to conclude that in few areas can any element of regularity in town distribution be discerned. The concept of the hexagonal lattice therefore holds little value to the student of the real world whatever may be its attraction to the student of social and economic geometry.

At the same time the idea of ranks of towns universal to the Western world, as sketched in Chapter 4, was also becoming less tenable as attempts to deploy such schema proved illusory (Davies, 1964). The concept of all settlements in all countries neatly dropping into appropriate slots could not be sustained and, as one commentator observed (Lukermann, 1966), there appeared a multiplicity of taxonomies rather than explanatory generalization. Moreover in the background of these classificatory studies was the 'rank–size' rule which states that empirical investigation in reality revealed not a stepped but a log-normal relationship between size of town (measured by population) and rank in the system (see p. 35).

This situation was greatly disturbing to urban geographers who in the early days had seen central place as a universal theory of town location. Berry, along with other authors, turned to these problems, first showing that the rank–size relation and a hierarchical structuring were not incompatible and could be subsumed under the same system. The result depended largely on the analytical approach for, by adjustments, either situation could be revealed. Moreover it was argued that any system was composed of two parts, one called *aggregative*, the other, *elemental*.

At the aggregative level the blending of many varying, but locally homogeneous areas, leads to the emergence of a continuum of centres. Within each small local area, the levels of the hierarchy are clearly represented. The theoretical postulate is that the levels of centres are a function of the grouping requirements of certain bundles of central functions, or, in other words, that classifications of centres and functions exist in which interaction effects are highly significant.

(Berry and Barnum, 1962: 46)

The last part of this statement is nothing more than a direct repetition of the earlier argument by Smailes for the 'trait complex' though the statistical procedures in identification were more objective and rigorous. But if the merging of these locally homogeneous areas is one reason why the hierarchy is obscured it also means that a clear limit is set to the universality of any ranking, for it will be a product of the particular economic, social and spatial condition of individual areas.

Berry takes this problem further. At the outset it can be demonstrated, as implied in the last paragraph, that local environmental constraints will act upon the system. Thus, for example, there is no direct relation between size of area (sphere of influence) and population density nor is there a simple expansion of area to compensate for falling density. It was observed that areas do expand as the density falls, in keeping with threshold demand, but the areas do not expand as fast as the densities fall so that the size of the population served falls. As a result, under these conditions, functions with the greatest threshold requirements at any level of the hierarchy move up to the next level; there is what Berry termed a phase shift (Berry, 1967).

If these several limitations are considered, it is apparent that the notion of a series of universal levels was over-optimistic and that the ranking of places is most meaningful within limited and local areas. To rank the towns of England and Wales is a useful descriptive device but has little meaning in central place terms because of the enormous variety of environmental conditions (using the word 'environment' in its widest and proper sense). The concepts of central place theory are invaluable in the analysis of the size and spacing, that is in the distribution, of cities, but the uniform deterministic setting in which they are implied to operate bears little relation to actuality.

All these problems gave the impression that central place theory led to a cul-de-sac and research work gradually petered out; what had been the leading edge of urban geography became a trailing tail-piece. In this context it entered what may be termed its post-modern phase. The simplicities and the certainties of the original theory were abandoned, but in dissolution it gave rise to a wide variety of almost disparate elements each owing something to the older formulation.

5.2 PRIMITIVE AND PERIODIC MARKETS

If modern urban growth has produced major metropolitan agglomerations which cannot be easily integrated into central place theory, then at the other end of the scale, in relation to the smallest settlements, problems also emerged. By its very adoption of the name 'place' or 'settlement' the theory envisaged a series of urban or semi-urban centres. But it is manifest that thresholds can be so limited as to be unable to generate permanently operational places of transaction. There are two closely related ways in which such a condition can be resolved. The first is by the reduction of the overhead costs of premises by the use of temporary and removable structures. At its most primitive this was achieved by the traditional seller of things such as matches or flowers in the nineteenth century; in fiction, by Eliza Dolittle, in reality by Michael Marks, the penniless Russian immigrant into Britain in 1881, with his tray of goods labelled, 'Don't ask the price, it's a penny'. At present in the markets of the developing world, the trader with just a medley of often recycled bits and pieces on a blanket is of the same order, while it is perhaps not fanciful to ascribe the car boot sale of the developed world to it also. At a slightly higher level, stalls and booths, which are temporary and movable, are used in both the developed and the developing worlds. These temporary conditions are associated with the second means of meeting lowered thresholds, the limitation of operations to specified days or times. The conventional system is the periodic market which adopts a cycle related to local culture and custom, hence, usually weekly in the Western world. The Sunday markets which have become popular in Britain demonstrate a characteristic association of a weekly cycle with temporary stalls. Indeed, many of these operations locally adopt a cycle moving from place to place during the week, with Sunday becoming no more than a day included rather than excluded.

Dawson defines periodic markets as 'Collections of itinerant sellers using non-permanent structures and operating in a designated place at a set time' (Dawson, 1979: 172). Quoting Forman and Rieglhaupt's (1970) study of northeast Brazil, Dawson notes three types of periodic market:

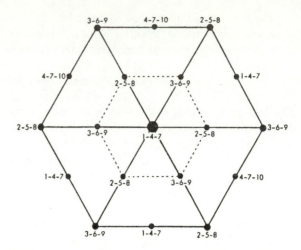

FIGURE 5.1 *Periodicity in a traditional Chinese 3-per-hsun cycle (after B. J. L. Berry, 1967). Source: B. J. L. Berry 1967: Geography of market centers and retail distribution. Englewood Cliffs, NJ,: Prentice Hall. Fig. 5.4, p. 94.*

1. **The Rural Consumer Market.** This is the classic one day a week set-up (or whatever time variant is apposite), mainly serving rural areas where there are limited populations with problems of access to towns. There may be no permanent structures and there is not necessarily a permanent settlement of any sort. There is a complex mix of producer–sellers, producer–sellers–buyers and full-time retailers.
2. **The distribution fair** where professional retailers purchase their stock, but where there is also associated retailing. These are usually located in larger towns.
3. **The urban consumer market** which is attached to a permanent market but which only operates on specified days. This includes the conventional 'market day' of the West.

There are some problems associated with this listing. The first category is by definition non-urban and though clearly of relevance to central place studies is somewhat out of place in an urban geography. The third is made up of two types, the single market in the smaller town and the series of markets on different days which can be found in the larger cities and which form part of the intra-urban hierarchy of shopping centres which has yet to be considered (see p. 66). In addition, there is a fourth variant which has most in common with the urban consumer market. This is where, in small towns in thinly peopled rural areas, shops, and especially services such as banks or surgeries, are only opened for limited times on specific days by personnel from the nearest larger town. More often than not the opening periods will correspond with the market day, but will possibly include other times also.

Within the literature of the rural consumer market the most widely quoted example is that presented by Skinner as long ago as 1964–5. He described a 10-day cycle in China within which each market was visited three times per cycle. The system is illustrated in Fig. 5.1, where a merchant can move between the central market and a pair of standard (lower order) markets in a 10-day cycle divided into units of three: the central market (day 1), first standard (2), second standard (3), central (4), first standard (5), second standard (6), central (7), first standard (8), second standard (9) and central on day 10,

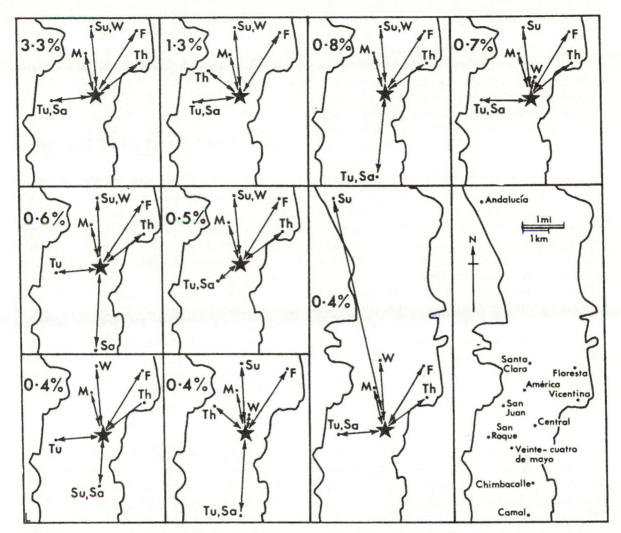

FIGURE 5.2   *Quito. The most popular combinations of markets and market days used by mobile traders, outside the permanent market buildings, who visit three or more different markets during the week. Home for all the traders is indicated by a star arbitrarily placed in one of the high density central residential areas where many market traders live. The percentages indicate the proportion of all the traders interviewed outside the market buildings who use each combination (after R. Bromley, 1974). Source: R. Bromley 1974: The organization of Quito's urban markets: towards a reinterpretation of periodic central places. Transactions of the Institute of British Geographers 62, Fig. 8, p. 66.*

when no business is transacted (Berry, 1967: 95). There are two crucial controls in the system. The first is that the merchant is truly itinerant not returning to the home centre each night and the second is that 'a well integrated regional system of periodic markets has an inverse relationship between spatial proximity and temporal proximity' (Bromley, 1974: 46).

Bromley has described the characteristic urban consumer markets in Quito and Fig. 5.2 reproduces the nine most popular combinations of markets and market days used by mobile traders. Of the total twenty-five urban markets, he identified nine which could be classed as stable, daily markets, the remaining sixteen showed considerable day-to-day variations in market activity with a clear element of synchronization so that the inverse relationship between temporal proximity and spatial proximity was preserved (Bromley, 1974: 67). It is worth noting that Bromley rejected the simple central place argument for periodic markets based on the accumulation of thresholds by the diminution of range, that is an issue of viability, and rather emphasized that traders opt for the flexibility of periodic marketing and consider that greater profits can be made from it. Such an argument would be equally applicable to 'Sunday markets' in contemporary Britain.

There is, however, a further complexity in developing countries. Milton Santos (1979) has proposed the notion of two circuits of activity within the urban economy. There is one which reflects the standard capitalist structure which in turn produces central place systems akin to those described by Christaller. But there is also a lower circuit relating to the urban poor outside the transactions of the upper circuit from which the dispossessed are excluded. Informal trade and barter dominates, especially in squatter settlements. Thus, for example, Nici Nelson describes the charcoal shops of the Nairobi squatter neighbourhood, Mathare valley, as follows:

> Charcoal shops are usually the most rudimentary of establishments, often consisting only of an awning to protect the charcoal from rain. They are supplied by lorries which come into Nairobi from the rural areas where the charcoal is prepared. Most charcoal sellers have an arrangement with a rural charcoal maker, often a relative or a friend from home.... Rental costs, if any, are low. Most operators merely store the coal in their rooms and set up a plastic awning each day on a piece of waste ground nearby, often near a well-frequented road.
>
> (Nelson, 1988: 193)

Here there is even no agglomeration and Santos's argument that threshold and range

have no relevance is possibly valid. Even so, what is probably nearer the truth is that there are two different systems superimposed on each other though controlled by the same principles.

5.3 THE INTRA-URBAN HIERARCHY

In the discussion of periodic markets, consideration switched in part from the inter-urban hierarchy, which Christaller introduced, to the intra-urban hierarchy, for all large and even smaller towns do not have one central retail core but a whole series of subsidiary retail business areas.

The earliest studies made of subsidiary business districts were not linked with central place ideas but were derived from empirical observation of city structure. The most widely known were by M. J. Proudfoot who identified five types of retail structures within the cities of the United States (Proudfoot, 1937a and b). These he defined as follows:

1. The CBD
2. The outlying business district
3. The principal business thoroughfare
4. The neighbourhood business street
5. The isolated store cluster.

Proudfoot noted that these displayed a 'progressive change' clearly indicating that he was identifying a ranked structure although making no link with the then nascent and little known central place theory.

Given the rapid development of central place notions, the growing number of empirical works being published and the existence of studies which identified ranks of business districts within the city, it was inevitable that the two lines of research should be brought together and attempts made to see in the types of business districts the equivalence of ranks in the general urban hierarchy. Hans Carol, writing of his work on Zurich, claimed, 'When this survey was begun in 1952, it was, to my knowledge, the first attempt to use the central place concept for analysing the pattern of central functions within the city' (Carol, 1960: 419). Carol recognized that distinctions in level were not simply related to numbers of shops but also to the variety of goods available for sale at a centre and the quality (price) range within each good (Fig. 5.3(a) and (b)). The extent of the service area was also considered.

From a survey of three centres of different 'levels' a hierarchical classification was derived which recognized four ranks of centre:

1. Central business district (CBD) serving the whole city.
2. Regional business district offering middle order goods (Fig. 5.3(a) and (b)) and serving some 90,000 to 100,000 population.
3. Neighbourhood business district offering low order goods, frequently needed and with a very limited range of higher order commodities. Such districts were usually situated towards the periphery of the city and served some 5000 to 10,000 people.
4. Local business district comprising a cluster of a few round-the-corner shops.

This was further advanced in the work of W. Garrison and others (Garrison *et al.*, 1959). In this it was pointed out that much of the existing work was based on the conventional distinction between 'convenience', 'shopping' and 'speciality' goods. A survey of definitions indicated the variety of different bases used to identify the ranks of centre which have been so far indicated in this chapter. In addition, Garrison emphasized the additional problem of string streets which were much more apparent in an American context. A study of Spokane, in which 49 business types located in 285 business centres were identified, was undertaken by means of a 49 × 49 correlation matrix of business types. By means of linkage analysis a series of nine groupings of businesses was derived and linkage analysis again applied to a 9 × 9 matrix of the average correlation for each group of business types. The result was a clear distinction of a nucleated component made up of three groups,

which were given no distinctive names, and an arterial component with an automobile set, a supplies set and a string street set.

This was undoubtedly the first study to demonstrate the contrasting characteristics of these subsidiary business districts in statistical terms and also to make the link with a wider country area. 'It is readily evident both empirically and theoretically that nucleated shopping centres *within* urban areas (densely built-up areas with concentrated purchasing power) and alternate nucleated urban centres (nucleations in the otherwise more sparsely populated areas of less concentrated purchasing power) are of the same nature' (Garrison *et al.*, 1959: 99). It also follows that intra-urban business districts are not only to be found within large metropolitan centres. If a small town has a central shopping array equal to that of the neighbourhood business district, using Carol's terminology, then it will be possible to identify local business districts at the lowest level, within that urban area. The intra-urban arrangement of business districts is not a problem related solely to the large city.

The reference above to Carol's terminology indicates one of the problems which arose in these studies, the variety of nomenclature employed for the subsidiary districts. This was met by Brian Berry who, drawing on work such as Garrison's, outlined a complete typology of the commercial characteristics of American cities outside the CBD (Berry, 1962) as shown in Fig. 5.4. It was based on a threefold division of these areas as follows:

*   A hierarchy of business centres – *centres*.
*   Highway orientated commercial ribbons and urban arterial commercial developments – *ribbons*.
*   Specialized function areas – *specialized areas*.

The hierarchy of centres included four ranks below the CBD, thus adding one to the pattern already established, largely it would seem by splitting the regional level into two and calling the lower segment the 'community centre'. This provided a parallel to the five ranks of towns within a central place system which Berry had

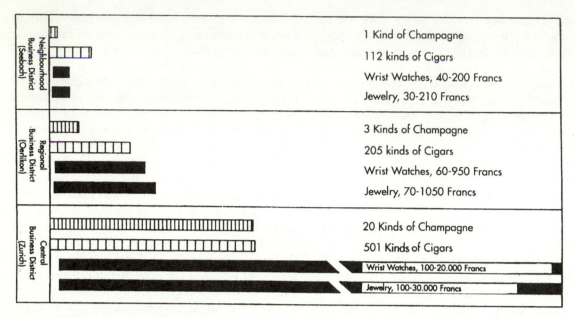

FIGURE 5.3A *Zurich. The range of goods in business districts at three different levels. This indicates the variation in the range and quality of similar goods at different levels of the hierarchy. (after H. Carol, 1960).*

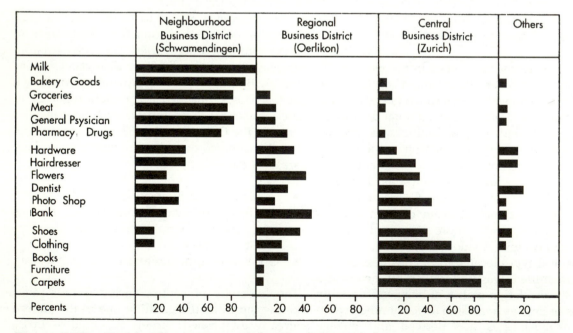

FIGURE 5.3B *Zurich. The percentage of shopping trips made by a person residing in a Zurich suburb (Schwamendingen) to three levels of business district. This indicates the differing orders of goods which are bought at the different hierarchical levels (after H. Carol, 1960). The data in both Figs 5.3A and 5.3B relate to the 1950s.*

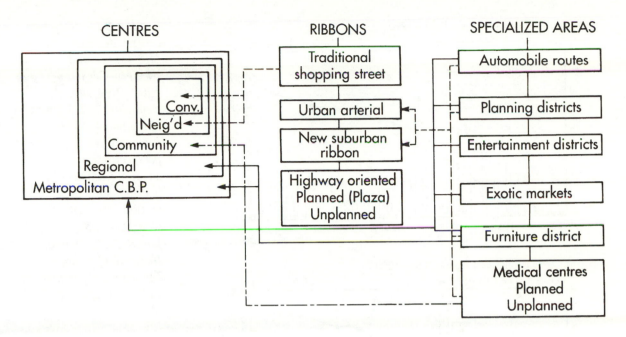

FIGURE 5.4 *A typology of business districts within the city (after B. J. L. Berry, 1967).*

indicated – hamlet, village, town, city and regional capital.

Under the broad heading of 'ribbons' a number of different types was indicated. The first is the well-established shopping street which usually leads from the nucleated centre or is substituted for the lower order centres. The second is the highway-orientated ribbon which is ubiquitous in the USA but much less common in Europe due to a somewhat less intensive use of the automobile and the more effective application of planning controls. It consists of a characteristic and garish assembly of motels, restaurants and filling stations and other car-related operations. Also included under this heading are the closely associated urban arterial locations made up of stores with excessive space requirements, such as furniture or appliance stores and building and lumber yards. They rely on special-purpose trips and provide easy access along the main routes together with good parking facilities. The third type, the specialized areas, is often closely related to the nucleated business district and provide grouped but highly specialized services, as for example medical districts with associated doctors, dentists, opticians and pharmacists.

This scheme has remained a definitive statement of the typology of subsidiary business districts. Two problems arise in a locational review of the centres. The first of these is derived from the social contrasts which exist within the city. These mean that purchasing power is not evenly spread and in consequence demand generated varies spatially. At the same time, and in part related, population densities and shopping habits vary. Where families do not own cars, refrigerators and deep freezers purchases will tend to be frequent and the 'corner shop' an important element. In well-to-do suburbs where two car families are common and fridges and freezers ubiquitous, shopping habits will be very different and most convenience goods will be purchased at the weekly visit to the supermarket so that the corner shop will not exist. The second problem is that the implied neat nesting of the various orders, and the discrete tributary areas that go along with

Shopping Centre Development UK 1964 – 1990

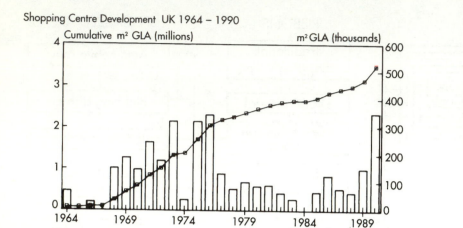

Source: Hillier Parker/URPI/Thorpe 1992

Shopping Centre Development France 1964 – 1990

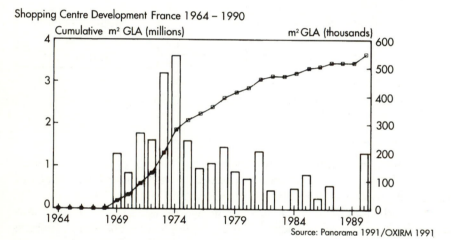

Source: Panorama 1991/OXIRM 1991

Shopping Centre Development Western Germany 1964 – 1990

Source: DHI Brehbung/OXIRM 1991

FIGURE 5.5 *Shopping centres developed in the UK, France and Western Germany 1964–1990 (after J. Reynolds, 1992) Source: J. Reynolds, 1992: International perspectives on shopping centre development. Paper read at IBG Annual Conference, Swansea, Jan. 8 1992. Oxford Institute of Retail Management, Templeton College, Oxford. OX1 5NY.*

that, are not borne out in studies of the real world. Influences such as public transport routes and the perceptions of shoppers as to range and value break across the nice academic resolution (Dawson, 1983; Dawson and Lord, 1985).

Hardly had some settled notion of a hierarchy of intra-urban sub-centres become accepted in the literature of urban geography, than a major revolution became apparent in the decline of the CBD (see Chapter 10) and the growth of out-of-town shopping centres. The difference between the earlier range of convenience, neighbour-hood, community, and regional centres and the new regional shopping centres or malls, a term which helps to avoid confusion of terminology, is obviously not a sharp one. But the older hierarchy was always subservient to the CBD, complemented it and generally offered lower order goods to specific suburban residential areas. The new breed replaced the CBD, or at least replicated its offerings, and by locations near to expressways sought to serve much larger sections of the city population than adjacent suburbs. Moreover, the new centres were planned and managed as unitary estates. Although there is an overlap with earlier intra-urban shopping provision, the new develop-ments were a distinctive addition to the city.

O'Brien and Harris (1991) see the change in Britain as taking place in three phases. The first was during the 1960s when discount stores initi-ated what they term 'an embryonic phase of retail decentralization' (O'Brien and Harris, 1991: 98), largely based on the conversion of existing premises such as garages and motor showrooms and auction rooms. This initial phase was dominated by grocery retailers but furniture and household goods soon followed, often by the modification of factories and warehouses. The second phase they term 'the entry of the multi-ples'. The lead was taken by the grocery multi-ples who sought new prestige premises and hence formally designed centres. The abandon-ment of resale price maintenance in 1964 provided one catalyst in the UK. Figure 5.5 shows the process of shopping centre growth in Britain and France with a massive extension after 1969

and continuing to the late 1970s. It was during this phase that the superstore (usually between 25,000 square feet and 50,000 square feet – 2325 to 4650 square metres) came on to the scene, together with the even larger hypermarket.

The third phase O'Brien and Harris call 'acceleration and agglomeration in the 1980s'. The early 1980s was a period of limited growth but the end of the decade saw an unprecedented consumer boom and a time of expansion on an even larger scale. This was accompanied by the move out of traditional town centre retailers: Marks and Spencer joined the trend in 1984. The first such major centre in Britain is usually regarded as Brent Cross, near London, which was fully opened in 1976.

The classification of these new centres is no longer based on the population thresholds of traditional central places, but on the area covered. However, variable bases, such as the actual retail space, the total shopfloor space which is normally equivalent to the gross leasing area or GLA, and the total space includ-ing car parks and service areas, do produce problems of comparison. Based on an examina-tion of European shopping centres Reynolds (1992) has proposed the classification which is set out in Fig. 5.6.

All these changes have undermined the idea of a reflection of the broad central place system within the city. But it can be argued that the intra-urban hierarchy still exists, as it most certainly does, and that the so-called regional shopping centres are somewhat different since they aim to be a replacement of the central business district itself. That is, they do not provide lower order goods at a point in subur-bia where a high enough threshold can be gener-ated over limited ranges which are shorter, in distance or in time, than the journey to the centre. Rather they provide all the facilities of the CBD. Thus restaurants, cinemas and other leisure provisions are directed towards enabling the shopper to spend the complete day in the out-of-town centre, just as in the past it would have been in the city centre. Perhaps at this point it is worth noting that the word 'regional' has two meanings. In the older intra-urban

I: Regional Shopping Centre (30,000 m²+)*

(Centres commerciaux régionaux, grandes centros periféricos, regionalen Shopping-Center)
(Two or more anchors)

Locational variants	Central area in traditional core	Eldon Square, Newcastle, UK
	Central area adjacent traditional core	La Part-Dieu, Lyon, France
	Non-central suburban growth pole	Vélizy 2, Versailles, France
	Green field site/transport node	Curno, Bergame, Italy
Compositional variants	Hypermarket-dominated	A6, Jonkoping, Sweden
	Department and variety store-dominated	Lakeside, Thurrock, UK
	Food, non-food and leisure anchors	Parquesur, Madrid, Spain

II: Intermediate Centres (10,000 m²–30,000 m²)

(Centres intercommunaux, centros intermedios)
(at least one anchor, integrated)

Locational variants	Non-central suburban community	Auchan, Torino, Italy
	Greenfield site/transport node	Cameron Toll, Edinburgh, UK
Compositional variants	Hypermarket-anchored	Euromarché
	Speciality non-food-anchored	BHV, Cergy, France

III: Retail Parks (5,000 m²–20,000 m²)

(Centres de magasins d'usine ou parc des entrepots, parques commerciales, retail warehouse parks)
(Not obviously anchored; not wholly integrated centres)

Locational variants	Non-central suburban community	
	Greenfield site/transport node	Lakeside Retail Park, UK
Compositional variants	Retail warehouse tenant mix	
	Factory outlet tenant mix	Direct Usines, Nancy, France
	Hybrid tenant mix	Fosse Park, Leicester, UK

III: Speciality centres (1,000 m²+)

(arcades, galeries marchandes, galerías commerciales, Galerien)

Locational variants	Central area in traditional core	Arcades, Lille, France
Compositional variants	Non-food specialist traders	Powerscourt Town House, Dublin, Eire
	Department store conversion	Centre Point, Braunschweig, Germany

*Floorspace figures are indicative only.
Note: Centre providing for local or neighbourhood needs are excluded.

FIGURE 5.6   *Models of shopping centres (after J. Reynolds, 1992). Source: As Fig. 5.5.*

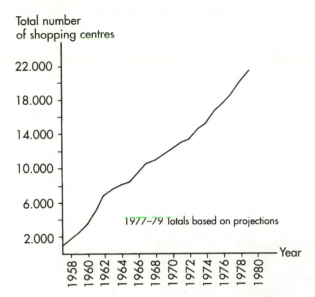

Total number
of shopping centres

1977–79 Totals based on projections

FIGURE 5.7 *The growth of shopping centres in the USA, 1957–1979 (after P. Muller, 1981). Source: P. Muller 1981: Contemporary suburban America. Englewood Cliffs NJ: Prentice Hall. Figure 4.1, p. 122.*

hierarchy it stood for a defined region of the city; in the newer context it implies a regional hinterland just like that of the whole city.

These major regional centres are, as might be expected, best developed where they first appeared, in North America. There new terms have been coined to describe the largest – mega-malls and mini-cities, although as will be demonstrated there is a difference between the two. Figure 5.7 shows the growth of shopping centres in the USA from 1957 to 1979. Muller (1981) has identified the immediate post-war period, 1945 to 1960, as one of 'leisurely outward drift' (Muller, 1981: 122) with that from 1960 to 1970 as the catalytic phase of retail decentralization. He then characterizes most of the centres built after 1970 as 'superregional' (adding to the complex meaning of the term 'regional') and quotes Kowinski (1978: 33): 'These meticulously planned and brightly enclosed structures ... have taken the concept of one-stop shopping, as old as the ancient public market, and turned it into a virtual one-stop

culture, providing a cornucopia of products nestled in an ecology of community, entertainment and societal identity'. A comparison between O'Brien and Harris's account of developments in Britain and Muller's for the USA will show the similarities, with Britain following about a decade behind America.

Two examples from North America can be briefly reviewed. The first is the West Edmonton Mall in Canada to which a special number of *The Canadian Geographer* was devoted. The introductory paper begins, 'with a gross leasable area of almost four million square feet, West Edmonton Mall (WEM) is the largest and most famous indoor shopping center constructed to date anywhere in the world, it is the prototype of the mega-mall' (Jackson and Johnson, 1991: 226). The location of the mall (Fig. 5.8(a)) in the affluent suburbs of western Edmonton reflects the comments made earlier (p. 69), although it is only some 5 miles or 20 minutes driving time from the CBD. Of the total floor space some 600,000 square feet (55,740 m^2) are devoted to recreation and 250,000 (23,225 m^2) to the Fantasyland Hotel. The remainder is conventional retailing, although it must be remembered that mall accessory areas, that is mall and service corridors and infrastructural space total nearly a million square feet (92,900 m^2) (Johnson, 1991: 252). As the so-called anchors which are critical to large centres, there are four out of the five Canadian department stores and nearly 600 other establishments (Fig. 5.8(b)). The relationship between WEM and the CBD is demonstrated in Fig. 5.8(a) and, as Johnson argues, the emergence of the mall challenges the conventional notion of the primacy of the CBD since regional shopping centres (or malls, the terms seem interchangeable) have become larger and larger and the CBD's retail component relegated to that of a local centre (Johnson, 1991: 250). But in the context of this chapter the most significant conclusion is that the study of the West Edmonton Mall 'provides further support for relying less on central place theory as a basis for understanding retail structure at the intra-urban scale'. As Jones and Simmons (1987) have pointed out, at least for Toronto, the intra-urban

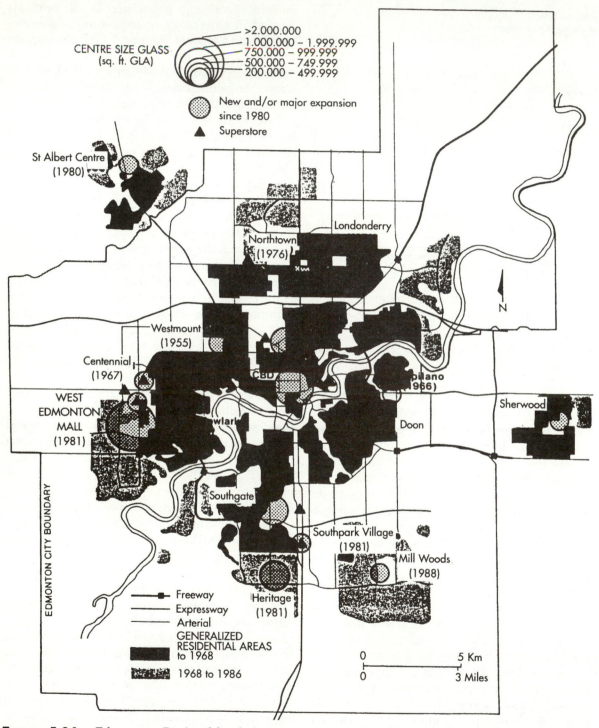

FIGURE 5.8A   *Edmonton. Regional level shopping centres (after D. B. Johnson, 1991). Source: D. B. Johnson, 1991: Structural features of West Edmonton Mall. The Canadian Geographer, 1991, 35(3), Fig. 1, p. 251.*

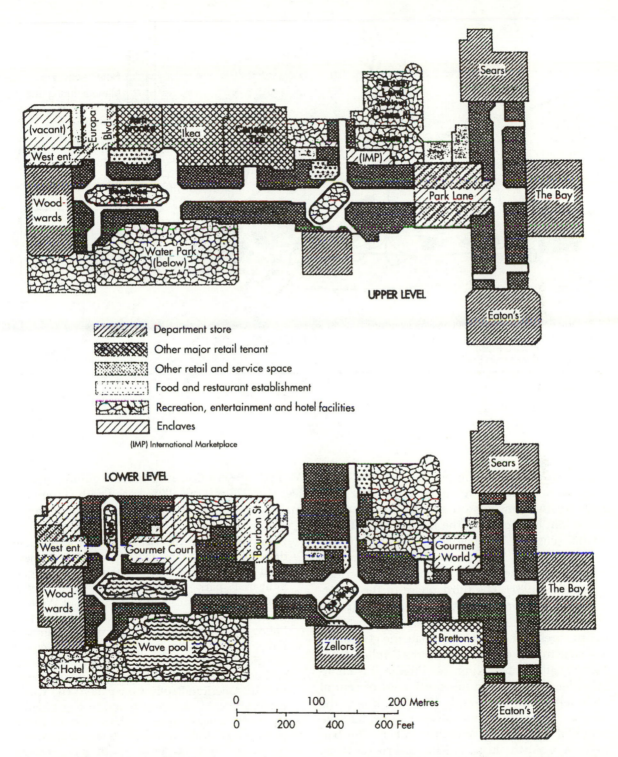

Figure 5.8B *The internal structure of the West Edmonton Mall (after D. B. Johnson, 1991) Source: As Fig. 5.8A. Fig. 4, p. 254.*

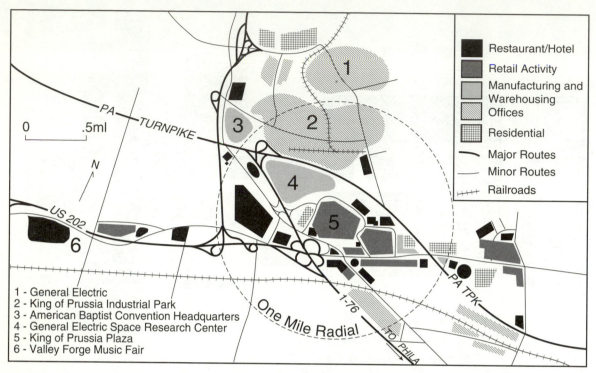

1 - General Electric
2 - King of Prussia Industrial Park
3 - American Baptist Convention Headquarters
4 - General Electric Space Research Center
5 - King of Prussia Plaza
6 - Valley Forge Music Fair

FIGURE 5.9 *A typical suburban mini-city in the USA. King of Prussia, Pennsylvania (after P. O. Muller, 1981). Source: P. O. Muller, 1981 Contemporary suburban America. Prentice Hall, Englewood Cliff, NJ. Fig. 4.13 p. 164.*

retail hierarchy accounts for a relatively small proportion of urban retail facilities. 'We may have to find concepts other than the hierarchy to explain how parts of the retail puzzle fit in a spatial sense' (Johnson, 1991: 260).

Akin to, but somewhat different from, the mega-mall is the suburban mini-city described by Muller (1981). His third phase of the super-regional centre in the USA has been noted (p. 73) and he contends that many were multifunctional urban cores that came to dominate the geographical structure of centreless suburbs (Muller, 1981: 162). Muller cites King of Prussia in Pennsylvania, some 20 minutes north-west of central Philadelphia as an example (Fig. 5.9). It is at the intersection of the Pennsylvania Turnpike, the Schuykil Expressway and three other suburban highways. At the heart is King of Prussia Plaza (yet another name to add to mall and regional shopping centre) which had, at the time of

Muller's description in 1980, 1.8 million square feet (167,220 m²) of selling space with some nine major department stores and 200 smaller shops.

Distributed about the vicinity within a 5-minute drive of the mall are dozens of highway-orientated retail facilities...: one of the region's largest industrial park complexes (over 750 acres), containing the plants and warehouses of Western Electric, GM-Chevrolet, Sears, Borg-Warner, Philco-Ford, three major pharmaceutical manufacturers and 30 smaller companies: one of General Electric's leading research and manufacturing facilities; a variety of office parks and buildings... and the American Baptist Convention. The Valley Forge Musical Fair, a year-round theater offering top-name entertainers weekly; five first-run cinemas, at least a dozen fine restaurants;... numerous superior quality high and low-rise apartment complexes; and six large motor hotels.
(Muller, 1981: 163)

This description emphasizes the contrasts between the mega-mall, which is a planned development mainly concerned with retailing but with added recreational facilities including restaurants and hotels, and the mini-city which is an agglomeration of a much wider range of land-uses more fully replicating the city centre and where the growth has been more organic and not totally planned *ab initio*.

The conclusion from this review is virtually to reiterate Johnson's comment quoted above (p. 76). Against all these developments the concept of an intra-urban hierarchy seems outdated. But that reaction can be taken too far. As in the consideration of the inter-urban condition, where underneath the new metropolitanism, and outside its immediate reach, lie the elements of a traditional hierarchy, so too in the intra-urban context there are still the basic layers of sub-centres which Berry identified some 30 years ago in 1967.

5.4 PROBLEMS OF CULTURAL AND BEHAVIOURAL VARIATION

A basic assumption of central place theory is, of course, that of economic man or woman seeking always to maximize economic benefits and able to deploy a full and complete knowledge of the market. That assumption is unjustified and the later development of central place studies was built about the analysis of cultural and behavioural modifications of spatial actions. There is certainly a common element between the two modifiers, indeed behaviour is conditioned by culture, that is by inherited and acquired ways of perceiving the world. But in central place studies a contrast can be postulated in that culturally moderated actions are conceived in terms of broad groups, while behavioural responses are regarded as those of individuals or aggregates of individuals within the culture realm.

Cultural variation

The analysis of culturally based variation has been relatively little pursued apart from an early and widely quoted paper by Murdie significantly titled 'Cultural differences in consumer travel' (Murdie, 1965). Murdie studied consumer travel in an area of southwestern Ontario where there were two contrasted cultural groups which he called old order Mennonites and modern Canadians. The old order Mennonites form a relict element equivalent to the Amish or Pennsylvania Dutch peoples of the USA. They are the strictest preservers of the traditional ways. In a standard analysis in central place terms a series of regression analyses was carried out where the dependent variable was distance travelled to first choice centres and where a series of independent variables was considered, such as the number of central functions in the first and second choice centres and frequency of purchase. The major explainer of distance travelled was the number of functions in the first choice centre in accordance with classical central place theory, but there were considerable differences between the two groups. These are best illustrated by Figs 5.10(a) and (b) and 5.11(a) and (b) demonstrating significant differences in travel for purchases of clothing and yard goods. Whereas the outer range of modern Canadians increased with the size of centre visited, for the Mennonites it did not, for the outer range was fixed at six miles. The old order Mennonites make no concession to changing fashion and still wear the same type of clothing as their ancestors did when they arrived in Canada. They buy their clothes locally, at the nearest point. New fashions coming into an area reach the regional capital first and are diffused down the hierarchy, usually related to price since the most exclusive and expensive remain available only in the higher order centres. But that whole process is meaningless to the Mennonite people. Effectively cultural mores have eliminated an ordering of goods on which a Christaller type hierarchy rests. 'One may conclude that the central place importance has virtually no effect on distance travelled by the old order Mennonites for these goods' and that 'socio-economic influences on relative mobility, as these are reflected in differences in the levels of space preference attained by different

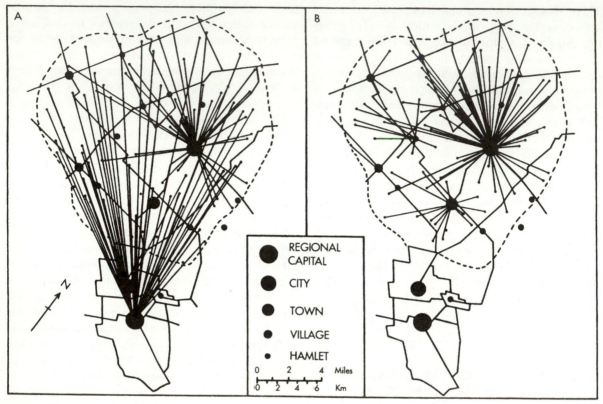

FIGURE 5.10 *Cultural determinants of the journey to retail centres. The journey to shop for clothing and yard goods. A: Modern Canadians; B: Menonites (after R. A. Murdie, 1965).*

consumers, must be considered in more hetero-geneous areas' (Murdie, 1965: 232–3). Or, in other words, the perceptions of the consumer are critical.

The lack of follow-up to Murdie's paper in central place literature was probably because most investigation was carried out within one cultural realm where such variation was irrelevant. There is one contemporary aspect, however, where such a cultural issue has arisen and that is the opening of shops on Sundays. The legal situation varies greatly (Kirby, 1992). It is admirably clear in Germany where by the Store Closing Act of 1956 hours of business are restricted to between 0900 and 1400 hours on Saturday and 0700 and 1830 hours on Monday to Friday. In France the situation, based on labour regulations, is extremely flexible. In England and Wales the controlling legislation

was the Shops Act of 1950 which had given rise to a complex and opaque situation. As a result of long agitation it was amended in 1994 so that shops are now allowed to open on Sundays, but there was a limit of six hours imposed on all large shops, where large was defined as over 3000 square feet. Effectively Sunday opening became universally possible. There was certainly opposition from the Lord's Day Observance Society and by shop workers. Thus, Kirby (1992) reports an unwillingness of some 40 per cent of female retail employees to work on Sundays. Though it is reported that the unwillingness and the opposition to Sunday work is more a consequence of the demand for leisure and family time, rather than for keeping holy the seventh day, nevertheless the underlying drive 'to keep Sunday special' is an inheritance from the Christian religion, and especially its protestant

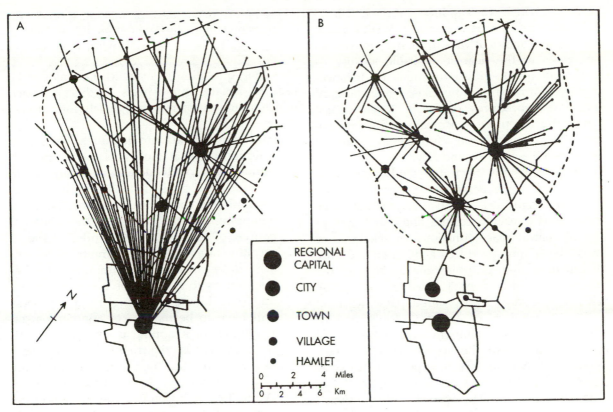

FIGURE 5.11   *Cultural determinants of the journey to retail centres. The journey to shop for shoes. A: Modern Canadians; B: Menonites (after R. A. Murdie, 1965).*

section. It is difficult to identify a specific spatial consequence of Sunday trading but the increased demand is at least partly related to the growth of out-of-town shopping centres and would probably reinforce their significance.

Behavioural reactions

One of the earliest observed discrepancies between the theory of central places and the real world was that of consumer behaviour. Among the protagonists of that criticism were Golledge *et al.* (1966) and Rushton (1966) who based their empirical investigations on the shopping behaviour of the dispersed rural population of Iowa, examining the assumption, critical to central place theory, that people will travel shorter distances for those goods and services which occur with the greatest frequency. It is argued that the above assumption, together with the hypothesis that many goods and services have similar thresholds and ranges (i.e. trait complexes occur), is the rationale for inferring consumer behaviour from the distribution and frequency of occurrence of central place goods.

The study dealt with two sets of data, the distance of travel for maximum purchase and the nearest purchase for a set of thirty-three goods and services. A considerable degree of flexibility in the purchase characteristics of some goods was shown by an analysis of standard deviations from mean distances travelled, but the crux of the study was a rank correlation between order of entry of goods by distance to maximum purchase town and by distance to nearest purchase town. The correlation coefficients derived were 0.23 and 0.32 respectively, neither of which was significant at the 0.01 level of confidence. It was apparent from the result that Iowans do not always make their maximum purchases at the nearest available locality. A further attempt to group 'functions' by distances travelled for maximum purchase using standard grouping procedures produced seven 'ranks' which did not provide a means of distinguishing the conventional functional contrasts between village, town and city.

The conclusions from the study were that deductions on expenditure patterns derived from the occurrence of functions did not agree with direct interview results, and that a 'grouping of central place functions on the basis of travel behaviour produces a different ordering of functions than grouping on the basis of the occurrence of functions' (Golledge *et al.*, 1966). This type of study leads to an impasse. Rushton writes 'Central place theory is only one of the many areas of human geography where assumptions about the spatial behaviour patterns of individuals are incorporated in explanations of spatial structure' (Rushton, 1969). But it is a 'chicken and egg' type of problem: which comes first, the city system or the patterns of behaviour.

These initial studies led to much wider investigations as the behavioural paradigm became high fashion in human geography and observed aggregate patterns were demoted in significance. This was accompanied by a fundamental change in scale of analysis from the city in the region to the individual in his or her environment; the new approaches were called 'cognitive behavioural'.

The cognitive behavioural approach to location theory and consumer behaviour is a synthetic framework, starting from the individual decision maker as the basic unit of analysis. The individual's behaviour is viewed as a function of the environmental situation and the decision-making processes with respect to the environment. In contrast to the deterministic location theory framework which makes a set of assumptions that factors out the processes of human decision making, the cognitive behavioural approach specifically focuses upon

the nature of the decision-making process and the parameters which determine its outcome.

(Downs, 1970a and b)

This long quotation from R. M. Downs can be used to isolate four aspects of this approach which although separate are closely interlinked.

1. **Motivation, goals and attitudes.** The whole question of what motivates the shopper is paramount for it determines consequent attitudes and goals. If the individual seeks to minimize costs then time will be spent 'shopping around' to find the cheapest source of goods. On the other hand a wealthy, or a lazy, shopper might be content with the nearest source, although a further decision as to the cost of travel in money and time in the shopping around process will have to be made. Yet another person might look on shopping as a basis for socializing and choose the friendliest shop where local gossip can be collected. In more general terms it is possible to adopt Wolpert's conclusion that the concept of the spatial satisficer appears more descriptively accurate of the behavioural patterns of populations than the normative concept of economic man. The individual is adaptively or intendedly rational rather than omnisciently rational (Wolpert, 1964).

2. **Decision-making and preference studies.** The motives and attitudes of the consumer have to be translated into action and this is done via a decision-making process in which various desiderata are played off against each other. Each shopper will have a preference structure derived from basic attitudes and this can be represented by a series of cognitive categories such as cost (cheap–dear); distance (near–far); variety (good range of goods–small range of goods); and so on.

3. **Perception.** Between the abstract preference structure of an individual and the actual choice in the real world will intervene the individual's perception of the various shopping centres. Apart from preference

Table 5.1 *Attributes of shopping centre structure and design*

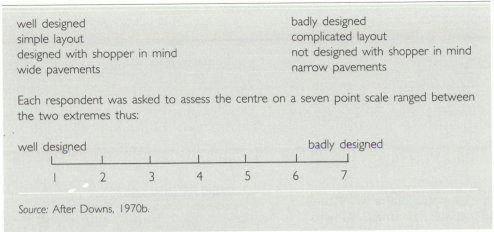

well designed	badly designed
simple layout	complicated layout
designed with shopper in mind	not designed with shopper in mind
wide pavements	narrow pavements

Each respondent was asked to assess the centre on a seven point scale ranged between the two extremes thus:

well designed   badly designed

1   2   3   4   5   6   7

Source: After Downs, 1970b.

there will be a wide range of other influences, including past experience.

4. **Search procedures and learning processes.** No individual will have perfect knowledge of an area on which to base decisions. There will be rather a continuing process of learning and of comparing centres. During that process successive searchings will probably be made to find those centres and shops most in accord with preferences. There then arises the problem that such a process will culminate in habit, that is the search will be given up, and eventually action will become out of phase with a changing reality. In such conditions habit may result in the behaviour of the individual no longer reflecting preferences as they would be expressed.

Given the four aspects of a cognitive behavioural approach it will be apparent that the critical issue becomes the techniques of measurement which have to be devised in order to evaluate behaviour. A real difficulty here is that as deeper comprehension of complex behaviour is sought so investigation merges into the work field of the psychologist concerned with human motivation and choice.

An example from the early work of Downs will serve as an illustration of the type of investigation. His basic problem was to assess the image of a shopping centre held by a group of shoppers. The hypothesis to be tested in more general terms related to the way a segment of the spatial environment was evaluated. In order to operationalize this Downs made use of the semantic differential which was devised in the 1950s by C. Osgood to measure the connotative meanings of concepts in what he termed 'semantic space' (Osgood *et al.*, 1957). In its application in this case it operated by asking respondents to assess the shopping centre in relation to a series of bipolar attributes. Nine cognitive categories were hypothesized with four attributes in each category giving thirty-six in the total scale. The nine categories were: price; structure and design; ease of internal movement and parking; visual appearance; reputation; range of goods; service; shopping hours; atmosphere. The breakdown of one of the categories into the four attributes can be given as an example (Table 5.1).

Respondents placed a mark at what they considered the appropriate point. A mark between 2 and 3 would indicate that the centre was adjudged to be only fairly well designed. Each response was given a value and accordingly means and standard deviations for each attribute were calculated. In addition the total responses for each scale were correlated with the responses for all other scales giving a 36 × 36 matrix. Downs then factor analysed the matrix to give the principal factors contained – that is,

Table 5.2 *Dominant categories in the image of shopping centres*

1. Service quality
2. Price
3. Structure and design
4. Shopping hours
5. Internal pedestrian movement
6. Shop range and quality
7. Visual appearance
8. Traffic conditions

Table 5.3 *The relative importance of factors in grocery shopping*

Group class and attributes	Average points	Within group rank
A. *Commodity traits*		
1 Price	5.791	1
2 Quality	5.754	2
3 Variety	4.866	4
4 Brand range	5.267	3
B. *Outlet factors*		
5 Friendly and courteous	5.487	3
6 24 hour shopping	3.267	9
7 Cleanliness	4.844	5
8 Attractive display	3.941	8
9 Pleasant surroundings	5.374	4
10 Quick service	5.770	2
11 Internal organization	4.283	7
12 Nearness of parking	6.024	1
13 Cheque cashing	4.684	6
C. *Locational features*		
14 Proximity to residence	5.689	1
15 Proximity to bank	2.331	4
16 Proximity to drugstore	3.129	3
17 Prox. to discount store	2.026	5
18 Prox. to bus stop	1.406	6
19 Prox to a route you follow to or from work	3.716	2

Source: Schuler, 1979.

how the image of the shopping centre was structured. Reviewing these findings Downs concluded that the image of a downtown shopping centre was composed of eight cognitive categories which, arranged in order of descending importance, were as shown in Table 5.2.

Downs then divided these categories into two groups: those related to the retail establishments themselves (1,2,4 and 6); and those related to the structure and function of the centre (3,5,7 and 8).

It is interesting to make a comparison with a later study by Schuler (1979) concerned with single shops rather than a shopping centre. He used a questionnaire answered by a comparatively limited number of respondents (110) in Bloomington, Indiana. A preliminary section ensured that all those whose answers were used were familiar with the major stores. Respondents were asked to rate 19 characteristics on a 7-point scale. The results, showing the average points scored for the 19 characteristics and the rank within a three group division of them, are given in Table 5.3.

The results indicate that quality and price of the merchandise, followed by speed of service check out and distance, were most important to the respondents. Schuler was considering single shops as against Downs's investigation of a shopping centre. Even so, it is significant that service quality and price head the list of desiderata in both. Economic motivation seems to dominate behaviour.

This type of investigation constitutes a significant movement away from the simplistic assumptions of central place theory and towards the scrutiny of the bases of the way in which people behave. Even so there is a considerable omission since the image needs to be examined in relation to the way it might vary with the socio-economic status of the respondents as well as age and the whole gamut of characteristics of the population which made up the group surveyed. This is attempted by Schuler but his two groups derived from the sample are not sufficiently differentiated to give a clear result.

One of the major problems of these sorts of study is the arbitrary selection of categories, characteristics or attributes, by the investigator. An answer to that lies in personal construct theory which is intended to derive the categories from the respondents (Bannister, 1962; Kelly, 1955; Harrison and Saare, 1971). In the sort of study outlined above the researcher presents the person being questioned with three shopping centres and asks that the two most similar be identified. The criteria of differentiation are then ascertained and used as the categories in the semantic differential, that is the categories are derived from the respondents themselves and not imposed by the investigator.

The outline of personal construct theory above implies that Downs's problem was relatively simple in that he was concerned with the evaluation of attitudes to one centre only, whereas in most studies the task is to unravel the attitudes which led to the decision to select one out of a number of possible centres. To achieve this geographers have resorted to attitude tests which were initially derived for work in the fields of social and educational psychology. The scaling of attitudes is usually accomplished by means of Likert (Likert, 1932) or Thurstone (Thurstone and Chave, 1929) scales based on the expression of agreement or disagreement to prepared statements.

This brief discussion has been directed towards indicating the nature of work on consumer behaviour which entered the realm of central place studies in the 1970s. Inevitably its development in depth moved away from the spatial basis of geographical study and became more psychological in nature and of more interest to the retail and business researcher.

This is perhaps best exemplified in the most detailed survey of shopping behaviour undertaken, that of the Cardiff Consumer Panel Survey (Wrigley *et al.*, 1985). The survey produced a 'continuous record on all aspects of daily food and grocery shopping ... for 454 households over a 24 week period' in 1982. For each store visited a wide array of detail was recorded by the participants. But the list of topics which formed the basis of published work indicate how retailing rather than the geographic relevance of consumer behaviour was the paramount concern. An illustration is the consideration of the interaction between store choice and brand choice.

If central place studies are beginning to consider consumer behaviour, relatively little attention has been devoted to the behaviour of the entrepreneur. Lewis attempted to demonstrate how entepreneurial interpretation of opportunities available can be seen in the context of part-time operation of functions (Lewis, 1970). Barton approached the same problem in a different way by examining the development of the central place system or the creation of centrality. She concludes the 'exchange and the entrepreneurial role are basic features of the achievement of centrality. The entrepreneur occupies an international niche created by the existence of comparative advantages and imperfect competition. As a result, places achieve excess importance, or centrality, by virtue of capturing a role in exchange' (Barton, 1978). However, the role of the entrepreneur, and also the constraints exercised by public policy, were to attract attention in a rather different way in the rapid growth of retail geography (Dawson, 1980).

5.5 RETAIL GEOGRAPHY

At this point in the consideration under the chapter heading of problems and developments from central place theory, retail geography presents a real difficulty. Certainly retail geography emerged from urban geography and its pedigree has to be traced back to the study of central places. Its present concerns are fundamentally urban, partly because of the truism that the vast proportion of retailing is urban based, but also because the prime interest has been that of the large retail store chains seeking advice on the choice of locations. There has been relatively little attention paid to the small village shop or to the lowest order towns. Geography, like so many other disciplines, has followed the sponsor's cash which becomes the determinant of development rather than intellectual paradigms. Moreover, as it has developed, retail geography has become the study of retailing with aspatial aspects of operation and internal store design more prominent. Many urban geographers have indeed moved to departments of business and management. It is unreal, therefore, to contain retail geography within an exposition of urban geography and, although it would be impossible to omit treatment, that treatment must be brief and, to a degree, superficial.

Probably the earliest contribution was that of Reilly in his Law of Retail Gravitation which, in spite of an overlay of great technical sophistication and development, still remains the basic notion of how to assess the drawing power of a town, or a store, or how to assess the potential drawing powers of a number of possible locations. Reilly's 'Law' is usually presented in the form

$$Bb = \frac{Dab}{1 + \dfrac{Pa}{Pb}}$$

where Bb is the break point between city A and city B in miles from B, i.e. B's sphere of influence; Dab is the distance between A and B in miles; and Pa, Pb are the populations of A and B respectively. Huff was responsible for restating this in probabilistic terms on the basis that the likelihood of a consumer going to any centre is based on the number of items required and the effort and expense which he has to make. From these two variables Huff (1963) derived an expression:

$$P(Cij) = \frac{\dfrac{Sj}{Tij^\lambda}}{\sum_{j=1}^{n} \dfrac{Sj}{Tij^\lambda}}$$

where $P(Cij)$ is the probability of a consumer at a given point of origin i travelling to a given shopping centre j; Sj is the area of selling space devoted to the sale of a particular class of goods by shopping centre j; Tij is the travel time involved in getting from a consumer's travel base i to shopping centre j; and λ is a parameter to be estimated empirically to reflect the effect of travel time on various kinds of shopping trips. If $P(Cij)$ is calculated for a series

of points of origin (i), then isopleths, or equiprobability contours, can be drawn for each of the centres (j). A trade area can be identified by isolating points where the same equiprobability contours from different centres intersect.

An early attempt to estimate the durable goods sales potential of a shopping centre was developed by Lakshamanan and Hansen (1965) and is presented in the form set out by the UK Department of the Environment (1973). The problem is to estimate the durable sales potential of a shopping centre. This is taken to be a function of its own size, the size and prosperity of its population catchment and its spatial relation to competing shopping facilities. In this context the size modified by income levels to give deployable expenditure can be equated with the threshold principle of traditional central place theory, while competition from other centres represents the idea of range. The basic formulation is that of the gravity model with the region divided into a number of arbitrary zones as opposed to spheres of influence:

$$SALj = \sum RSP_i \sum_k \frac{\dfrac{ATRj^b}{Tij^a}}{\dfrac{ATR^b_k}{Tik^a}}$$

where $SALj$ represents durable sales at centre j; $RSPi$, the retail spending power in zone i; $ATRj$ $ATRk$, the attraction indices for centres j and k; Tij, Tik, the interzonal travel time between zone i and centre j and zone i and centre k; and a, b are constants.

Before this can be run the various terms need to be defined; for example a measure of the attraction index has to be devised and travel times determined. The problems these cause will not be discussed since the concern here is with the nature of the concept rather than the technical problems of making it operational.

Further developments in this same area can be represented by the forecasting model for Tesco, a British grocery chain, as set out by Penny and Broom (1988). As the authors note, the initial focus was to provide an accurate model for sales forecasting and the starting point was the basic spatial interaction model which dominates this section.

$$EXP_{is} = \beta_o(\sum_k E_k H_{ik}) \cdot \frac{W_s \cdot e^{-\beta_1 t_{is}}}{\sum_m W_m \cdot e^{-\beta_1 tim}}$$

where EXP_{is} is the expenditure from zone i to site s; β_o and β_i are parameters; E_k is the mean food expenditure by household category k; H_{ik} is the number of households of category k located in zone i; W_s is the measure of the attraction of the proposed store s; t_{is} is the travel time from zone i to the site at s; W_m is the attractiveness of competitor m; tim is the travel time from zone i to competitor m.

The authors state 'the aim was to calibrate this model successfully using customer survey data; the result could then be used as a sales-forecasting model as well as being applied systematically to nodes in the network in order to scan areas for optimum store location' (Penny and Broom, 1988: 115). Thus it is noted that the term *ExPis* gives a measure of available expenditure in each zone which in itself provides a valuable analytical device. 'The plotting of available expenditure as a map, overlaid by the locations of trading and planned stores, provides a means of "eyeballing" areas in order to identify locations with a high expenditure potential but low superstore availability' (Penny and Broom, 1988: 116).

This outline of procedure simply stresses the basic interactional content. It can be made more sophisticated by the addition of the disposable income and the tastes of those within a catchment area. While all these analyses shift more into the field of retail geography rather than urban geography, it is critical to observe that they are largely behind the creation of out-of-town shopping centres which have already been discussed. Here, perhaps, is the perception and behaviour of the entrepreneur which has been noted as being absent from behavioural studies which have concentrated on the consumer. Geographers should perhaps examine geographers as consultants and as the creators of patterns!

As a final comment on retail geography it is worth recalling that there are two trends which take the issue beyond spatial interaction, at least in the conventional sense:

1. **Mail order sales.** These are an increasing element in purchasing and are unrelated to local central places.
2. **Information technology and retailing** (Guy, 1988). The electronic bases of selling are related to Cable TV and armchair shopping. Where deliveries are local these do not necessarily impact upon central place patterns, but where mailing from central warehouses takes place then there is a clear influence.

This very brief excursion into retail geography has demonstrated that the active policies of entrepreneurs have to be incorporated into the consideration of the process of shopping centre location and development. So do the policies of government, both local and national, for all developments are subject to planning approval. There are very many relevant forces at work which are well beyond the naive view of central places as set out by Christaller in the 1930s (Wrigley, 1988).

5.6 HISTORICAL CHANGE AND CENTRAL PLACE SYSTEM

The view of a wide range of forces beyond the initial statement of central place theory is re-echoed in another central weakness for it is a deterministic economic model which creates from first stated principles an instantaneous spatial pattern. But manifestly towns grow and decline, and the analysis of such processes is essential. Christaller's chapter on dynamic processes did nothing to meet that point. So that if central place theory was welcomed in that it replaced narrative and descriptive accounts of particular towns, it nevertheless could not replace the concept of status as a thing acquired over time. There are, however, three scales on which change can be considered. One is the long-term consideration of secular change, the second is the medium-term and the third the short-term review of contemporary change. These are clearly part of the same process, but they are best examined separately.

Long-term change in the city system

The clear implication of central place theory is that cities are ordered into a system, or to put it in another way, cities are related to each other systematically, although the area over which that relationship appertains is problematic. But at the largest scale in space and the longest scale in time that notion of a system of cities has become linked with the postulation of the evolution of a world order by Immanuel Wallerstein (1974, 1979, 1980). If a broad process be envisaged of the emergence under capitalism of a world order divided into core, periphery and, possibly, semi-periphery, then that process must have been intimately associated with urbanization and hence with the systematic relationship between cities. If urbanization be a component of the capitalist mode of production, then its specification must be visible in the ordering of cities. It follows that if there be a world order then there should be a world order of cities; certainly that is implicit in the very notion of 'world cities' (Hall, 1966). However, attempts to present and interpret a world order through the character of the city system have been relatively unsuccessful. Thus Chase-Dunn (1985) attempts to calculate a Standardized Primary Index which attempts to express the deviations of city-size distribution from the log-normal (rank–size) rule (see Chapter 3, p. 35). It is zero when the city-size distribution conforms to the rank–size rule, positive when the distribution is primate and negative 'when the distribution is less hierarchical than would be predicted by the log-normal rule' (Chase-Dunn, 1985: 275). Calculations reveal that the hierarchy taking the ten largest cities of the capitalist world, that is assuming they constitute an ordered system, is always less hierarchical than the log-normal rule would predict. Chase-Dunn concludes that 'this finding might be interpreted as indicating that the notion of a world system is mistaken', but 'instead it could be postulated ... that the globe

is composed of many separate urban systems, and that comparing the sizes of the largest cities in this way combines things that ought to be analysed separately' (Chase-Dunn, 1975: 275). That might be considered as a rather self-evident truism were it not for the fact that any supranational, world organization must necessarily imply supranational relationships.

Chase-Dunn does, however, provide some intriguing facts by extending his analysis over time. Especially notable is a flattening of the rank–size graph of these largest cities over recent decades due to the increasing presence of the cities of the semi-periphery in the rankings, a feature already noted in Chapter 2 where world urbanization was considered. Thus by 1975, Mexico City was ranked fourth, Shanghai fifth and São Paulo seventh. But this was at a time when all Latin American countries, with the exception of Colombia, had primate systems, while the countries of the core had begun to move towards less hierarchical national city systems. Indeed, the entry of these large cities of the developing world or the semi-periphery is not seen as marking any shift of power and resources to the periphery, an interpretation which may be true but which has implications for the validity of the concept of a world order of cities by rank–size.

At a level lower than the rather speculative interpretation of the world system in urban hierarchical terms has been the attempt, derived from comparisons of its various parts, to establish some rationale for the contrasts between countries which show primacy and those which show log-normal distributions, a matter already raised in Chapter 3. There is a long-standing assumption that the rank–size rule relationship is an end or steady-state distribution and, therefore, represents a standard outcome of the long-term operation of an unfettered market economy, given a fairly large geographical area and the simultaneous effect of a variety of cross-cutting influences (Walters, 1985: 72). In essence that is no more than a restatement of Zipf's original argument. It follows that a primate distribution marked underdevelopment and characterized Western countries at earlier times and 'Third World' countries at the present. However, as Walters comments, attempts to demonstrate such a simple relationship have been at best inconclusive (Walters, 1985: 73). It is also germane that the simple contrast of a primate as against a log-normal pattern is too stark. Thus it can be proposed that in countries where there is a distinctive primate city, the remainder may be disposed either in a fairly flat curve, which is the usual assumption, or in a log-normal curve thus combining two systems of ordering. Vapnarsky (1969) argued that Argentina demonstrated the former state during the nineteenth century, but the latter at the present. The implication is that the pattern which is usually associated with primacy is related to a condition where the country is marginal to the world economy so that there is a lack of internal integration, whereas later movement into the semi-periphery and internal sorting engenders a log-normal relationship beneath the primate city. But that is very close to standard interpretation of development as the conditioning factor. It is certainly a replication of the points made in Chapter 3 that the determinant of system organization is the degree to which a rationalized Western system of economic organization and activity is deployed.

Carol A. Smith has examined the notion of primacy against the empirical evidence in Guatemala. She presents the three well established explanations for urban primacy (Smith, 1985a and b: 99–113). The first is colonialism, based on the view that colonial powers usually established one dominant centre of control, leaving the rest of the territory in a traditional and backward condition, although 'backward' has a pejorative meaning in this context. The second explanation is export dependency, largely devised to explain primacy where there was no direct colonial inheritance but rather a state of dependency generated by the dominance of primary products for export via a single primate port/city. The third explanation has been called rural collapse and is in contrast in that it envisages the growth of industry in the largest city and the consequent collapse of the rural peripheries which are dependent on plantation products for export. It is one of the standard interpretations of urbanward migration and hence of the exaggerated growth of the target city and its elevation to primacy. These do not seem to be discrete explanations and are little more than exegeses of development.

Smith offers her own explanation epitomized by the title under which she presents her material – class relations and urbanization in Guatemala: towards an alternative theory of urban primacy (Smith, 1985b). The whole process of shift in the city system organization is seen as part of the transition to capitalism, and in particular the transformation of class relationships. The élites of a peripheral economy gather in one city and, tied to world commerce as capitalism impacts, need a free and mobile labour force, that is one removed from what can loosely be called feudal obligations. This labour force is created locally and in the primate city solely, leaving the rest of the countryside with traditional relations intact. 'Thus, at the same time that free labour is released and encouraged to move to some cities, it is discouraged from entering others, especially provincial cities' (Smith, 1985b: 129). Where the attracted labour cannot be employed a large informal sector is generated; hence the creation of 'overlarge' or primate cities with their shanty towns.

Certainly Smith's interpretation adds depth to the more generalized explanations, but itself seems mainly an extension of the development theme. It offers no convincing explanation in itself of why the dominance of a single city should develop. Thus in the same book as Smith's papers appear there follows a study which contrasts the distinct primate system of the Philippines with that of South Korea where no such condition emerged (Nemeth and Smith, 1985). The basic explanation is one of historical contrast. The colonial Philippines had no existing urban system so that the colonial power developed a predominating centre much after the fashion already noted in the colonial explanation of primacy above. In contrast in South Korea there was a pre-existent and indigenous urban system which was simply taken over and hence no primate pattern developed. 'The major argument of this essay has been that an understanding of the historical context of urbanization – and especially key junctures in history such as the initial incorporation into the capitalist world economy – is critical to explain adequately the shape of the present urban structures of South Korea and the Philippines' (Nemeth and Smith, 1985: 204–5).

It is apparent from the material which has been briefly reviewed that there are two issues concerning the longest term evolution of city systems at the largest scale which is relevant to urban geography. Both are related to the development of a world system. The first is the extent to which that system, primarily economic and political in its definition, has generated a world system of cities related at the highest level to each other and overriding regional and national boundaries. The second issue is the way in which different organizations of city systems within regions or nations reflect the growth of the world capitalist order, and especially variations within the core, periphery and semi-periphery. Here are considerable grounds for analysis and debate for an overarching interpretation has yet to be presented. Urban geography awaits its Wallerstein.

Medium-term change in the city system (Carter, 1983: 96–113)

Under this heading of medium-term change modifications within established systems will be considered rather than the largest-scale nature of the systems themselves. Inevitably there is some overlap. The earliest attempt at a modification of central place substituted random process for the simple determinism of Christaller. To some extent this was envisaged by a change to the employment of a probability (stochastic) approach to the generation of settlement patterns in the late 1960s (Morrill, 1962, 1965a; Olsson, 1967). But the 'explanatory' element in these studies remains slight and little further exploration has taken place.

The actual process by which the city system is built up, however, is an area of enquiry which appears to offer a fair possibility of generaliza-tion. In most cases it can be argued that at an early stage in the development of a city system, the competition and interaction between the centres, which is postulated as a necessary condition of central place systems, did not obtain. What did rule was the principle of separation whereby each centre served a surrounding area but lack of communications prevented competition and each centre was at a level locally determined.

Guttenberg characterized this as a situation with 'distributed facilities' (Guttenberg, 1960), that is, where the various urban facilities are distributed according to population in a system where little interaction takes place. This same situation was envisaged by Webb (1959) who postulated two theoretical viewpoints under which the phenomena of urban function may be developed. These he termed 'isolated urban society' and 'integrated urban society'. In the first, there are differentiated functions but no contact; in the second, full communication and free exchange of goods. As an example of the first condition, Von Thunen's isolated city is used and it is suggested that if this is cut off from all others that each city would of necessity duplicate the same services, all other things being equal.

But even where economic organization was rudimentary, there was still a need for political and administrative control which in a 'pre-industrial' stage was usually dominant over economic relations. This political control was often exercised from a single centre which was the epitome of the folk culture and which thus attained a pre-eminence above a multiplicity of cities of the same low order, that is, it occupied a primate position, a somewhat different interpretation of primacy from that offered in the previous section, although in keeping with the theme of Nemeth and Smith (1985). But economic development, the thickening of the whole web of economic activity which accompanies the process of urbanization, together with advances in communication techniques, led to interaction between these distributed points and to the possibility of a selected few being vested with higher order services by virtue of their nodality.

Within the above process there are two conditions which must be noted:

1. Markets are not necessarily economically based, they can be and were ordered and maintained by central authority. That is, centrality can be created by government, particularly by despotic government.
2. Christaller stressed centrality which, although it is a term denoting function, carries locational implications. But often the most significant locations were those of the so-called gateway cities or ports of entry where the vital external trading links were concentrated. Such a situation could generate a dendritic rather than a centrally orientated system (Smith, 1976).

At this point one can introduce the ideas of primary and secondary urbanization. It is possible that this move towards discrimination between centres had already been partly anticipated by the creation of a hierarchy of administrative centres related to the primate capital and the two emergent town systems one based on economic activity, the other on political organization, do not mesh. The result is a haphazard system of towns, 'rotten boroughs' and 'ancient cities', with important administrative but negligible economic functions and large 'urban areas' important economically but not otherwise. Further developments result in a sorting out of this situation, mainly, but not solely, in economic terms, though nearly all countries show that this sorting is far from complete.

This analysis is in line with that of Guttenberg who argues that transportation is the means by which the distributed features become 'undistributed' or related to a different ordering principle. Transportation decisions 'will result in a constantly changing structure with the emphasis shifting along the continuum between the situation with highly distributed centres to the situation with one major undistributed function' (Guttenberg, 1960: 109). Lukermann argued in a similar way, identifying three elements in urban systems analysis: hierarchy, nodality and circulation. Nodality is identified as 'a behavioural

act of man', not simply a geometric point or a circulation intersect. In behavioural terms a nodal location is that place where the individual has the greatest freedom to interact. Such a definition involves both population density and areal accessibility, as well as functional availability. Expressed in locational terms, nodality and hierarchy are conceptually analogous. A spatial hierarchy is the specification of a nodal system (Lukermann, 1966: 22).

The generalized pattern of city system development is now revealed as that of a series of isolated centres being brought into interaction through the operation of the circulation manifold and from that interaction a hierarchical situation appears.

This is in accord with the present author's work on Wales (Carter, 1965: 1970b). There the Anglo-Norman conquest introduced the castle town or bastide into a non-urbanized folk society where the move towards urbanization was but incipient. There was no primary urbanization, no epitome of the folk culture, but only secondary urbanization and this created a series of military centres isolated amid an alien population and in an area of fragmented topography. But with the breakdown of isolation through the increase of transport facilities, differentiation became apparent. At first, this is represented by administrative terms, for government was the prime urban function and a hierarchy of administrative centres emerges, but one clearly reflecting the older military situation rather than the newer evolving economic one. The result is a period of transition or change until, by the end of the eighteenth century, a sorting out has taken place and a discernible 'economic' hierarchy is present.

It is clear that this process of development is seldom uninterrupted and complete, and any nation or area can go through the same process more than once. Thus in Wales a degree of equilibrium had barely been attained before the beginnings of iron working and coal mining in the mid-eighteenth century (or, to put it another way, the beginning of the Industrial Revolution) resulted in the generation of a whole new set of unifunctional settlements. The mining village is the epitome of the unifunctional 'node' dominated by the principle of separation. But interaction was now quicker and the patterns of circulation exercised a discriminatory influence on these nodes, creating the degrees of nodality which have resulted in the size arrangement of contemporary centres.

The present urban arrangement of Wales is a complex mix, resulting from the partial integration of two phases of development, neither of which had been completely consummated, worked out in terms of contemporary technological, economic and social characteristics. But if the complexity be disregarded, then the basic sequence can be understood in the terms outlined above. In this context, the 'primate situation' is not regarded as inherent in the scheme: at no time in Wales has there been a primate city. It is, however, interesting to speculate that the tendency to equilibrium which is envisaged in the climax phase need not lead to a situation exclusively hierarchical in terms of the definition given above, and depending upon economic and cultural conditions, one could envisage the process leading to an arrangement in accord with either the 'law of the primate city' or the 'rank–size rule'. This would depend on the exact nature of the interaction between primary and secondary urbanization and between the developmental sequence outlined above. The aim here is not to postulate an abstract but unreal situation but to generalize the processes that have produced the wide range of variant condition which now obtain.

The failure of an orthogenic city to develop in Wales and the complete dominance of secondary urbanization is simply revealed in the fact that Wales had no capital city until 1955 when the opposition of the folk culture was finally overcome by the reality of the present economic situation.

The accompanying diagram (Fig. 5.12) attempts to portray a conceptual framework which summarizes the various influences on the urban system. The urban system (S) is conceived as being created at time t_1 through a phase of genesis (P_1). The system itself becomes rapidly composed of two elements, one a set of towns

owing their origins to special functions with peculiar locational demands, the other a set of towns owing their growth to central place functions derived from general regional demand. These are intimately related for their complex interaction produces an urban set (S_1). This net is then itself subject to the continuing modification brought about by two sets of influences. The first of these is the economic, technological changes which progressively bring new pressures to bear. Thus changes in transport techniques or the productivity of agriculture will influence the system. The second set of influences is closely related and could be included under the same general heading. These are the series of minor exogenous influences which, except in very unusual cases, continually affect the system.

The system is, therefore, never a closed system but remains open (S_2). The minor or longer-term slow changes can be absorbed into the system without signs of discontinuity but in most areas there occur major changes which are clearly related to specific new urban roles with specific and new locational demands. This can involve the creation of an unconformable urban net (P_2). There follows a period of rapid adjustment with marked fluctuations in the rank position of towns in the system until, through the revised pattern of regional demand, a modified system (S_3) is produced. Once more this is subject to minor interpolations and will react to technological and organizational change. The present city system, therefore, is seen as a momentary still in a moving picture in the true context of 'sequent occupance'. It is the product of the interaction of a series of time- and place-related acts of growing and declining towns.

There are other theoretical formulations which can be linked with the scheme outlined. The whole process of town founding can be conceived as an innovation and the subsequent development as a diffusion process. Hagerstrand's three stages in the process (Hagerstrand, 1952) are very similar to the stages of 'town genesis', 'transition and sorting', and 'climax' that have been outlined for the Welsh system and indicated in a more generalized

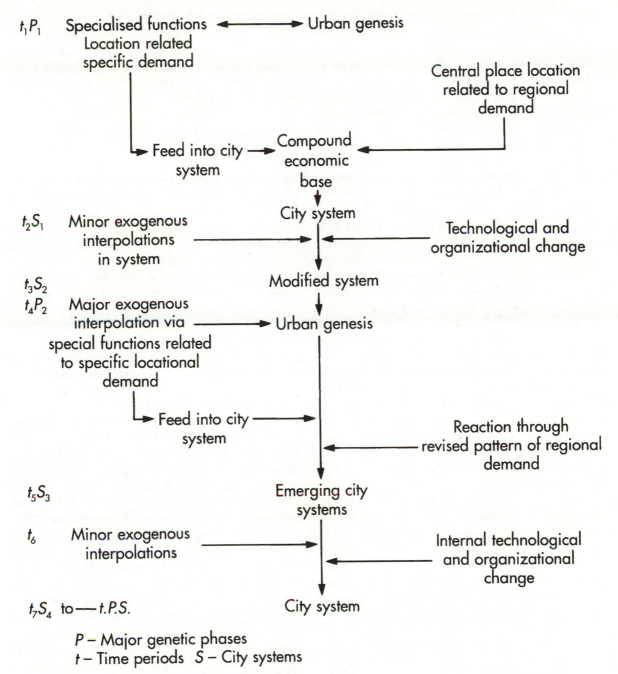

t_1P_1 Specialised functions — Urban genesis
Location related
specific demand

Central place location
related to regional
demand

Feed into city → Compound
system   economic
                  base

City system

t_2S_1 Minor exogenous
interpolations
in system

Technological and
organizational change

Modified system

t_3S_2
t_4P_2 Major exogenous
interpolation via — Urban genesis
special functions related
to specific locational
demand

Feed into city
system

Reaction through
revised pattern of regional
demand

t_5S_3 Emerging city
systems

t_6 Minor exogenous
interpolations

Internal technological
and organizational
change

t_7S_4 to — t.P.S. City system

P – Major genetic phases
t – Time periods   S – City systems

FIGURE 5.12 *The growth of a system of cities.*

form on Fig. 5.12. Each major phase has many analogies with the way in which the available space and economic potential are eventually taken into the urban service network and the way in which this network increases in density. Again the patterns of urban-industrial growth in a cumulative fashion set out by Pred (1977) and and developed in Section 6.3, can be associated with the way in which the initial creations of a genetic phase can effectively remain the major foci by the positive influence of self-sustaining growth rather than by the sole operation of external circulation.

Lampard (1968) in considering urban systems introduces the two notions of morphostasis and morphogenesis, Morphostasis represents the result of 'deviation-counteracting feedback networks' or 'a most probable state under constraint'. Morphogenesis is a deviation-amplifying process, giving an open-ended development with no evident tendency to entropy. Lampard quotes Maruyama (1963) as concluding that every system, including city systems, is made up of sub-systems variously manifesting deviation-correcting or deviation-amplifying processes whose prevalent effect will tend either to return the whole to morphostasis or to transform it by morphogenesis. To some extent the genetic phase, the major exogenous interpolations, envisaged in Fig. 5.12, represent transformations by morphogenesis, the emerging city systems within phases of transition represent morphostasis. Generalizations which argue for increasing entropy in a city system – that is the movement of the system to a more uniform and less random form – are only acceptable for certain selected phases or periods: no universal process can be isolated.

The conclusions set out above are, to a degree, confirmed by Robson in his book *Urban growth* (1973). After examining the growth of British towns during the nineteenth century, Robson concludes 'the details of the spatial patterns can only be interpreted in terms of the particular factor endowments and historical events which underlaid the growth of certain towns and the decay of others ˈRobson, 1973: 126). These endowments and events epitomize the consequences of major exogenous interpolation into the system of cities. The remainder of Robson's book seeks to demonstrate that the 'diffusion of entrepreneurial innovations through a set of nineteenth century cities may well be characterized by a simultaneous process of hierarchical space jumping from larger to smaller cities and of localized spread outwards from regional

centres' (Robson, 1973: 186). In essence this is a version of the cumulative character already considered as part of the way in which the city system grows. Robson's study added both depth and elegance to previous work on system development at this scale.

Short-term changes in the city system

To a degree the short-term changes in the city system have already been considered in a review of most recent developments which have seen the emergence of metropolitan and megalopolitan dominance. These are no more nor less than the most recent modifications of the urban system. It is also true that if one looked for the area in central place studies where most contemporary discussion had taken place it would be in the concept of change under the stimulus of the International Geographic Union's Commission on Urban Systems in Transition (Bourne *et al.*, 1989; Bannon *et al.*, 1991).

The nature of change is critical to contemporary planning problems. For example, at the end of the consideration of long-term change in the Welsh city system it was noted that a significant factor is continuing readjustment with the end of the nineteenth-century interlude of coal mining and the settlements it created. Like the redundant fortified towns of the Middle Ages, these coal-mining towns, for example, have lost their *raisons d'être* and their collapse, as against new patterns of circulation and growth along motorway and expressway, creates a prime planning problem. It is in relation to policies related to such problems that Wayne K. D. Davies (1991: 228) has set out a skeletal tabulation of the nature of government intervention set against the components of the settlement pattern upon which it can act (Fig. 5.13).

The sort of problem that arises can be illustrated briefly from Ireland. There the central problem is the dominance of Dublin. In 1986 Dublin metropolitan region had 1.13 million people or 31.8 per cent of the national population (Hourihan, 1991: 144). It was, and is, the

Settlement components affected	INCREASING GOVERNMENT INTERVENTION			
	Persuasion	Inducement (Incentives/Disincentives)	Controls	Active intervention
Population: Nat. Inc. Internal migration Internat. migration	Child policy 'To the frontiers' Advertising	Child cost payment Tax relief Free passages	Pass laws Immigration laws	Sterilization Forced removal Forced removal
Land use	Early stages of development control	Transfer of development rights	Green belts	Land banking
Generators of employment	Location offices bureau (UK)	Development grants	Industrial expansion certificates	Office re-location
Settlements (complete)		Provision of market charter. Tax incentives	Key settlements	New towns
Infrastructure		Capital provision possibility	Electricity equalization schemes	Hospital and school building, roads, rail, etc.

FIGURE 5.13 *A conceptual model of urban systems development policies (after W. K. D. Davies, 1991). Source: W. K. D. Davies, 1991: Urban systems development policies: the Alberta experience. in M. J. Bannon, L. S. Bourne and R. Sinclair eds. Urbanization and urban development. Recent trends in a global context. Dublin: Service Industries Research Centre, University of Dublin.*

absolute epitome of the primate city. It was true that during the 1970s small towns with a population of between 5000 and 10,000 showed the greatest growth, but that was no more than the universal Western trend of central city loss and suburban gain, for most were commuter settlements (Fig. 5.14). Dublin was not growing at a slower rate, its growth was simply spread over a wider area (Bannon, 1983). Moreover, Dublin has one of the most youthful populations in Europe with some 49.2 per cent below the age of 25 in 1981. Not only is that likely to lead to increased dominance but it effectively eliminates any of the measures under Davies's population heading. Hourihan's conclusion is that Ireland's regional policy was one of 'missed opportunities, of what might have been if Dublin's growth had been more balanced, if a national settlement policy had been adopted' (1991: 149). And 'in the absence of some form of national plan, it is

difficult to foresee the future role of medium and smaller sized towns as many of the more recent initiatives favour a strengthening of the functional base of Dublin, thereby, its role in the national urban system' (Hourihan, 1991: 149).

These brief considerations of change in the Welsh and Irish city systems can be elaborated by considering a review of recent changes in the French city system by Pumain and Saint-Julien (1989). The centrepiece of their analysis is a measure of attractiveness where attractiveness is defined by net migration rates. A principal components analysis of 107 urban agglomerations against 63 variables is undertaken. The variables encompass economic activity, socio-professional groups, age groupings, salary levels, income groupings, unemployment rates and house prices. The analysis produces two components which account for 40 per cent of the variance. These are identified as 'image de

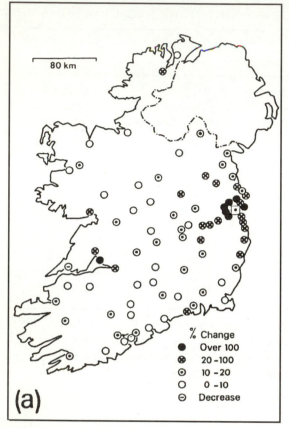
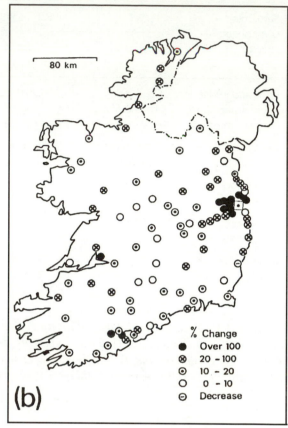

FIGURE 5.14   *Ireland. Percentage population increases in towns of over 2500 population. A: 1961–71, B: 1971–81 (after D. G. Pringle, 1991). Source: M. J. Bannon, L. S. Bourne and R. Sinclair eds, 1991: Urbanization and urban development, as Fig. 5.13 and Fig. 14.2, p. 156.*

marque' which is seen as an inheritance from industrialization and arranges cities from the most industrialized to the least industrialized, and 'modernity' which epitomizes the emergence of new service industries and growing socio-economic groups, such as managers and executives. The distribution of the variables is shown in Fig. 5.15(a) and of the cities in Fig. 5.15(b) while changes across two analyses, one in 1975 the other in 1982, are shown in Fig. 5.15(c). The authors conclude that urban growth has taken place in those cities where the socio-economic structure was dependent more on free enterprise, and it has been less in those cities dependent on old industry, salaried staffs and

public services. It is significant that the trajectory of all the cities on Fig. 5.15(c) is the same, one part being parallel to modernization, the other to 'image de marque', or in other words, deindustrialization. No city moves in the opposite direction. The 'attractiveness' with which the study began is closely linked with the growing cities of the Rhones-Alpes region, as well as the regional capitals, but further, 'it may also mean that the most recent cycle of urban development (tied for instance to the new forms of the integration of information and communication in economic activity) is entering a stage of geographical concentration, perhaps not so marked but analogous to those of the previous

A – Variables

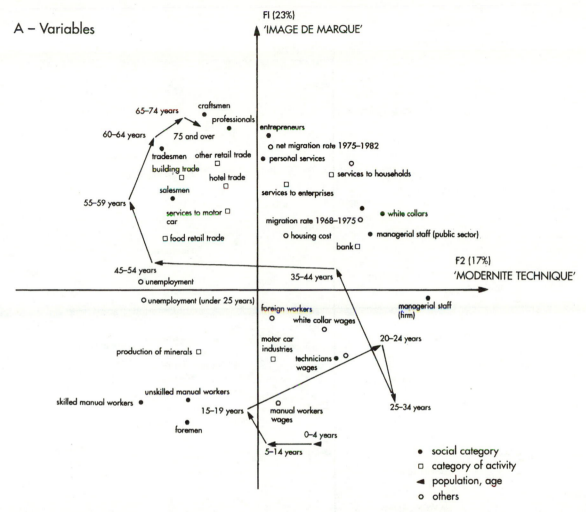

FI (23%)
'IMAGE DE MARQUE'

65–74 years   craftsmen
•
professionals
•   entrepreneurs
•
60–64 years   75 and over   ○ net migration rate 1975–1982
• personal services
tradesmen   other retail trade   ○
building trade   □   □ services to households
□   hotel trade   □
salesmen   □   □
•   services to enterprises
55–59 years
services to motor   □   white collars •
car   migration rate 1968–1975 ○
□ food retail trade   ○ housing cost   managerial staff (public sector) •
bank □

F2 (17%)
'MODERNITE TECHNIQUE'

45–54 years
○ unemployment   35–44 years

○ unemployment (under 25 years)   managerial staff
•   (firm)
foreign workers
○   white collar wages
○   20–24 years
motor car
production of minerals □   industries
□   technicians • ○   25–34 years
wages
unskilled manual workers
•
skilled manual workers •   ○   manual workers
15–19 years   wages
•   0–4 years
foremen   5–14 years

• social category
□ category of activity
◀ population, age
○ others

FIGURE 5.15(a)   *France, A: Structure of the urban system in 1982: variants.*

industrial "revolution"' (Pumain and Saint-Julien, 1989: 248).

There are clear parallels with what has been observed in Wales and Ireland, indeed in Wales the 'image de marque' component and modernity have very clear equivalents while the dominance of Dublin is closely linked to the development of all those characteristics of modernity, those variables indeed on Fig. 5.15(a) which characterize 'modernity'. But it is possible to take this argument much further and relate it back to those significant transforma-

tions which were observed in the city systems of Western Europe (p. 57 et seq) and to the now hackneyed comparison of 'sun belt' and 'snow belt' in the United States. Thus the medium-term and the short-term changes in the city system merge.

5.7 CONCLUSION

The very disparity of elements which have been treated in this chapter is indicative of the situa-

B – Cities

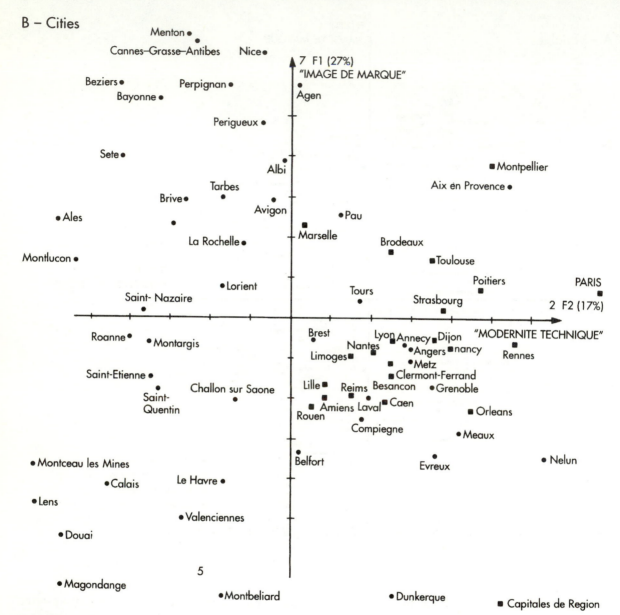

FIGURE 5.15(b)   *France, B: Structure of the urban system in 1982: cities.*

tion that central place theory is no longer the central postulate even in the examination of city systems. Christaller's work is revealed as not establishing some universal spatial condition, but as being time bound and related to particular conditions at a particular scale when free-standing discrete settlements were universal.

Even then it only operates within the confined assumption of economic man/woman. The limitations of the theory, therefore, pose a question as to whether this section of the book, that is this chapter and the one preceding it, should have been organized about the theory and begun with it. Is not the theory now

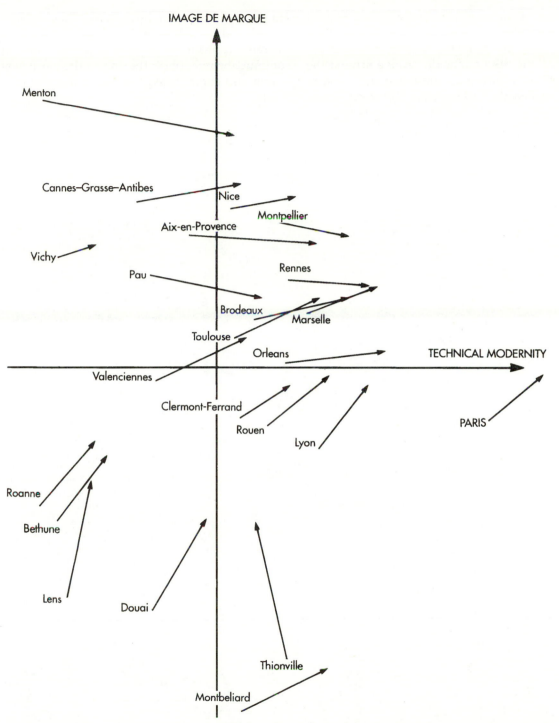

FIGURE 5.15(c)   *France, C: Urban changes 1975–1982 (after D. Pumain and T. Saint-Julien, 1989) Source: D. Pumain and T. Saint-Julien 1989: Recent changes in the French urban system, in L. S. Bourne et al. ed., The changing geography of urban systems. Dept. of Geography, University of Navarra for IGU Commission on Urban Systems in Transition. Figs. 3A and B and Fig. 5. pp. 243, 244 and 248.*

irrelevant, or only of historical interest? Should not a new starting point be established? To a degree these are proper questions; but even so, nearly all the discussions do revolve around the two notions of threshold and range and a consequent systematic relationship between cities. However these are reinterpreted they remain crucial. Central place theory is still, therefore, the most appropriate point of departure. But, to continue the metaphor, the end of the journey is a long and complicated way from that point of origin.

URBAN FUNCTIONS AND TOWN CLASSIFICATION

6.1 INTRODUCTION

At the beginning of Chapter 3 it was pointed out that a town has a dual role. It is on the one hand a central place serving its surrounding area, on the other it is the location of specialized activities which are not necessarily found in all towns. Having considered the more general area-serving role it is now essential to graft on to that the specialized or resource-orientated functions for the light they will throw on three closely related aspects of urbanism:

1. The urban role in the national or regional economy.
2. The particular aspects of urban society which correlate with types of specialism
3. The relation between functional and locational patterns which is at the heart of the geographer's interest.

Along with centrality, therefore, it is necessary to consider those aspects of a town's activities which are not directly related to it in the sense of having been derived from it. Although, to add complexity, some of the special activities will have been established because of the size engendered by the central role. This involves a separation, therefore, which is artificial but necessary. Later in this chapter a discussion of the logic and pertinence of divorcing the study of resource orientated and manufacturing activities from central place functions will review the possibility of producing a unified theory of city size and location.

It is significant, however, that in recent urban geographical literature but little attention has been devoted to urban functional analysis. Whereas in the early stages of urban geography it was one of the prime themes, now it has all but disappeared. This is because little progress has been made from the initial process of town classification which has remained little more than a minor advance from the common usage of terms such as port, industrial town or market town. But it is also a consequence of the nature of urban developments reviewed in the last chapter. As the extreme conditions which were produced by the Industrial Revolution are modified, then functional differentiation is itself diminished. Even when at the present comparable developments take place there is a determined attempt to avoid the creation of highly specialized, single function towns. The title of a study of Australia's biggest gold mine is sufficient – 'Commuter mining and the Kidston gold mine: Goodbye to mining towns?' (Jackson, 1987). As a result of these trends two rather different procedures, with somewhat different ends, have become more frequent. The first is the use of multivariate analysis which is based on a much wider range of variables than the older considerations of function which were based on occupational data alone. The resultant classes or groups are much broader in connotation. The second is even further removed from the traditional field, though a derivative from its multivariate development. It is the measurement of the quality of life in cities. It seeks a different range of variables in an attempt to rank towns in terms of their attractiveness as places in which to live. Even so, role and location, which are the basic elements of functional study, are two critical arbitrators of quality of life.

Since no overarching model exists, the approach in this chapter will be somewhat different from that in Chapter 3. Instead of starting with a theoretic model, successive attempts which have been made to understand functional character will be reviewed, leading to contemporary assessments of urban quality. The starting off point is classification since that is always the first, if primitive, stage in attempting to derive meaning from the array of data which purports to identify the functions of towns.

6.2 SYSTEMS OF CLASSIFICATION

From earliest times, chorographical works have included descriptions of town functions and have carried the implication that towns fall into classes or groups by virtue of the functions they perform. The simple designation 'market town' or 'seaport' is a form of functional classification. With the increasing diversity of town function that followed the industrial developments of the

eighteenth and nineteenth centuries such descriptive classifications became more frequent and more elaborate. In Chapter 3 it was noted that in Britain the Committee on the Health of Towns which reported in 1840 (*Report of Select Committee on the Health of Towns*, 1840, vol. iv) found that a prerequisite of their work was some orderly arrangement of the towns to be considered. Five groups were identified:

1. The metropolis
2. Manufacturing towns
3. Populous seaport towns
4. Great watering places
5. County and other considerable inland towns not being the seats of particular manufactures.

The distinction is made between central place cities (5) and special activity cities (2, 3 and 4) and the variable of size is clearly included in the words 'populous', 'great', and 'considerable'.

Since the time of this classification attempts to suggest groups of towns linked by common functions have become more elaborate and statistically more sophisticated. At the same time, greater efforts have been made to understand the logical bases of classification and the nature of town function. It can be suggested that these attempts can be arranged in progression from the simple general statement to the contemporary multivariate analysis. In general, this is a chronological arrangement and one in which statistical analytical tools become increasingly used.

General description

This is the earliest stage in the analysis of town functions. Classes are established in descriptive terms only and towns are allocated to a class on a subjective basis. There are innumerable schemes of this sort appearing at the most elementary level but they all owe a great deal to one of the earliest of such classifications to appear in a geographic context. This was included by M. Aurousseau in his paper – 'The distribution of population: a constructive

Table 6.1   *A classification of cities*

Class I Administration	Class II Defence	Class III Culture
Capital cities	Fortress towns	University towns
Revenue towns	Garrison towns	Cathedral towns
	Naval bases	Art centres
		Pilgrimage centres
		Religious centres

Class IV   Production
Manufacturing towns

Class V   Communication

(a) Collection	(b) Transfer	(c) Distribution
Mining towns	Market towns	Export towns
Fishing towns	Fall line towns	Import towns
Forest towns	Break of bulk towns	Supply towns
Depot towns	Bridgehead towns	
	Tidal limit towns	
	Navigation head towns	

Class VI   Recreation
Health resorts
Tourist resorts
Holiday resorts

Source: After Aurousseau, 1921.

problem' (Aurousseau, 1921). The scheme recognized six classes of active towns which were then subdivided (Table 6.1).

Criticisms of these classes are not hard to find. There is a confusion of functional and locational terms; for example, 'Tidal limit' town cannot in any way be a functional term; it is descriptive of a location which may or may not result in certain definable functions. The group labelled 'Communications' seems oddly elaborate compared with the other groups; moreover, its constitution appears open to criticism, for it includes two of the three basic types of urban activity identified earlier. 'Market towns' and 'Mining towns' are hardly of the same functional order and the promotion, at least, of the three sub-groups to major groups would seem logical and necessary.

Aurousseau's scheme, although subject to criticism, marks an important stage in the development of functional study. It brought together many diverse ideas into one comprehensive scheme which was at once the climax of a long period of purely descriptive work and the springboard for new methods.

Statistical description

This stage in the consideration of town functions introduces objective, statistical material into the problem of classification. The most consistently used data have been occupation or employment ratios. There is a clear link between an employment group and a town's function. Numbers employed are not immediately important, but rather the proportional place which an employment group takes in the whole range of groups in the town. If mining employs 30 per cent of the total occupied population in a town, then this is a clear diagnosis of an important mining function and such figures can be referred to as the diagnostic ratios. But there are difficulties which arise in the equation of employment category and town function.

The first is the obvious problem of reducing the many thousands of occupations into a limited number of significant groups. In this process, a large number of arbitrary decisions are made, which the classifier has to accept on the principle that all the results will be similarly affected. The second major problem is that the same occupation group can mean different things in different places. To cite a single example, in the British census the group entitled 'Personal service', including hotel and cafe proprietors, barmen and domestic servants, is usually taken to be diagnostic of a resort function. But if the occupation structure of Oxford is examined it will be found that the largest single group is 'Personal service', accounting for some 14.6 per cent of the total. The next highest group is 'Metalworking and engineering' employing nearly 13 per cent. The conclusion, that Oxford is a resort with an important engineering sector, might have an uncomfortable element of truth but it is hardly the whole truth. The source of error is the substantial employment by the Oxford colleges of those occupations grouped under 'Personal service'. This is an obvious example, but much more subtle confusions can arise and the data have to be handled with care.

Studies based on the principle of statistical description can be found in the earliest stage of urban geography, as for example in Olinto Marinelli's 'Dei tipi economici dei centri abitati a proposito di alcune citta italiane ed americane' (1916). But the most widely quoted example was put forward in 1943 by Chauncey D. Harris in which a functional classification of the cities of the USA was outlined. Eight classes of towns were recognized: Manufacturing, retail, wholesale, transport, mining, university, resort and retirement and diversified.

One example will be sufficient to indicate the principle used. Transport centres are defined as towns where 'Transportation and communication contain at least 11 per cent of the gainful workers and workers in transportation and communication equal at least one third the number in manufacturing and mechanical industries and at least two thirds the number in

trade' (Harris, 1943). This example illustrates the problem of diagnosis. 'Communications' should include workers engaged in telephone and telegraph services, and yet the class 'transport centre' seems to imply something rather more limited.

If the criteria for the definition of this town type are considered then two bases for recognition are apparent:

1. A certain minimum proportional employment, the diagnostic ratio in this case of 11 per cent.
2. A certain degree of dominance of this group measured by comparison with other groups.

Both these criteria were set by simple empirical means. The experience of the classifier was used to assign towns to groups which were then converted into functional classes by means of a definition which fitted with a minimum degree of disturbance, that is with a minimum number of accessions and deletions. This was a logical step forward from Aurousseau's system for the groups there named were now given precise statistical definitions. But the definitions were the result of subjective decisions, and, although dispersion graphs and other aids may be used, the ultimate allocation of town to group is a personal one.

This was clearly accepted in another classification which although made some 10 years later was of the same broad period and demonstrated some parallel features. Duncan and Reiss in their book *Social characteristics of urban and rural communities* (1956) included a large section dealing with 'functional specialization'. In their classification the lowest value of upper decile or quintile groups was used as a criterion of definition, although it was varied in application. For specialization in transport, to compare with Harris's definition above, the critical value was determined thus: 'since the distributions of places by the percentage of persons employed in transport are quite leptokurtic, although positively skewed, the upper decile was taken as the criterion' (Duncan and Reiss, 1956: 244). This gave a ratio of between 8 and 12 per cent dependent on size, for allowance was made for the different size classes in the classification. This is in line with the proportion used by Harris. The most appropriate name for these types of classification is 'statistical description'.

Statistical analysis

The next step forward in functional classification was linked with the attempt to offset criticism of the sort directed at Harris's scheme. This necessitated the classes being derived statistically from the raw data. The critical basis of Harris's classification is the comparison of the proportion employed in a city with the national average, virtually on the same basis as the calculation of location quotients. This remained the theme in most attempts at development of these early schemes. A good example dating from 1955 is that by H. J. Nelson. In his paper, 'A service classification of American cities' (Nelson, 1955), he posed the question, 'How large a percentage of the labour force must be employed in a particular service to make the performance of the service far enough above normal to warrant separate classification?' Once again the critical issue is the definition of what is to be taken as normal and Nelson answered his question by defining normal as the mean or average for the whole country and the degree above normal was identified by using the standard deviation.

Diagnostic occupational groups were selected from the census returns. Those related to manufacturing; retail trade; professional service; transportation and communications; personal service; public administration; wholesale trade; finance, insurance and real estate; mining. For each occupational group the standard deviation from the mean for all towns was calculated. Any town which then showed a percentage employment of more than mean plus one standard deviation was said to be significantly characterized by the function diagnosed by the occupation group. That was further developed by recording how many times the employment ratio in a town was above the mean for all towns in terms of the standard deviation. For comparison's sake the average in the broad transport

and communications group as calculated by Nelson was 7.12 per cent and the standard deviation 4.58 giving the diagnostic ratio of 11.6, compared with Harris's 11 per cent and Duncan and Reiss's 'between 8 and 12'. Nelson's classification was updated for the 1980 census by Archer and White (1985) when the threshold was 8.7 per cent. However, the array of settlements was somewhat different being limited by Archer and White to SMSAs.

A further example can be quoted in the identification of mining towns where the mean for all urban areas in the USA was 1.62 per cent and the standard deviation (SD) was 5.01 per cent. Thus the diagnostic ratio was 1.62 + 5.01 = 6.63 per cent. In Archer and White's update the diagnostic figure of mean (0.97) and standard (2.36) deviation was 3.33 per cent. Nelson measures further degrees of specialization by mean plus two SDs and mean plus three SDs. Since the standard deviation is properly only valid when the distribution about the mean is normal, no more than three SDs were measured. Butte, Montana, with 32.1 per cent employed in mining was characterized as Mi3, indicating that it was a mining town with an employment proportion over three standard deviations above the mean.

R. S. Dick working in Queensland (Dick, 1961) adopted a similar procedure but expressed his results more fully by including the percentage as well, so that Butte would have been shown as Mi32[+3], indicating that it was a mining town with a total employment of 32 per cent as well as the three SDs above the mean.

These schemes produced no exclusive classes. Any town could show a number of different occupations above the mean plus one SD, indeed a town could show no occupation in that category and to meet that problem Nelson had to introduce a further functional class called 'diversified'. The forcing of multifunctional towns into unifunctional classes is of necessity an unsatisfactory process which both Nelson and Dick avoided, but at some sacrifice of simplicity, for the purpose of classification in the first place is to reduce complexity to a comprehendible form.

Two further points need to be added here. In these procedures functional classes are determined by the occupational groups of the census. Thus in Britain it would not be possible to identify, as Nelson does, a wholesale function, since no appropriate occupational group is recognized in the census of occupations. The second point is that diagnostic ratios will vary from census to census and from place to place. There is no universal proportion to identify a specific type of town so that international comparisons are vitiated.

Urban economic base studies

These studies have developed during the whole of the period covered by the classifications reviewed and indeed, there is an overlap between the two. In practice, the studies introduce little that is very different, but there is a more extended background of principle.

A review of principle could start with Chauncy D. Harris's classification already considered. Under that classification, transportation centres are defined as having 11 per cent of their gainful workers employed in transport and communications. If this is analysed, it means that Harris regards 11 per cent as the critical value at which transport employment becomes diagnostically significant in terms of the country as a whole. But this is not necessarily a useful measure, for what is required is some estimate of the point at which employment in transport becomes critical in the life of the town, not merely keeping it going but making a distinctive and generative contribution to its economic well-being. This is the essence of the concept of the 'economic base', around which an elaborate theory has accumulated and which seems to offer at least the possibility of model building.

The earliest suggestion of the concept appears to have come in 1902, when W. Sombart in *Der moderne Kapitalismus* identified a dual function in towns, which he characterized as 'stadegrunder', or basic; and 'studefuller' or complementary or non-basic. These concepts have come into English mainly via the work of planners,

particularly in the USA. In a work published in 1928, Robert M. Haig distinguished between 'primary occupations', or the producing of goods for external purposes, and 'auxiliary occupations', or the producing of goods and services for the convenience of the primary group (Haig, 1928). This concept was developed, and the terminology of the economic base was introduced by Homer Hoyt in 1939 (Weimer and Hoyt, 1939). Here the idea is propounded in its modern form. Economic activity can be broken down into two components:

1. **Basic or city forming:** that which meets non-local demand. This is the contribution to the national economy.
2. **Non-basic or city serving:** that which meets local, internal demand. This keeps the city going but makes no contribution to the national economy: 'We cannot live by taking in each other's washing'.

It follows from the above definitions that the basic component is 'city forming' because it induces growth. It creates a centripetal flow of income into the city which is available for distribution and circulation. But some qualification is necessary. The dictum quoted in 2 above, 'We cannot live by taking in each other's washing', is only partly true. By extending the boundaries of the unit for which the economic base study is to be made, one eventually arrives at an area where there is no export of goods or services at all, at which point all activities would, by definition, be non-basic. Such a unit would be hard to find, but nation states might approach the condition in varying degrees. Nevertheless within the limited unit of the city, it is argued that if it is possible to isolate the basic component of a city's economy, then we have isolated the growth-inducing or 'city forming' element. This is the obvious key to the study of locational advantages and growth, and should be the basis of classification.

The practical problem remains in the method by which this basic component is to be isolated. Hoyt himself proposed a cumbersome procedure by which local and non-local destinations of goods and services sold were determined by questionnaire, and the trade and professional population apportioned to basic or non-basic activities by using the proportion of the national income obtained by the city. Any such procedure is clumsy, unreliable and hardly practicable when a large number of towns is being considered. Most methods subsequently proposed have aimed at simplification and have fallen back on the old idea of comparing local with national, particularly with the mean or some other condition. The national proportion employed in any occupation is obtained and the number expected to be so employed in a town of any size is calculated. If the actual employment, when compared with this expected total, shows a surplus, then this represents basic employment. This was the principle used in the clearest statement of this concept by J. M. Mattila and W. R. Thompson who proposed an 'index of surplus workers' (Mattila and Thompson, 1955).

$$S = ei - et/Et.Ei$$

where ei = the regional employment in the industry concerned; et = the total regional employment in all industries and economic activity; Ei = the national employment in the industry concerned; Et = the total national employment in all industries and economic activity.

This index can be used in a classification of towns by the following procedure. The index is calculated for all occupational groups in the town and all positive values are totalled giving an aggregate of surplus workers for the town. The percentages of the aggregates formed by each occupational surplus are calculated and ranked. These rankings are then used as a basis for functional classification. An application of this Index of Surplus Workers can be examined in *Provincial Metropolis* where it was used in a 1959 study of the economic base in south-east Lancashire (Green, 1959: 43)

A somewhat different approach was that of G. Alexandersson in a study of the industrial structure of cities in the USA (Alexandersson, 1956). The problem, as he stated it, remained the same

– to identify the value above which employment is significant. In this case, the value attempted to answer the question – 'What ratios in different industries are a necessary minimum to supply a city's own population with goods and services of the type which are produced in every normal city?' The clear difference between the term 'normal city' and the idea of a 'national average' is at once apparent. The national average does not measure an economic base and in so far as such figures are used, as in the Index of Surplus Workers, these methods are little different from those of Nelson *et al.* described in the previous section.

Alexandersson, therefore, asked the more pertinent question in terms of economic base analysis. Accordingly, he ranks the employment ratios in each industry for all the towns he is considering and constructs a cumulative distribution diagram of percentage of cities against percentage employment. By empirical means two points (K_1 and K) were chosen, 1 per cent and 5 per cent from the point of origin, and employment structures set up using these values to provide a diagnostic employment ratio. That is, in a sample of 500 cities, the 5th and the 25th would be chosen and in Alexandersson's 864 cities of over 10,000 population in the USA, the 9th and 43rd cities respectively were chosen. In the final analysis the 5 per cent point was selected so that for any functional category employment in the 43rd city in rank order from the minimum became the diagnostic proportion and these proportions were used to set up a 'normal' city with a balanced occupational structure. Specialization was then measured by proportional employment above the K value.

This scheme has the merit of using an established normal structure rather than an 'average' structure, but the setting up of this normal structure is marked by empirical decisions, while a very large sample is needed to make it worth employing.

Perhaps the logical conclusion of these analyses was reached in the suggestion by Ullman and Dacey (Ullman and Dacey, 1962) whose work, in principle, parallels that of Alexandersson. They argued that the *minimum* percentage employed in *any* city provides the non-basic ratio since this must identify the smallest proportion for a viable city. This is an argument clearly open to serious challenge. Indeed, the use of minima had been considered and rejected by Alexandersson on the grounds that anomalous cases would become the decisive ones. His choice of the *K* points was specifically made to avoid that problem. To some extent Ullman and Dacey discounted the difficulties since the variation of the minimum with city size is taken into account. This non-basic ratio is subsequently used to classify cities not by comparing one employment sector with another but by measuring the extent to which a city departed from the various minima, thus providing an Index of Diversity (*D*).

$$D = \Sigma i \left[\frac{(Pi - Mi)^2}{Mi} \right] \frac{[\Sigma i Pi - \Sigma i Mi]^2}{\Sigma i Mi}$$

where *i* = each employment group; *Pi* = percentage employed in each *i* group; *Mi* = minimum requirement for each group; *Σi* = the sum of all the groups.

The higher this index the more specialized is a town, the lower the more diversified. This index demonstrates how statistical measures were increasingly employed and how the way towns differed from each other became the basis of classification.

But however refined these assessments of urban function became during the 1950s and 1960s they were never regarded with great respect by urban geographers and were always assigned to the periphery of the discipline. This was probably due to the fact that to some degree they simply confirmed what was obvious; but possibly more to the view that processes of classification were perceived, at a time when theoretical advance and model building were *de rigueur*, to be essentially descriptive and outmoded. Moreover, it was apparent that the classifications were seldom effectively used in the fields where they were supposed to tell, for they remained little more than classifications with illustrative distribution maps rather than

interpretations of location. However, as statistical methods during the quantitative revolution became more sophisticated so too inevitably did the procedures for classification. Pre-eminently multivariate forms of analysis provided a clear path for advance. But once variables other than the standard employment categories were introduced then the very basis of classification, for grouping or clustering was still the end product, itself changed. And as that happened so the purpose shifted in response. One of the earliest and most widely quoted examples, that by Moser and Scott (1961) for British towns was carried out by social statisticians. No longer was the prime end the comprehension of geographical location but rather the identification of common characteristics such as social composition which is more in the form of social typology than of economic role. In turn, this shifted interest to quality of life and eventually away from functional classification to ranking in terms of attractiveness as places in which to live. In spite of attempts to drag this field of research back into the initial geographical demand for the understanding of location, attempts which will be considered later in this chapter, a destination very different from that envisaged at the point of departure has been reached. That redirection has been partly due to the the fact that classification seemingly added little to the central concerns of urban geography, but also to the much greater public interest in ranking by quality of life which is of immediate concern and of media interest as against the somewhat remote and academic grouping of towns according to functions performed.

Multivariate analysis

The prime example of multivariate classification appeared in 1961 in *British towns: a statistical study of their social and economic differences* (Moser and Scott, 1961). The reasons for and the objectives of the work were succinctly stated. It was contended that 'no systematic and general research has been done into the ways in which British towns differ or resemble one another'. Accordingly the authors tried to assemble and

Table 6.2 *Principal components analysis of British towns, 1951*

Component	Interpretation	Variation explained North	South
I	Social class	24.0	27.4
II	Population growth 1931–51	14.3	19.0
III	Development after 1951 and the working population 1951	8.7	14.6
IV	Housing conditions	6.7	5.1

Source: After Moser and Scott, 1961.

collate material relevant to that purpose and then to classify towns on the basis of their demographic, economic and social characteristics. They confined their work to towns of over 50,000 population and for them assembled a total of 57 variables grouped under the headings Population size and structure; Population change; Households and housing; Economic character; Social class; Voting characteristics; Health; Education. This was clearly a much broader study of difference than the more limited functional analyses which have so far been discussed. There was also a dynamic element introduced by using changes between 1931 and 1951.

The principal components analysis which was the basis of the investigation produced the results shown in Table 6.2 with the percentage of total variance explained being divided into northern and southern groups. These are taken to be parameters of difference among the towns of England and Wales and a process of classification is set up based on the weightings recorded on the first two components with the additional evidence of the next two used when needed. The procedure aimed at producing ten groups of towns and ended up with fourteen. From these London and Huyton were excluded since they were 'too different from other towns to be included in any group'. Although the allocation to groups was not based on function but on a wide array of criteria, nevertheless the final groupings are presented in essentially functional terms. In outline the scheme is given in Table 6.3.

Table 6.3 *Classification of British towns*

Mainly resorts, administrative and commercial towns

1. Mainly seaside towns
2. Mainly spas, professional and administrative centres
3. Mainly commercial centres with some industry.

Mainly industrial towns

4. Including most of the traditional railway centres
5. Including many of the large ports as well as two Black Country towns
6. Mainly textile centres in Yorkshire and Lancashire
7. Including the industrial towns of the north-east seaboard and mining towns of Wales
8. Including the more recent manufacturing towns.

Suburbs and suburban type towns

9. Mainly 'exclusive' residential suburbs
10. Mainly older mixed residential suburbs
11. Mainly newer mixed residential suburbs
12. Including light industry suburbs, national defence centres and towns within the sphere of large conurbations
13. Mainly working class and industrial suburbs
14. Mainly newer industrial suburbs.

Not allocated: London and Huyton

Source: Moser and Scott, 1961.

Table 6.4 *Sample profiles of American cities*

Profile item	New York	Chicago	Cincinnati	Arcadia*
Total populations, 000s	7781	3550	0502	0041
Per cent single dwellings	0	0	0	0
Density	9	9	7	3
Median income	4	7	3	9
Deprivation index	7	7	7	0
Per cent non-White	7	8	8	0
Per cent foreign born	9	8	3	5
Median age	8	7	6	9
Per cent population increase 1955–60	1	1	1	8
Per cent same house 1955–60	8	4	3	3
Per cent migrants	2	0	1	4
Education centre	6	3	7	5

Source: After Hadden and Borgatta, 1965.
*Arcadia (Calif.) is included as a small town example.

It is interesting to compare the results shown in Table 6.3 with the initial identification of urban functions which was suggested at the beginning of this chapter:

1. Central place activities are directly comparable with the 'mainly administrative and commercial towns' which Moser and Scott identify as Group 3. But the 'resorts' which they also include (Groups 1 and 2) would have to be placed with the special activities and it is unfortunate that this major class straddles distinctive central place and special functions.
2. Transport activities are not directly shown to produce a distinct set of towns, though Group 4 and 5 in the classification are identified broadly in these terms. But these groups include other towns as well with no distinctive transport function.
3. Special activities include the remaining groups, although 'suburbs' have no distinctive functional connotation. They are parts of larger units which appear in the classification due to the use of local government units.

This detailed investigation can be effectively compared with Aurousseau's scheme for it is particular to an area, precise in definition and

'the criteria of classification emerged from the analysis itself' (Moser and Scott, 1961: 18).

Another attempt at investigation into urban character was made by Hadden and Borgatta (1965) in relation to American cities. Sixty-five variables were used and separate analyses for different city sizes were carried out. From the matrix sixteen factors were extracted and the first of these, 'socio-economic status', was the same as that in the British study. The second, third and fourth, however, were linked to 'non-white population', 'age composition' and 'education'. Subsequently the factors were used to select variables in order to construct a profile for each city by a simple decile division and though this does not provide a classification as such, it does provide a socio-economic profile (Table 6.4).

This study again emphasizes that general social character had become the end rather than the functional type of the narrower sort with which this chapter began.

6.3 SPECIALIZED FUNCTIONS AND URBAN GROWTH

At this point it is necessary to reinforce the notion of the influence of specialized functions

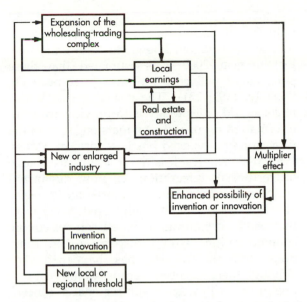

FIGURE 6.1 *The circular and cumulative feedback process of local urban size–growth for a single large US mercantile city, 1740–1840 (after A. Pred, 1977). Source: A. Pred 1977: City systems in advanced economies. London: Hutchinson & Co. Fig. 2.8, p. 72.*

on urban growth and to demonstrate them as something more than a mere distortion of a basic regularity given by central place functions. Perhaps the simplest model of these influences has been put forward by Pred (1966) although his work owes something to the earlier schemes of 'cumulative causation' put forward by Myrdal (Keeble, 1967) in his studies of economic growth on a regional basis.

Pred concerns himself more directly with the growth of American cities between 1860 and 1910 and proposes a model of 'the circular and cumulative process of industrialization and urban-size growth'. If the introduction of factory industries into a mercantile city is envisaged then the chains of reaction illustrated in Fig. 6.1 are evoked. 'New manufacturing functions, whether or not they primarily serve local markets, will have an initial multiplier effect', that is, new services will be demanded, while linked industries will be encouraged. The result

will be 'an alteration of the city's occupational structure (with a more even balance struck between the industrial sector and the wholesale-trading complex), an increase in population, or growth in urban size, and the probable attainment of one or more new local or regional industrial thresholds' (Pred, 1977: 90). This attainment of new thresholds will, in turn, support new manufacturing functions and encourage invention and innovation, and so the whole circular and cumulative process continues, until interrupted or impeded by diseconomies or by competition from other growing centres.

To some extent Pred's work was a more sophisticated interpretation of the old elementary notion of 'geographical inertia' for which he provided a convincing rationale. Nor has Pred been alone in attempting to suggest generalizations of this sort. Smolensky and Ratajczak (1965) proposed a sequence of three stages which they called 'elemental settlement', 'conforming city' and 'urban agglomerate'. The 'elemental settlement... originates because of economies of specialization in performing what would otherwise be ubiquitous economic activities' (Smolensky and Ratajczak, 1965: 108). This explanation is directly in line with central place theory at least as set out by Lösch. Such a settlement 'becomes a "conforming city" when a factor specific to that site, giving an absolute cost advantage to entrepreneurs locating in that town, becomes economically relevant to profit-maximizing entrepreneurs' (Smolensky and Ratajczak, 1965: 91). This implies the sort of transformation suggested by Pred which initiates cumulative growth to an urban conglomerate.

Wilbur R. Thompson proposed what he called 'stages of urban growth' (Thompson, 1965, 1968) beginning with the 'stage of export specialization' where the local economy is dominated by a single industry or even a single firm. Presumably one could envisage this as equivalent to the town in former times, when it was dominantly a defensive, military strongpoint. This is followed by a 'stage of export complex' where a broadening of the local economy takes place possibly by extending forward or backward stages in production or by adding suppliers or consumers of intermediate products. Presumably, in the parallel quoted above, this stage would be achieved by the broadening of primarily military functions into commerce and administration. Thomson next proposes a 'stage of economic maturation' or 'local service sector puberty' where local activity replaces imports with its new 'own use' production and business and services are expanded. This leads to the 'final stage of regional metropolis', where the local economy is seen as a node connecting and controlling neighbouring cities, once rivals but now satellites. It is interesting to observe that Smolensky and Ratajczak envisage a regular pattern of 'elemental settlements' transformed by the discriminating advantages of particular site characteristics, while Thompson seems to think in terms of an initial stage, where site characteristics discriminate, transformed to one where the regional metropolis 'organizes' its satellites. It is this latter view which now seems most appropriate and links back to much that is already established.

One phrase used by Pred can here be isolated. The changes which have been contemplated are accompanied by 'an alteration in the city's occupational structure' and this has been the concern of the present chapter. Industrialization, or the growth of any special function, affects urbanization and vice versa so that the size and spacing of cities is in part the product of the way in which the specialized tasks which cities perform – mining coal, making cars or providing rest and relaxation – are carried out. This means that consideration of these roles has to be married with that of general regional functions in any total explanation of the urban pattern.

This becomes even more apparent when the role of the modern metropolis is more closely scrutinized. The switch of employment out of manufacturing and into services characterized the 1970s, but the 1980s and 1990s have been dominated by the rise of information-based employment. In the USA in the mid-1950s, 37 per cent of the working population were employed in manufacturing, commerce and industry, 20 per cent in other services, and 9 per cent in information, knowledge and education. By the year 2000 it is estimated that these percentages will be 29, 17 and 50 respectively (Blakely, 1991: 231). 'The principal function of major world cities today is to provide access to information users and providers engaged in the provision of advanced business services' (Moss, 1991: 184). The result is that the primary source of territorial integration is now organized technology rather than transportation (Wheeler and Mitchelson, 1989). Many predictions were that such organization based on electronic communications would disperse urban functions, but the reverse has happened. For although there has been 'because of internal cost pressures... a process of sub-regional deconcentration of information handling industries and activities from cores to peripheries of such regions' (Hall, 1991: 20), at the highest level advantages of agglomeration economies in the acquisition of information have reinforced the absolute control of the largest centres. Holly has demonstrated the migration of banking functions up the urban hierarchy, leading to an increased spatial concentration (Holly, 1987). Information generation and diffusion, therefore, have a clear spatial dimension which favours large urban centres. That was the theme of a study by Wheeler and Mitchelson (1989) of information flows among major metropolitan areas in the USA.

The data they used were the sources and destinations of letters, packages and boxes transmitted via the Federal Express Corporation, among 48 large metropolitan areas. A principal components analysis of the 48×48 data matrix gave five components – New York, Los Angeles, Chicago, Atlanta, and Dallas-Fort Worth. Table 6.5 reproduces some of the analysis.

From their data Wheeler and Mitchelson conclude that there is a concentration of information upon the American landscape and they note New York's unique centrality in the information network. Los Angeles emerges as the second leading national centre for information dissemination, but it is adjudged emergent in

Table 6.5 *Information flows among metropolitan areas*

Metro area	D-values	No. of centres controlled
New York	15,264	46
Los Angeles	7,857	44
Chicago	5,508	42
Dallas Ft Worth	3,608	42
Boston	3,599	45
San Jose	3,476	45
San Francisco	549	39
Philadelphia	540	38
Houston	514	36
Atlanta	461	35
Miami		35

Source: Wheeler and Mitchelson, 1989.
Notes:
1. The D-value is the absolute difference between messages sent and received.
2. Number of centres controlled is the number of centres for which a flow origin is greater than a flow destination
3. The high figures for San Jose are probably related to the shipment of computer components.

that context and included with Dallas, Houston, Chicago, Atlanta, Boston, San Francisco, Philadelphia and Miami as making up the predominating regional centres. The remainder of the 48 areas are assigned to a third level where flows are directed either to New York or one of the regional centres.

Here, then, is both specialized function of the late twentieth century and hierarchical ordering. But neither is simple cause or effect for the ability to capture innovative functions is derived from past role and size. Considerations of function and hierarchical status merge into one, as they must inevitably do. Thus if one looks back at a paper published in 1984 on 'shifts in the wholesale trade status of US metropolitan areas' (Lord, 1984), then although the highest echelons were losing in terms of dominance of wholesale trade, in rank order the top ten were

New York, Chicago, Los Angeles, Detroit, Dallas-Fort Worth, Houston, San Francisco-Oakland, Philadelphia, Atlanta and Boston. That is, they were, with the exception of the special case of San Jose ranked 38th in wholesale trade and with the understandable exchange of Detroit for Miami, ranked 30th in wholesale trade, those which lead in information flows. Specialization leads to specialization based on established advantages of size. It is, however, also apparent that the larger the city, or the more advanced the economy in terms of Western industrial capitalism, then the more multifunctional the city becomes. Functional contrasts become diminished over time even though new and innovative ones are added. It is worth noting, however, that Archer and White's (1985) replication of Nelson's classification referred to earlier did not show these trends. There was an equivalent representation of specialized centres, but since the qualification is based on relative rather than absolute figures then that might be expected. Moreover there was a fall in the qualifying proportion for manufacturing from 43 per cent to 32.7 per cent and the main conclusion was a swing from manufacturing to service centres. Even so, the most interesting finding was that there was no statistically valid relationship between diversified cities and size, a conclusion which seems to go against the interpretation of trends presented here.

Another attempt at the sort of classifications which have been discussed in this chapter is that by B. J. L. Berry considering what he terms the latent structure of the American urban system (Berry, 1972). A factor analysis with a varimax rotation was carried out on a 1762 (cities) by 97 (variables) data matrix. The factors abstracted are shown in Table 6.6. The conclusions from this exercise suggest that, if the factors are uncorrelated, then the economic base of urban centres tends to act independently of other urban structural features. As multifunctional towns lose distinction in economic specialization it is the broader socio-economic dimensions which emerge as bases of contrast. The distinctive towns specialized by their economic bases

Table 6.6 *Latent dimensions of the American urban system in 1960*

Factor no.	Factor description
1.	Functional size of cities in an urban hierarchy
2.	Socio-economic status of the city residents
3.	Stage in family cycle of the city residents
4.	Non-White population and home ownership
5.	Recent population growth experience
6.	Economic base: college towns
7.	Population proportion foreign born or of foreign stock
8.	Recent employment expansion
9.	Economic base: manufacturing
10.	Extent of female participation in the labour force
11.	Economic base: specialized service centres
12.	Economic base: military
13.	Economic base: mining
14.	Extent to which elderly males participate in labour force

Source: After B. J. L. Berry, 1972.

are small and unimportant. It follows that the traditional economic approach to city classification is of minimal and declining relevance. 'The one exception is the market-orientated activities, for every urban system is hierarchically structured, the structure resting on aggregate economic power', and, as Berry concludes, the functional size of centres in an urban hierarchy is a universally latent dimension. Two conclusions follow:

1. Into the pattern of urban growth so far set out needs to be added a process of functional change by which, as divergences of economic base are progressively diminished, contrast is developed in the general and separated contexts of socio-economic status and life cycle stage even lifestyle character, and possibly a racial or ethnic characteristic as the society becomes culturally heterogeneous. Again this relates to the basic theme at the end of Chapter 5.
2. The one exception is the universal size dimension.

6.4 QUALITY OF LIFE IN TOWNS

At this point the conclusions reached in Chapter 5 and in this chapter converge. The nature of contemporary urbanism reveals a trend towards metropolises, or metropolitan regional systems, no longer distinguished by functional character but rather by their reaction with service and administrative roles revealed in socio-economic structure, life cycle stages and possibly immigrants and ethnicity. But even beyond that, their accumulations of functions and history, set against the background of location, determine the quality of life of their inhabitants. The logical sequence of Berry's 'latent functions', as set out earlier in this chapter, is the assessment of attractiveness which has become a distinctive, if not a dominant, strand in urban geography.

The attempt to identify and measure indicators of economic and social well-being has a long history, As long ago as 1939 and 1940 E. L. Thorndike published two volumes, *Your city* and *144 smaller cities* (Thorndike, 1939, 1940) which considered the quality of life in American cities. But such work never took off in urban geography although analysis became both more intensive and more sophisticated during the 1960s and 1970s. In 1975 in the USA the Midwest Research Institute based in Kansas City published *The quality of life in the US 1970. Index rating as statistics*; but it was based on the State as territorial unit and reactions suggested that metropolitan areas would be both more meaningful and useful. Accordingly in 1976 Ben Chie Liu published *Quality of life indicators in US metropolitan areas* (Liu, 1976). The spatial bases were the SMSAs although they were divided by size into 'Large' with a population of over

500,000, 'Medium' with 200,000 and 'Small' with under 200,000. Although urbanists in general were somewhat slow to take up these assessments, as might be expected, public interest rapidly grew. The result has been the publication annually of the *Places rated almanac* with the significant title *Your guide to finding best places to live in America* (Boyer and Savageau, 1989). It is not surprising that a parallel British venture should have appeared in the Glasgow Quality of Life Group linked to the Applied Population Research Unit at the Department of Geography at Glasgow University.

There are two basic and immensely complex problems at the heart of all measurements of quality of life, assuming in the first place that it is possible and meaningful. The first is the identification of acceptable and available criteria as indicators. The second is the statistical procedures by which the many diverse indicators are resolved into one overall score for a city. In his book Liu proposes five separate components of quality of life. They are: economic, mainly related to employment and earnings; psychological, which encompasses the satisfaction which people derive from living in an area; environmental, which is merely linked to pollution and county access; political, which is based on the presence of a democratic system and hence the ability to participate in and influence government; and sociological, which is mainly expressed in the presence/absence of discrimination on any grounds. But when these are converted into obtainable measures, compromises have to be made. The psychological component is dropped, since it is deemed impossible to measure, and it is replaced by health and education. Liu thus sets out five components – economic, political, environmental, health and education, and social. In turn each of these is divided into a number of factors which in turn are expressed by the actual input variables. The make-up of the economic and political components is shown in Table 6.7. As might be expected, the remaining components are not as easily measured. Thus in the political component an item such as 'informed citizenry' is measured by variables such as the local

Sunday newspaper circulation per 1000 population; 'political activity' is measured by such variables as participation in Presidential elections; local government efficacy or 'professionalism' is measured by such variables as the salaries of teachers and firemen; local government 'performance' is a factor in which crime rates are used; and 'welfare assistance' includes per capita local government expenditure on public welfare. In all, some 120 variables are assembled for the five components.

The problem of deriving a combined score for each component is met by using an adjusted standardized additive method, usually known as the z score method, where the z score is a linear transformation of the original data, such that the mean of the z score becomes '0' and the standard deviation '1' (Liu, 1976: 83). Each variable is then weighted so that the input to the main factors in the component is equal even though the number of variables per factor differs.

The results given by two of the components are shown in Figs 6.2 and 6.3. It should, however, be remembered that 1971 data were used and considerable change has taken place since then. The economic component shows highest ratings in the East North Central region of the USA, while most of the lowest ratings occur in the Middle Atlantic and the South. In contrast, on environmental grounds the western areas especially are highly ranked as opposed to the traditional centres of heavy industry, no doubt an explanation of the dominant trends in population movements which have characterized the last quarter of the twentieth century.

Liu does bring the five components together into a composite index on the basis of relationship to group means and the standard deviations from them deriving five categories called 'outstanding', 'excellent', 'good', 'adequate' and 'substandard'. These are shown for the large SMSAs in Table 6.8. Liu's conclusions in locational terms are somewhat muted.

> This study ... found that although the Pacific, the East North Central, the Mountain and the New England regions had relatively more

Table 6.7 *Quality of life indicators in the USA*

Factor effect and weight	Factors
Economic component	
I. *Individual economic well-being*	
+ (0.25)	A. Personal income per capita ($)
	B. Wealth
+ (0.05)	1. Savings per capita ($)
+ (0.05)	2. Ratio of total property income to total personal income
+ (0.05)	3. Per cent of owner-occupied housing units
+ (0.05)	4. Per cent of households with one or more automobiles
+ (0.05)	5. Median value, owner-occupied, single family housing units ($1000)
II. *Community economic health*	
+ (0.07)	A. Per cent of families with income above poverty level
− (0.07)	B. Degree of economic concentration, absolute value
	C. Productivity
+ (0.014)	1. Value added per worker in manufacturing ($1000)
+ (0.014)	2. Value of construction per worker ($1000)
+ (0.014)	3. Sales per employee in retail trade ($1000)
+ (0.014)	4. Sales per employee in wholesale trade ($1000)
+ (0.014)	5. Sales per employee in selected services ($1000)
+ (0.07)	D. Total bank deposits per capita ($)
	E. Income equality index
− (0.035)	1. Central city and suburban income distribution
− (0.035)	2. Per cent of families with incomes below poverty level or greater than $15,000
− (0.07)	F. Unemployment rate
+ (0.07)	G. Number of full-time Chamber of Commerce employees per 100,000 population
Political component	
I. *Individual activities*	
	A. Informed citizenry
+ (0.083)	1. Local Sunday newspaper circulation per 1000 population
+ (0.083)	2. Per cent of occupied housing units with TV available
+ (0.083)	3. Local radio stations per 1000 population
+ (0.025)	B. Political activity participation-ratio of Presidential vote cast to voting age population
II. *Local government factors*	
	A. Professionalism
+ (0.02)	1. Average monthly earnings of full-time teachers ($)
+ (0.02)	2. Average monthly earnings of other full-time employees ($)
+ (0.02)	3. Entrance salary of patrolmen ($)
+ (0.02)	4. Entrance salary of firemen ($)
+ (0.02)	5. Total municipal employment per 1000 population
+ (0.02)	6. Police protection employment per 1000 population
+ (0.02)	7. Fire protection employment per 1000 population
+ (0.02)	8. Insured unemployment rates under state, federal, and ex-servicemen's programmes
	B. Performance
− (0.03)	1. Violent crime rate per 100,000 population
− (0.03)	2. Property crime rate per 100,000 population
+ (0.03)	3. Local government revenue per capita
+ (0.03)	4. Per cent of revenue from Federal Government
+ (0.03)	5. Community health index
+ (0.03)	6. Community education index
	C. Welfare assistance
+ (0.053)	1. Per capita local government expenditures on public welfare ($)
+ (0.053)	2. Average monthly retiree benefits ($)
+ (0.053)	3. Average monthly payments to families with dependent children ($)

Source: After Liu, 1976

Note: The factor effect sign indicates whether the factor is a positive or negative contributor to the component.

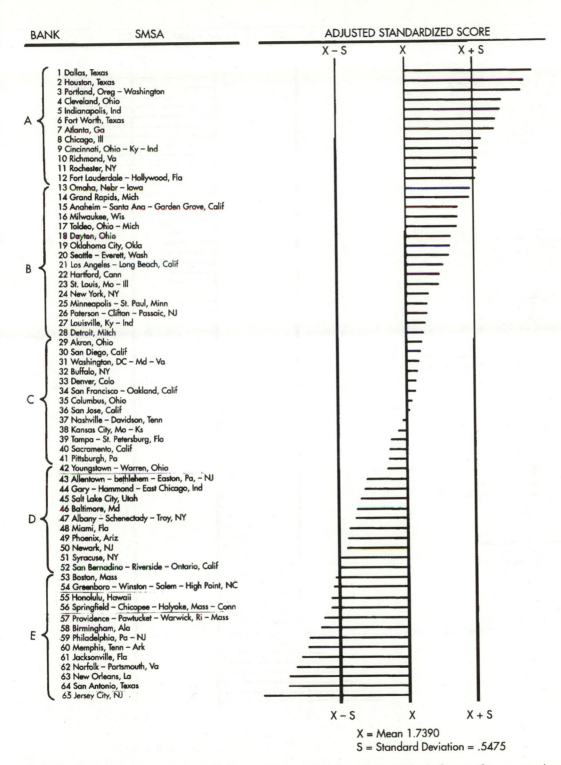

BANK	SMSA	ADJUSTED STANDARDIZED SCORE

1 Dallas, Texas
2 Houston, Texas
3 Portland, Oreg – Washington
4 Cleveland, Ohio
5 Indianapolis, Ind
A 6 Fort Worth, Texas
7 Atlanta, Ga
8 Chicago, Ill
9 Cincinnati, Ohio – Ky – Ind
10 Richmond, Va
11 Rochester, NY
12 Fort Lauderdale – Hollywood, Fla

13 Omaha, Nebr – Iowa
14 Grand Rapids, Mich
15 Anaheim – Santa Ana – Garden Grove, Calif
16 Milwaukee, Wis
17 Toldeo, Ohio – Mich
18 Dayton, Ohio
19 Oklahoma City, Okla
20 Seattle – Everett, Wash
21 Los Angeles – Long Beach, Calif
B 22 Hartford, Conn
23 St. Louis, Mo – Ill
24 New York, NY
25 Minneapolis – St. Paul, Minn
26 Paterson – Clifton – Passaic, NJ
27 Louisville, Ky – Ind
28 Detroit, Mitch

29 Akron, Ohio
30 San Diego, Calif
31 Washington, DC – Md – Va
32 Buffalo, NY
33 Denver, Colo
34 San Francisco – Oakland, Calif
C 35 Columbus, Ohio
36 San Jose, Calif
37 Nashville – Davidson, Tenn
38 Kansas City, Mo – Ks
39 Tampa – St. Petersburg, Fla
40 Sacramento, Calif
41 Pittsburgh, Pa

42 Youngstown – Warren, Ohio
43 Allentown – bethlehem – Easton, Pa, – NJ
44 Gary – Hammond – East Chicago, Ind
45 Salt Lake City, Utah
46 Baltimore, Md
D 47 Albany – Schenectady – Troy, NY
48 Miami, Fla
49 Phoenix, Ariz
50 Newark, NJ
51 Syracuse, NY
52 San Bernardino – Riverside – Ontario, Calif

53 Boston, Mass
54 Greenboro – Winston – Salem – High Point, NC
55 Honolulu, Hawaii
56 Springfield – Chicopee – Holyoke, Mass – Conn
57 Providence – Pawtucket – Warwick, Ri – Mass
58 Birmingham, Ala
E 59 Philadelphia, Pa – NJ
60 Memphis, Tenn – Ark
61 Jacksonville, Fla
62 Norfolk – Portsmouth, Va
63 New Orleans, La
64 San Antonio, Texas
65 Jersey City, NJ

X = Mean 1.7390
S = Standard Deviation = .5475

FIGURE 6.2 *Quality of life indicators. USA regional variations in indexes: the economic component (after Liu, 1976). Source: B-C Liu 1976: Quality of life indicators in US metropolitan areas: a statistical analysis. New York: Praeger. Chart 1, p. 102.*

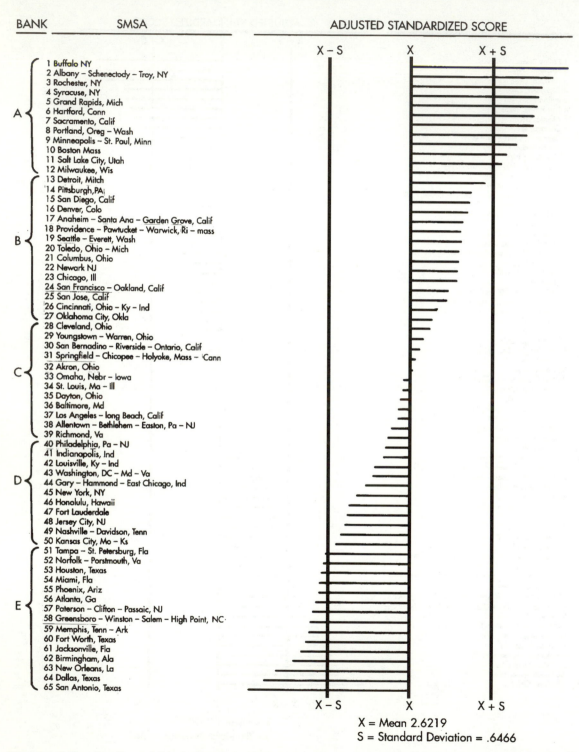

FIGURE 6.3 *Quality of life indicators. USA regional variations in indexes: the political component (after Liu, 1976). Source: As Fig. 6.2. Chart 2, p. 106.*

Table 6.8 *Quality of life indexes and ratings in large SMSAs in the United States*

SMSA	Economic Value	Rating	Political Value	Rating	Environmental Value	Rating	Health & Education Value	Rating	Social Value	Rating	Overall Value	Rating
1. Akron, Ohio	1.8786	C	2.6319	C	-0.9667	C	1.1250	C	0.1835	E	0.9705	C
2. Albany-Schenectady-Troy, NY	1.3286	D	3.7431	A	-1.2917	D	1.8625	B	0.5836	B	1.2452	B
3. Allentown-Bethlehem-Easton, Pa-NJ	1.4286	D	2.4792	C	-0.6167	A	0.3875	D	0.2173	D	0.7792	D
4. Anaheim-Santa Ana-Garden Grove, Ca.	2.1786	B	3.0486	B	-1.0500	C	2.0125	A	0.4762	C	1.3332	B
5. Atlanta, Ga.	2.4714	A	1.8750	E	-1.2833	D	0.8375	D	0.2806	E	0.8362	D
6. Baltimore, Md.	1.3429	D	2.5278	C	-1.2667	D	0.3625	D	0.1392	E	0.6211	D
7. Birmingham, Ala.	1.0500	E	1.6944	C	-1.4250	E	-0.0250	E	0.0931	E	0.2775	E
8. Boston, Mass.	1.1786	E	3.3889	A	-1.2500	D	2.0125	A	0.6036	B	1.1867	B
9. Buffalo, NY	1.8357	D	3.8819	A	-1.2000	D	1.4250	B	0.7019	B	1.3289	B
10. Chicago, Ill.	2.3643	A	2.9653	B	-.8167	E	0.6625	D	0.3056	D	0.8962	C
11. Cincinnati, Ohio-Ky.-Ind.	2.3429	A	2.8403	B	-1.0333	C	0.6250	D	0.0711	E	0.9692	C
12. Cleveland, Ohio	2.5143	A	2.7847	C	-1.4250	E	1.0875	C	0.5837	B	1.1090	B
13. Columbus, Ohio	1.7857	C	3.0208	B	-1.0917	C	1.4875	B	0.7621	B	1.1929	B
14. Dallas, Texas	2.7571	A	1.4653	E	-0.9083	B	0.7625	D	0.4585	C	0.9070	C
15. Dayton, Ohio	2.1214	B	2.5625	C	-1.3167	D	1.0625	C	0.3421	D	0.9544	C
16. Denver, Colo.	1.8357	C	3.0903	B	-0.9917	C	2.5000	A	0.9604	A	1.4789	A
17. Detroit, Mich.	1.8929	B	3.2222	B	-1.7250	E	0.9625	C	-0.0248	E	0.8656	D
18. Fort Lauderdale-Hollywood, Fla.	2.3143	A	2.1319	D	-1.0833	C	0.2000	E	0.5823	B	0.8290	D
19. Fort Worth, Texas	2.4786	A	1.7986	E	-0.8583	B	0.3500	D	0.4372	C	0.8412	D
20. Gary-Hammond-East Chicago, Ind.	1.3929	D	2.2778	D	-1.1750	D	0.7000	D	0.2106	D	0.6813	D
21. Grand Rapids, Mich.	2.2643	B	3.6319	A	-1.0333	C	1.5375	B	0.5527	C	1.3906	A
22. Greensboro-Winston-Salem-High Point, NC	1.1571	E	1.8333	E	-1.3000	D	0.1000	E	0.2337	D	0.4048	E
23. Hartford, Conn.	2.0357	B	3.6181	A	-1.1250	C	2.2750	A	0.5981	B	1.4804	A
24. Honolulu, Hawaii	1.1357	E	2.1458	D	-0.4583	A	1.5375	B	0.4496	C	0.9621	C
25. Houston, Texas	2.7000	A	1.9167	E	-1.0000	E	1.0875	C	0.5573	C	1.0523	C
26. Indianapolis, Ind.	2.5143	A	2.4236	E	-1.5250	E	0.6500	D	0.4303	C	0.8969	C
27. Jacksonville, Fla.	0.8929	E	1.7569	E	-1.2500	E	0.1125	D	0.3169	D	0.3658	E
28. Jersey City, NJ	0.5857	E	2.1250	D	-1.0167	C	-0.5250	E	-0.1694	E	0.1999	E
29. Kansas City, Mo.-Ka.	1.6857	C	2.0486	D	-1.1250	C	1.1125	D	0.8089	A	0.9061	C
30. Los Angeles-Long Beach, Ca.	2.0500	B	2.5278	C	-1.0583	C	1.7375	B	0.8315	A	1.2177	B
31. Louisville, Ky.-Ind.	1.9071	B	2.3403	D	-1.4167	E	0.3125	D	0.2603	D	0.6807	D
32. Memphis, Tenn.-Ark	0.9429	E	1.8264	E	-1.2083	D	0.6125	D	0.1198	E	0.4587	E
33. Miami, Fla.	1.2857	D	1.9097	E	-0.4167	A	0.6000	D	0.7634	B	0.8284	D
34. Milwaukee, Wis.	2.1786	B	3.2708	A	-1.0417	C	1.7000	B	0.8453	A	1.3906	A

continued

Table 6.8   *Quality of life indexes and ratings in large SMSAs in the United States (Continued)*

SMSA	Economic Value	Rating	Political Value	Rating	Environmental Value	Rating	Health & Education Value	Rating	Social Value	Rating	Overall Value	Rating
35. Minneapolis-St Paul, Minn.	1.9357	B	3.4722	A	-.9000	B	2.2375	A	.8329	A	1.5157	A
36. Nashville-Davidson, Tenn.	1.7286	C	2.0833	D	-1.0833	C	.6375	D	.7218	B	.8176	D
37. New Orleans, La.	.7857	E	1.5625	E	-1.2667	D	.4250	D	.1783	E	.3370	E
38. New York, NY	1.9500	B	2.2014	D	-1.3333	D	1.2125	C	.5179	C	.9097	C
39. Newark, NJ	1.2571	D	2.2931	B	-1.2000	D	1.2625	C	.1000	E	.8825	D
40. Norfolk-Portsmouth, Va.	.8500	E	1.9306	E	-.8667	B	.0625	E	.2507	D	.4454	E
41. Oklahoma City, Okla.	2.1143	B	2.8056	B	-.8250	B	1.3750	B	.8852	A	1.2710	B
42. Omaha, Nebraska-Iowa	2.2786	B	2.5833	C	-1.3083	D	1.7500	B	.9966	A	1.2600	B
43. Paterson-Clifton-Passaic, NJ	1.9357	B	1.8542	E	-1.0000	C	1.4625	B	.1371	E	.8779	D
44. Philadelphia, Pa.-NJ	.9500	D	2.4306	D	-.0250	C	.3000	E	.2234	B	.5758	C
45. Phoenix, Ariz.	1.2786	D	1.9097	E	-.5917	A	1.6000	B	.7246	B	.9842	C
46. Pittsburgh, Pa.	1.5929	C	3.1181	B	-1.8667	E	.7875	D	.3510	D	.7966	D
47. Portland, Oreg.-Wash.	2.6786	A	3.5486	A	-.6500	A	2.1375	A	1.0273	A	1.7484	A
48. Providence-Pawtucket-Warwick, RI-Mass.	1.0786	E	3.0347	B	-.7667	B	-.1750	E	.1606	E	.6664	D
49. Richmond, Va.	2.3357	A	2.4722	C	-1.1333	D	.4500	D	.1123	E	.8474	D
50. Rochester, NY	2.3214	A	3.6667	A	-.7000	B	2.0000	A	.2196	D	1.5015	A
51. Sacramento, Ca.	1.5929	C	3.6181	A	-.2000	A	2.1875	A	.9576	A	1.6312	A
52. St Louis, Mo.-Ill.	2.0357	B	2.5833	C	-1.5833	E	.5625	D	.1583	E	.7513	D
53. Salt Lake City, Utah	1.3714	D	3.3542	A	-1.0250	C	2.5625	A	.5728	B	1.3672	A
54. San Antonio, Texas	.7857	E	1.3403	E	-.8333	B	.2875	E	.2463	D	.3653	E
55. San Bernadino-Riverside-Ontario, Ca.	1.2000	D	2.6944	C	-.4750	A	1.3625	B	.6042	B	1.0772	C
56. San Diego, Ca.	1.8786	C	3.1111	B	-.5333	A	1.8125	A	.9020	A	1.4342	A
57. San Francisco-Oakland, Ca.	1.8357	C	2.9444	B	-.7000	B	2.3750	A	.8189	A	1.4548	A
58. San Jose, Ca.	1.7500	C	2.9167	B	-.5333	A	2.7250	A	.7364	B	1.5190	A
59. Seattle-Everett, Wa.	2.1071	B	3.0347	B	-.2667	A	2.2625	A	1.0144	A	1.6304	A
60. Springfield-Chicopee-Holyoke, Mass.-Conn.	1.1357	E	2.6667	C	-.6167	A	.7000	D	.4634	C	.8698	D
61. Syracuse, NY	1.2071	D	3.6458	A	-.1500	D	1.8500	B	.6157	B	1.2337	B
62. Tampa-St Petersburg, Fla.	1.6214	C	1.9514	E	-.0583	C	.0000	E	.5526	C	.6134	E
63. Toledo, Ohio-Mich.	2.1714	B	3.0278	B	-.1833	D	.9375	C	.5617	C	1.1030	B
64. Washington, DC-Md.-Va.	1.8571	C	2.3403	D	-.8333	B	2.1000	A	.6848	B	1.2298	B
65. Youngstown-Warren, Ohio	1.5857	D	2.7222	C	-.9667	C	.6375	D	.3634	D	.8684	D
Mean ($\bar{x}$) =	1.7390		2.6219		-1.0342		1.1252		.4809		.9865	
Standard Deviation (s) =	.5475		.6466		.3452		.7868		.2928		.3688	

Source: After Liu, 1976

A = outstanding ($\geq \bar{x} + s$); B = excellent ($\bar{x} + .28s \leq B < \bar{x} + s$); C = good ($\bar{x} - .28s < C < \bar{x} + .28s$); D = adequate ($\bar{x} - s < D \leq \bar{x} - .28s$); E = substandard ($\leq \bar{x} - s$)

Table 6.9 *Places rated almanac: a selection of overall profiles and ranks*

Metro area	Cost	Jobs	Crime	H@E	Tran	Educ	Art	Rec	Clim	Cum	Rank
Seattle	261	36	270	33	12	30	11	1	12	666	1
Washington	296	4	245	13	3	6	4	80	51	702	4
Cincinnati	214	75	138	104	63	31	41	56	95	820	14
San Antonio	130	31	284	172	70	71	60	183	302	1303	100
Tuscaloosa	53	327	244	201	130	143	297	242	256	1943	292
Pine Bluff	1	311	203	301	305	288	293	327	264	2293	333
New York	319	71	332	2	1	1	1	15	44	786	7

SMSAs with outstanding and excellent ratings, they also had substandard areas, though relatively few in number. In contrast, the southern states did show relatively larger numbers of low rated SMSAs with quality of life measures which were beyond the 'adequate' or 'good' category, e.g. West South Central and South Atlantic although these regions showed up fairly strong in the economic component. In other words, the hypothesis is inconclusive with respect to this test of regional differentiation.

(Liu, 1976: 219)

There is nothing here of 'sunbelts' and 'snowbelts'.

It is not surprising that assessment by quality of life is perhaps the only academic classification of cities which has achieved a popular status. The *Places rated almanac* (Boyer and Savageau, 1989) has already been noted and it has become an accepted measure of residential desirability. It presents ratings for 333 metropolitan areas on nine categories: cost of living; jobs; crime; health and environment; transportation; education; the arts; recreation; climate. There are clear links with Liu's work, but also a move to more immediate criteria which are more easily understood. The measurement procedure is to select diagnostic criteria for each category. Thus crime is assessed by two criteria, the violent crime rate and property crime rate divided by 10, the sum of the two being the 'metro score'. Cost of living is measured by indices for the costs of housing,

food, health care and transportation set against the US average of 100. Each area is then ranked 1–333 to give a 'places rated rank' for each criterion. These ranks are used both to present an overall profile and to give a cumulative score which is the basis for an overall ranking. Table 6.9 gives a brief selection of examples.

There are two immediate conclusions to be drawn fron the overall rankings. The first is that the profile is more important than the cumulative score. Thus Pine Bluff, Arizona, which is the worst ranked area overall actually ranks first for cost of living. This point will be taken up later. The second is that, unlike in Liu's ranking, no allowance is made for size and, although many of the measures are 'per head of population', the higher thresholds of the larger metropolitan areas, and in central place terms the consequent generation of higher order services, means that size is a significant determinant of rank. But there are anomalies in that, as might be expected, the smaller areas score better on cost of living and on crime.

The study by the Glasgow Group in the UK in many ways produced comparable results. Some fifty measures were taken but they were weighted by the importance given to them derived from a nation-wide questionnaire with 1200 respondents (Rogerson, 1989). The weightings of various dimensions are shown in Table 6.10. They range from violent crime which was regarded by 93.3 per cent of respondents as important or very important to quality of life to some 44.2 per cent who regarded the cost of

Table 6.10 *Ranking of highest 19 out of 50 indicators of quality of life. UK sample of 1200*

Indicator	Percentage recording it as very important
Violent crime	93.3
Health facilities	93.5
Non-violent crime	92.9
Pollution	88.8
Cost of living	87.3
Shopping facilities	86.7
Racial harmony	80.2
Scenic quality	78.2
Cost of owner-occupied housing	73.9
Wage level	70.2
Education facilities	69.7
Employment prospects	69.2
Unemployment levels	67.0
Sports facilities	66.5
Leisure facilities	62.7
Travel to work time	62.4
Quality of council housing	53.0
Access to council housing	48.3
Cost of private rent accommodation	44.2

Source: Rogerson *et al.,* 1988.

Table 6.11 *Rank order of UK towns according to an aggregated index of quality of life*

Edinburgh	Derby
Aberdeen	Norwich
Plymouth	Birkenhead
Cardiff	Blackpool
Hamilton–Motherwell	Luton
Bradford	Glasgow
Reading	Bournemouth
Stoke-on-Trent	Leeds
Middlesborough	Sunderland
Sheffield	Bolton
Oxford	Manchester
Leicester	Liverpool
Brighton	Nottingham
Portsmouth	Newcastle
Southampton	London
Southend	Wolverhampton
Hull	Coventry
Aldershot–Farnborough	Walsall
Bristol	Birmingham

Source: Rogerson *et al.,* 1988.

private rented accommodation in the same light (Rogerson *et al.,* 1988).

Standardized scores were again used as the basis for the derivation of an 'aggregated index of quality of life' for towns of over 250,000 population. The rank order of towns is given in Table 6.11.

In the UK as in the USA there is difficulty in deriving any generalization. The so-called Celtic fringe seems to come out very well, presumably because of high scores on low living cost and on environmental, countryside access and limited commuting-time grounds. The north–south divide does not appear; the highest ranked town in the South East is Reading at seventh, a place below Bradford. But the inclusion of London as a single entity rather than its separate parts probably confuses the issue.

6.5 BOOMING TOWNS AND SUNBELTS

The apparent contradictions between towns ranked highly for quality of life and the more general image of successful towns both in the USA and in the UK calls for comment. The CURDS work already referred to in Chapter 3 has also been concerned with the identification of what are called 'booming towns' in Britain (Champion and Green, 1987, 1991). The procedures are very similar to those used in assessing quality of life, the derivation of an aggregate index from a number of selected variables. In this case, however, the variables are wholly economic and demographic (Champion and Green, 1991) and akin, therefore, to only one of Liu's components in the US study. In the latest assessment ten variables were used related to unemployment, type of employment and house prices, with changes over time in the categories

Table 6.12 *The ten highest and ten lowest scores of British LLMAs on an amalgamated index of economic standing and growth/decline*

Top 20 LLMAs			Bottom 20 LLMAs		
Rank	Name	Score	Rank	Name	Score
1	Horsham	0.796	280	Greenock	0.253
2	Newbury	0.788	279	Peterlee	0.271
3	Milton Keynes	0.713	278	Kilmarnock	0.274
4	Haywards Heath	0.708	277	Kirkcaldy	0.281
5	Bishop's Stortford	0.704	276	Castleford	0.282
6	Tunbridge Wells	0.700	275	Sunderland	0.283
7	Bracknell	0.698	274	Liverpool	0.292
8	Reigate & Redhill	0.696	273	Paisley	0.293
9	Reading	0.695	272	St Helens	0.293
10	Stratford-on-Avon	0.690	271	Thurso	0.294
11	St Albans	0.685	269	Stranraer	0.294
12	Andover	0.675	270	Mexborough	0.298
13	Dereham	0.673	268	Alloa	0.299
14	Bournemouth	0.667	267	Ayr & Prestwick	0.302
15	Woking & Weybridge	0.667	266	Peterhead	0.313
16	Winchester	0.666	265	Bishop Auckland & Aycliffe	0.315
17	Guildford	0.666	264	Dunfermline	0.317
18	Basingstoke	0.661	263	South Shields	0.319
19	Maidenhead	0.661	262	Llanelli	0.320
20	Northampton	0.660	261	Mansfield	0.322

Source: Champion and Green, 1991.

also used. A similar additive procedure was employed. The highest in each variable was given the score of 1 and the lowest 0, the remainder being scored accordingly. The total scores for each area were then divided by the maximum possible – 10 without any weighting – to give scores in the range 0.000 to 1.000. The highest and lowest scores are set out in Table 6.12. It must be noted that the areal units (Local Labour Market Areas or LLMAs) are the same as those used by the Glasgow Group so that comparability is possible.

It is at once apparent that the contrast with the quality of life index is absolute. All the highest ranked scores on this index are in the South East, the majority of them in a halo about London. The worst performing areas are all northern, ten being accounted for by Scotland.

The highest ranked Scottish area is Perth at 158, and the Welsh area, Carmarthen at 55. The only conclusion that can be drawn from these figures is that economic benefit is by no means the most important criterion in the assessment of quality of life. This can be the only reason for the complete contrast between these two indices. The north–south divide may be one based on wealth; it does not seem to have much relevance to the perceived quality of living.

A similar pattern is found in the USA as derived by Liu. Table 6.13 lists the twelve highest ranked SMSAs on the economic component and sets alongside the ranks on the environmental component. The lack of agreement is clear, indeed of the twelve only Portland and Rochester appear in the aggregated 'outstanding' category.

Table 6.13 *The top SMASs in the USA on the Economic Component and their ranks on the Environmental Component*

SMSA	Economic rank	Environmental rank
Dallas	1	21
Houston	2	26
Portland	3	11
Cleveland	4	60
Indianapolis	5	61
Fort Worth	6	18
Atlanta	7	52
Chicago	8	64
Cincinnati	9	30
Richmond	10	41
Rochester	11	13
Fort Lauderdale– Hollywood	12	36

Source: Liu, 1976.

It is significant again that consideration of these issues leads back to the changes in the city system which were discussed in Chapter 5. The analyses of Pumain and Saint-Julien (1989) referred to in that chapter could well be included at this point as 'image de marque' and 'technical modernity' become critical influences in the determination of both the economic success and attractiveness of towns. A point is reached where all the aspects treated in the last three chapters come together so that hierarchical rank, functional character and the quality of life offered become integrated parts of a complete whole, which is the role and character of towns.

6.6 CONCLUSION

It would of course be ideal if in this conclusion it were possible to weave all the separate strands which have been followed into one whole cloth – hierarchical ranks, functional specialisms, quality of urban living, the transformation of free-standing towns into metropolitan regions and the booming towns, all set into one integrated

schema. Over the last three chapters the inter-linkages that occur have been noted and stressed, but that is a long way from integration. But the topic has hardly been addressed by urban geographers. It has always been there but pushed to the fringe, possibly because of its intractable nature, possibly because it lacks the immediate relevance of studies of the internal structure of cities. But it has had some recent attention. Thus Preston and Mitchell (1989) have published 'notes on a combined economic base – central place theory'. The common ground is found in the assertion that 'the part of a town's total production of central goods and services accounted for by centrality could be considered basic or export, while the part accounted for by local consumption could be considered service or non-basic' (Preston and Mitchell, 1989: 93). Although interesting and of conceptual value it would seem of limited significance, especially as economic base studies have virtually disappeared from the urban scene.

On a far more ambitious scale is W. K. D. Davies's 'conceptual model of urban differentiation' (Davies, 1989) which is shown in Fig. 6.4. It is difficult to present a brief account of this complex schema, but as Davies himself suggests, 'the easiest way of understanding the range of information included is to begin with the concept of the city system as a mode of concentration in space' (Davies, 1989: 70), which is the central column 5 in Fig. 6.4. Immediately to the right are placed seven concepts of what constitutes urban or 'types of differentiation'. They are location, aggregation or size, functional complexity, nodality and interaction, social heterogeneity, permanence and identity. Of these the first four have been the subject of the last three chapters, while social heterogeneity will be discussed later. Permanence and identity are perhaps not as clear. Permanence simply denotes the difference between places of temporary assembly, as, for example, fairs or periodic markets held at unsettled points, and permanent settlements permanently occupied. Identity refers to the notion, which at one time dominated Western urban history, that a town could only be a settlement with a recognized political identity via a charter or some such instrument. To the right of these

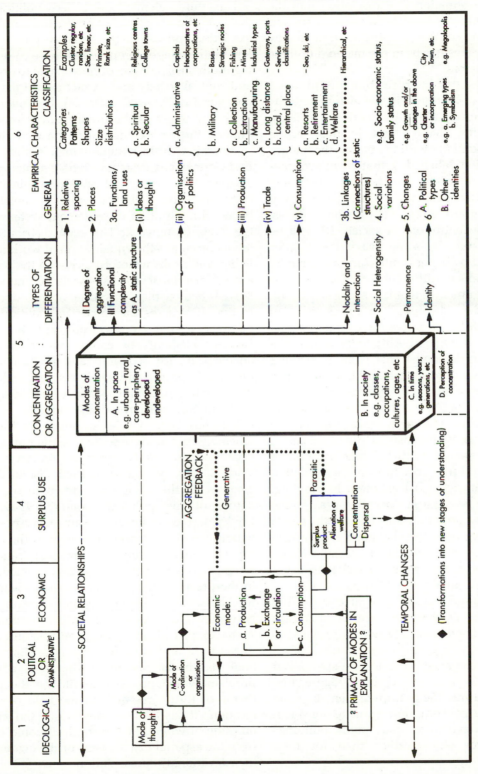

FIGURE 6.4   *A conceptual model of urban differentiation (after W. K. D. Davies, 1989). Source: W. K. D. Davies, 1989: Towards a conceptual model of the urban systems literature. As Fig. 5.15 above. Fig. 3, p. 71.*

types of differentiation is a resolution into the general and thence into the more specific and empirical means of their identification mainly by classification. This has been the material of these last three chapters.

Davies sees the interpretation of all these observations to be found in the elements to the left of the central column of modes of concentration. In this chapter and in this section concern is essentially with the part of the 'modes of concentration' labelled *A*, that is with concentration in space, for the time factor (*C*) is largely one for historical geography (Carter, 1983) even though there are daily and seasonal fluxes, while the 'concentration' in society (*B*) will be dealt with later. In some ways these latter aspects would be better omitted from the model since they lead in a different direction.

Davies's argument is that it is necessary to look at the broader, theoretical interpretations of urbanism in order to derive the common explanation which all the complexity to the right of the central column demands. That is, he is carrying out a reductionist process parallel to those which appear in interpretations of the internal structure of cities. Accordingly, the first of his explicands is derived from Harvey's insistence on the derived surplus as the basis of urban existence; towns are concerned with the manipulation of that surplus and the procedures in manipulation determine character and location. It is tempting, for example, to suggest that the greater the degree of alienation of the surplus into the hands of an élite, the greater the likelihood of a primate city; the greater the devotion to social welfare, then the nearer to the presumed socialist ideal of equally spaced, equally sized towns.

But the surplus has to be generated, as well as manipulated, so Davies introduces what he calls the economic level of society which is related to production, exchange and consumption, which have direct relevance to cities as specialized centres, central places and built, aggregated forms. To continue the example given above, in a primate city consumption is made conspicuous by being devoted to costly public buildings constructed for show rather than for their function.

The third element is called political, for 'behind all aspects of the production, exchange, extraction and use of the social surplus product lies another stage, composed of those co-ordinating mechanisms or loci of power able to organize and control society and extract or use the available surplus' (Davies, 1989: 74). It is interesting to review many of the urban studies produced under former communist East European regimes. F. Grimm, summarizing his study of the East German city system in 1984 (see this volume Chapter 17), wrote, 'the present administrative division of the GDR has led to a stable, well-balanced integration of settlement systems with the Kreis framework. The even distribution of towns promotes the equalisation of working and living conditions in accordance with the political and social aims of socialist society' (Grimm, 1984: 381). Whether that was true or not does not matter. It stresses the impact of political ends upon city systems. It also leads on to the last of Davies's elements of control which is called 'ideological', for economic and political procedures and controls ultimately rest on an ideological foundation, the ideals and values in society, as the quotation from Grimm suggests.

This has been a brief summary of Davies's schema which undoubtedly is setting an appropriate and necessary direction for city system studies. It is, of course, reductionist in that it resolves all the complexity down to the basic nature of the political economy. Inevitably, Davies derives and uses the same bases of explanation which have been deployed in the search for the deeper, sub-structural explicands of internal city structure. But the real problem rests in the precise tracing of the aspatial theoretical mechanisms to the left of his central column and how they are translated into the locational or spatial empirical observations and classifications to the right and with which this section of the book has been concerned. There lies a major problem which possibly should be the prime theme of future research in the field of urban location. It also, of course, raises the critical question of where the geographer should begin, from the aspatial mechanisms or the evidence of their locational representation.

URBAN LAND-USE: GENERAL PROBLEMS

7.1 INTRODUCTION

The last chapters have been concerned with the evaluation of the related characteristics of the location, size and character of cities. Much of it has been concerned with giving both precision and depth to what are everyday concepts. Thus most people use terms like hamlet, village, town and city, unconsciously deriving a settlement hierarchy. Likewise descriptions such as resort or port are widely used making an elementary functional classification. In the same way the internal structure of the city is part of everyday perception. City centre or downtown, shopping centre, industrial estate, housing estate, all these order the city into a spatial framework in order to make sense of the environment in which people live and work. In this context too, the academic analysis of the internal structure of the city attempts to formalize, and in turn interpret, the elements which make it up and their spatial interrelations.

Like the work on central places, most of the development which came with the growth of urban geography was based on ideas which were generated by non-geographers before the Second World War. Indeed, the first attempt at land-use generalization must be that of Von Thunen as early as 1837, in relation to agricultural land-use about the city (Von Thunen, 1837). Even the early attempts to rationalize urban land-use occurred early in the present century (Hurd, 1924). But urban geographers, in seeking to offer the most general interpretation of city land-use, used two sources: one was the work of human ecologists in Chicago during the 1920s; the other was the work of land economists which had been developed earlier in the century. Undoubtedly human ecology had a greater initial impact upon what was a specialism developing in the field of human geography, and it is appropriate to consider that source first although its origins are later in time, and in some respects a reaction to the earlier more mechanistic explanations of the land economists. There was, however, a reaction to these more formal, mechanical interpretations, one which viewed the city as derived from the complex activity systems of its inhabitants. This can be added as a third approach to give the following aspects:

- Ecological
- Economic
- Activity systems.

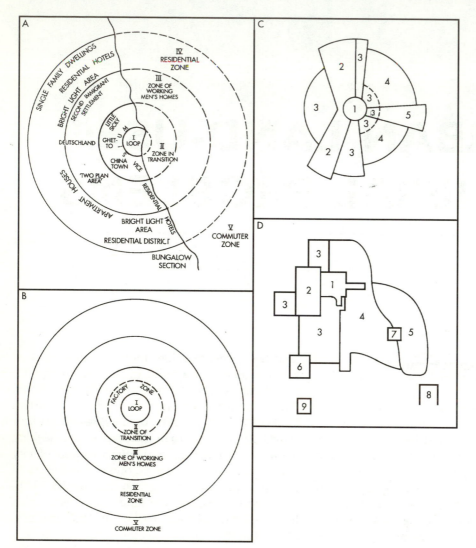

FIGURE 7.1 *General schemes of urban land-use. A: is Burgess's interpretation of the structure of Chicago in 1925, whilst B: is the generalization derived from it for all rapidly growing industrial cities. C: is the sector scheme of H. Hoyt where 1: CBD; 2: Wholesale and light manufacturing; 3: low-class residential; 4: middle-class residential; 5: high-class residential. D: is the multiple nuclei scheme of Harris and Ullman. The areas are numbered as in C with the addition of 6: heavy manufacturing; 7: outlying business district; 8: residential suburb; 9: industrial suburb.*

7.2 HUMAN ECOLOGY AND URBAN LAND-USE

E. W. Burgess's thesis, first set out in 1925 (Burgess, 1925) that urban land-use tended to display a zonal organization concentrically arrayed about the city centre (Fig. 7.1(a) and (b)), has for the remainder of the century been the starting point for most considerations of the utilization of land in cities. Each succeeding author has outlined the concentric zone scheme, or theory, as it is sometimes called, subjecting it to what is now a fairly standard critique. Since this process continues it seems that the concentric zone scheme must either present something of value or be the only generalized statement available, and is, therefore, presented *faute de mieux*.

The development of model building in the social sciences has to some extent put Burgess's scheme upon a more elevated plane. It can quite justifiably be considered a normative model, 'a simplified structuring of reality which presents supposedly significant features of relationships in a generalized form' (Haggett and Chorley, 1967: 22). Comparison is at once suggested with Christaller's statement of central place theory as another early model concerned with spatial and locational aspects within the social sciences. But here little progress can be made, for Burgess did not deduce a structure from a series of initial assumptions. His model was explicitly inductive and, presumably, intuitively derived from the observation of a large number of American cities in general, and of Chicago in particular; though the process was not made clear in his chapter on 'The growth of the city: an introduction to a research project' in the book *The city* written with R. E. Park in 1925.

The scheme was baldly introduced. 'The typical process of the expansion of the city can best be illustrated, perhaps, by a series of concentric circles, which may be numbered to designate both the successive zones of urban extension and the types of areas differentiated in the process of expansion' (Burgess, 1925: 49). Moreover, there was a strong historical element in the scheme, for when process was introduced (in terms of invasion and succession and concentration and decentralization) it was envisaged in the historical context of the growth of the American city and not in terms of contemporary forces creating by their interaction an equilibrium condition which is the model land-use pattern. Neverthless it is a model in the proper sense of the term. It was to Burgess:

an ideal construction of the tendencies of any town or city to expand radially from its central business district... Encircling the downtown area there is normally an area of transition which is being invaded by business and light manufacture, a third area is inhabited by the workers in industries who have escaped from the area of deterioration but who desire to live within easy access of their work. Beyond this zone is the 'residential area' of high class apartment buildings or of exclusive 'restricted' districts of single family dwellings. Still further, out beyond the city limits, is the commuters' zone – suburban areas or satellite cities...

(Burgess, 1925)

This sketch of the concentric zones was immediately followed by a section which firmly linked it with an historical process, 'the tendency of each inner zone to extend its area by the invasion of the next outer zone', so that in the early history of Chicago all zones were included within the inner zone but emerged as distinctive parts of the city during the process of growth and expansion. Undoubtedly Burgess's presentation was sketchy and muddled. The zones were related partly to generalization based on evidence from Chicago and other cities, partly to the processes of city growth, partly to the sub-social ecological forces of competition then dominating the thought of the 'Chicago School' but which are very difficult to distinguish from more standard economic competition for land.

Reactions to the concentric zone model can be considered under three headings. Most critiques contain elements of all three and the division is one of convenience of review. The divisions are:

- Destructive criticism which rejects the thesis.
- Criticism which lays emphasis on factors which were either excluded by Burgess or to which he gave little weight.
- Attempts to exclude historical process and to evaluate the implicit assumptions on which the thesis is based and hence recast the whole argument in deductive form.

These can be considered in turn.

Rejection of the Burgess thesis

Intensive examination of Burgess's scheme reached a peak in the decades after the Second

World War when it was taken more literally than it is now. This was partly the result of the way in which Burgess presented it but even so its generality was ignored and attacks directed at specific detail. Thus the concentric circles were attacked as being transition gradients rather than sharp breaks and the zones as having more heterogeneity of use than the scheme implied. But these were irrelevant attempts to match specific detail with a general model. However, broader criticisms were developed which had more substance since they argued that the concentric zonal structure was particular in time and place, that by the late twentieth century it was anachronistic and it was limited to large Western industrial cities.

1. **The scheme is anachronistic.** Since Burgess's model was stated in a specific historical context, it is only to be expected that it would become outdated. Though the zonal scheme might have been apposite for the American city of the 1920s, it is certainly not useful for world-wide application at the end of the twentieth century.

2. **The scheme lacks universality.** It is only just to record that Burgess made no claim that the model had relevance other than to rapidly growing, industrialized American cities, but inevitably it has been taken as universally applicable to all cities. The examination of what he called the 'pre-industrial city' by Gideon Sjoberg (Sjoberg, 1965: 95) effectively disposed of this extension. Sjoberg demonstrated that the city prior to industrialization, in both an historical and a spatial context, displayed a structural pattern which was a negation of the concentric scheme in all ways. Three major aspects of this contrast can be emphasized.

   (a) 'The pre-eminence of the centre over the periphery especially as portrayed in the distribution of social classes' (Sjoberg, 1965: 95). In this context of the 'pre-industrial' city the privileged classes, the élite, gather at the centre because of their need for the closest association with the governmental and religious buildings which physically and symbolically represent political power.

   Because political and religious activities in feudal cities have far more status than the economic, the main market, though often set up in the central sector, is subsidiary to the religious and political structures there... the commercial structures in no way rival the religious and political in symbolic eminence; typically these tower above all others... These land-use patterns refute the still widely accepted proposition of the Chicago School that the 'central business district' is the hub of urban living, a generalization fulfilled only in industrial cities.

   (Sjoberg, 1965: 97)

   (b) 'Finer spatial differences according to ethnic, occupational and family ties' (Sjoberg, 1965: 95). There is a characteristic segregation along these various lines. A prime example is the ghetto of the medieval European city or indeed the monopolization of some areas by particular crafts which is still visible in the street or 'quarter' names of contemporary cities. It should be stressed that in chart 2 of his chapter on the growth of the city, Burgess clearly showed a 'black belt' in Chicago which cut across the concentric zones. 'Wedging out from here is the black belt' he wrote, almost using a sectoral terminology.

   (c) 'The low incidence of functional differentiation of land-use patterns' (Sjoberg, 1965: 90). The multiple use of single plots is a further characteristic of the pre-industrial city, as for example in the use of church land for market places, even the use of the churchyard itself; those who were buying and selling in the outer court of the Temple at Jerusalem were doing so in a typical location. Again, more often than not merchants and artisans live at their place of work in quarters above or behind their shops so that there is not that essential division of place of work and place of residence on which the concentric zone scheme is based. This, of course, is partly a consequence of a technology unable to provide the necessary rapid transport to make this feasible.

'To reiterate, the feudal city's land-use configuration is in many ways the reverse of that in the highly industrialized countries' (Sjoberg, 1965: 103). Although it is not necessary to accept Sjoberg's terminology, his views have been expounded at some length since they so effectively demonstrate that Burgess's model depended not only on those processes which human ecologists called sub-social (but which seem to have simply been economic competition for a scarce commodity, that is, central city land) but also upon a whole array of social and technological conditions which were never made explicit and which were often ignored when the model was seized upon as being universally applicable.

However, the most significant development has been the association of the schemes of Burgess and Sjoberg and their presentation as representing an historical continuum (Carter and Lewis, 1983, 1990). By the processes and consequences of nineteenth-century industrial urbanism, the feudal, or pre-industrial, city of Sjoberg was transformed into that described by Burgess. In this way criticism based on Sjoberg becomes incremental rather than destructive and seeks to offset the particularity of time and place which marked Burgess's scheme. Since a parallelism can be argued between pre-industrial Western cities and the cities of the developing world, both criticisms on the basis of anachronism and of the lack of universality can at least in part be met.

There is no need, however, to look at widely diverging areas and times to make the point that Burgess's model is limited in universality. The location of British municipal housing estates on city margins shows that the simple notion of a rise in social status with distance from the city centre is not true. These estates are the consequence of an entirely different set of socio-political forces than those moulding the American

city in the 1920s, when American municipal authorities were effectively not in the house constructing business. Once one moves outside the 'American culture province' and tries to interpret city patterns created by different social–economic–political–technological (i.e. cultural) conditions, then the applicability of American-based concepts and models becomes less than complete.

The whole tenor of the lines of criticism which have been outlined is to reject the Burgess model because it is limited historically and culturally to a particular situation at a particular time in a particular country, and second, even then it quite arbitrarily emphasizes clear-cut boundaries which cannot be justified by gradient study and assumes a natural sorting of distinctive areas, whereas the city is in reality a most complex patchwork which defies the massive generalization. In some ways these various attacks make elaborate parade of a simple theme: that the model of concentric zones is too simple and too limited in historical and cultural application to carry one far into an understanding of land-use patterns. With this clearly understood the model remains useful as a first approximation and as a pedagogic device; it is no longer a springboard for research.

Extension of the Burgess model

Since the concentric zone model excludes a wide variety of factors which affect urban land-use it is possible to introduce many of these in extension of the model. Four of these can be considered here, the first and the last perhaps not standard inclusions but the middle two now invariably sketched alongside the Burgess model.

The introduction of building height

There is an implication in the concentric zone model that the city is a two-dimensional feature and little attention was paid in its formulation to the height of buildings and the variation of use with height. It is quite simple, however, within the overall limits of the model itself, to introduce this feature. Accessibility not only

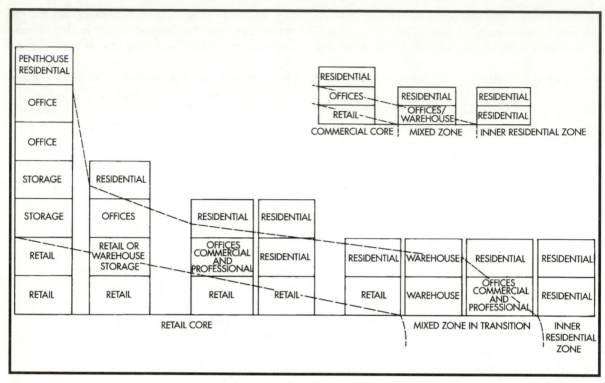

FIGURE 7.2 *The relation of land-use to building height. The upper diagram shows a more simplified form of the larger diagram. These are only generalized, unscaled sketches to demonstrate the principle of the way in which the respective zones of the concentric scheme overlie each other at the city centre.*

diminishes outward from the centre but upward from the ground. If the vertical changes in use are examined they are often found to parallel the horizontal changes. This is depicted in Fig. 7.2. Characteristically in the smaller town the ground floor retail shop has office space above, frequently occupied by legal or insurance firms, while above this is an apartment or maisonette. This parallels the horizontal sequence. Types of use which cannot effectively compete for a desired central location because of the high costs are relegated to the transitional or mixed fringe which surrounds the CBD *or* they are relegated to the upper floors of centrally located buildings.

The whole model can, therefore, be thought of as a cone reaching its highest point at the peak land value and with a series of bands of land-use taking off progressively from the outer edges, thinning markedly towards the centre where they overlay each other at the central vertical axis. Thus logically the outermost commuter zone, where there are the most valued residential locations, would form the uppermost layer of the cone at the centre where, indeed, it is found in the form of penthouse flats and apartments. Its extension into the vertical has only been sketched here and would involve a number of problems if strictly interpreted. It suggests, however, an area of extension of the model in need of more thorough and logical exposition.

The emphasis of sectors

The earliest constructive criticism of Burgess was best presented in a volume by Homer Hoyt published in 1939 by the United States Federal

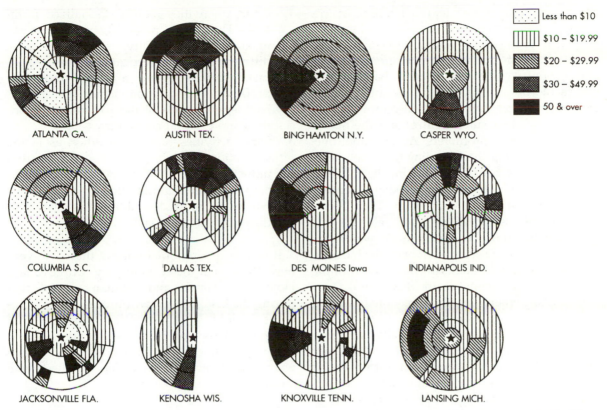

Less than $10
$10 – $19.99
$20 – $29.99
$30 – $49.99
50 & over

ATLANTA GA. AUSTIN TEX. BINGHAMTON N.Y. CASPER WYO.

COLUMBIA S.C. DALLAS TEX. DES MOINES Iowa INDIANAPOLIS IND.

JACKSONVILLE FLA. KENOSHA WIS. KNOXVILLE TENN. LANSING MICH.

FIGURE 7.3 *The theoretical pattern of rent areas in twelve American cities (after H. Hoyt, 1939). These are 12 of the 30 examples depicted by Hoyt. He arranged the rent areas to correspond to an idealized pattern of concentric circles and concluded from the evidence – 'rent areas in American cities tend to conform to a pattern of sectors rather than concentric circles'.*

Housing Administration and entitled *The structure and growth of residential neighborhoods in American cities* (Hoyt, 1939). After conducting a factual examination of residential rent patterns in twenty-five widely distributed cities in the USA (see Fig. 7.3), Hoyt concluded that:

there is, nevertheless, a general pattern of rent that applies to all cities. This pattern is not a random distribution. It is not in the form of sharply defined rectangular areas, with blocks in each rental group occupying completely segregated segments. It is not in the form of successive concentric circles .... Even when the rental data are put into a framework of concentric circles there is revealed no general gradation upward from the centre to the periphery. From

the evidence presented, therefore, it may be concluded that the rent areas in American cities tend to conform to a pattern of sectors rather than of concentric circles [see Figs 7.1(c) and 7.3].

Thus Hoyt presented what has been called the 'sector theory'. As the title of the volume where it appears suggests, he was concerned primarily with residential uses and the key to the sector arrangement was to be found in the location of the high quality areas. These, Hoyt maintained, 'do not skip about at random in the process of movement – they follow a definite path in one or more sectors of the city' (Hoyt, 1939: 14); that is, they are extended outward along communication axes thus producing the sector; they do not encircle the city as its outer limits. In this process certain features exercise a dominant influence. These are, in addition to communication lines, high ground free from flood danger, open country with no physical barriers, the homes of community leaders, the attraction of office blocks and stores which also move, established outlying business centres and the operation of real estate promoters. At the same time, as is evident from the foregoing list, some land-users will attract each other, as in the move of quality residence to outlying business centres, while some will repel each other. Thus heavy industry will also probably extend along a line of communication, be it railway or waterway, and also occupy a sector – but one which due to repulsion will be located directly away from the high quality residential sector (Fig. 7.1(c)).

Essentially, therefore, the directional element will control land-use contrasts rather than distance, and accordingly, the city structure becomes sectoral in character. Hoyt did not discount the distance variable but rather added the further directional element which he considered vital in any attempt to give more reality to Burgess's model: he was adding a further degree of verisimilitude at the expense of simplicity.

The emphasis of multiple nuclei

A model made up of a number of separate nuclei was proposed by C. D. Harris and E. L. Ullman (1945). It was conceived as a further move away from the massive generalization and towards reality. It is an observed fact that many towns and nearly all large cities do not grow simply about a single central business district but are formed by the progressive integration of a number of separate nuclei into the urban fabric – hence the suggestion of a multiple nuclei model. These nuclei, and further districts which become specialized and differentiated in the growth process, are not located in relation to any generalized zone or distance attribute, but rather they are bound by a number of controls which produce a pattern of characteristic associations betwen the nuclei. These are:

1. Certain activities require specialized facilities. Thus the retail district demands maximum accessibility, something quite different from geometrical centrality.
2. Like activities group together since they profit from association, hence the specialized legal districts or theatrelands. In short, there are external economies.
3. Some activities repel each other, as in the separation of high quality residences from industry.
4. Some activities cannot afford the high rents which the most desirable sites, relative to their interests, demand.

It will be apparent that none of these was very new (they had been noted by Burgess and Hoyt), but Harris and Ullman argued that during the historical growth process these influences would condition the emergence of separate areas, neither zones nor sectors, forming in effect a patchwork but where there were clear controls as to which 'patches' were joined together (Fig. 7.1(d)).

The introduction of a size variable

In presenting his model Burgess used the words 'great cities' and 'our largest cities' and he clearly had in mind a process which applied to the largest cities. But how large are 'great' and 'largest'? Burgess himself stated that all the zones were contained in one when Chicago was but a small and primitive settlement. The concentric zone model does not apply to the large metropolis where the simple concept of a CBD has fragmented into a large number of specialized tracts – financial, legal, retail, theatre areas to identify only the beginnings of specialization.

The conclusion from this is that if the evolution of the zones from a 'primitive cell' to a great city can be traced historically, then this same process should be identifiable at the present moment by examining different size classes of town. It should be possible to trace a progression from the smallest settlement where there was no apparent differentiation of uses, to the next class where a fairly well-defined residential ring has become segregated and surrounds a core of mixed but mainly non-residential uses. The next size class is characterized by the crystallization of a retail core with a mixed surround which is partly the product of the expulsion of non-retail uses as the core developed and of their taking over of premises in the first residential ring. And so on through to the metropolis. It must be emphasized that these are identified as the characteristics of size classes not stages in the growth process.

Nevertheless this immediately suggests the attempt by Griffith Taylor (1949) to formalize the sorts of contrasts which have been outlined. His scheme recognized the following towns:

- **Infantile towns:** haphazard distribution of shops and houses, no factories.
- **Juvenile towns:** differentiation of zones begins, shops are separated.
- **Adolescent towns:** scattered factories, but no definite zone for first-class houses.
- **Early mature towns:** a clear segregation of first-class houses.
- **Mature towns:** separate commercial and industrial areas, four zones of houses, ranging from mansions to shacks.

This is an inadequate table in almost every way. The use of terminology adapted from the cycle of erosion as propounded by W. M. Davis is ill-considered and inappropriate; above all it is unnecessary since the characteristics outlined are correlates of size and possibly of economic base.

Indeed, Taylor wrote size into the text indicating, for example, that the population of the 'juvenile town' would be about 20,000 but then arguing with some justification but little proof that 'stage' and 'population' need not be closely associated. There are problems; in its early stages a crude mining town may have a large population yet remain still little more than a 'camp', but these are problems partly associated with the 'economic base'. The progressive differentiation of uses in Taylor's 'cycle' was but thinly indicated and not rigorously pursued. His

identification of 'the various ages of towns' was, therefore, not very impressive and has been seldom quoted in the literature of urban geography, but it pointed in the right direction and posed questions which have yet to be effectively answered. If the appearance of segregated land-uses is part of the urban growth process, at what town sizes or at what hierarchical grades and in what manner does this differentiation take place and under what functional conditions? Perhaps it is sufficient here merely to point out that Burgess's model clearly involved a size variable which was barely considered, although held constant by implication.

This brief review of models which have aimed at generalizing the total land-use pattern of a city has demonstrated some of the ways in which Burgess's original scheme has been extended. These extensions are the product of an inductive approach, the thesis being that knowlege of land-use in a large number of cities will show that there is little correspondence between model and reality. If, therefore, further variables are introduced – size, height, axial growth, multiple nuclei – then progress will be made by making the generalization 'more real'. But as this is done, so too some of the grand simplicity of the original is lost. Indeed one questions whether the 'multiple nuclei' model can rightly be so called. It is based on well-known statements regarding uses and makes no marked contribution to their complex association in area in a locational sense. At an earlier stage a quotation from Hoyt was introduced denying that land-uses formed a patchwork; multiple nuclei theory argued that they do and because of this ought perhaps to be associated with destructive criticisms rather than with those which modify.

Perhaps it is worth repeating that Burgess's original formulation of a concentric zone model remains as still providing stimulating insight into land-use patterns and, if it has lost the relevance to the formulation of research projects and designs which it had in the 1920s, it is still a most useful pedagogic device for approaching the complexities of the problem it seeks to illuminate. Perhaps this is why the more frequently it is dismissed as no longer of relevance, the more consistently it appears in texts such as this.

Presentation of the Burgess model as a deductive theory

As dissatisfaction with the concentric zone scheme of urban land-use grew, so, sometimes explicitly, sometimes implicitly, attempts were made to reframe the model as the end product of a deductive argument. In this process a statement of premises becomes essential, making explicit the assumptions on which the model rests. The clearest exposition of these assumptions was that by Leo F. Schnore (1965) who listed five:

1. **Assumption of heterogeneity.** This was a restatement of Quinn's argument that widely contrasted population types are a prerequisite. Quinn envisaged contrasts in race and in degree of cultural assimilation as well as in social class and occupations (Quinn, 1950: 120). This would hardly seem essential particularly if the model were only to take in 'large' Western cities when occupational and class differences are implicit.

2. **Assumption of an economic base.** Certainly a mixed industrial commercial city was implied by Burgess, but a much greater clarity is to be obtained if a certain economic functional character is specified.

3. **Economic and cultural assumptions.** Burgess assumed those conditions which were appropriate in the America of the 1920s, such as private ownership of property, economic competition and efficient transport, equally easy, rapid and cheap in all directions.

4. **Assumptions as to the geometry of space.** Schnore (1965: 353–4) maintained that it is necessary to assume a single centre (hence disposing of the multiple nuclei problem) and that since physical area increases as the square of radial distance, then space is in shortest supply at the centre. In addition there is the further necessary condition that

central areas are most highly valued by virtue of short supply and of accessibility to all other areas. Competition for these central locations leads to a 'sifting and sorting' with physical positions largely determined by economic ability to compete for space.

5. **The assumption of occupancy patterns.** Schnore argued that for the Burgess model it is necessary that social classes with superior economic status will be able to pre-empt the newer and more desirable areas, or will at least have a high degree of freedom of choice as against a very restricted choice for the poor.

But even starting from this list, comparatively little progess has been made in the presentation of a convincing argument. Schnore proceeded to list variables rather than deduce an urban arrangement. To some extent a somewhat limp end is partly bound up with the way human ecology itself developed in this particular context.

The classical ecologists had been concerned to demonstrate parallels with plant ecology and hence they looked upon urban land as subject to competition from rival interests carried on at a 'sub-social' level. 'The basic process in human relationships is competition largely involving a struggle for space'. In this process 'human society was seen as organized on two levels: the biotic and the cultural. The biotic level involves basic, non-thoughtful adjustments made in the struggle for existence. The level is regarded as sub-social ... The cultural level is seen as a super-structure and excluded from human ecology' (Theodorson, 1961: 3). This framework might be applicable to the study of diseases or the spatial arrangement of a primitive population but if the land-use pattern is to be interpreted in this way two problems immediately arise.

1. The more the concept of sub-social competition for a scarce resource is examined and refined the more the ecologist is taken directly into the field of land economics, and this is particularly so when a deductive model of the city is desired excluding those cultural and

social influences which the ecologist eschewed. When Quinn wrote 'Ecological interaction underlies many aspects of community life. Merchants who want strategic commercial sites at the heart of the city compete for the limited supply of land in this location' (Quinn, 1939: 166), and then went on to argue that urban residents 'bid for the limited supply of residential space', one is forced to conclude that 'sub-social' is merely another term for 'economic', for he was making a straightforward statement which is almost a definition of economics – the allocation of scarce resources to alternative ends. It follows that the vigorous attempts to present a theory of urban land-use have passed to the land economists and that the stream of work started by Burgess, while still of value in its empirical side, has not been a source of rich theoretical construction.

2. The clear rejection by the classical human ecologists of social and cultural influences while making for simplicity in terms of economic theory has led to severe criticisms in the inductive sphere for, as the economic implications of competition become apparent, it was only to be expected that protests would arise against these mechanistic theorems.

Two such protests can be noted:

• **Sentiment and symbolism as ecological variables.** In a paper with this title Walter Firey (1945) called attention to the difficulty outlined in the last paragraph – 'A different order of concepts, corresponding to the valuative, meaningful aspect of spatial adaptation, must supplement the prevailing economic concepts of ecology.' Firey made his objection to the classical position on two grounds arguing that there were two kinds of human adaptation to the urban environment.

    The first he called 'volitional adaptation' since 'values which express primarily the identity or solidarity of a small cultural system may acquire spatial articulation'. His work on central Boston emphasized, for example, the value system derived from a cultural heritage, which gave Beacon Hill its distinctive character and determined its residential status unrelated to any distance or directional controls (Firey,1947).

    This objection may be restated in the form that culturally rooted values exert a 'causative' influence on urban land-use patterns.

    The second sort of adaptation was termed 'rational' since 'interests dominate the spatial adaptation of certain social systems'. This is nearer the economic approach of his precursors in human ecology but even so, Firey argued, 'these interests come directly from broader and larger cultural systems'. This objection to economic determinism may be stated in the form that market forces themselves stem indirectly from larger cultural controls; they are not self-generated.

• **The place of social power in the determination of land-use.** This aspect of land-use determination was earliest stated by W. H. Form (1954). He prefaced his case with a plea for models which consider social reality and not solely economic abstraction. In particular he argued that it is necessary to discard the idea of a free unorganized market in which individuals compete impersonally. This done it becomes imperative to identify who are the largest consumers of land, which organizations deal in land and which associations mediate conflicts of land-use. In a way this anticipates later and much more intensive Marxist critiques of traditional urban land-use analysis.

    He identified four organizational complexes which are dominant; the real estate and building businesses; larger industries, businesses and utilities; individual home owners; local government agencies. The problem becomes the exposure of the resources or influence which each of these can deploy, their functions in terms of the land-use market, the nature of their internal organizations, the accountability of the groups and the 'image of the city' within which each group works and the 'values' they consider paramount. Finally it is necessary to consider the relations between them much as one would study a collective bargaining process in industry. This provides a very different view of the sorting process from that of the classical ecologists.

With the rise of a radical human geography in the late 1960s these sorts of interpretations of urban land-use became much more common, even predominant. A good British example is Ambrose and Colenutt's book *The property machine* (1975). It was specifically written – and given a title – to demonstrate the influence of one of the complexes Form had identified the real estate and building businesses. As the authors write in relation to the development of the 25 acre site of Brighton railway station, the problem was 'whether a "windfall site" in a town short of land should be used for commercial purposes or to accommodate a large addition to the town's stock of low cost housing' (Ambrose and Colenutt, 1975: 105). Here, however, are much larger issues which will be taken up in a later chapter.

In conclusion to this section dealing with human ecology and land-use it would seem reasonable to infer that work on ecological lines had reached an impasse by the 1960s. Its predominant economic bias has little to offer besides the more rigorous theoretical constructs of the thorough-going economist, while its rejection of cultural and social influences has hampered a vital line of enquiry. Moreover, human ecology had given rise to social area analysis and factorial ecology (see Chapter 12) which were still couched in zonal and sectoral terms but took the analysis of internal urban structure into a new dimension.

7.3 LAND ECONOMICS AND URBAN LAND-USE

The earliest work of importance in this field predates that of the human ecologists by many years for Richard M. Hurd's (1924) *Principles of city land values* appeared in its first edition in 1903. Much of this enquiry, as the title of Hurd's book implies, is directly concerned with land value, but even so, it is inextricably bound up with land-use for 'the patterns of land-uses and land values will be mutually determining' (Alonso, 1964a: 16).

Hurd drew upon earlier work concerned with land value and adapted the principles of Ricardo, propounded for agricultural land, to an urban setting. Thus he began from the principle that 'value in urban land, as in agricultural land, is the resultant of economic or ground rent capitalized' (Hurd, 1924: 1). Further, 'in cities, economic rent is based on superiority of location only, the sole function of city land being to furnish an area on which to erect buildings' (Hurd, 1924). With the growth of a city more remote, and hence by definition inferior, locations are brought into use so that rents at the most accessible points rise. For this accessible and consequently more valuable land competition will take place; 'any utility may compete for any location within a city and all land goes to the highest bidder' – who therefore obtains maximum convenience or economy in time and effort by being most accessible.

Hurd summed up his argument in a succint sentence, 'Since value depends on economic rent and rent on location, and location on convenience, and convenience on nearness we may eliminate the intermediate steps and say that value depends on nearness' (Hurd, 1924: 13). But Hurd notes that 'nearness' is a relative term and has to be evaluated in terms of the growth and physical structure of the city on the one hand and the nature of the use required on the other. He concludes by emphasizing the necessary limitations imposed by an economic investgation which does not take into account 'individual and collective taste and preference as shown in social habits and custom' (Hurd, 1924).

These ideas have remained as basic to the mainstream of land economics. The major restatement took place in the 1920s by Robert M. Haig (Haig, 1926) but in much the same terms. Haig saw rent as the charge for accessibility or the saving in transport costs and invoked a bidding process to determine the occupancy and the use of land. Undoubtedly his major contribution was the introduction of the notion of the 'friction of space', that is, that hindrance to perfect or immediate accessibility, for without such 'friction' there would be no transport costs and all locations would be perfect. The purpose of transport is to overcome this imperfection or friction, but while transport partly offsets friction, site rentals and transport costs represent the charge for that friction which remains. Rent thus appears as a payment for the saving in the costs of transport and these (rent and transport costs) will vary with site since 'the theoretically perfect site for the activity is that which furnishes the desired degree of accessibility at the lowest cost of friction' so that 'the layout of the metropolis tends to be determined by a principle which may be termed minimizing the costs of friction' (Haig, 1926). This involves a three-way relation of rent, transport costs and the costs of friction which are interdependent.

Ratcliff, in a volume specifically called *Urban land economics* (1949), adopted an argument that is basically similar. 'The utilization of land is ultimately determined by the relative efficiencies of various uses in various locations. Efficiency in use is measured by rent-paying ability – the ability of a use to extract economic utility from a site. The process of adjustment in city structure to a most efficient land-use pattern is through the competition of uses for various locations. The use that can extract the greatest return from a given site will be the successful bidder (Ratcliff, 1949: 369). From this there emerges an 'orderly pattern of land-use, spatially organized to perform most efficiently the economic functions that characterize urban life' (Ratcliff, 1949: 369). Ratcliff follows further enquiry into the competition of users for sites by the bidding process and into the minimizing of costs of friction, by an analysis of the different demands each user will have in locational terms, for not all users are in direct competition. Nevertheless 'in summary one might say that the structure of the city is determined through the dollar evaluation of the importance of convenience' (Ratcliff, 1949: 375).

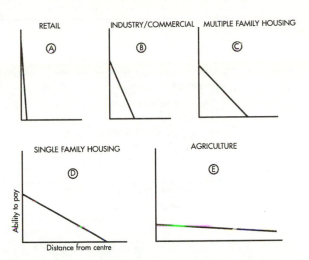

FIGURE 7.4 *Bid rents made by different users of urban land in relation to distance from the city centre (after B. J. L. Berry, 1959). In each of these graphs the ability to pay rent of a potential user of urban land is plotted against distance from the city centre.*

This sort of argument can be used in the derivation of a city structure closely analogous to that proposed by Burgess. In Fig. 7.4 distance is used as a measure of 'convenience' in the sense used by Ratcliff. Each crude and broad category of land-user can be examined by ability to pay rent against distance from a single most accessible core. Retail uses demand greatest accessibility to the whole city population in order to maximize profits, while away from the city centre this quality falls off very rapidly as does the willingness to pay high rents (Fig. 7.4(a)). Offices, such as those of legal or insurance firms or of doctors and dentists, also require accessibility and a central location but the very nature of their businesses means that the very high rents of the centre can be avoided by marginal locations. The slope for rent/distance is, therefore, less steep (Fig. 7.4(b)). Multiple housing schemes, such as apartment blocks, will give larger return per unit area and can, therefore, obtain a greater degree of accessibility than single family housing, but will not need the central locations which the retail and commercial offices have to obtain (Fig. 7.4(c)).

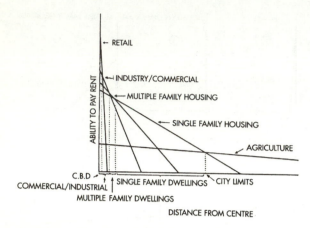

FIGURE 7.5 *Superimposition of the graphs in Figure 7.4. Where the graph lines intersect, working from the city outskirts to the centre, then the innerward use can afford to pay a higher rent and takes over the land. Those points are indicated by dotted lines. If the base line is taken as a single radius then the total city pattern will be one of concentric circles.*

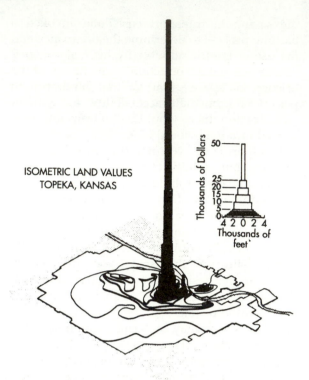

FIGURE 7.6 *The distribution of land values in Topeka in 1962 (after D. Knos, 1962).*

The advantages of accessibility are, therefore, sacrificed in the interest of lower costs, although it must be emphasized that to these costs in terms of land must be added the costs of travel for, as has been emphasized, value or the capitalized expectation of rent, transport costs and location are interdependent. Finally, agricultural uses are the least intensive and although they, too, would derive advantages from location at the centre they are outbid by all the urban uses and the rent/distance slope is the least steep (Fig. 7.4(e)).

If all these relationships, that is, all the slopes, are now superimposed (Fig. 7.5) then it can be seen that, moving from the city margin, where the slopes intersect the user nearest to the core will be outbidding the user further from the core and consequently that type of use will succeed. If this is converted into a two-dimensional diagram then a series of rings will be generated as in Fig. 7.6. There are a large number of assumptions which are not proven and a number of serious flaws in the argument which has been presented. For example, all users are assumed to require

equal quantities of land and no allowance is made for the possible substitution of a larger land area for a less convenient location or, in crude terms, for buying cheaper land further out. Several attempts have been made to provide a more sophisticated basis and a more complex model in place of the simple concepts so far considered (Wingo, 1961, 1963; Alonso, 1964a).

The initial outline of one of these models can be briefly considered in order to indicate its rationale rather than to trace its full development which is impossible in the present limited context.

William Alonso, in his work *Location and land use*, introduced two additional variables: first, the quantity of land which each user will wish to acquire; and second, the amount of disposable income which will be devoted to land and travel costs on the one hand, and all goods and services, including savings on the other. The equilibrium of the individual household in the urban situation is then seen as a consequence of the relation of all the possibilities in expenditure open to the individual and the patterns of preference which can be exercised. The possibilities open define the 'locus of opportunities' which is seen as a surface generated by the interrelation of the three variables noted above: the amount of land; the amount of goods and services; and distance from the centre.

To relate this back to Fig. 7.5, it will be observed that the ability to pay rent depends partly on the size of the land area required and partly on the desired expenditure on other things. A complex surface is generated since the relations are not simply linear. Thus if the amount of goods and services is held constant and the relationship between quantity of land and distance considered then:

> the price of land ... decreases with increasing distance from the centre. Therefore, the quantity of land that may be bought increases with the distance, since land is becoming cheaper. On the other hand, distance enters ... in the form of commuting costs. As distance increases, so do commuting costs and consequently the amount of land that may be purchased decreases.

The resultant curve of quantity against distance:

> rises to the point at which marginal increases in commuting costs are equal to the savings realized from the decreasing price of land. Thereafter, the amount of land which may be bought with increasing distance decreases.
> (Alonso, 1964a: 23)

The locus of opportunities surface defined along the three axes – quantity of land as against distance, quantity of land against quantity of goods and services and quantity of goods and services against distance – is set against the pattern of preference, defined as a surface derived from the possibility of substitution among the three variables so that the individual household would be equally satisfied by different combinations of land, goods and accessibility (distance). From this the equilibrium of the household is derived, the desires in terms of preferences being linked to the possibilities in terms of the locus of opportunities. This individual equilibrium solution is reformulated in the terms of market equilibrium, but this is a complex process for land and distance are so intertwined that the individual demand curves cannot be amalgamated into a demand curve for the market.

It will be apparent that this model is working, as the sub-title of the volume states, 'towards a general theory of land rent'. The assumptions on which the model is built are clearly set out. These include a featureless plain, all land of equal quality, all land ready for use without improvement, land freely bought and sold with both buyers and sellers having perfect knowledge of the market, no legal or planning restraints, no social restraints, sellers who wish to maximize revenue and buyers their profits or satisfaction, and the city regarded as a two-dimensional unit with no vertical element. The result is a deductive argument with clear premises explicitly rejecting all but purely measurable economic factors. It is rigorous, internally consistent and far removed from the reality of the city as it comes closer to pure economic theory. Again to a large extent within urban geography, the rise of radicalism turned attention away from these mechanistic interpretations.

7.4 ACTIVITY SYSTEMS AND URBAN LAND-USE

The discussion so far has revealed that work on the structure of towns in human ecology has reached an impasse, while economic models contributed but partially to the solution of the whole problem. There is a third approach which, while presenting no model, seems to provide a stimulating way of looking at urban land-use structure and leads to a coherent framework for interpreting it (Chapin, 1965: 244). This is the approach via activity systems which can be defined as 'behaviour patterns of individuals, institutions and firms which occur in spatial patterns' (Chapin, 1965: 29). F. Stuart Chapin sketched a conceptual framework to describe some of the major elements and dynamics of

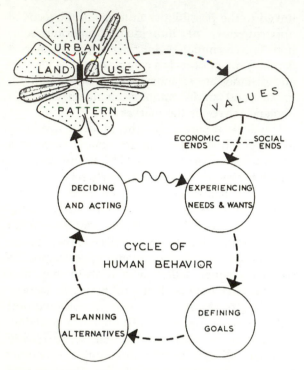

FIGURE 7.7 *The sequence of action and the influence of values in bringing about a change in the urban land-use pattern (after F. S. Chapin, 1965).*

human behaviour as they relate to land-use and ultimately to the approach which has dominated urban geography.

Figure 7.7 indicates the sequence which Chapin invoked to provide a framework consisting of values, behaviour patterns and outcomes. 'Certain individual or group-held values concerning the use of a particular parcel or area set in motion a four-phase cycle of behaviour which culminates in the parcel or area being put to a particular use' (Chapin, 1965: 30). These phases are part of the 'cycle of human behaviour' or of the behavioural pattern indicated above and are classified as: experiencing needs and wants; defining goals; planning alternatives; deciding and acting. Because of his particular interest, Chapin put this in the context of planning, but it will be seen that he accepted implicitly the views of Firey in starting with culturally determined values which find no place in any other

construct. From these values, both conscious and subconscious, or explicit or implicit, of individuals or groups of individuals, patterns of behaviour are generated leading to those actions in the urban setting which are the determinants of the land-use pattern. The behaviour patterns can be interpreted as constituting systems of activity and this is as equally applicable to the firm or the individual. Rannels classifies these activities into three states of elaboration (Rannells, 1956):

1. **Routine activities.** Standard patterns of movement of the individual, for example, the journey to shop.
2. **Institutionalized activities.** These are concentrated on to particular points by institutions as, for example, onto a theatre which necessarily focuses a number of individual systems upon it.
3. **Organization of processes.** This is the most complex situation where there is a whole series of patterned cross-relationships. For example, the 'activity system' of a bank by the very nature of its business will be made up of a wide array of associated but different links.

The word 'link' used above indicates that the study of these systems leads directly into the examination of linkages, for few 'establishments', or indeed individuals, are isolated for they are associated with others by a series of linkages where a linkage is defined as a 'relationship between establishments characterized by recurrent interactions which require movement of persons or of goods or the exchange of information'.

Thus, for example, the emergence of a well-defined legal area in a city can be seen as a consequence of a system of activities and linkages which bind legal firms together. There is a relation to the general public demanding accessibility in the sense which it has previously been used in this chapter; there is a relation to the law courts, the location of which may be due to essentially historical circumstances; there is a relation to standard sources of information such as a registry of wills and births, marriages and deaths; there is a relation of firm to firm in negotiation which involves face-to-face situations; there is a relation in Britain of solicitor to barrister. For any such area these links can be traced, indeed they could to an extent be measured and quantified. Goddard's work on London offices which will be referred to later in Chapter 11 attempted to do that.

The result is a land-use element which can be seen as the product of a complex of linkages which are the creation of the activity systems of the various components. The tracing of the systems and the identification of the linkages is a prerequisite to the understanding of the components or the pattern of land-use.

The crucial diversion from this sort of interpretation, however, was again derived from the rise of radicalism and the revision of process to be viewed not as some equitable product of multiple activity systems, but rather as one of ruthless exploitation by those with access to power. It was this type of study which dominated urban geography in the last quarter of the twentieth century. As a result, models of city structure, complex or simple, were marginalized, for the complex of land-uses was seen as the product of an exploitative capitalist system and it was in the operation of that system that explanation was to be found. Moreover, that system in post-modern times, in what Harvey called 'its carnival mask' was directed towards the creation of 'spectacular spaces', the means 'to attract capital, people (of the right sort) in a period since 1973 of intensified inter-urban competition and urban entrepreneurialism' (Harvey, 1989: 92). There resulted the city as a 'collage of highly differentiated spaces' (Knox, 1991: 188), separate parts planned and developed with no attention to the city as a whole, what has been called 'the Galactic Metropolis' (Lewis, 1983).

7.5 THE GALACTIC METROPOLIS

One of the problems in presenting urban geography is that all the complex aspects of the city impinge upon one another so that it is difficult to isolate treatment of any theme. It is evident from the distribution of out-of-town shopping described in Chapter 5, and its impact upon the idea of an urban hierarchy, that the models discussed in this chapter, themselves largely derived from the inter-war period however subsequently modified, bear little relevance to the nature of the modern metropolis. Thus, Pierce Lewis has written, 'the residential subdivisions, the shopping centres, the industrial parks seem to float in space; seen together they resemble a galaxy of stars and planets, held together by mutual gravitational attraction, but with large empty spaces between the clusters' (Lewis, 1983: 35). If this be an effective interpretation then the relation to concentric zones and sectors is tenuous indeed and a pattern of multiple nuclei would be implied.

Paul Knox, in a paper on what he calls 'the restless urban landscape', has identified a series of characteristic elements of these post-modern formations (Knox, 1991). They include what he calls Multi Use Developments (MUDs) and Mixed Use Developments (MXDs) making up 'edge cities' and linked to private master-planned suburban and exurban communities, 'high-tech corridors' and 'festival settings'. As a result he advances the notion of a new urban geography with a radically different form and ecology from that of the classic American city depicted by factorial ecologies (see Chapter 12) and explained, with varying degrees of success by bid-rent theories, theories of residential mobility and non-Marxist theories. 'The spatial patterns associated with the [above] landscape elements ... do not fit comfortably within the sectors, zones or mosaic patterns that have been the focus of academic debate surrounding the classic American city' (Knox, 1991: 203) and he concludes, 'fragmentation, multinodality, fluidity, plurality and diffusion are more in evidence than homogeneity, nodality and hierarchy' (Knox, 1991: 204). In short, as the references in the previous sentence suggest, as with the study of the urban hierarchy there are contrasted forms of what is urban. The first is the smaller towns and cities which have characteristics which still are amenable to analysis by what can be called the traditional approaches of urban geography. The second is the great metropolises where such traditional modes of analysis are no longer appropriate or tenable. Even then, as has been stressed, contemporary mobility is such that even those who live in the small towns use the varied facilities of the surrounding countryside much as if they constituted a dispersed city.

It is evident that many of the ideas which have been noted here will be developed in succeeding chapters, but their introduction in this chapter, just as discussions in previous chapters of contemporary developments in relation to the urban hierarchy, illustrate the tensions within urban geography which have developed between traditional models and modes of explanation, initially developed in relation to the free-standing town, and the reality of the modern metropolis. However, just as there are still large areas of the earth's surface where traditional hierarchies are to be found, so there are very many small to medium-sized towns where the inter-war models have at least some relevance.

7.6 CONCLUSION

An attempt has been made in this chapter to survey in general terms those studies which have attempted either to construct a model of the entire city or to introduce a conceptual framework in relation to which it can be understood. It is apparent that generalization on such a vastly complex topic must inevitably be unreal – hence the flight away from classical ecology, either to the discipline of theoretical economics or the rationality of an approach via the detail of the deployment of power through individual or corporate activities, and still further to the 'floating spaces' of the contemporary metropolis. The deployment of power immediately invokes competing ideals and interests involved in the creation of urban patterns. It is appropriate, therefore, to turn aside briefly to consider the impact of planning in the geography of towns.

8

PLANNING THE CITY

8.1 INTRODUCTION

It may seem somewhat strange to introduce a chapter called 'planning the city' at this point in the book; even more so when it is manifest that a few pages cannot hope to review a literature which would fill a large library. But the purpose of this brief chapter is not to survey that literature, it is far less ambitious than that but, nevertheless, crucial to the development of the proper study of urban geography. There has been in the presentation of all the material to this point, and especially of central place theory and of the standard models of urban land-use, an assumption of a completely free market economy. Modifications have inevitably been introduced but now need to be more explicitly considered. Equally, complete knowedge of the market has been assumed and that too will have to be reconsidered at a later stage.

The untrammelled operation of a free market has never been true. Indeed, over most of urban history the converse has applied. Towns were as much the creations of absolute rulers as they were of merchants; the history of urbanism is largely the narration of the eras of town founders, from the origins of the city down to contemporary new towns and capitals such as Canberra, Chandigarh and Brasilia. Even with regard to the mercantile town the main definition of urban status was taken as legal identity, and the town community so defined acted through its burgesses or citizens to constrain the independent actions of the individual inhabitants. The close and intimate living of the urban condition demanded control. The idea, therefore, of uninhibited economic forces as the creator of either the urban hierarchy, or of the internal structuring of towns, although it has been the basic assumption within urban geography as it has been set out in texts, is really out of keeping with reality. It follows that there are many reasons why this issue should have formed an introductory chapter rather than the eighth. It is, however, interpolated at this stage with a necessary reference back to the nature of the urban system (see Davies's diagram in Chapter 5, Fig. 5.13), and a reference forward to the discussion of the internal structure of the city with which the rest of the book is largely concerned.

There are two associated but somewhat different aspects of planning the city. They can be called, in fashionable terms, the proactive and the reactive. The proactive would include the planning of whole cities or parts of cities. That is, it takes in the determined actions of individuals or communities to create settlements, envisaging their hierarchical ranks and laying out internal structures, both physical and functional. The proactive is concerned with acts of creation. The reactive, in contrast, defines the more limited controls by which the operations of the market are constrained and made to take account of the interests of the community, or of sections of the community. These are acts of control. These two aspects can be taken as separate concepts of planning, or regarded as the poles of a continuum. Post-modernism is perhaps more sympathetic to the anecdotal. The present author, as a consultant, has been involved in public inquiries in Britain concerning on the one hand the establishment of a new town at Llantrisant, near Cardiff in South Wales, and on the other hand a proposal by Gwynedd County Council in North Wales to demolish two early nineteenth-century houses in the vicinity of a medieval town gate (Porth yr Aur) in the town of Caernarfon. These two inquiries, legitimately classified under the heading of urban planning, were very different in their scale and nature and illustrate the degree of variation which can be covered by the term urban or city planning.

As has already been stressed, this chapter cannot be a review, still less a history, of city planning. It is, in some respects, no more than an interlude to call attention to the fact that a volume constructed about the notion of a free market is one which can be accused of starting at the wrong point. The concept of a free market so that models can be constructed on the economic basis of the maximization of profits, is the product of the nineteenth century and its contemporary inheritance, perhaps of the growth of urban geography itself; of a time when urban industrial entrepreneurs, symbolized by Henry Ford, had broken free from the control of monarch and aristocrat on the one hand and from the mercantile bourgeoisie on the other. The new entrepreneurs were able to use their freedom to create the 'insensate industrial town' (Mumford, 1938) and the succeeding conditions which became the progenitors of most urban models.

It is essential at this stage, therefore, to identify the various roles of planning so that any subsequent analyses of urban structure can be related to them, and so that the range of modifiers of 'the market' can be made apparent. To this end a classification of planning interventions can be suggested, although the very word 'intervention' is pejorative assuming a 'natural' free market and antagonistic restrictions. These 'planning interventions' can be considered as ranging between the two poles of proaction and reaction which have been identified above.

8.2 THE TOWN PLANNING REVOLUTIONARIES

These make up the most extremely proactive as well as the most theoretical group. Their argument is that throughout the history of the town in the West, the characteristic has been either planning by absolute rulers to demonstrate their power, extend their control and increase their wealth, or, and this aspect is the more immediate, exploitation by capitalist entrepreneurs for their own ends of accumulation, private profit and economic dominance. The main practitioners in the field of urban geography have been Bunge (1971) and Harvey (1973), although it has been the fashionable stance adopted by the majority. Thus Harvey concluded *Social justice and the City* by stating:

> the old structure of industrial capitalism, once such a force for revolutionary change in society, now appears as a stumbling block...cities – 'those workshops of civilization' – are founded upon the exploitation of the many by the few. An urbanism founded upon exploitation is a legacy of history. A genuinely humanizing urbanism has yet to be brought into being. It remains for revolutionary theory to chart the path from an urbanism based in exploitation to an urbanism appropriate to the human species. And it remains for revolutionary practice to accomplish such a transformation.
>
> (Harvey, 1973: 314)

The real problems with writings of this order is that they never become translated into reality

– using Harvey's words, 'it remains'. Presumably the exemplification to which one would look must be, or must have been, the cities of communist regimes. But they have hardly been celebrated as typifying the good urban life, or even the egalitarian urban life. Most studies have been more concerned with explaining why that should be so than in presenting the cities of communist states as exemplars. Indeed, more characteristic of the extreme left are quite violent anti-urban attitudes. Moreover, in the 1990s communism seems to be in total disarray, and largely because of a failure to deliver the quality of urban life expected. Specific features of the socialist city will be considered later (Chapter 17) and all that needs to be noted here is the failure of the revolutionary reaction to become realized, other than as one representative of a whole range of pressure groups seeking modification rather than revolution.

It is possible within the West to identify groups who do adopt revolutionary attitudes and lifestyles, although unconventional might be a better description than revolutionary. These are the 'New Age' communities who first appeared as drop out 'hippy' or 'flower children' in the USA in the 1960s. As Ravetz comments, 'the relative ease of going on social security...made it possible for thousands of young, educated people to survive at the expense of the taxpayer while they themselves might live rent-free in squatted houses.' (Ravetz, 1980: 329). But Ravetz identifies clear positive elements too. 'In other vital respects the new life-styles challenged conventional values: for instance in their refusal to distinguish work and leisure, and their consequent repossession of time, which was no longer divided into time "sold" to other people for money, and time retained for oneself for leisure' (Ravetz, 1980: 331). And as she goes on to point out, part of the environmental contribution was to observe the 'reinhabitation of places'. 'This was not so much a question of being permanently rooted in a locality, for as the people concerned were mainly youthful their stay in any place was often temporary; but it represented a commit-ment to place that was quite contrary to "cosmopolitan" and mobile middle-class life' (Ravetz, 1980: 331). This comes very close to implying an anti-urbanism which has been so traditionally a part of radical Western philosophy. Certainly, the 'New Age' travellers of Britain in the 1990s are more associated with the countryside than the town. And it is difficult to envisage 'new age' communities other than as living on the infrastructures and the social provisions of the capitalist city; they can possibly be innovative and creative in country communes, but in the city they are wholly parasitic.

There is, at the end, a conclusion which the revolutionaries refuse to face, one endemic in the failure of revolutionary theory to chart the path which Harvey sought in 1973. The very core of urbanism is based on spatial advantage and competition; a hierarchy is the product of competition, the very word hierarchy is an opposite of equality. As long as there is the friction of distance, places, and places within places, cannot offer the equality which is at the heart of the revolutionary message. Communities, acting in a communal way, are another matter; but their specification in urban terms is moot.

8.3 THE URBAN UTOPIANS

It is significant in itself, even anomalous, that the name given to this group comes from an attempt to describe the ideal state, because the basic characteristic is not the overturning of the state systems of the West but, rather, enhancement by the implanting into them of idealized solutions to the urban problems which have arisen. It is, however, necessary to define the meaning of 'idealized' and of the ideal in urban terms. The 'citta ideale' of the Renaissance was the epitome of a ruler's wealth and dominance made manifest in an urban form. It was a piece of sculpture, a monument, rather than a provider of ideal living conditions for the inhabitants. But taking a view of planning from the perspective of the end of the twentieth century, Utopian schemes are more directly related to attempts to offset the worst conditions generated by the free market economy of the nineteenth century. The aim has not been to change the responsible politico-economic system, but rather to regenerate a debased form. The precursors were the more direct attempts to control building and housing standards. There were model villages, such as those of Robert Owen; and model industrial towns, such as that of J. S. Buckingham (Carter, 1983: 126–9). But the apogee must be the Garden City associated with the name of Ebenezer Howard; it was certainly the greatest in its world-wide impact (Fisherman, 1977).

There is little purpose in presenting Howard's scheme in detail since it has been discussed with extensive critiques in every volume on urban planning, for example in Purdom's book *The building of satellite towns* of 1925 and more recently and authoritatively in Peter Hall's *Cities of tomorrow* (1988). But there are two aspects which are directly relevant to the theme of this chapter: the nature and the consequences of planning interventions on urban development.

The first of these is the manner of presentation. Howard's basic theme was that both the countryside and the town had advantages and disadvantages in terms of the richness of living, or the quality of life. It was, however, he argued, possible to combine the advantages of both in the garden city. Hence the demonstration by the famous three magnets – town; country; town–country (Fig. 8.1). Quite explicit in the presentation is an acceptance of the socio-economic system as it was; there is no suggestion of overturning it. The garden city was the means by which the better characteristics of what already existed could be more effectively exploited. It was utopian rather than revolutionary.

The second aspect is the quite enormous influence which the garden city was to have on subsequent planning. It was the progenitor of the new town movement not only in Britain but throughout the Western world and beyond (Hall, 1988: 112–32). It was widely adopted, however modified in form, as the solution both for the internal restructuring of cities by decant-

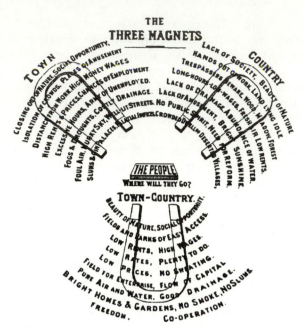

FIGURE 8.1 *The three magnets of attraction to the people as set out by Ebenezer Howard in 1898.*

ing population out, and for the extension of cities by planned satellite settlements.

But the major contribution to the nature of urbanism by the garden city movement was made not directly by garden cities as such but through the popularization of low-density residential developments. The saddest aspect of the movement was its degeneration into the garden suburb, an altogether different concept which lacked the urban quality of the garden city expunging all the complex of land-uses apart from housing. The consequence was a widespread type of suburban development – large, low-density estates of houses which represented the eventual debasement of a Utopian ideal. True, these estates can be looked upon in market terms as representing a switching of capital investment, at a time of over-accumulation, into a different circuit – 'the housing sector became a Keynesian "contra-cyclical" regulator for the accumulation process as a whole' (Harvey, 1978: 126–7). But whatever plays the primary role, both the planner and the market are crucial in the interpretation of form.

8.4 CREATIVE PRAGMATISTS IN TOWN PLANNING

With this group one is still dealing with the proactive, but there is a considerable reactive element. These are the planners who are concerned with the drawing up of large-scale national or regional plans, as well as structural plans at a lower level. They are clearly influenced by the Utopians and operate against a background of planning theory, such as it is. Their main tasks are pragmatically to solve identified problems. In many ways they have the greatest influence in directly determining the bases of urban development. In the consideration of this group the centre of interest shifts from any broad revolutionary or Utopian basis towards the more technical means of procedure. The most obvious examples are the demands made by the great metropolises where planning becomes a major element of national policy since extensive and vital regional areas are involved. London, Paris, Randstat in The Netherlands, the Ruhr, these are the prime examples in Europe (Hall, 1977; Jones, 1990).

Paris is the example cited by most analysts as the most successful case of determined planning. It also demonstrates the way in which direct political influences are made manifest in the city. After the Second World War, exercised by concern over the dominance of Paris as a primate city and by low demographic projections, the first reaction was to seek to limit extension. The 'Plan d'Aménagement et d'Organisation Générale de la Region Parisienne' (PADOG) of 1960 envisaged the limitation of Paris to the area already urbanized, thus invoking one of the oldest and least successful concepts of an urban ring fence. Internal restructuring, however, was included, with a new node at La Défense-Montesson and the shifting of the old central market – Les Halles – to a new suburban site at Rungis.

By 1965 it was clear that the plan had failed. One consequence had been rising land prices within the ring fence and another speculation at its margins. Again, the idea of decentralizing industry and commerce from Paris to the provinces had not been successful; Paris was far too strong an attraction. The reponse was the formulation of a major new plan – The Schema Directeur d'Aménagement et d'Urbanisme de la Region de Paris – set out in 1965 with a first revision in 1969. The Schema Directeur abandoned the idea of population stabilization and in its place accepted an increase of some 6 million, to a total of 14 million by the year 2000. Also abandoned was the notion of a restricting perimeter. The idea of imitating London with a surround of new towns and a defined green belt was considered but dismissed on the grounds that it was too artificial and difficult to enforce. The central theme adopted was to give structure to the dominant pattern of growth which was occurring and to use new towns for that purpose. Such a scheme would give coherence to development and would give unity to the whole region. At the same time transport costs would be minimized.

The structure proposed (Fig. 8.2) was based on two tangential axes to the north and south of the central city. Initially eight new towns were to give strength to the axes though they were subsequently reduced to five – Cergy-Pontoise; Evry; St Quentin-des-Yvelines (originally Trappes); Vallée de la Marne (originally Noisy le Grand and now the site of Euro Disneyland); and Melun-Senart (originally Tigery-Lieusaint). With successive revisions to 1990 the broad outlines of the scheme have been retained and implemented. In 1990, with greater sensitivity to green issues, open or green zones have been more clearly defined. In addition, the development of Charles de Gaulle airport at Roissy to the north had somewhat modified the northern axis pulling it northward and out of line. However, that is to be balanced to the south with a new 'technopole' at Sarclay-Palaiseau which is to be based on higher education establishments.

In relation to direct political impacts on the centre of the city one can do no better than quote Burtenshaw *et al.* (1991: 267):

Presidents in France have long patronised and supported those aspects of urban change in

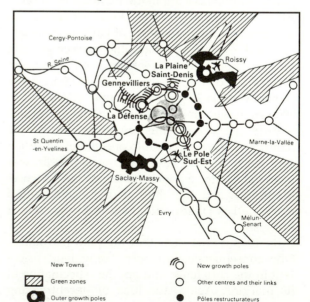

FIGURE 8.2 *Regional development plan for the Paris region. A: The scheme in the late 1960s with two dominant axes to the north and south of central Paris based upon new towns. B: Proposals 1990 (Livre Blanc) (after Burtenshaw et al., 1991). The concept of extending La Defense further to the west, balanced by a pole to the south-east. In the outer suburbs a new technopolis will be developed at Saclay-Paliseau and there will be further growth around the airport at Roissy. These proposals represent urban planning at its largest scale. Source: Fig. 8.2B is from D. Burtenshaw, M. Bateman and G. J. Ashworth 1991: The European city. A western perspective. London: David Fulton Publishers. Fig. 11.6, p. 268.*

Paris which have best reflected their vision of France and the ideology of the times.... De Gaulle (La Défense and Roissy Airport), Pompidou (Centre Pompidou) and Giscard d'Estaing (Forum Les Halles) have all made their mark on the townscape. Mitterand's impact has succeeded all these in that he has enabled the construction of several 'grands projets', ranging from the Pei Pyramid in the Louvre forecourt to the vast museum complex on the site of the old abattoirs at La Villette.

This brief review of regional planning about Paris has been undertaken not so much as an end in itself, for much has been omitted, but rather to demonstrate, in line with the purpose of this chapter, the extent to which not only urban planning has to be kept in mind, but even the remoter issues such as the ambitions of presidents. Perhaps the contrast with the age of absolute rulers is not as remote as might be thought. It is too simplistic to dismiss the self-esteem of political leaders from the realm of urban geography.

8.5 DEVELOPMENT ENGINEERS AND TOWN PLANNING

With this group the transition is made from the proactive to the reactive, although it can be argued that many of the operations of the civil engineers are derived from Utopian notions but are debased in application. The problems remain the same but the solutions are more pragmatic than idealistic. The design and building of large housing estates at low densities were the developers' response to the garden city ideal. But the best and most quoted example is the widespread use of the high-rise residential block as a solution to poor housing and overcrowding. Logically it was an eminently sensible solution and, it is worth commenting, one which has been mostly criticized from the experience of hindsight rather than during the period of construction. Subtopia (Nairn, 1955) had been extravagantly condemned both for its consumption of valuable land, with predictions that

countries such as Britain would be covered by concrete by the middle of the twenty-first century, and for the despised quality of its middle class lifestyle. Nothing gratified the arrogance of the intellectual more than criticism of suburbia. Moreover, the epitome of luxurious city living was to be found in skyscraper blocks, themselves the products of maximum exploitation of limited land.

The result of these positive forces, as well as the reaction to low-density extension, led to the widespread building of high-rise blocks, and not only in the West but also in the socialist countries where the tower block provided the most rapid and cheapest solution to the problem of housing large populations. Why they have proved so disastrous will be investigated at greater depth in a later chapter (Chapter 14), all that needs to be noted here is that their building was a clear response of architects and civil engineers to a very specific problem – how to house low-income populations rapidly and at high densities so as to conserve land.

The categories of planning and planners which have been put forward in this chapter are clearly not separate and this phenomenon of high-rise building has both its Utopian protagonists and its capitalist free market interpretations.

The idealist progenitor was the Swiss architect/planner who used the name Le Corbusier. He firmly believed that urban living could only be effectively brought about by concentration; sprawl was anathema. The city had to retain the qualities of medieval and post-medieval settlements, compactness and liveliness, with a vital street life. Le Corbusier's model was the monastery where the cell maximized privacy when it was required, but where there was also a dominant pattern of communal living and cooperation. He designed what he called a 'unité d'habitation' where in one large block all the necessities of urban living were present – shops, banks, restaurants, theatres. Time has not been kind to Le Corbusier, but it must be said in his defence, as of the garden suburb, that his ideals were corrupted by practitioners. High-rise blocks were built without the facilities and services which were an essential part of his scheme. Hall calls the consequences of his work 'at best questionable, at worst catastrophic' (Hall, 1988: 204). But it is the corruption of Le Corbusier's concepts which led to catastrophe, and that was due to the development engineers looking for quick and cheap solutions to the urgent problems which faced them.

As more detailed analysis will reveal (Chapter 14) it is also possible to argue that civil engineering firms moved to high-rise building in order to develop a new technology which was to be used later in more prestigious projects, and to practise mass production methods of industrial building to generate large profits. The market also had its role. As ever in urban geography no simple interpretation is wholly satisfactory; Utopianism, idealism, pragmatism and profitability, all have a part to play.

8.6 PRACTISING PRAGMATISTS OF TOWN PLANNING

With this group one comes to the core of the reactors – though hardly, even figuratively, in the nuclear sense! These are the planners employed by local governments to oversee the character of development and to ensure that it is in accord with whatever local, regional or national legislation or planning regulations are in place. It is, therefore, not surprising that the bulk of the work can be summarized in the procedure called in Britain 'development control'.

In most democracies there is a hierarchy of planning with national government setting out guidelines in relation to which local plans are produced. Ideally such plans are agreed after local consultation and then approved by central government. This means that even at this level there is proactive planning in the drawing up of local plans, but clearly initiatives are limited.

8.7 ADVANTAGE PRESERVERS IN TOWN PLANNING

In the sphere of the definition of constituency boundaries these advantage preservers would be called gerrymanderers; it is a term which could still be used. There is an erroneous idea, possibly implicit in this chapter, that planning is solely concerned with the preservation of a public interest as against the selfish operation of the free market. Such is not the case. Planning decisions are not made in the abstract by disinterested parties, but by competing interests. Hence planning regulations can be used to preserve or develop the advantage of minority groups who can gain access to power. One of the best examples is the use of zoning in the USA. F. B. Williams in *The law of city planning and zoning* (1922) derived zoning from the attempt to control the spread of Chinese laundries in California, first in the city of Modesto and then in San Francisco (Hall, 1988: 58). Then from 1909 onwards Los Angeles developed comprehensive zoning. Reyner Banham (1971) provides the key; 'Bending the zoning regulations is reckoned to be a bigger area of graft than the vice industry, and nowhere has it been more significant than in the preservation of white suburbia.' The most extensive, accessible study of the legislative decisions relating to zoning can be found in Johnston's *Residential segregation, the state and constitutional conflict in American urban areas* (1984). Essentially, the advantage preservers used zoning to define areas as properly limited to high quality, low density, expensive housing, thereby excluding lower income groups and especially blacks. Johnson quotes Popper in the introduction to his chapter on the challenges of zoning:

> The basic purpose of suburban zoning was to keep Them where They belonged – Out. If They had already gotten In, then its purpose was to confine Them to limited areas .... The advocates of exclusionary zoning justified it with euphemisms and technical jargon that sometimes even provoked [*sic*] protection of the environment ... zoning gave every promise of continuing to keep many suburbs closed to all but affluent, acceptable whites.
>
> (Popper, 1981: 54–5)

Johnson outlines the long and complex series of test cases on zoning, many of which have gone to the Supreme Court. But the outcome has been no clear rejection of the use of zoning for exclusionary purposes and Johnson concludes, 'zoning is a powerful tool for the protection of suburban interests' (1984: 104).

The manipulation of zoning regulations is a clear demonstration of the way in which laws relating to planning can be used to preserve advantages and so structure the city in a way favourable to groups in whom power is vested. It operates at all scales. In Britain the ability of local councillors to influence the allocation of houses on municipal estates can be a factor in the determination of social areas.

The simplest epitome of this group is the acronym NIMBY – 'Not in my back yard'.

8.8 THE CORRUPT IN TOWN PLANNING

It might be maintained that having considered the use of planning to preserve advantage, one has come close to the characteristics of the free market. But the advantage preservers work within the law, however it is bent. There is still one futher stage, though hardly justified as an aspect of planning. This is where operations are undertaken outside the law and corrupt practices are involved. The blatantly corrupt, as against the bending and manipulation of the law, is difficult to prove, though few would doubt its significance.

8.9 CONCLUSION

This very brief review of what has, somewhat grandly, been called 'planning the city' obviously cannot claim to be an account of planning principle and practice. That was not its intent. But it is crucial to stress the necessary corrective to those urban models, elaborated by urban geographers, which assume without question the operation of a free, and freely comprehended, market. Not only is that market subject to the constraints and distortions of planning, it is itself a myth, at least in the way it is sometimes interpreted as having some abstract legitimacy of its own. The so-called 'free market' is the creation of competing interests, both individual and corporate. All those interests, while primarily governed by money motives, are also subject to a whole range of political and social ends. Whatever democracy may be in theoretical terms, in the real world it is no more than a means of regulating and controlling selfish interest and reconciling a wide spectrum of concerns. Any outcome is most often an untidy consequence of the compromises made in the process. That is often greatly to the concern of those who are ruled by the absolute convictions of their own beliefs, but it is the stimulating heart of the geography of urban places.

9

TRANSPORT IN TOWNS

9.1 INTRODUCTION

There are two closely associated paradoxes which are immediately generated by an approach to transport in towns. The first is that the movement of people and goods is a consequence of the segregation of land-uses, but, at the same time, it is the ease of movement which creates and reinforces segregation. Thus, for example, the divorce of workplace and residence for shopkeepers in the nineteenth century was due to the increase in the possibility of daily movement, of commuting. On the other hand, the specialization of land-uses in the industrial city of necessity brought about movement including daily fluxes of population transference which were limited at first but gradually extended in reach. The second paradox is that cities are essentially agglomerations dependent on easy movement between the various parts, effectively they offer reductions in transport costs by association and nearness, a point already noted as crucial in Chapter 7. But the continued increase of that movement results in congestion and rising costs; it is no longer 'easy'.

To a large extent, all the problems of traffic in towns, however complex their resolution may be, are derived from those two simple

paradoxes. But there is still a third for although transport is concerned with movement, in itself it is a major consumer of static space, a fact often ignored in land-use studies.

> Above all it is profligate in its use of space, precisely where space is scarcest, both when in use and also for 98 per cent of its existence when it is stationary. Estimates of the numbers of cars parked for periods up to eight hours on working days in the central areas of cities such as Paris or London are around a million, and whether on- or off-street, legal or not, the result is an enormous consumption of the scarcest of urban resources, space. Most European cities devote between 10 and 20 per cent of their total land area to accommodating cars. In Cologne, for example, road and parking space took up twice the area of the city's parks and almost half as much as was devoted to buildings of all types.
>
> (Burtenshaw *et al.*, 1991: 97)

There is a further issue which, although it characterizes every aspect of urbanism and recurs throughout this book, is particularly acute in terms of transport. This is the extent to which movement in the city should be a matter of private choice or public provision, and hence the degree to which it is class based. The determination of that extent becomes all the more contentious since it impinges upon another area of interest, that of environmentalism, for it is contended that public transport is far less damaging in terms of exhaust emissions than private transport in the form of the motor car; not of course the bicycle (Tolley, 1993). Thus three partly competing interests come into play. The first is direct cost efficiency, including convenience, in moving goods and people. The second is social desirability in moving all the people in relation to the goods and services they desire. The third is environmental benefit in minimizing the impact of transport both on the physical, visible townscape and on the atmosphere locally and globally. It is not surprising, therefore, that urban transport is a contentious issue. Perhaps more than any other it is in the public eye, even if only in daily radio reports of problems.

Given the background which has been sketched then the starting point must be the demand from city dwellers, and from the transport industry, for free and unhindered movement, preferably door to door, by the most convenient medium. Clearly in that context compromises have to be made. But demand is the key element and that is broken down in most transport models into four (Pas, 1986: 60). The first is the basic reasons why people need or wish to move, usually termed trip generation. The second is the means or mode of transport used, usually called the modal split. The third is the destination to which the journeys are directed or trip distribution, while the fourth is the particular routes chosen, or traffic assignment. These four can be considered in turn.

9.2 ANALYSES OF MOVEMENT

Trip generation

Table 9.1 is a representation of urban movement broken down into standard categories. Similar figures are generally recorded. A survey by Burns in Newcastle upon Tyne in 1967 gave percentages of bus journeys with destination purposes of work at 22.8, business at 4.7, education at 3.7, shopping at 10.9, recreation at 15.4,

Table 9.1 *Trip purpose categories – Chicago 1976*

Purpose	Percentage of trips
Work	15.0
Shopping	13.1
Social and recreational	13.4
School	2.4
Business (Work)	3.2
Business (Personal)	6.9
Home	42.8
Other	3.2

Source: Chicago area transportation study of 1976.

other purposes at 4.2 and home at 38.3. An accord between bus journeys in Newcastle and all trips in Chicago indicates the universality of the recurring pattern. The category 'home', which dominates, occurs because purposes are identified for single trips and since many transport uses are from home and back, then it is logical that something approaching 50 per cent would be in that category. Again, in Britain the National Travel Survey in 1978/9 recorded that 21 per cent of all journey miles consisted of journeys to and from home (Whitelegg, 1985: 32). That presumably includes all travel, not just urban travel, as does the Department of Transport's 'Transport statistics'. The latter source in terms of journeys per person per annum records 153.7 for commuting, which was only outdone by social and recreational purposes at 178.3. Shopping recorded 144.9 and other personal business 146.5. In percentage terms the figures are 19.9, 23.1, 18.7 and 18.9 respectively. Given the quite large differences in the bases of these data, there is a good deal of accord between them, certainly in their ordering of categories.

These generalized data are the combined product of the space–time budgets of individual people and the stimuli which produce them. The notion of a space–time budget was popularized by Hagerstrand (1975). It concerns the relationship between the use of time and the use of

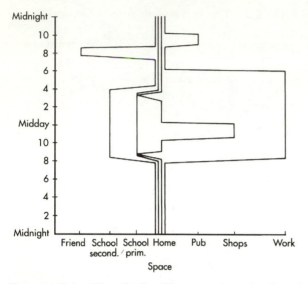

FIGURE 9.1 *Time budget for an urban family of four.*

space. A typical example is displayed in Fig. 9.1 for a family of four, parents and two children. It effectively brings to the fore the major urban problem which is derived from the patterning of trip generation the aggregate form of which is illustrated in Fig. 9.2. It is the concentration of movement into clear peak periods, one in the morning related largely to commuting journeys to work and to school, and one in the late afternoon/evening again linked to the same flows, with shoppers returning home being added, which is the problem. Figure 9.2(b) shows the movement in three German cities, broken down into six population groups, by activity profiles or the proportion away from home during a given period of the day (Kutter, 1973: 247). There are differences. Thus school children under 15 are generally less active and are more likely to return home at midday and are less likely to go out in the evening, while retired people tend to have a later peak of movement. But generally all contribute to the overall tendency to generate peaks of movement and it is these tidal surges which are responsible for the bulk of congestion. Thus the average traffic speed at morning peak in the centre of London in 1990 was 10.2 miles per hour, having declined

from 12.7 in 1968/70 (Dept. of Transport, 1993: 4.13).

There are essentially three ways in which the problems brought about by the unequal time distribution of journeys can be resolved:

1. The first is simply to alter the times of arriving at and leaving work. Such flexitime working is supposed to level out the graph of flows against time of day. While such schemes do help some individuals they have made little impact upon the aggregate pattern.
2. The second is to redistribute land-uses or, more directly as a consequence, to redistribute employment. This was very much the favoured approach after the Second World War. The creation of a circle of new towns around London was in part an attempt to reduce congestion by shifting centres of employment. Further, planning permissions were used, as in the Location of Offices Bureau, to inhibit the growth of some types of employment. As will be seen later in the chapter shifts from centre to periphery have now become universal.
3. The third is the placing of constraints on private transport, which is considered the least effective means of moving people about the city, and the provision and promotion of mass transit systems. Here, however, an economic problem intervenes. If mass transit systems have to provide accommodation adequate for effective movement at two limited peak periods, then for the rest of the day they are either underused or not used, with consequent costs to the operator. Any consideration of urban transport necessarily brings all aspects together and the comparison of the private car with other means of movement introduces the concept of the modal split.

Trip means, the modal split

Table 9.2 is derived from the US Bureau of Census, Current Population Reports which indicates characteristics of the travel to work for

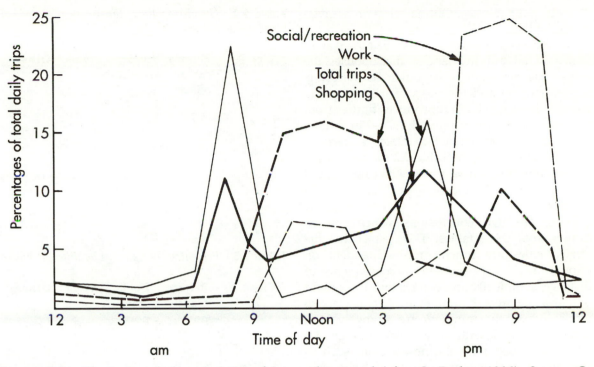

FIGURE 9.2A *The temporal characteristics of intra-urban travel (after G. Barber, 1986). Source: G. Barber 1986: Aggregate characteristics of urban travel, in S. Hanson ed. The geography of urban transport. New York, The Guilford Press. p. 77.*

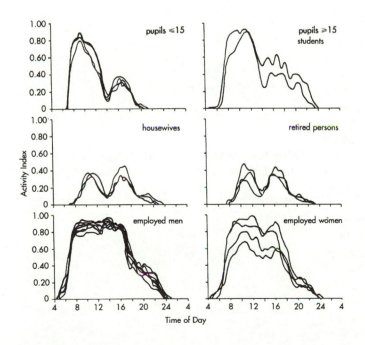

FIGURE 9.2B. *Activity profiles of six population groups in a West German town (after Kutter, 1973). Source: Kutter 1973: A model for individual travel behaviour. Urban Studies, **10**: 233–58.*

20 metropolitan areas. Only a selection is given in the table.

The figures in Table 9.2 emphasize the way in which in all but the largest metropolises the private car has come to dominate as the prime mode of travel and the basis of the journey to work. Whitelegg (1985) reproduces figures from the National Travel Survey of 1978–9 in Britain. They can be summarized to match as near as possible those given in Table 9.2 for the USA.

The 1992 data for those entering central London during the morning peak are given in Table 9.4.

There are two conclusions which can be drawn from these tables. The first is that the largest cities are much more dependent on public transport, presumably a consequence of distance, complexity and congestion. The second is that a far greater number in Britain walk to work, indeed the numbers are strangely high when compared with New York, though not when compared with the newer cities such as Houston. 'Smaller cities tend to have very limited public transportation systems with a transit share usually less than 5% while larger cities tend to have more developed public transit systems with possible transit shares in the 15–30% range; newer fast growing western cities such as Los Angeles and Denver have dominantly auto trips' (Barber, 1986: 79). The totally dominant use of the car in these newer cities is related to their greater adaptation to it because of their later growth; they are largely cities whose structure has been dictated by the car. But even in Britain the growing numbers owning a car bring pressure to bear on urban road systems. In 1951 there were 86 per cent without a car and only 1 per cent with two cars. By 1991 those figures had changed to only 32 per cent with no car and 19 per cent with two. Given these changes, and although a later chapter will be devoted to cities in the developing world, it is useful to make a comparison of the model split in non-Western cities (Table 9.5).

The bases of the surveys were so different that it is perhaps misleading to set them alongside each other. Even so Table 9.5 reveals interesting comparisons and contrasts with Western cities.

Table 9.2 *US metropolitan statistic areas – mode of travel of commuters (%)*

City	Private car or truck	Public transport	Walk	Other
New York	47	44	8	1
Baltimore	83	12	5	1
Houston	93	4	3	1
Denver	97	5	5	2

Table 9.3 *Percentage of journeys by different modes*

City	Private transport	Public transport	Walk	Other
Greater London	44	17	37	1
Birmingham	40	18	42	1
Manchester	38	17	44	1
Glasgow	22	26	49	3
All urban areas	44	13	40	22

Source: Whitelegg, 1985.

Table 9.4 *Mode of transport of those entering central London during the morning peak, 1992*

Mode	Percentage
Public transport	40
Underground	34
Bus	6
Coach	2
Private car	15
Cycle or motor cycle	2

Source: Department of Transport, 1993.

Table 9.5 *The modal split in three non-Western cities*

Mode	Shanghai	Karachi	Baghdad
Bus	35.8	63.2	47.3
Bicycle	13.2	20.1	—
Car	—	10.4	35.3
Taxi	—	—	17.4
Taxi rickshaw	—	5.5	—
Walk	43.2	—	—
Other	1.0	0.8	—
No travel	6.8	—	—

Source: From Heraty, 1991.

Notes

1. The Karachi data were derived from household survey hence the absence of walking as a mode.
2. The Bagdad data were derived from household survey and traffic counts. Again absence of walking is probably a result of the data collection where it was excluded.
3. In Karachi and Baghdad the figure for 'bus' is an amalgamation of a variety of bus modes, e.g. minibus, office bus.

Baghdad is clearly comparable with a significant use of the private car which is very limited in Karachi and totally absent in communist China. The 'bus plays a significant role simply because it is the cheapest form of mechanized transport and the most flexible. The bicycle, which is both cheaper and still more flexible, is clearly of considerable significance – one which many would wish to restore to Western cities (Tolley, 1993). Over all the 'modal split is ... likely to be a function of city size and density (affecting travel speeds) but is also affected by bicycle ownership levels, the availability of public transport services and the extension of these to serve new outlying residential and industrial areas' (Kneebone, 1991: 152).

One problem with the modal split is that all data are given by one category – that covering the greatest distance or taking up most time – whereas in reality two legs to a journey are more common. Thus a walk to a railway station covers two modes, although conventionally only the longer is returned. Likewise, 'park and ride' schemes are omitted, as are the so-called 'kiss and ride' where one partner drives the other to rail or bus station and so retains the use of the car for the day. These schemes, of course, hide a greater use of the private car than tabulations by the longest leg of a journey imply. And it is, above all, the ever growing use of the private car which has generated much of the congestion in urban areas. It is to solve that problem that much transport planning is directed mainly by the provision of mass transit systems. This introduces the issue of traffic assignment, but before that is discussed it is necessary to consider traffic destinations since the assumption so far is that movement is directed towards the city core.

Trip distribution, the destination of movement

The whole of the discussion to this point has referred to journeys to work or to shop with the implication that these are all movements from suburbs to city centre. But destinations are much more diverse. Whitelegg (1985) identifies four types of commuting trips: from outside the urban core into the Central Business District (CBD); from the CBD to the suburbs or reverse commuting; from one suburb to another or cross-suburb commuting; from outside the metropolitan area to either the suburbs or the CBD or long distance commuting. Using data from the late 1960s he demonstrates that whereas the first two categories have declined, the others have considerably increased as a function of the changing distribution of employment which will be considered in later chapters. Some specific figures from the United States (Table 9.6) will demonstrate that in consequence the notion of the dominance of flows from suburbs to city centre is simply not true.

There is some variation in the naming of types of commuting but Hanson's set seems more complete. She quotes Plane 1981 to emphasize how wrong is the idea that commuting patterns are dominated by suburb to central city flows and how complex the actual patterns are. They are made even more complex by the uninhibited

Table 9.6 *Commuting flows by type for selected urban fields 1970*

Flow type	Per cent of all flows	
	Worcester (Mass.)	Median for 28 New England urban fields
Within central city	4.5	4.8
Suburb to city (inward)	2.7	3.4
City to suburb (reverse)	0.8	1.7
Suburb to suburb (lateral)	60.5	61.4
Between urban fields (cross)	31.5	25.4

Source: After Hanson, 1986.

use of shopping malls and out-of-town shopping centres, as well as specialized shops and centres. All this is a consequence of the decentralization of employment and retailing. But the car can respond much more easily and rapidly to change than public transport and hence its greater and increasing use. In the car the full potential of the city for any purpose – employment, shopping, entertainment – can be more easily and immediately exploited. But this also contributes greatly to congestion, especially as in the exploitation of total city space, expressways designed for through-city movement become used over short stretches by suburb-to-suburb movement.

Trip assignment or route and modal choice

In every large city on every day there are broadcast messages or flashes dealing with congestion on particular roads or with delays caused by accidents. Their purpose is to warn and presumably to encourage the selection of an alternative route or mode of transport. They are directed at influencing trip assignment. The basis of assignment eventually becomes a behavioural study

which at one time in the late 1960s became popular with urban geographers who rode as passengers in cars in order to determine how decisions were made as to the route taken. The general conclusion was usually that habit was the dominant control and having established a route then familiarity took over and it was adhered to. Apart from habit, the logical control is time taken, for travellers will choose the quickest route between two points and that in turn will depend upon traffic volumes.

Assignment, however, introduces a further point in the choice of travel mode, and that is its affordability. The first constraint upon any journey in the city is its cost and for the deprived and the very elderly or unfit that immediately rules out the car. The dominance of the car and the decline of public transport is often seen in the context of the isolation of the rural poor, but it is equally applicable to the urban poor as well. The characteristics of the contemporary city, such as out-of-town shopping centres and the elimination of the small-scale providers of services, are generated by accessibility and if the capacity to move about the city easily is diminished then it is the deprived who suffer a diminution in their quality of life. Even further, where only one car is owned and it is used by one partner for commuting, the other partner can be left isolated. Trip assignment, therefore, involves a great deal more than the construction of models to be used in the prediction of traffic densities. It is a central issue of social concern.

Out of the discussion of the characteristics of urban transport two issues arise for further consideration. The first is the array of solutions which have been offered to relieve congestion and the second is the impact of transport problems and facilities on the post-modern city.

9.3 URBAN TRAFFIC PROBLEMS AND SOLUTIONS

The problems generated by traffic in towns are so extensive and universal that no brief discussion can do them justice; a library would be needed rather than a few paragraphs. Two aspects can be considered however. The first is the provision and promotion of mass transit systems so that as much movement as possible by private cars is removed. The second is the discouragement or even the forbidding of the use of the car in the most congested central areas.

Bus and rail systems are the obvious components of mass transit. The bus is the cheaper and the more flexible way but even with designated bus lanes contributes to road congestion and has a lower capacity. Rail travel is much the more expensive, but offers higher capacity at greater speeds. The London 'Underground', with the Metropolitan Line begun in 1863, was one of the earliest attempts to develop a metropolitan mass transport system and by putting the track below the surface avoided the need for disturbance at street level. Since then many cities have adopted similar schemes but usually by surface rail which is cheaper, or by combinations of surface, underground and elevated track. They are now usually identified by an acronym – DART (Dublin Area Regional Transport) and MARTA (Atlanta's Metropolitan Area Regional Transportation Authority's rapid rail system) are examples. Perhaps the best known and the first to be funded in the USA by the federal Urban Mass Transit Authority is the San Francisco Bay Area Rapid Transit system or BART (Fig. 9.3). It began operations in 1972 and was completed in 1974, with a track length of 71 miles. Unfortunately it did not achieve times for equivalent journeys better than the car; 'operation problems affected service frequency and reliability, and although BART improved transit times to the major employment centres, auto travel to these cities remained on average 35 per cent shorter than transit' (Guiliano, 1986: 214). The share of total movement of such transit systems varies greatly. Thus in 1980 it accounted for 60.2 per cent of CBD-bound work trips in Boston and 58.2 per cent in New York, but only 13.7 per cent in San Diego and a mere 3.4 per cent in Anaheim – Santa Ana – Garden Grove (Fielding, 1986: 232). The best known of the British schemes is the Tyne and Wear Metro

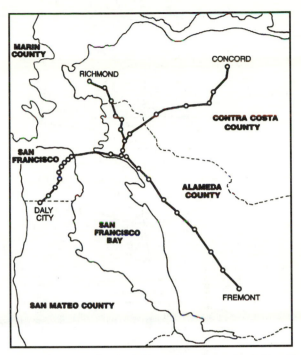

FIGURE 9.3 *The San Francisco Bay Area Rapid Transport System (BART). Source: G. Guiliano 1986: Land use impacts of transportation investment: highway and transit. As 9.2a: p. 274.*

which was completed in 1984 and 'is Britain's only example of a fully integrated public transport system' (Whitelegg, 1985: 28) for bus and British rail services are linked in.

The crucial problem of such schemes is their costs both in capital terms – the Tyne and Wear Metro at £5 million per kilometre 'has been one of the cheapest ... in the world to build' (Whitelegg, 1985: 28) – but also in running costs. For most the problem of the twin daily peaks remains. Even so, 'after more than 50 years of little or no investment, no fewer than eleven west European cities have built new underground systems in the last two decades' (Burtenshaw *et al.*, 1991: 109) and of the fifteen pre-existing systems many have been extended, including London's. In spite of such development, however, because of the cost, metro type solutions are only feasible in the largest and most prosperous cities and are unlikely to be possible

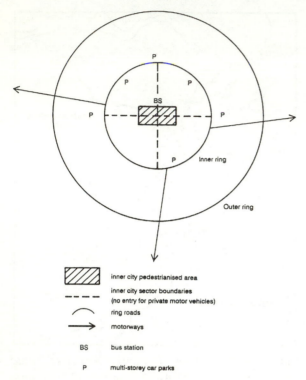

inner city pedestrianised area

inner city sector boundaries
(no entry for private motor vehicles)

ring roads

motorways

BS   bus station

P   multi-storey car parks

FIGURE 9.4   *Groningen: traffic planning scheme. Source: D. Burtenshaw, M. Bateman and G. J. Asworth, 1991: The European city. A western perspective. David Fulton Publishers. p. 112. Fig. 5.6.*

in smaller cities, and especially in the larger cities of the developing world where buses operating along exclusive busways are the most likely solution (Armstrong-Wright, 1991: 23). Thus in Porto Alegre in Brazil some 30 kilometres of median lanes were built in the early 1980s along five radial roads. The costs were just over US $1 million per kilometre (Lindau, 1991: 214). However, there are six metro systems in Latin America: in Mexico City (118 km); Sao Paulo (285 km); Rio de Janeiro (26.9 km); Buenos Aires (35.4 km); Santiago and Caracas (11.7 km). The number carried per day ranges from 4.1 million in Mexico City to 0.35 million in Caracas (Figueroa and Henry, 1991: 234).

The second approach to problems of congestion is to prohibit or limit the use of the private car, and also the lorry which has not been mentioned in this chapter although juggernauts of 36 tonnes cause major difficulties in city centres. The earliest and simplest measure was exclusion by means of the creation of pedestrian precincts, or pedestrianization as the process was known. Such precincts now characterize nearly every city in the Western world, though varying greatly in scale with Mainz at 610 hectares 'as a threshold beyond which there would be an unacceptable loss of accessibility and thus commercial viability' (Burkenshaw *et al.*, 1991: 106). In Western Europe these types of schemes have been extended to the extreme exclusion of the private car from town centres completely, as for example at Groningen (Fig. 9.4). If exclusion is not politically acceptable then the alternatives are first to charge for parking at a true economic level (meters were introduced into London in 1958) and then to devise a scheme for charging for entry into the central area. Meanwhile, park and ride schemes also make their contribution.

The other long-standing solution to keep the car, and especially the lorry, out of the centre is to divert through traffic by means of bypasses. Again these range in scale from a mile or two around a small town to the major circular roads with their huge planning and construction costs, such as the Boulevard Peripherique about Paris and the M25 about London. The problem of such schemes is that they pay the price of their own success by attracting intra-city movement, as well as through-city travel, and by generating additional traffic lead once more to congestion.

This has been but a brief survey only to indicate some of the attempts to solve the problem of providing easy, unhindered and rapid movement within cities. That they have had limited success is clear by the fact that such movement remains unachieved in the contemporary city.

9.4 TRANSPORT AND LAND-USE

The growth of urban expressways and of beltways, or 'circumferential limited-access highways' to give them their formal definition in jargon, has had a major impact upon city development. To some extent this has been considered in Chapter 7 with reference to 'the galactic city' where the composing elements were seen to 'float in space' rather than have a structured relationship with the historic city. If that is the case, then the enabling basis is the expressway and the creation of what Sudjic (1993) calls 'edge cities'. The nature of these developments about Washington DC along the Washington beltway has been noted (Chapter 7), but Sudjic's description is worth quoting at some considerable length since it so admirably epitomizes the nature of these developments:

Tyson's Corner, the fastest growing office concentration on the East Coast, represents the new edge city at its starkest. Twenty years ago it was still a tract of undeveloped Virginia countryside. Half an hour's drive from Washington's Dulles airport, through thickly wooded forests where you can still smell the leaves and hear the birds as the sun goes down, I-7 suddenly jolts you into another world. There is a high-rise hotel with a conical pointed roof protruding through the trees. Then you see a Nieman Marcus store, housed in a brick palazzo – windowless, but four storeys high, beautified with the liberal addition of pastel stripes, the most universal symbol of the area. This is a place which grew in parcels of little private estates, their individual presence marked by vestigal gates bearing names such as Tysons 2. From the humblest beginnings – distribution depots, start-up software companies, professional buildings to house lawyers and dentists – the scale has ballooned. Suddenly here in the middle of nowhere is Bloomingdales and Taco Bell, and Holiday Inn, next to Woolworths and Nordstrom.... Philip Johnson was not too proud to turn his attention to a thirty storey office tower here. In the context of Tyson's Corner, Johnson and Burgess's banal design, with its detached columns rising the whole height of the building and its venetian window motif, provides some sort of identity to a placeless address that is otherwise defined only by work and mobility.

(Sudjic, 1993: 121–2)

The last word of the quotation is the significant one – 'mobility', derived from the transport system. If the title of Sudjic's book *The 100 mile city* is properly descriptive of the post-modern metropolis, with its spatial diffuseness and its scattering of elements 'floating' in space, then its enabler, if not neccessarily its creator, is the transport system. Sudjic goes on to quote other examples, 'Las Collinas, once a ranch near Dallas (Fort Worth Airport) and now boasting a collection of towers that rivals that of the self-styled Metropolex itself, is another, more premeditated version of the same phenomenon' (Sudjic, 1993: 128). Southfield Office Park in suburban Detroit has more office space than the city itself. Stockley Park is a London example. It is also significant in transport terms that a following chapter in Sudjic's book is called 'The airport as city square' which develops a discussion of the airport as workplace and meeting place; Stockley Park is adjacent to Heathrow, conference centre and shopping mall. At Frankfurt 'the terminal area is organised like a small city, with more than a hundred shops selling everything from mink coats to hi-fis, twenty six restaurants, three cinemas, a chapel which offers wedding ceremonies, a medical centre with five doctors equipped to handle child birth and heart attacks, and a battery of resident social workers' (Sudjic, 1993: 151). There is a proposal to develop a major shopping centre complex at Gatwick. Here then is another piece of floating space which is now one of the prime determinants in the spatial organization of the post-modern city and with links to the city centre which are also crucial in patterning metropolitan structure.

This chapter must, however, close with two points of reservation. It is notable that a US Department of Transportation study of the impact of belt ways on land-use came to the conclusion that there was no consistently significant impact. Developments that did occur were linked to high income residential areas, available land for development and favourable land zoning policies. These universal factors rather than the specific belt way itself generated development (Guiliano, 1986: 270).

The second point is that most writers on urbanization live in the great metropolises where, for example, the most prestigious universities are located and house most commentators. Small towns where a good proportion of the people across the world live are consistently ignored. There is the danger, therefore, that some writing on urban transport is obsessed with the spectacular and the grandiose. Though perhaps, it is also true that as procedures filter down the urban hierarchy, the metropolises are the models for the future of all urban settlement.

URBAN LOCATION OF ECONOMIC ACTIVITY: CBD

10.1 INTRODUCTION

Once the general patterning of the city has been established, be it zonal, sectoral or multiple nuclei, then there follows the need to consider the components of that patterning in some further detail, bearing in mind the interaction of both market forces and planning. There are two ways in which that can be accomplished, The first is area based, being concerned with identification and analysis of the distinctive parts which make up the city, a kind of urban regional geography. The second is user based and concentrates on the different competitors for urban land across the whole city, such as shops, offices or residences. In general, and perhaps more and more in recent times, geographical analyses have tended to concentrate on users, practitioners regarding themselves as specialists, say in the location of offices, rather than as urban geographers *per se*. But the one exception, and even then only to a degree, has been the central area which is still discussed as a distinctive region of the city with a complex web of associated and interwoven land-uses (Murphy, 1972; Sim, 1982). Some of the difficulty of treatment is epitomized in the names which are variously used – central area or city centre, central business district (CBD) from the Burgess model, and shopping centre, although the last has acquired quite a different connotation. The terms are, of course, in part a reflection of size. It is possible to envisage the shopping centre of a small town, but it is meaningless when applied to a metropolis. Size is, therefore, a significant factor, for with its increase the city centre becomes much more complex in the range, distinctiveness and association of land-uses. In all this an assumption is made as to what constitutes this central part of the city and to this attention must be directed.

10.2 DEFINITION: PROBLEMS AND CONSEQUENCES

Studies of the city centre, or CBD, did not play a significant part in the initial development of modern urban geography which, at least in the early stages, was more concerned with central place theory. When such studies did emerge they were directed not at evolution or process but somewhat obsessively at problems of definition. A quite elaborate literature arose (Carter, 1972) concerned with providing a standard means of putting a boundary around the CBD. Since, as always, a single line limit is less than satisfactory, more often than not two boundaries were identified and the CBD presented as a core with a surrounding frame (Horwood and Boyce, 1959). The most elaborate procedures for definition were developed by Murphy and Vance (1954a and b) based on a series of indices which related the area of arbitrarily defined central business district uses, such as retailing, to the area taken up by those regarded as non-CBD, such as residence. These are shown for Worcester, Mass., in Fig. 10.1 where the Central Business Height Index (CBHI) is derived from Total Central Business Use Space divided by Total Ground Floor Space, and the Central Business Intensity Index (CBII) is derived from Central Business Floor Space divided by Total Floor Space and converted to a percentage by multiplying by one hundred. The critical defining line is the CBI or Central Business Index which demands a CBHI of 1 and a CBII of 50 per cent. However, a series of operational rules are needed in order to arrive at a single line boundary.

Although an arbitrarily defined boundary, based on an arbitrary distinction between CBD land-uses and non-CBD land-uses, was never very satisfactory, studies using the criteria devised by Murphy and Vance became popular in the late 1950s and the 1960s. But they ran into the ground and out of fashion largely because nothing was done with them and as an end in themselves they lacked any point or purpose.

One of the few attempts to carry the study forward was that by Hartenstein and Staack in 1967. Six German cities (Fig. 10.2) were studied and by using Murphy and Vance's methods a 'hard core' and 'fringe' were identified. Data relating to these methods were then used for a series of analyses, from which generalizations were derived, such as: the larger the total floor

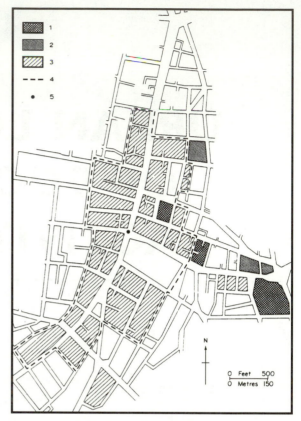

FIGURE 10.1 *The definition of the Central Businss District (CBD): Worcester, Mass. (after R. Murphy and J. E. Vance, 1954a). 1: Central business height index (CBHI) of 1 or more; 2: Central business intensity index (CBII) of 50 or more; 3: CBHI of 1 or more and CBII of 50 or more; 4: CBD boundary 5: Peak land value intersection.*

area of the urban core, the larger the floor area for all uses; with growing size of the core area, the floor area for retail use did not grow at a parallel rate – probably due to the fact that larger cities had decentralized their retail functions; with growing size of the core area, the floor area in office use grew at a higher rate – probably due to the fact that larger cities fulfil a number of regional and international functions which smaller cities do not.

Many of these conclusions were not very startling or unexpected, but they are important, for if analysis is to be based on the locational decisions of individual firms then this study showed that they are not simply concerned with making decisions in the abstract in relation to location. The different combinations of users in cities of different size must inevitably be part of the decision-making process. Undoubtedly Hartenstein and Staack made good use of a uniform method of definition, but even so their conclusions were linked to a number of fairly obvious statements about the relative density of uses in the core and the rate of diminution of those uses away from the centre. To a large extent in other studies the fixing of a boundary became an end to itself, devoid of purpose and hence academically barren. In some ways the search for regional boundaries in the city centre followed the earlier search for regional boundaries in geography and in doing so fell into all the same traps.

Murphy and Vance in their CBD studies were aware of this problem and seem to have realized that definition for its own sake was hardly a stimulating end product of their research. A consideration of the internal structure of the CBD (Murphy *et al.*, 1955) revealed its dynamic and changing nature, for the edges were seen to be either advancing or contracting, and zones of assimilation and of discard were identified and related to surrounding areas. This is an appropriate development, for, as assimilation and discard are considered, attention is directed towards process and away from definition. But these are still only a part of the whole complex of processes operative in the city centre and determining land-use. Inevitably one is led towards a more realistic evaluation of the central area not being made up of a CBD (with or without a hard core) but of a number of closely associated areas or sub-cores, constantly subject to pressures and, in consequence, with changing boundaries. These areas have emerged and have crystallized out in the long historical process of the town's growth.

This was implicit, although not fully developed, in the core and frame organization (Fig. 10.3) proposed by Horwood and Boyce where a series of characterizing uses such as wholesal-ing, transportation terminals and auto-sales and services are seen to make up the frame.

One of the more elaborate attempts to demonstrate the complexity of the city centre was that by D. H. Davies (1965) in his study of Cape Town. He attempted to identify distinctive areas by means of cluster analysis. This is carried out for each category of use by using floor space per lot (the basic areal unit). Each lot with a particular use is plotted and the centre of gravity is found by the standard method of drawing two axes to contain the distribution and by finding the mean of all the distances from each lot to each axis (Fig. 10.4). Davies weighted the distance values by amount of floor space in each use by multiplying each measurement by the floor space devoted to the use in each lot and then dividing the totals by the total floor space in that use, so that the weighted centre of gravity takes areal extent into account as well as locational pattern.

The conventional method of establishing the cluster boundary is to draw a circle of radius equal to the mean of all the distances from the individual lots to the centre of gravity, or in more sophisticated studies to draw in standard deviational ellipses. This was rejected on account of its arbitrary areal character. Instead, each lot distance was expressed as a percentage of the distance of all lots from the centre of gravity, and likewise the floor space in any use was expressed as a percentage of all the floor space in the use under consideration. These were then plotted against each other. In order to bring together strong representation, increasing distance was associated with decreasing size by plotting $1/area$. On this graph, the lots which were to be regarded as forming the cluster were identified by a factor of inclusion (F), where $F = A/D$ where A is percentage area and D percentage distance. A series of curves for F, expressed as values of $(1/A)/D$, were used to identify inclusion by totalling the percentage of floor areas of lots to the left of the curve (Fig. 10.4). Eventually the very high figure of 97 per cent was chosen to identify clusters and a line was drawn which encompassed all the lots identified as making up this proportion. The superimposition of these

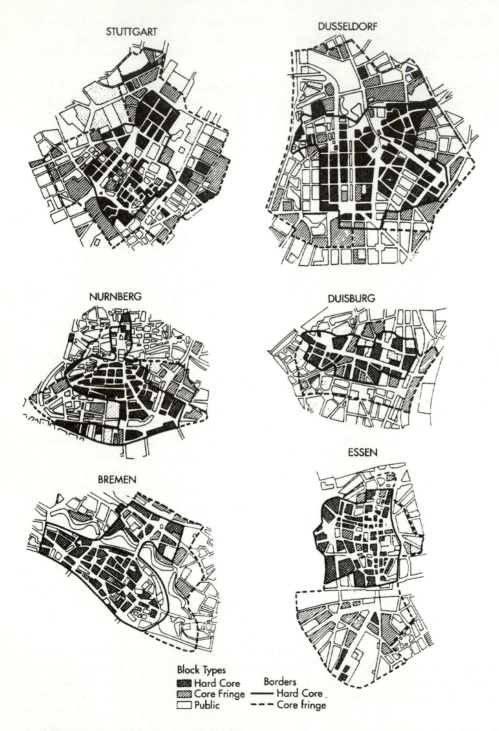

Block Types
- ▨ Hard Core
- ▧ Core Fringe
- ☐ Public

Borders
- —— Hard Core
- - - - Core fringe

FIGURE 10.2 *The hard core and the fringe area in six German cities (after W. Hartenstein and G. Staak, 1967). The figure illustrates the application of the core and fringe (or frame) concept to a series of large German cities in the 1960s.*

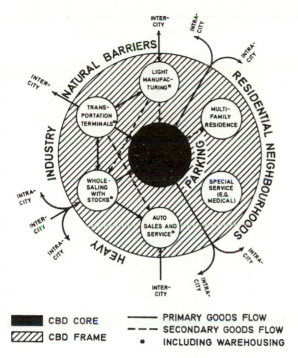

CBD CORE
CBD FRAME
PRIMARY GOODS FLOW
SECONDARY GOODS FLOW
* INCLUDING WAREHOUSING

FIGURE 10.3 *The core and frame interpretation of the CBD (after Horwood and Boyce, 1959). Source: E. Horwood and R. Boyce, 1959: Studies of the CBD and urban freeway development. Seattle: University of Washington Press.*

cluster boundaries produced a complex map of city centre 'regions' (Fig. 10.5). These were then analysed as part of the structural make-up of the city centre.

The conclusions which Davies drew were related to the complexity of the spatial patterning, which has to be set against the discrete CBD concept, and also the relation of this patterning to concentric zones, sectors and nuclei, a backward look to the generalizations of whole city structure which is not illuminating. Rightly, he commented, 'it is suggested that currently there may well be diminishing returns in any delimitation studies *per se*', but he let his study end rather inconclusively, neither following up the concept of linkages which he introduced, nor the historical emergence of the areas identified, that is, the way present linkages have come about; for he stated 'cause and effect are interwoven in a manner too

complex to unravel without historical investigation beyond the scope of the present study'.

Three conclusions emerge from a consideration of these studies by Hartenstein and Staack and by Davies:

1. The CBD, even during the 1960s, was the subject of quite considerable change as land-uses, considered as defining criteria by Murphy and Vance, began to move away. Subsequent developments, which were introduced in Chapter 7, have further modified the nature of the contemporary Western city centre.
2. The CBD is in no way a uniform central region of the city, rather it is made up of a complex of use areas, interdigitating it is true, but even so quite different in character. The larger the city then the greater the segmentation and the clarity of the segments. In essence, there is a complex of uses even in small cities but in the largest those uses become segregated into distinct areas.
3. Since points 1 and 2 are a consequence of change operating over time, the study of the evolution of the central area is a crucial background to the patterns of the present. (Carter and Rowley, 1966).

From these three conclusions it follows that some consideration of development is essential, to be followed by a review of the traditionally predominant use, retailing, and its present status in the city centre.

10.3 HISTORICAL PROCESS AND THE CBD

Historical investigation is best exemplified by Ward's study of the evolution of central Boston (Ward, 1966). Three maps (Figs 10.6(a), (b) and (c)) indicate the basic pattern of developments as successive specialized areas crystallized out as the city grew rapidly.

From an early nineteenth-century pattern where specialized business was restricted to a small section of the waterfront and the market

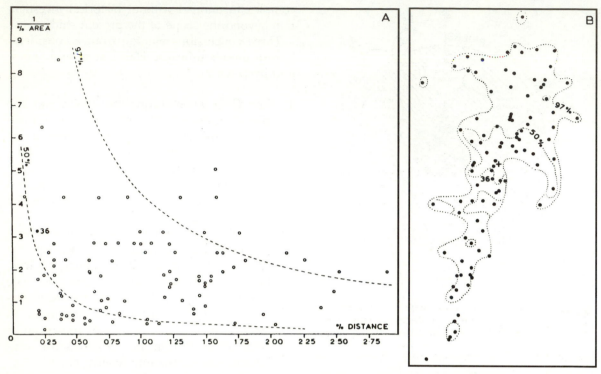

FIGURE 10.4 *Identification of use clusters in central Cape Town: household furniture sales (after D. H. Davies, 1965). The process of identification is explained in the text, p. 165–7. A: Inclusion process. B: Limits of cluster.*

halls, there was a progressive segmentation of specialized activities into financial, administrative, retail and wholesale commercial functions, occupying by the early twentieth century an area larger than the original town.

The first distinctive area to emerge was that devoted to financial and insurance services which were valued by the merchant community, and warehouse accommodation developed nearby. The development of 'mass consumption' and 'mass transit' resulted in the emergence of distinct retail and wholesale sections while manufacturing was displaced. At the same time financial and administrative functions were enlarged to form an extensive segment. Thus Ward depicts how, from a small nucleus, and related to changes brought about in the nineteenth century, the distinctive parts of Boston's CBD emerged. At no point in his illustrative maps is the CBD as such defined nor is

there any attempt at objective definition, indeed no direct evidence is produced at all to support the identification of areas or directions of growth. The prime concern is with process rather than with definition.

It is, however, possible to generalize the situation to indicate the way in which the composing elements of the central area crystallize out from the amorphous general central area of the pre-nineteenth-century city (Fig. 10.7)

In Fig. 10.7 which relates mainly to Britain, diagram A represents the situation early in the eighteenth century with a regionally undifferentiated kernel surrounded by a fringe belt, eventually the Inner Fringe Belt, IFB (see Chapter 19). By 1850 (diagram B) the segregation of central city land-uses had appeared with the emergence of specialized financial warehouse or wholesale districts (F and W on Fig. 10.7). These overlap the retail core which was

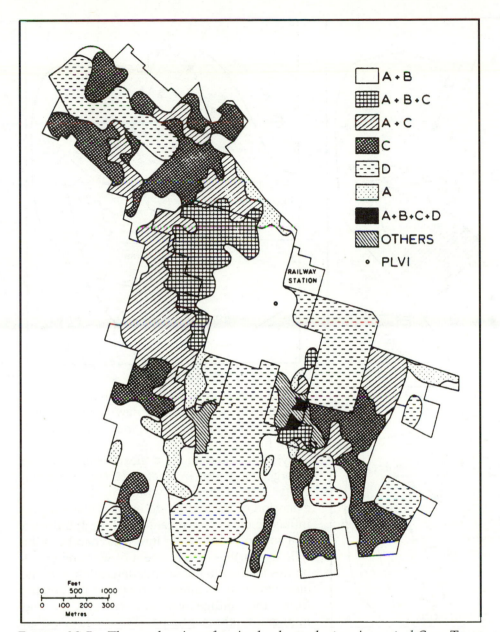

FIGURE 10.5   *The overlapping of major land-use clusters in central Cape Town (after D. H. Davies, 1965). A: retail uses; B: office uses; C: automobile; industrial wholesale and commercial storage uses; D: public and government uses. PLVI : peak land-value intersection.*

still in an 'incipient' form. The new industrial areas (I) had much in common with the warehouse district and there was considerable overlap. Industrial housing (Ih) of low quality was closely associated with the industrial region

itself and spread on to the estates of the IFB. The core still retained upper class housing (U) but, under the pressure of changes, a sectoral extension was underway. Vance would argue that the merchants still needing rapid access, especially

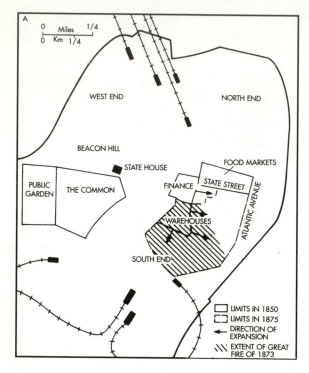

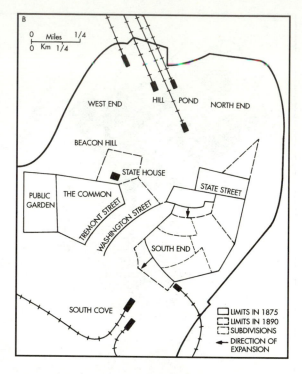

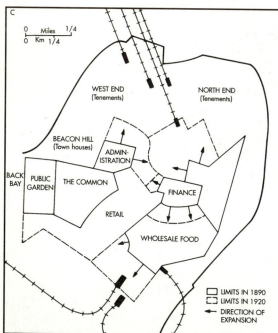

FIGURE 10.6 *The development of the central area of Boston Mass. from 1850 to 1920 (after D.Ward, 1966) A: 1850 to 1875; B: 1875 to 1890; C: 1890 to 1920.*

to the financial area, tended to remain at or near the centre often in imposing newly built squares, but the non-working élite were beginning to move further out.

By 1880 (diagram C) a clear retail core had emerged and also by that date the first chain shops were appearing within it. The sectoral extension of the well-to-do exerted a significant influence on the town centre pulling it outward in a zone of assimilation into what had been the IFB. The further development of working men's houses had led to the further deterioration of the oldest inner areas into which filtered the derelict from the indigenous society (WCS) and immigrant ethnic groups (E), in the case of Britain mainly the Irish. At the margins a middle class area had developed, deteriorating both zonally into the centre and sectorally towards the industrial areas. A Middle Fringe Belt (MFB) had come into being characterized by public utilities, including cemeteries (C), waterworks (W) and asylums (LA).

By the end of the nineteenth century the modern city, before Second World War destruction and renewal, had come into being (diagram

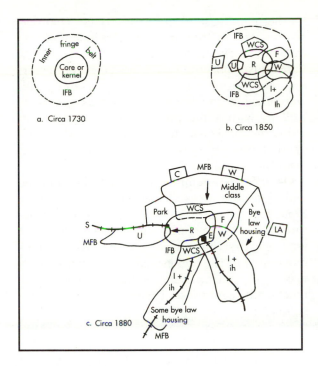

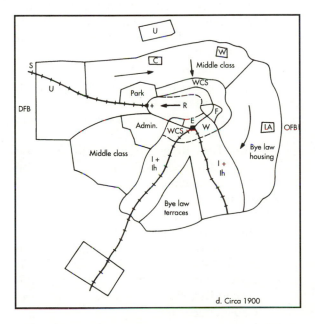

FIGURE 10.7 *Generalized development of the central area of an industrial town 1730 to 1900. For explanation of the abbreviations see text pages 170–171. WCS: Working Class slum. E: Immigrant ethnic area.*

D). The distinctive retail area continued to be dominated by chain and specialist stores and its continued migration towards the sector of upper class housing had created zones of assimilation (+) and discard (–). The remaining areas were largely as they had been twenty years earlier. The MFB had been assimilated by intermediate class housing in which the public utilities remained lodged, but an Outer Fringe Belt was coming into being.

This simple descriptive generalization of the internal structural changes in British cities during the nineteenth century indicates the way in which an approach via the evolutionary sequence can be made to the modern city centre. To it must be added post-war change which in many cases has left the old core isolated by the conversion of the surrounding zone by comprehensive redevelopment programmes into an inner ring road or urban motorway with new high-rise municipal housing closely related to it.

The trend towards high specialization, as well as the more general themes, can best be illustrated by a review at greater length of the evolution of central London.

There the separation of functions had begun as early as medieval times, for the City of London to the east was quite distinct from Westminster to the west. The former had an early history common to many of the medieval successors of the Roman towns of Britain; the latter was unique. The Westminster site was first used, ignoring fragmentary Roman evidence, as the location of an Abbey on what was Thorney, or Thorn Island, with the first reliable record dating from 970. Royal associations date from the eleventh century for the Abbey was rebuilt by Edward the Confessor and the royal residence was moved there from Winchester in the same century. This addition of the royal palace and the seat of 'national' government transformed the path of development. At a later date the Tudor monarchy added considerably to the area. In the early 1530s Henry VIII built St James's Palace as well as beginning the development of Whitehall.

From early medieval times, therefore, the functions of central London were located in two separate areas which even by the late sixteenth century, as Hogenberg's map of 1572 demonstrates, were only linked by the thin line of the Strand, the name indicative of its location. To the west lay the Court and the centre of national government. Its nature inevitably attracted prestige residence, spreading along the Strand at first, and all that was to become associated with it in the way of shopping and personal service and entertainment. These all became mutually promoting to produce the 'West End'. In contrast, to the east was the 'city', the base of the independent London burgesses and of the merchants, out of which was to grow the financial centre, the trading interests and the warehouses and industries of the river front and the constructed docklands, the 'East End'.

Even in the Middle Ages a further distinctive area had appeared. Already in the fourteenth century the Order of St John, which had succeeded the Templars on the Temple site in 1324, had leased it to students of law. Lincoln's Inn was also left as a residence for lawyers by the Earl of Lincoln. Thus, appropriately located at the divide between the Court and the City, the royal power and the merchant interest, a legal area was early apparent.

Those strong forces which were to create modern London are perhaps first apparent in the Tudor period. The increase in trade and the generation of wealth led to population growth and an extension which the monarchy tried vainly to contain. The dissolution of the monasteries made available large amounts of land in and around the city. Characteristic was the Convent Garden, given to Sir John Russell in 1553 and in 1660 translated as Covent Garden into the first British piazza, a significant symbol of the westward extension of prestige residence. It was followed by Soho Square in 1681.

Economic changes also began to operate. The trader or the dealer began to play a bigger role, especially related to overseas trade, and consequent increased pressure on city sites meant that land values rose. Craft industry was not able to meet the rising costs and was gradually displaced to the surrounding villages, especially Spitalfields and Clerkenwell, thus creating a northern and eastern industrial fringe. That was accentuated by the creation through maritime trade of a riverside area from the Tower to Limehouse concerned with the handling, repairing and provisioning of ships, and with warehousing and the storage of bulk goods, as well as with some processing of imports. An industrial and warehousing East End developed, based in part on industries 'extracted' from the city.

Oscar Spate attempted a reconstruction of the main land-use areas of London as they were in 1750. This is reproduced in Fig. 10.8. The structure reflects the early development together with the first impacts of the massive western movement of the highest class residences which was to be characteristic of the eighteenth and nineteenth centuries. There were three constituents:

1. East London
   (a) The City: a core of merchants and traders grouped around the Bank, the Exchange and the houses of the great companies, East India, South Seas and Hudson Bay.
   (b) Industrial areas, with three sections:
       (i)   Clerkenwell, characterized by metal working and jewellery.
       (ii) Spitalfields, characterized by silk weaving.
       (iii) The River Front, with a wide variety of activity with maritime connections.
2. West London
   (a) St James and Westminster: the seat of the Court and government.
   (b) Aristocratic residences, largely on the great estates which had been developed before 1750, including the Cavendish-Harley estate and the Grosvenor estate.
   (c) Shopping: the beginning of the extraction of retailing from the city, at this period mainly collected about Charing Cross and the Haymarket.
3. The Strand. This was the connecting link between west and east, but with its

Land-use areas in London in the eighteenth century (after Spate).
1. Aristocratic residential sector
2. Government offices
3. Middle class and professional residential areas
4. Amusement and vice area
5. Legal area
6. Industrial areas and artisan dwellings
7. Wharfs, warehouses, waterside trades, including labourers' dwellings
8. 'The City' – commerce and finance
9. Boundary of the City Liberties

10. West End shopping and hotel centres about the Haymarket and Charing Cross
The principal markets are shown
S – Smithfield (meat, hay)
L – Leadenhall (meat, provisions, leather)
G – Covent Garden
Q – Queenhithe (corn, meal, malt)
B – Billingsgate   R – Roomland (coal)
The boundaries are obviously approximate and the built-up area is about 1750.

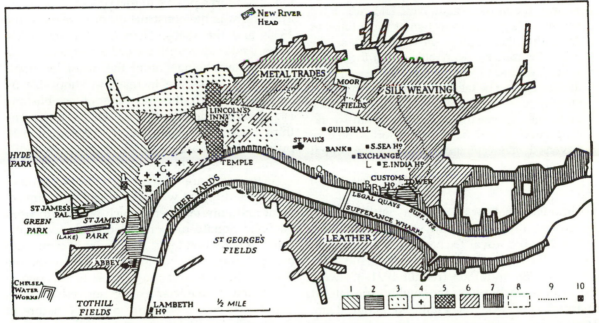

FIGURE 10.8   *Land-use areas in London in the eighteenth century (after O. H. K. Spate, 1948). Source: O. H. K. Spate, 1948: The growth of London* AD *1660–1800 in H. C. Darby ed. An historical geography of England before 1800. Cambridge: Cambridge University Press.*

own distinctive associations which were largely related to its intermediary location:

(a) The legal area, which as has been indicated had grown about the Temple and extended northward.

(b) Amusement and vice: this role is partly related to the older 'parade' areas such as Lincoln's Inn Fields which had fallen into disrepute as sites for what would now be termed 'pick-ups'; partly to the transformation of Covent Garden into a market; and partly to the development of the theatre area epitomized by Drury Lane, one of the minor alleys off the Strand.

(c) Shopping: again this was an intermediary area between the city and the developing new shopping areas of the west.

It was on to this situation that the massive growth of the nineteenth century was impressed, driving forward both eastern and western growth on their separate lines and accentuating the specialization of areas. Kellett in his study of the impact of railways on Victorian cities wrote:

Because they set in motion a whole chain of operations in manufacturing, warehousing, transporting and selling, and because they depended to such a marked extent, even in the days of the telegraph and efficient postal service, upon personal meetings, the dealers on the Exchange, and their associated offices and credit agencies, were able to bid for central locations against all comers. Land values in the City of London were six to eight times as high as even the most elegant West End residential address. They were even two or three times as great (for the smaller sites required) as the prices the railways were able to bid for central land. The core of the central business district, therefore, must be taken as immovable by direct railway pressure. Residential areas, historic buildings, graveyards, hospitals, craft workshops, even, where necessary, factories could be traversed or swept away, but not the central Exchange area.

(Kellett, 1969: 298–9)

Although that was written with specific relation to railways there are wider implications. In London, the Royal Exchange rebuilt between 1841 and 1844 and the Bank of England built between 1788 and 1808 (rebuilt 1921–37) were the critical anchors of the highly specialized financial core which crystallized about them. Trading in commodities was filtered out and characterized appropriately the area to the east more immediately in contact with the warehouse and industrial dockland which had been extensively developed after 1800 with the beginning of the West India Dock. That again is typical of the micro-specialization of the largest cities.

Between the mixed margins of the financial core about St Paul's and the legal area to the west, a further specialized section emerged deriving its name from an east–west street which crossed a small tributary of the Thames, the Fleet. The growth of a literate public and the increasing ease of national distribution, rapidly enlarged the printing and publishing business, while legal publishing, an even more specialized activity, had already occupied the intersection of the areas. To quote Kellett once more, 'Mercery

... was virtually chased into the West End by the expanding printing business; first the professional publishers of legal and other documents in Paternoster Row, then the newspapers of Fleet Street' (Kellett, 1969: 300).

Continued growth and extension completely transformed the western margins after about 1815 when modern retailing emerged. The leader was the department store. Shaw has identified three phases in this process (Fig. 10.9). The first was the construction of bazaars in the 1830s and the 1840s. There were ten in early Victorian London with Oxford Street, a new axis of extension, being one of the major locations. The second was the increase in shop size by amalgamation with the intention of establishing island sites which dominated whole blocks. The third phase after 1880 was the construction of purpose-built stores reaching its apogee with the opening of Selfridges in 1909.

As the retail area extended so, too, were other land-uses dragged west. Foremost were amusement and entertainment which became focused on Soho Square and included the area to the south where London's theatre-land came into being. Again, to the north the section around the British Museum and the University of London developed a character partly derived from its intellectual environs, partly from the nature of the theatre-land to the south. It is best summarized by the name of one square, Bloomsbury.

National administration pushed north from Westminster along Whitehall to meet the western extension at Charing Cross, thus establishing an extensive governmental and administrative complex in the area between the river and the royal parks.

As these various extensions took place, core areas of high specialization became clearly identifiable, although the interstitial areas were much less clearly characterized and great mixtures of uses and occupations occurred. The general picture at the end of the nineteenth century is summarized in Fig. 10.10. Indeed, this has remained the basic structure throughout the present century and is the key to contemporary spatial organization. But there have also been very considerable modifications. Oxford Street

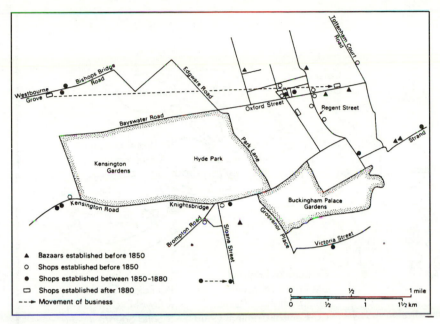

Bazaars established before 1850
○ Shops established before 1850
● Shops established between 1850–1880
□ Shops established after 1880
---→ Movement of business

FIGURE 10.9 *The development of department stores in London's West End (after G. Shaw and M. T. Wild, 1979). Source: G. Shaw and M. T. Wild, 1979: Retail patterns in the Victorian city. Transactions of the Institute of British Geographers, N. S. 4, pp. 278–91.*

has developed as the major 'popular' shopping axis, although not unaffected by more recent trends in retail location (see p. 184). Publishing in the last two decades of the twentieth century has moved significantly away from Fleet Street as a response on the one hand to the demands of new technology and on the other as an escape from the deeply entrenched restrictive practices which had become associated with an activity so long lodged in the same area. The move has been mainly to east London. There, too, perhaps the most significant developments have occurred as a response to the decline of the old docklands as maritime trade and industry have moved away. This type of change will be discussed later (p. 187). At this point two aspects of the central districts of large cities can be reiterated.

1. As city size increases so the notion of a central business district becomes inappropriate in the context of the complexity of specialized areas which come into being.

2. These highly specialized areas are most likely to be affected by short-term fluctuations in fashion. Today's high profile area is tomorrow's problem. The larger the city the greater the volatility of areal significance, unless there is a strong underpinning of established buildings. Carnaby Street with its 'swinging sixties' image which has long degenerated is a good London example. Again, Soho has long been associated with raffish entertainment and what urban geographers once euphemistically called 'vice'. By the last quarter of the twentieth century it had reached a low point of trashiness in all respects. But the 1990s have seen yet further change. Figure 10.11 is a journalist's map of Soho in 1993. Stuart Wavell writes, 'London's Soho, stigmatised for decades as the exclusive haunt of drunks, pimps and whores, has been transformed beyond recognition ... the area has emerged as the vibrant outpost of cafe society' (Wavell, 1993: 1–7). He continues

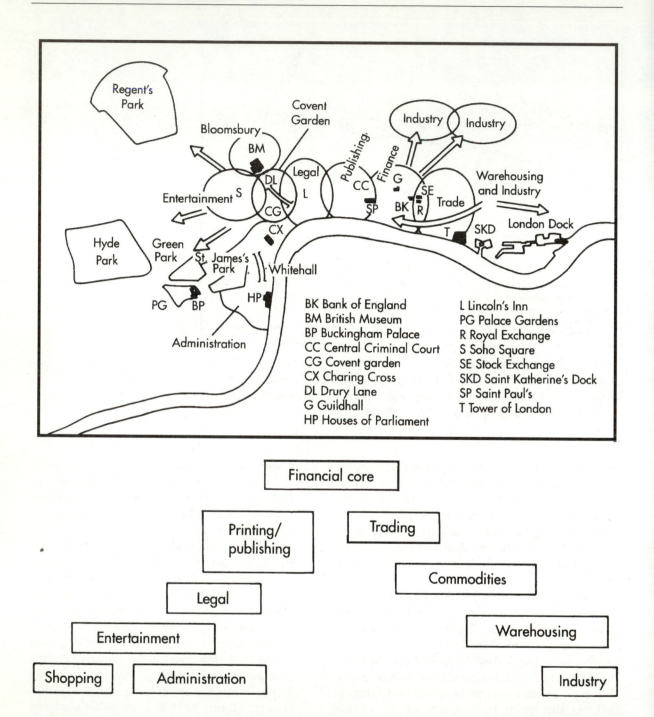

FIGURE 10.10 *A generalized diagram of the development of specialized areas in central London. A: An outline of areas shown in a most generalized fashion. There is a great deal of overlap of all the uses. B: An indication in diagrammatic forms of the way in which specialized areas progressively broke away from the commercial and financial core of 'the city', which itself developed internal specialisms (see Fig. 11.7).*

to note that the 'vice merchants' have not gone but 'they are tacky irrelevancies to most of the bright young things who are now flocking to the area'. An explanation of the change is linked to the growth of an even more specialized area. 'There is also the gay community, whose "pink pound" spending power was largely responsible for the initial resurgence and continues to underpin the mushrooming phenomenon. A popular theory is that once gay entrepreneurs had revived Old Compton Street as the central hub, with straight businesses prospering in their wake, the knock-on effect began to fan out along adjoining streets' (Wavell, 1993: 1–7). But whatever the reason, micro-specialization is characteristic of the central parts of great cities but it is also unstable in its nature by virtue of the strong influences at work in such cities.

At this point, after a review of the development of the largest central areas it is necessary to return to consider the predominant use in all central business districts – retailing.

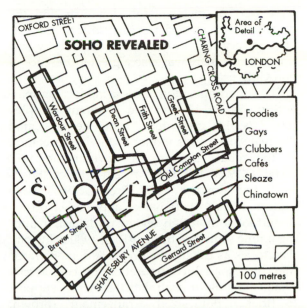

FIGURE 10.11 *Soho, London. The major uses of land in Soho. Source: S.Wavell: Soho is reborn as the centre of the universe. Sunday Times, 15 August 1993, 1–7.*

10.4 RETAILING IN THE CBD

It has already been made apparent that the central business district is a complex of uses, nearly every land-use that can be found in reality is found to some degree there. But distinctive single-use-dominated areas do emerge and the most significant, and to some extent, the defining use is retailing. The standard model of retailing within the city centre was developed by Ross Davies (1972a) some twenty years ago. The crux of the model was the assemblage of three models or interpretations which were commonly available. The first was Garner's simple model of the structure of a shopping centre which was based on the standard bid–rent curves of urban land-users (Garner, 1966, 1970). Thus the stores with the highest thresholds, those which are either large in size, or specialized in character, will demand the most accessible and hence expensive sites and, best able to compete in the bidding process,

will be located at the centre or peak land value.Those with the lowest thresholds, relatively small and unspecialized will be the least able to compete and will take up marginal locations. Between these there will be an intermediate group. But that simple structure fails to take into account the greatly complex configurations of shopping centres which have already been considered in Chapter 5, where Berry's basic division into centres, ribbons and specialized areas was illustrated (Fig. 5.4). The only observation that is needed here is that there is a contrast between the single high specialism which is centrally located, and the grouping of specialized outlets which benefit from association and need not be central. The third model which Davies introduced was that of the CBD divided into core and frame as identified by Horwood and Boyce (1959) which has again been noted (Fig. 10.3). This does little more than replicate the centre–periphery contrast which Garner had demonstrated, although the wider varying land-use types, apart from retailing, were part of their model, demonstrating the presence of lower order specialisms, such as car or auto-servicing in the frame; in Murphy and Vance terms there was also a greater presence of non-CBD uses.

The superimposition of these models is apparent in Fig. 10.12 and the resultant complex model is also presented. Davies demonstrated the relevance of his model to the interpretation of the retail structure of Coventry in 1972 (Davies, 1973) but more recent studies (Shepherd and Rowley, 1978; Brown, 1987) have indicated that it remains applicable. Brown undertook a thorough review of Belfast, including a time sequence from 1910 to 1980 using city directories as the data source. He demonstrated that 'the spatial configuration of commercial establishments in central Belfast lends support to [the] hypothesized arrangement' (Brown, 1987: 11), that is, that lower order establishments tend to be peripheral, higher order establishments central. But not only is this so for 1980, but for all the other dates from 1910 investigated. There seems, therefore, to be considerable spatial stability in the pattern.

The clustering of specialized activities is demonstrated by examining the spatial association of retail types. To some degree this would be expected regardless of distinctive locational expressions but, even so it reveals that the basis of specialized association exists. It is even more difficult to demonstrate distinctive arterial ribbons from the data collected. Belfast was subjectively divided into core, frame and ribbon elements:

> when the commercial compositions of these areas are compared, chi-square reveals that every pair-wise combination (core-ribbon, ribbon-frame, core-frame) is significantly different at every single period under consideration... [suggesting]...that retail ribbons represent distinctive elements of the city centre retail system and, moreover, have done so throughout the twentieth century. Whereas the 'core' has been composed consistently of high order retail outlets such as general stores, outfitters and specialists, and some service establishments, the frame has persistently been the

NUCLEATED CHARACTERISTICS

SHOP TYPES	EXAMPLE CLUSTERS
1 Central Area	A Apparel Shops
2 Regional Centres	B Variety Shops
3 Community Centres	C Gift Shops
4 Neighbourhood Centres	D Food Shops

RIBBON CHARACTERISTICS

1 Traditional Street	E Banking
2 Arterial Ribbon	F Cafes
3 Surburban Ribbon	G Garages

SPECIAL AREA CHARACTERISTICS

1 High Quality	H Entertainments
2 Medium Quality	J Market
3 Low Quality	K Furniture
	L Appliances

THE COMPLEX MODEL

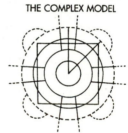

FIGURE 10.12 *A structural model of central area core retailing facilities (after R. L. Davies, 1972b). Source: R. L. Davies 1972b: Structural models of retail distribution. Transactions of the Institute of British Geographers, 1972, 57, 75.*

domain of service activities and vacant premises. Ribbons, on the other hand. have tended to be retail rather than service orientated with food stores being particularly prominent.
> (Brown, 1987: 14–15)

The only points one would add, are that vacancy suggests areas of discard from the CBD, while the traditional ribbon would have a wider range than food stores as such, including, especially, car sales and services and restaurants.

If Davies's schema is accepted as a valid representation of the structure of city centre retailing, then its key lies partly in the three patterns or systems which he integrates, those of Horwood and Boyce, Garner and Berry, but also in related forces. These include the relative amount of space which any user needs – compare the demands of a furniture store with those of a jewellery shop – and particularly the necessity for association or repulsion with regard to other retailers and other land-uses. The latter is, indeed, the criterion which Davies uses to demonstrate specialization, by showing that similar types of retailing cluster together. It is, as he admits, a far from convincing basis to prove the evidence of specialized areas. Indeed, quality and image can become more important bases of association. As long ago as 1958 Nelson, in his book *The selection of retail locations*, put forward eight principles operative in the selection of a retail site:

1. **Trading area potential.** A minimum volume of business is an obvious first necessity. In central place theory terminology it must be possible to tap sufficient custom to pass the minimum threshold requirement.

2. **Accessibility to trading area.** Maximum accessibility to the population of the trading area is needed. Presumably this operates on a regional and on a local city centre scale. It results in peak land value intersections or, in more everyday terms, competition for corner sites at main cross roads or squares.

3. **Growth potential.** Access to areas with growing populations and rising incomes is needed.

4. **Business interception.** A site between the main concentration of working population in the downtown area and the main shopping centre will enable customers to be intercepted along the main lines of daily movement.

5. **Cumulative attraction.** A number of similar units in a small specialized area can exert a much greater influence on potential customers than an isolated store. The possibility of inspecting several shops offering similar goods and providing a wide choice, that is a number of complementary units, will generate a pull of its own.

6. **Compatibility.** The location in an area of compatible uses will generate maximum customer interchange.

7. **The minimizing of competitive hazard.** This is self-explanatory for it is common business sense to avoid a location where competition is critical.

8. **Site economics.** A site which provides maximum economies through ease of access and shape is obviously an advantage.

Not all these principles work together. Thus cumulative attractions and compatibility may well bring a store into an area of competitive hazard. But nevertheless they form a basic group of tenets relative to store location.

At this point, and with reference to the above eight principles, the crucial transformation of the CBD in recent times has to be introduced. A review of the principles will make it clear that they can best be met not in the CBD itself but in the outlying residential areas. In those areas there is easier access to population without involving the hassle of travelling to the centre (Principles 1 and 2). Suburbs are the fastest growing areas (Principle 3) and are very well placed to intercept the flow from home to city centre (Principle 4). Groupings in suburban malls or shopping centres, as outlined in Chapter 5, can exert a significant cumulative attraction (Principle 5) of compatible users well away from all the downtown hazards from congestion to crime. Finally, and most emphatically, the maximum advantage of site economies (Principle 8) is in open greenfield sites where single storey building costs are considerably lower. Add to this the crucial benefit of plentiful free parking, again related to land values, and the pressures for retailing to move out of the city centre are evident. It is apt to quote as an example, Houston, a city of the American sunbelt. Even by 1982 only 2.3 per cent of the city's retail sales were accounted for by the CBD and in the preceding decade it had lost 27 per cent of its retail establishments (Feagin and Beauregard, 1989: 175). 'Houston went from a downtown-centered metropolitan pattern just after World War II, to eighteen "centers" spread

(a)

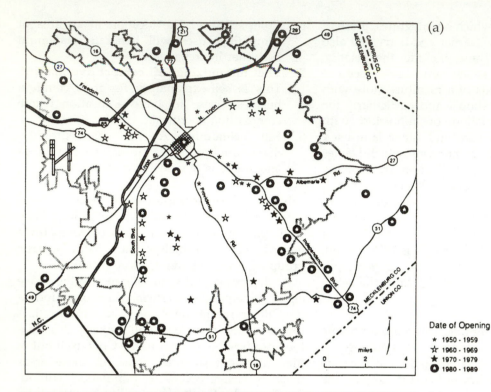

Date of Opening
★ 1950 - 1959
☆ 1960 - 1969
★ 1970 - 1979
◉ 1980 - 1989

(b)

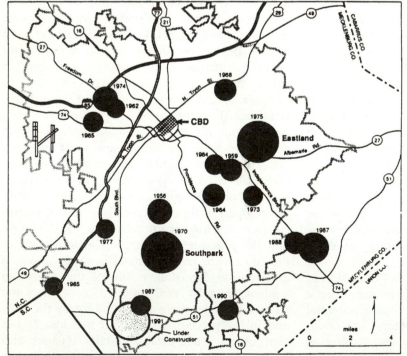

* Number on map next to center indicates year center opened.

Square Feet of Floorspace

1,200,000
800,000
400,000
200,000

FIGURE 10.13 *Shopping Centres in Charlotte, N. Carol. USA. (after J. D. Lord and C. M. Guy, 1991). A: Location of shopping centres by date of opening 1950–89. B: Planned shopping centres larger than 200,000 square feet of floorspace, 1989. Source: J. D. Lord and C. M. Guy, 1991: Comparative retail structure of British and American cities: Cardiff (UK) and Charlotte (USA). The International Review of Retail, Distribution and Consumer Research 1(4), 1991, pp. 422, 423 and 425.*

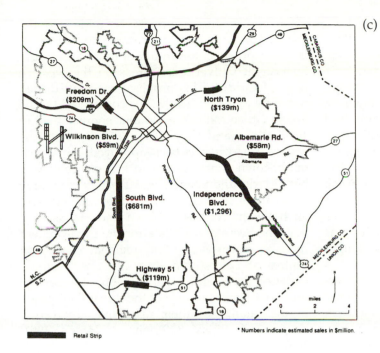

(c)

FIGURE 10.13 *continued*
Shopping Centres in Charlotte, N. Carol. USA (after J. D. Lord and C. M. Guy, 1991). C: Retail shops, 1989. Source: J. D. Lord and C. M. Guy, 1991: Comparative retail structure of British and American cities: Cardiff (UK) and Charlotte (USA). The International Review of Retail, Distribution and Consumer Research 1(4), 1991, pp. 422, 423 and 425.

across wide expanses of flat undifferentiated landscape with various land-uses juxtaposed in complex mixtures.' (Feagin and Beauregard, 1989: 161).

Here the issues discussed in Chapter 8 come into play. The discussion of structure and change in the CBD has assumed the untrammelled action of entrepreneurs in a free market economy, seeking to gain maximum profit and willing to modify location in the process. But as against that many cities have made strenuous efforts to prevent the growth of shopping centres at the periphery in order to retain the viability and significance of the city centre, and also to retain significant elements in their tax bases. The main instrument in this attempt to regulate market orientated processes has been planning permission for out-of-town developments (see Chapter 8).

The contrasts brought about by differences in attitude and planning control have been admirably illustrated in a comparative study of Cardiff in the UK and Charlotte in North Carolina in the USA by Lord and Guy (1991). Salient facts of the retail pattern of Charlotte are summarized in Fig. 10.13(a), (b) and (c). These illustrate the archetypal American process of the rapid decline of the CBD retail sales. Fig. 10.14 shows a fall in GAF sales from a peak of some $78 million in 1967 by as much as $22 million by 1972. By 1987 the total has declined to some $17 million. The acronym GAF refers to the three shopping goods categories of general merchandise stores, apparel and accessory stores and home furnishings and appliance stores. In the mid-1950s the CBD at Charlotte accounted for between 80 and 90 per cent of GAF sales in Mecklenburg County (largely Charlotte and its immediate fringe). By 1982 the CBD accounted for only 8 per cent of the sales. Again, in rounded figures, GAF sales in the CBD in 1972 were $71 million but the equivalent figure for Southpark Mall, a shopping mall to the south (Fig. 10.13(b) and (c)) was $48 million. But by 1982 a complete reversal had occurred with $45 million in the CBD and $90 million in Southpark. The authors' comment, 'by the late 1970s or early 1980s, Charlotte had two suburban retail clusters which exceeded the CBD in size ... given the trend of continuing retail decline in the CBD, it is likely that several other MRCs [Major Retail Centres] had surpassed the CBD in GAF sale values by the late 1980s' (Lord and Guy, 1991: 417).

There have been compensatory changes mainly in the growth of office employment; office space in the CBD doubled between 1974 and 1989. But, even so, retail sales declined in spite of the demands of the office workers for food stores and restaurants. Charlotte very effectively encapsulates the problems of a city losing the retail function from its CBD and, thereby, finding the traditional city centre deprived of its main function and facing decline – what has been aptly called in American terms, the dough-nut syndrome.

It is difficult to produce a comparative numerical assessment of Cardiff since the data are not available, but although there has been a growth of decentralized retailing it has been of a very limited nature and largely of the sort of operation which requires large floor space and is a comparatively new entry into the retail system, such as do-it-yourself warehouses. There is the expected intra-urban hierarchy of shopping centres. But they remain that – part of an older, traditional hierarchy. The authors write, 'the city centre dominates retailing in Cardiff, apart from sales of convenience goods and "bulky" household items such as furniture and do-it-yourself materials which have largely been decentralized' (Lord and Guy, 1991: 399). The reasons for this continued dominance are to be found in the support of local authority councillors and planners.

Since the 1950s they have taken a consistent stance in favour of city centre retailing and against major developments outside the established suburban shopping centres .... This support has included not only forward planning in a general sense but also purchase of land for subsequent leasing ... and provision of several multistorey car parks. Retail development has also been underpinned by council policies to make the city centre an attractive and convenient shopping environment.

(Lord and Guy, 1991: 401)

One addendum needs to be added to the above. The authors of the paper note that 'there is also a prospect for an out-of-town Marks and Spencer at Culverhouse Cross on the western edge of Cardiff. This proposal has been turned down by planners but will be considered by the Welsh Office following a public enquiry held in 1990' (Lord and Guy, 1991: 405). By 1992 the proposal had in fact been approved and a new out-of-town centre seems to be emerging with both Marks and Spencer and Tesco having large stores there, as well as some other retailers, and an hotel. In contrast O'Brien and Harris note that 'the Secretary of State for the Environment refused permission for regional shopping centres at Leicester (Centre 21), Southampton (Adanac Park) and Bristol (Cribb's Causeway) (but see below) because of the threat they posed to the vitality of their respective city centre shopping facilities' (O'Brien and Harris, 1991: 106).

Any simplistic interpretation of the contrasts between Cardiff and Charlotte must be avoided. There are complexities in relation to standards of living, the culture of shopping and land values which make the direct reference to differences in planning controls questionable. But there is little doubt that the crucial contrast lies in the degree of legislative control through the planning system.

This has been stressed in a study by Rowley of prospects for the central business district (Rowley, 1993: 110). An analysis of the central area of the city of Sheffield is used to illustrate the decay of that area as against the Meadowhall complex of some 1.25 million square feet (116,000 square metres) which was opened in 1990. Rowley comments, 'trade in the city centre appears to have fallen by more than one-third, and the entire northern third of the CBD...has been devastated. Streets are lined with boarded-up shops, whitewashed windows and down-market discounters' (Rowley, 1993: 118). While the central part of the CBD 'retains a certain ebullience, there are signs of decay there as well', while the southern section 'is seeking a specific market niche in what can be termed a bazaar economy with cheap street-market stalls

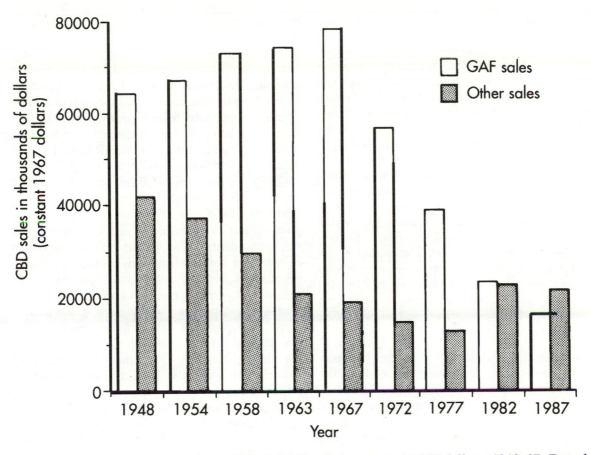

FIGURE 10.14   *Charlotte, N. Carol. CBD Sales Trends in constant 1967 dollars: 1948–87. Data from US Bureau of the Census: Census of Retail Trade. Source: As Fig. 10.13, p. 414.*

... the general air is of a non-CBD local shopping district that seems set to survive at a reduced threshold capacity level. Marks and Spencer closed its store ... when it opened its Meadowhall Branch' (Rowley, 1993: 120–1). Thus Sheffield has moved further towards Charlotteville than has Cardiff, although the out-of-town development at Culverhouse Cross which has been noted suggests that Cardiff, too, is in a process of change.

Another example of an out-of-town development in Britain is the Merry Hill centre outside the town of Dudley in the West Midlands. In late 1993 it employed some 3370 and turned over some £350 million. It is worth quoting at some length an article from *The Sunday Telegraph*

(28 November 1993), not only for the factual detail but also for the attitudes revealed. Adam Nicolson writes:

Most people you talk to in the town of Dudley, in the heart of the Black Country, will say that it has been devastated, decimated, filleted, ghettoised, forced to find a new level, dumped in the spiral of decline, reduced to a rump. 'All that we are left with', according to Sean Nethercott of the borough's planning department, 'is low-value, second-hand junk shops. That basically. We've now got a pawnbroker's occupying a prime site in the Market Place. That is a measure of what this town has sunk to'. Marks and Spencer, British Home Stores, C & A,

Littlewood's, Curry's, Next, Burton and Rumbelow's all used to have shops on the Market Place. Within the last four years they have all left. Tesco's 70,000 square foot site on Flood Street now stands empty. Sainsbury's has abandoned its supermarket in the Trident Centre, which is now half-occupied by a low quality indoor market. At the top end of the High Street what used to be the Cooks department store, for decades one of the Dudley landmarks, now stands vacant and visibly rotting. ... Shop after shop is boarded up and empty. ... In 1986, Dudley made a colossal mistake. The council gave planning permission – it must be said by a single vote – for one million square feet of retail shopping to be developed at Merry Hill, two and a half miles outside the town. In 1989 the new mall opened ... with 10,000 free car parking spaces, a monorail with three stations, and an overriding air of the future.

Nicolson concludes that 'it is a strange situation in which nobody, including the big retailers, the small town independents and the environmentalists, likes what is going on but where none of them can bring the process to a halt'. Significantly, he does not mention the shoppers themselves who presumably are voting with their feet. 'Perhaps', he concludes, 'it is up to the Environment Secretary... . He could freeze out-of-town developments ... and relieve everyone of their misery' (Nicolson, 1993: 22), though one might wish to question the pejorative 'everyone'. It would seem that in February 1994 the Environment Secretary did react to Nicolson's request. He refused permission for the development of the Duxford centre which was to be built nine miles south of Cambridge, and under the commitments of the Rio de Janeiro environmental summit in 1992 has indicated that no more permissions for out-of-town centres will be given. This directly reverses what has been government policy. Thus Rowley's consideration of the planning response noted the ambiguity of central government policy accompanied by a reduction in the powers and discretion of local government. The ambiguity he based on formal expressions of support for central areas

but decisions on appeal which go in the opposite direction with developers gaining ministerial consent for massive out-of-town complexes after having successfully appealed against earlier planning rejections at both local and national levels. This he illustrated with the case of Cribbs Causeway outside Bristol near to the M4/M5 motorway junction, long regarded as a location with major attractions for developers. In 1987 planning permission was refused by the then Minister on the basis of damage which would 'be done to the vitality and viability of the city centre of Bristol'. In 1990 the case was cited by government as an example of the removal of a threat to a central area, But in 1991 the Secretary of State for the Environment granted planning permission. Rowley concluded 'The new large regional shopping centres and other out-of-centre developments could have quite a dramatic and possibly catastrophic effect on the commercial futures of established British CBDs' (Rowley, 1993: 125). Thus the contrast referred to earlier between the USA and the UK which came about certainly in part by planning attempts to preserve the CBD, was slowly being eroded. Now, however, the government seems to have announced another change of policy and presumably emphasis will swing back to developments in the city centres.

The question as to which of the two – *laissez-faire* development of out-of-town centres or the protection of the CBD – is to be preferred is another and different question, although it should not be put in that simple way for there are other functions the centre could take over. But it is necessary to question the easy assumption, in Europe rather than in America, that the maintenance of the CBD is manifestly the better policy. The controlled environment and the undoubted convenience of the large shopping malls, with their free and convenient parking, offer a considerable improvement over congested city streets which are open to the weather and where the elementary segregation of vehicular traffic and pedestrian shoppers is not always achieved by pedestrianization. Moreover, proposals to limit access to the city centre by some form of cost levied on private

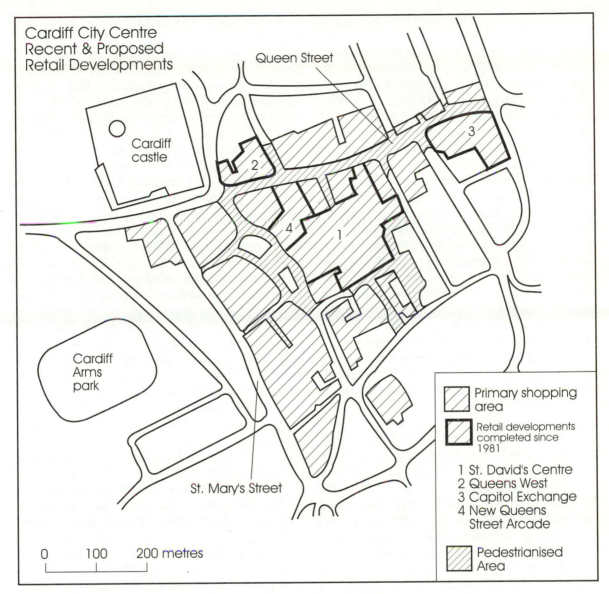

Cardiff City Centre
Recent & Proposed
Retail Developments

Queen Street

Cardiff castle

Cardiff Arms park

St. Mary's Street

Primary shopping area

Retail developments completed since 1981

1 St. David's Centre
2 Queens West
3 Capitol Exchange
4 New Queens Street Arcade

Pedestrianised Area

0   100   200 metres

FIGURE 10.15 *Cardiff City centre (after Lord and Guy, 1991)*

cars (see Chapter 9) will further disadvantage the CBD. The efforts to maintain the CBD can be regarded as a characteristic negative reaction to innovation, a rearguard reaction to change, moreover to change which the public demands. There is, too, more resistance in Europe where town centres have historic buildings with long associations with civic dignity and pride. Also, the intensive development of

small countries poses problems, which do not arise in the larger and extensive countries, of the preservation of open land. Again, the degree of car ownership and the quality of public transport are relevant where the poor will be the most disadvantaged by the loss of accessible shops (Bromley and Thomas, 1993). In cultural terms too, a 'throw-away' society is more ready to dispose of the outdated concept of the city centre as the retailing core. The real defenders of the CBD are theorists, planners and city fathers and, to a degree, those committed to public rather than private initiatives; the general public, although limited to that part with the greatest spending power, has other ideas, demonstrably so by their actions. It is, perhaps, appropriate that in the transactional city, retailing should be abstracted from the core leaving it to offices related to business, finance and administration.

This is symptomatic of a general character change which most commentators identify (Mackeith, 1985; Dawson, 1988; Davies and Champion, 1983) and which is related to the changing mix of land-uses within the city centre. As extensive retailing has moved out – usually in the order of food supermarkets, retail warehousing, clothing and quality comparison goods (Schiller, 1988) – so city centre retailing has become more specialized even to the extent of large companies breaking up their operations into smaller, more specialized ventures, with separate names. As retail uses have declined in area, so services of all kinds have increased, personal services, financial and business services, medical and particularly leisure services. Even in Oxford Street, at the heart of London's shopping, the area of retail uses declined between 1971 and 1982 whereas the largest gainers were offices and hotels (Dawson, 1988). It is possible, therefore, to envisage a different CBD, rather than one necessarily in decline as against out-of-town shopping centres. As Mackeith writes, 'the inhabitants will still require a meeting place, a cultural and social centre, and already the concept of the medieval market place is re-emerging in "speciality centres" which are designed to bring the pleasure and the fun back into shopping'(Mackeith, 1985: 10). A rather different view relative to contemporary Britain is that the current danger is that an uneven retreat from planning will leave many existing centres as decaying parts of the inner city' (Thorpe, 1988: 19). All this has generated a fight back by city administrators unwilling to accept decay, even if only to retain a tax base.

The measures taken by many city authorities to preserve the vitality of their city centres are based on the creation of conditions similar to those in out-of-town malls. The first and earliest reaction was pedestrianization which is now characteristic of even the smallest town. It sought, in the most elementary fashion, to segregate traffic and pedestrians. Along with that came the multistorey car park – multistorey in order to maximize a low quality use of expensive city land. Unfortunately such car parks are usually environmentally unpleasant places, bleak and crude and with a growing association with petty crime. The next stage in the process of city centre rehabilitation has been the creation of shopping precincts or malls within the CBD so that the environmental benefits of protection from weather conditions, together with warmer, friendly surroundings can be achieved. In addition decoration and embellishment can increase the attraction. In many cases direct pedestrian access to shopping mall from multistorey car park gives complete protection from weather and isolation from traffic. Figure 10.15 illustrates Cardiff's CBD, which has already been discussed, and the extensive pedestrianization which includes a number of nineteenth-century arcades, for protected shopping areas are no twentieth-century innovation. In addition, there has been a very substantial restructuring of the city centre, using both older commercial land and former stores, to create four substantial shopping centres. The largest is the St David's Centre with 450,000 square feet (42,000 square metres) of gross retailing space but also a major concert hall. The extent is modest indeed, compared with the Southpark Mall at Charlotte with 1 million square feet (93,000 square metres) and the later Eastlaw Mall with 1.2 million (111,000 square metres). But city centre developments are inevitably constrained by both existing building and the cost of land. Along with this restructuring at Cardiff has come the attempt to make the CBD attractive by statuary, one is of Aneurin Bevan, and by encouraging street theatre and entertainment. This again is a characteristic parallel to that of out-of-town malls.

The example of Cardiff has been continued in order to demonstrate the attempt to revitalize the city centre. But it is of modest dimension compared with attempts elsewhere. Thus in Britain between 1965 and 1978 four characteristic developments were the Brunel Centre at Swindon of 574,000 square feet (53,000 square metres), the Victoria Centre at Nottingham of 622,000 square feet (58,000 square metres), the Eldon Square development at Newcastle of 780,000 square feet (72,000 square metres) and the Arndale Centre at Manchester with 1,240,000 square feet (115,000 square metres). But much of this, including the Cardiff Bay development at Cardiff, follows earlier attempts to reinvigorate the centres of American cities which had suffered even more drastically for planning constraints hardly existed.

The initial move towards pedestrianization, as a measure to solve the CBD problem of the mixing of incompatible uses, has already been noted; but the most dramatic way of achieving that end was not to designate streets for pedestrians only, but physically to separate traffic and people by lifting walkways to the first floor (or the second floor in American terminology). Thus a major downtown development scheme in Cincinnati, which included the clearance and redevelopment of the river front – characteristically with the Riverfront Stadium – was based on the construction of a Skyline Walkway, partly enclosed and air conditioned, without the need to descend to ground level from the extensive multistorey car parking. The cost was $7 million and the Inter Downtown Executive Association called it 'an outstanding example of the kind of approach and commitment on the part of city centre leaders'. In contrast to raising pedestrian areas the Ste. Marie Centre at Montreal was constructed underground in order to provide an effectively controlled environment in the cold Canadian winters. A more usual alternative is to put the car parking underground.

The redevelopment of Cincinnati's downtown area was accompanied by the complete refurbishment of its core in Fountain Square, but that type of revitalization has characterized many US cities. One of the earliest and most well-known schemes was the Faneuil Center (Quincy Market) in Boston, a classic case of an historic but run-down part of a city, for the Faneuil Hall, known as the 'Cradle of Liberty', had been built in 1742, although rebuilt in 1763 and enlarged in 1805. It was flanked by two later Market Buildings. These were acquired and renovated by the Boston Redevelopment Authority which then leased them to the James Rouse Corporation for one dollar a year plus a share of the profits. The first phase was opened in 1976 and with extensive retail and office space, along with cafés and public open spaces has been very successful (Mackeith, 1985: 11). The development is illustrative of the 'Public–private partnership in American cities' (Foster and Berger, 1982) which has been crucial in most schemes of downtown rehabilitation. The most frequently cited are Baltimore and Pittsburg.

The origins of the movement to redevelop downtown Baltimore can be traced back to a non-profit-making organization, the Citizens' Planning and Housing Association which had been formed out of an even earlier organization in 1941, and which eventually gave rise to the Greater Baltimore Committee (GBC) (Lyall, 1982: 21 *et seq.*). 'By 1957 alarm at the eroding business climate and declining property values was so strong that the Committee for Downtown (an offshoot of the Retail Merchants Association) and the GBC commissioned a master plan for renewal. Alongside these private initiatives the public Redevelopment Commission with federal Urban Renewal Funds undertook large-scale clearance and redevelopment. By 1968 a process of merging had produced the Department of Housing and Community Development so that the city's physical development powers were consolidated in a single department (Lyall, 1982: 23). From the joint efforts of both private concern and finance and the city's public commitment two major projects followed.

The first of these was the Charles Center, opened in 1973, a unified concept based on three self-contained plazas with, significantly, a Federal Building.

By 1973, the Morris Mechanic Theater, the Baltimore Gas and Electric Addition, the Hilton Hotel, Mullen Towers, the center plaza, overhead pedestrian walkways and underground parking garages had been completed. Both the Hamburger's Store and Vermont Federal Savings and Loan successfully relocated within the area, and Sun Life cancelled a previously planned move to Baltimore County opting to stay in its original location. Charles Center became Baltimore's first successful public–private venture.

(Lyall, 1982: 37)

It covers some 33 acres with 2 million square feet (186,000 square metres) of office space. The ratio of private to public finance was 86:14.

The second element is the Inner Harbour, initiated in a 1964 document, which takes in the whole of the waterfront and its neighbouring area. It includes major public initiatives such as a World Trade Center, a Convention Center, the Maryland Science Museum and a National Aquarium. In 1980 Harbour Place, which mirrors the Quincey Market in Boston, was opened with its two shopping pavilions and leisure facilities as well as residences. It reverses the Charles Center ratio of private to public finance being something like 10:90, attracts some 20 million visitors a year and is estimated to have created 30,000 jobs.

The redevelopment of the so-called Golden Triangle at Pittsburg, with its two phases usually called Renaissance I, which saw the creation of Point Park and the Gateway Center, and Renaissance II, characterized largely by great office tower blocks within the CBD, is a second example of private–public partnership with the respective inputs showing great variation. It also marks the considerable use of offfice space in the CBD. Nearly every large city can now show a parallel venture. London Docklands is the prime British equivalent.

10.5 CONCLUSION

The discussion of the CBD in this chapter can be summarized under a number of headings:

1. **Definition.** In formal terms this has virtually been abandoned by urban geographers and most areal definitions are purely subjective. It is rather akin to the ranking of central places by a centrality index; it is such a long and time-consuming task that a surrogate is used, population in central place studies, arbitrary definition in CBD studies. Even so, exercises in definition, such as Davies's in Cape Town, emphasize that the CBD is not a single spatial unity, but is made up of a complex set of interlocking use regions.
2. **Retail structure.** The retail structure in spatial terms seems to be effectively encapsulated by Davies's model, itself an amalgam of bid–rent principles, core–frame contrasts, and the emergence of distinctive shopping environments.
3. **Decline.** The growth of out-of-town shopping, especially large complexes, has greatly diminished the retail viability of the CBD. However, the increase of offices of all kinds and services has contributed to its continued significance, although with a change of character.
4. **Redevelopment.** That change of character is best symbolized by the very many major schemes which have sought to offset decline and put in place large mixed complexes in which leisure provision plays a major part, tourists as well as citizens are attracted.
5. **Planning.** The patterns and changes which have been reviewed are the product of the operation of the free market but greatly constrained by planning decisions and determination. Renaissance at Pittsburg was largely due to the prime mover, the Allegheny Conference on Community Development. But after Renaissance I a new mayor, Peter Flaherty, was less amenable to being led and progress lapsed. 'Clearly the major distinction between the Flaherty years [1969–77] and those of Richard Caligviri, who followed him, is not between development and no development. Rather the difference lies in the spirit and flurry of activity' (Stewman and Tarr, 1982: 94). Stewman and

Tarr write, 'When Richard Caligviri was elected his primary goal was to make a second Renaissance happen on two fronts, the neighborhoods and the central business district. This view constituted a major change in philosophy, especially in terms of priorities and organization' (Stewman and Tarr, 1982: 95). Here then is not simply a planning constraint but an overt political influence of the sort discussed in Chapter 8, in operation determining the nature of the urban scene.

As these headings are reviewed it is evident that more than the CBD is involved. Even the core–frame notion brings into consideration the zone in transition as it was identified by Burgess, or what is now called the inner city. Uses other than retailing, such as offices and residence, need to be considered apart before it is possible to turn to consider the CBD margins and the inner city.

URBAN LOCATION OF ECONOMIC ACTIVITY: INDUSTRY AND OFFICE USES

11.1 INTRODUCTION

The last chapter was sub-titled 'The Central Business District' but it was devoted largely to the consideration of retailing within the city. Retailing rightly claims a priority since, residence apart, it is the most universal of land-users in towns whatever their hierarchical rank. But the predominant theme of the chapter was the flight of retailing from the CBD, as it was traditionally conceived, to out-of-town locations, with the consequent implication of significant change within the CBD. When Burgess used the description 'central business district' in 1925 he was undoubtedly envisaging and presenting the notion of a unified city centre as represented by 'The Loop' in Chicago.

In all cities there is a natural tendency for local and outside transportation to converge in the central business district. In the downtown section we expect to find the department stores, the skyscraper office buildings, the railroad stations, the great hotels, the theatres, the art museum and the city hall. Quite naturally, almost inevitably, the economic, the cultural and political life centres here.

(Burgess, 1925: 52)

Clearly, there was a diversity of uses made up of the winners in the bidding process for city centre land. Even so, there were two zones where users of city land other than residence, had established themselves.

The first is implicit in the designation 'zone of workingmen's homes'. It was clearly the creation of heavy industry which had grown about the core in the late eighteenth and nineteenth centuries. Burgess wrote, 'a third area is inhabited by the workers in the industries who have escaped from the area of deterioration but who desire to live within easy access of their work' (Burgess, 1925: 50). The word

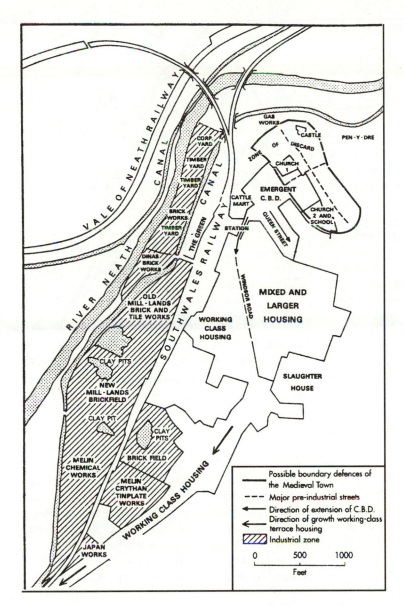

FIGURE 11.1 *Neath, W. Glamorgan. The basic structure of the town in the last quarter of the nineteenth century.*

'desire' was an unfortunate choice for, given the limitations of transport, workers had perforce to live close to their place of work; the zone was the creation of capitalist industry and limited mobility, rather than the free choice of the residents. It was, in consequence, seldom a concentric zone since the particular needs of industry, especially access to a variety of transport facilities, determined specific locations. Figure11.1 shows the form taken by the indus-

trial area in a small British town. River frontage, canal and railway created an elongated area. Closely related, even classically interdigitated, was the 'zone of workingmen's houses', or more directly, the working class residential areas.

The second variant from an all-embracing CBD was what Burgess called the Zone in Transition, although the name has been generally criticized since all parts of the city are in transition. It can be considered as the enlarged frame of the central core (see Chapter 10), the location of those users seeking but unable to afford a city centre site. The work of Murphy and Vance on the definition of the CBD was followed by others attempting to define the Zone in Transition, or the Tz as they summarized it. Consideration of the patterns revealed by an analysis of three American cities led Preston and Griffin to propose that 'within the zone the clusters (of use) are generally arranged in sectors and are separated from one another by areas of less intensive and less specialized land-use' (Preston and Griffin, 1966: 344). The outstanding clusters identified were as follows:

1. Wholesaling, private and commercial storage, light industry and transport
2. Public organizational and headquarters office establishments
3. Automobile sales and services and parking
4. Financial establishments, general offices, variety stores and transient residences
5. Food, household, service trade and various retail establishments.

At this juncture two further points can be made concerning the CBD and the Tz.

• In relation to the CBD Murphy and Vance (1954b) proposed a dynamic where there were zones of assimilation, into which the CBD was expanding, and zones of discard, that is areas being abandoned by CBD uses (see p. 165). These can be clearly seen in the late nineteenth century in Fig. 11.1 where the active town core was migrating away from a deteriorating medieval nucleus and, in sympathy with the direction of growth of the industrial/residential areas, was expanding into a newer area. The area which was discarded became part of the Tz, indeed it was classically an area in transition.
• The range of uses given by Preston and Griffin for the Tz is very similar to those set out by Horwood and Boyce for the frame. Frame and Tz appear to embrace the same areas.

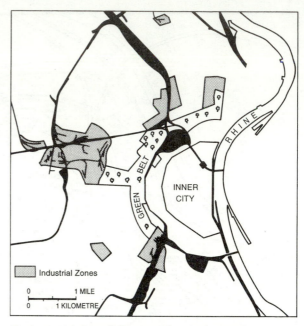

FIGURE 11.2 *Cologne. The industrial areas in 1939 (adapted from R. E. Dickinson, 1951). The black areas were railways and goods marshalling yards, Areas to the east of the Rhine have been excluded.*

But whatever the situation when Preston and Griffin developed their work on the Tz, the succeeding quarter century has seen the major transformation in the role of cities which has already been discussed, as well as the changes in the CBD itself. Thus, on the one hand there has been the move out of the city centre of retailing, as well as of industry, and on the other hand the coming of the transactional city where the crucial activities are those of data and information handling. Those changes have meant a virtual revolution in the role of those uses which Preston and Griffin described as 'financial establishments and general offices' as well as 'public organizational and headquarters office establishments'. The consequence is that city centres are now dominated by multistorey office blocks which are the dominant features of contemporary city skylines. It is to these users of city land to which attention must now be directed.

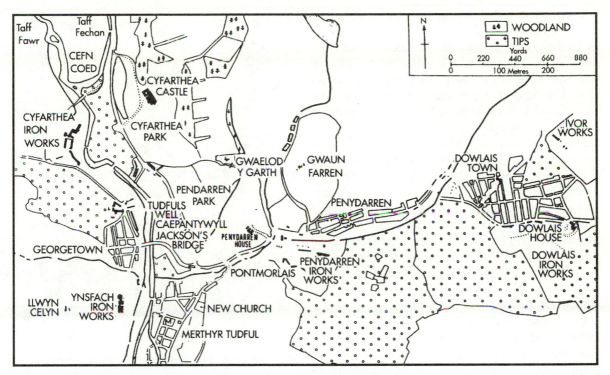

FIGURE 11.3 *Merthyr Tydfil in 1850. The separate iron working nuclei can be easily distinguished. To the northwest lay the Cyfarthfa Iron Works, to the east the Penydarren Iron Works and the Dowlais Iron Works. The fourth nucleus is not shown on this map but lay to the south of the town. The old pre-industrial village was located on the extreme southern edge of the map and new developments in the form of St David's Church (1847) and the market hall (M, 1835) indicate the growth of the industrial town. Note the extensive spoil heaps or 'tips'. This is an admirable example of a 'multiple nuclei' structuring.*

11.2 INDUSTRY IN THE CITY

It has already been noted in Chapter 7 and at the beginning of this chapter that E. W. Burgess identified a zone which was described as being dominated by working class homes and implicitly by heavy industry. This was largely the product of historical forces and, as Fig. 11.1 demonstrates, relevant to the 1920s when Burgess was writing. The rapid growth of industry during the nineteenth century meant that it took up a location, when associated with pre-existing towns, outside and on the margins of the old nuclei. This resulted, in some cases, in the neat concentric zone

envisaged by Burgess, as for example in Cologne, prior to the Second World War (Dickinson, 1951: 81–90). The construction of defensive walls on an even site had constricted the city into a series of zones while the area *non aedificandi* outside the 1880 defences was preserved as a green belt. Outside this the industrial developments of the latter part of the century took place, since they were barred from the river frontage for military reasons. Along with them were built the industrial suburbs to produce a classical zonal pattern (Fig. 11.2). The major industrial areas themselves were associated with the main radial railway lines, while the working class residential areas filled in the areas between.

In contrast to the arrangement exemplified by Cologne, most of the towns which were created by industry, and had no previous existence, grew around the formative elements, the factory and the mine. In consequence, multiple nuclei schemes tended to emerge, as suggested by Harris and Ullman (1945). The point location of exploited resources produced a number of nuclei in relation to which the residential areas were disposed. These often coalesced (the origin of the concept of conurbation) to form urbanized areas dominated by the separate nuclei and often in the process creating another, the central business district. A good example of this is Merthyr Tydfil in South Wales where four early ironworks established between 1750 and 1790 created four distinct nuclei, while the central area emerged as the fifth (Fig. 11.3) (Carter and Wheatley, 1982).

But these schemes are at the most general level of interpretation and it is progress from that stage which is crucial in resolving the complexity of traditional urban locations of industry. Three attempts by Loewenstein, Hamilton and Pred can be considered in turn and in conclusion some attempt can be made to establish what ground there is in common in order to provide a basic interpretation of industrial location at the start of the last quarter of the twentieth century.

Loewenstein (1963) based his survey on a wide array of empirical evidence where manufacturing uses were considered alongside a number of other characteristic uses. A selection of cities was examined in which a particular use, as demonstrated by employment, was dominant: 'Only that employment in which a particular city excelled was reproduced from the basic land-use maps' (Loewenstein, 1963: 409) on to a standard format consisting of five distant rings (Fig. 11.4). From this evidence, synthesized into one map representing manufacturing land-uses in total, it was concluded that 'manufacturing activities typically tend to be dispersed away from the core' (Loewenstein, 1963: 413) and to have a linear appearance due to extension along transport lines. This dispersal away from the core Loewenstein explains in a conventional

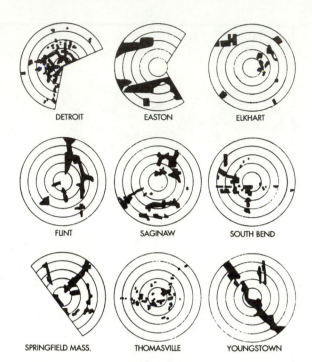

FIGURE 11.4 *The distribution of manufacturing in selected American cities (after L. K. Loewenstein, 1963). The areas in black denote manufacturing land-use. Five distance rings are superimposed in each case by taking the major CBD intersection as centre and drawing an outer ring at the mean limit of the metropolitan area. The radius of the outer ring is divided into four to give the circles shown.*

way as the consequence of the high costs and congestion in the theoretically optimum location in the city centre. There are also a number of other factors:

1. Labour or market-orientated firms are likely to be found in the city centre. This is because market-orientated firms, such as those engaged in newspaper publication, can save in terms of transport costs since distribution from the centre is much easier, while labour-orientated firms can draw effectively from a central location on the widest ranges of skills.

2. Manufacturing plants which are engaged in what Loewenstein calls organized, inte-

grated industries, cluster along 'radial and belt railroads and highways'. This is particularly true of assembly industries such as the automobile industry.

3. Large, basic processing industries, such as petroleum refineries or steel mills require large areas and create a great deal of noise and pollution, They, therefore, repel residential development and will be removed from it. This is only partly true: certainly high quality residential areas are unlikely to develop adjacent to heavy industry but historically this is by no means true of working class residential areas.

4. Large new plants are often located in a suburban context where land is most easily and cheaply available.

This is not a very impressive breakdown of industrial distribution but it does illustrate the point which Burgess never made clear. Industry is not confined to any one zone or any series of areas, it is located throughout the city because different types of industry require different locational attributes so that any attempt at generalization, unless linked to a convincing typology, becomes extremely hazardous.

This can be illustrated from Loewenstein's own work. He calculated for each use an index of concentration (Loewenstein, 1963: 418). It measured the percentage distribution of land-use of each of the activities, such as retail or manufacturing, in each of the five rings as a percentage of the total land area within the ring. In the computation of the index the percentage of the total land was accumulated and multiplied by the amount of land in a particular use. The formula is defined as:

$$\text{Index} = 1 - \sum(X_i - X_{i-1})\,(Y_i + Y_{i-1})$$

where Y_i is the accumulated proportion of total area through the ith ring (the fifth if the total city is under survey) and X_i is the accumulated proportion of the area for the given kind of employment activity, i.e. for a given type of use. The index varied between −1.0000 for a completed dispersal, through zero for even

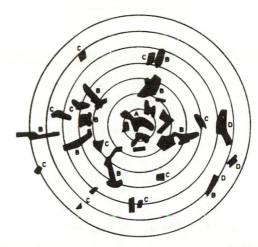

FIGURE 11.5 *A model of the spatial industrial structure of a metropolis (after A. E. Hamilton, 1967). The model is based on London. The four categories indicated are A: central location; B: port locations; C: radial or ring transport artery locations; D: suburban locations.*

distribution throughout the rings to +1.0000 for complete concentration. An analysis of the results reveals that manufacturing had the lowest index of 0.2800 compared with 0.6896 for finance, insurance and real estate, 0.7344 for retail and 0.4816 for personal service and 0.8400 for public administration. Of all the uses considered, manufacturing showed the least tendency to concentration and the highest tendency to an even distribution throughout the city. One example (Elkart, Ind.) even shows a minus index of −0.0352, the only occurrence, which demonstrates a tendency towards complete dispersal.

The various characteristics of intra-urban industrial location derived by Loewenstein are reflected in Hamilton's attempt (Hamilton, 1967), in considering models of industrial location, to reduce the map of London's industry to a generalized scheme (Fig. 11.5). On this four characteristic locations are identified:

A **Central locations.** These are taken up by industries which require access to skilled labour, for example, the clothing industry, and to the whole urban market for distribution, for example, newspaper publishing. It is in this context that the 'swarming' of closely associated activities occurs which gives rise to sharply defined industrial quarters.

B   Port locations

C   Radial or ring transport artery locations. Both these (**B** and **C**) are taken up by larger enterprises seeking cheaper land in larger quantities and good locations for assembly and distribution. In addition larger amounts of unskilled or semi-skilled labour can be obtained without calling on city-wide sources.

D   Suburban locations. Such positions are sought by industries requiring very large amounts of land for assembly or production lines and for storage or if the industry has repellent features, such as fumes or noise which compel it to seek isolation from residential areas.

The interpretation of this pattern is presented in the context of the conventional explanatory factors. 'If the metropolis is large, then substantial cost differentials exist between alternative locations within the city, especially with regard to land, labour and transport. The model of the metropolitan spatial structure of industry, therefore, comprises differing localizations of associated industries in different optimum conditions' (Hamilton, 1967: 408). There is clearly a very close parallel between the generalizations of Loewenstein and Hamilton

Finally Pred presented an array of what he termed seven flexible types of manufacturing each of which he argued should be characterized by a distributional pattern with a unique set of attributes, including in some cases randomness (Pred, 1964). Pred adopted the rationale derived from Chinitz's study of industrial plants in the New York Metropolitan Area. Chinitz proposed three broad types:

1.   Those serving markets predominantly local.
2.   Those serving national markets – subdivided by size and value of product.

3.   Those plants localized by external economies and not included under 1 and 2.

Pred wrote 'since each type and subtype has different transportation requirements or different abilities to absorb transport outlays, singular but not exclusive, locational tendencies emerge' (Pred, 1964: 173). This is comparable to Hamilton's statement above, but made in relation to a useful subdivision of industries with an emphasis on markets rather than on raw materials. Pred's seven types were as follows:

1.   **Ubiquitous industries concentrated near the CBD.** These are industries whose market is co-extensive with the metropolis. They often have linked wholesaling functions and can maximize distribution from the centre. Pred quoted bread, cake and pie plants as an example.

2.   **Centrally located 'communication economy' industries.** In these cases locations are determined by external economies derived from immediate accessibility to the purchaser prior to the process of manufacturing . An obvious example is job printing where the most frequent demand arises in the city centre and where face-to-face contacts are necessary. Pred characterized the then extant New York garment centre as archetypal.

3.   **Local market industries with local and ubiquitous raw material sources.** Within this group are industries for the local market and using ubiquitous raw materials, such as ice manufacturing plants, and also those using raw materials which are the by-products of other industries or semi-finished goods. These often appear to have random locations since the constraints operating on site selection are not great, even nearness to railheads is not vital where intra-urban access is required. The building of freeways has resulted in some dispersion of these industries: presumably the freeways should also influence those industries in group 1.

4.   **Non-local market industries with high-value products.** Where the market is greater

than the metropolis and where the finished product has a high value-to-weight ratio, then transport costs become relatively unimportant and a random location pattern can result. Fortuitous factors can govern site selection. A typical example is the manufacture of calculating machines. Other things being equal, when he wrote, Pred noted that such industries would tend to locate near the CBD, especially the smaller firms.

5. **Non-centrally located 'communications economy' industries.** Pred defined these as 'those industries which imperatively cluster in non-central locations to realize "communication economies"' (Pred, 1964: 177). He suggested that these are highly technical industries which need to cluster to keep abreast of innovation but are nationally orientated and hence not related to the CBD in any specific way. This results in strip development along major expressways. Space-age electronic industries are exemplars.

6. **Non-local market industries on waterfronts.** This is a well known group of port industries and requires no comment other than to make the reservation that all industries on waterfronts do not necessarily belong to the group.

7. **Industries orientated towards the national market.** These have extensive markets with locations strongly influenced by the bulk of the products and transport rates. The main point that Pred made is that these locations often show distinct biases in the sector of the city in which they are located which is characterized as one looking towards the regional or national market.

At the beginning of his study Pred set out to answer two questions : 'What kinds of industry remain in or near the core of the metropolis?' 'What patterns, if any, are to be distinguished among the decentralized industries?' He answered the first question with some success but made little headway with the second, and indeed admitted 'a considerable amount of evidence remains to be culled before any real understanding of intrametropolitan locational preference of industries oriented towards non-local markets is attained' (Pred, 1964: 180). The examples which he used were mainly derived from San Francisco and to some extent the characteristic types and locations were representative of that metropolis rather than general in concept. But it is evident that there is a great deal in common between the suggestions of Loewenstein, Hamilton and Pred.

The list set out by Pred is a representation of urban industrial location as it had evolved by the middle of the twentieth century. But even in such a listing the suburbanization of industry is consistently noted and that has continued and evolved to become the dominant theme. But the process has not been one of the simple transfer of operations, although that has occurred. Largely it has been associated with the structural change of industry, and the greater proportion of closures or 'deaths' as they are called in the centre and the greater proportion of new investment or 'births' at the periphery.

> The picture of considerable industrial change, but clear net decline, involving our conurbations, shows that most job losses in manufacturing have been due to complete closure or 'deaths' of plants, and that comparatively few plants have migrated to the suburbs, [or] the satellite towns.... Quite simply this growth has been largely through the location of new firms in the smaller towns and rural areas, not through any major geographical movement of existing firms: the decentralization thesis must be rejected.
>
> (Lawless and Brown, 1986: 190)

That situation is largely due to the fact that the nature of industrial activity has itself changed: one is not dealing with the same industries taking up different locations, but rather with new industries demanding new locations. There are, therefore, two elements involved, the one, the change in the nature of industry; the other, the emergence of novel factors in industrial location.

There is an initial problem that since most industry is urban then it is difficult to restrict a discussion to what is seemingly relevant to urban geography. Basically, deindustrialization, which has been largely the loss of the old heavy, so-called smoke-stack industry due to falling demand, high costs of production and competition from new producers where labour costs are lower, has created extensive tracts of derelict land. Those characteristically were part of Burgess's zone of heavy industry or the large basic processing industries of the classification presented above (p. 193). Thus the areas devoted to iron, steel and coal mining at Merthyr Tydfil shown on Fig. 11.3 have long been derelict and redeveloped for newer, modern industry such as Hoover washing machines at Pentrebach to the south of the mapped area.

Significant change has also affected those industries which were classified as 'centrally located'. It has been brought about by a whole range of factors which have made central locations, effectively those in the inner city, wholly unattractive. It is possible to make a long list of the drawbacks of the inner city. Land is expensive while the extensive areas needed are difficult if not impossible to assemble. The more intricate road system makes access difficult on the local scale, while the accessibility of motorways or throughways is limited, causing time loss and higher costs. Changes in the modes of production mean that cramped sites are inappropriate and physical extension with business expansion impossible. A classic example is the shift of newspaper production in London from Fleet Street, although to a revised inner city area rather than to the city fringes. Here too, the entrenched position of labour unions was another factor which can be set alongside the other drawbacks. But undoubtedly the major problem of old industrial areas is simply the image of place. The inner parts of the city are regarded as run down or derelict, characterized by every variety of anti-social behaviour and hence totally unattractive to new industry. And these new, so-called 'high-tech' industries demand locational criteria which have little in common with traditional manufacturing. Thus Marshall M. A. Feldman (1985) in a study of biotechnology based on the

Biotechnology Industry Survey of the California Commission on Industrial Innovation (Feldman and O'Malley, 1982) identifies the significant locational factors. The three most important are 'proximity to major research centres (for their research libraries, graduates, students and academic staffs), the quality of residential life (in order to attract and retain high calibre researchers) and the potential for growth (space for physical expansion, transportation facilities, etc.)' (Feldman, 1985: 76). In contrast factors which were commonly used to explain location are seen as being of minor or no importance. Among them are proximity to markets, availability of low cost labour, availability of reliable energy supplies, government incentives, proximity to major suppliers, availability of utilities and proximity to major competitors (Feldman, 1985: 76).

Any review of these criteria relevant to the location of 'high-tech' industries will immediately suggest those behind the growth of the archetypal example, Silicon Valley in Santa Clara County in California as outlined by Saxenian (1985). The area was until the Second World War an essentially agricultural county with the largest city, San Jose with a population of 68,000 and Palo Alto, the location of Stanford University, with some 20,000. Stanford was to become the generative centre of development as wartime research on electronic components for military use became an element in its work. It was the very specific post-war growth of Stanford's electrical engineering programmes which ensured continued growth. One of the major innovations was the development of the 600 acre Stanford Industrial Park in the 1950s, one of the first in the USA. 'No sooty smokestacks or shabby old factories mar the scenery. The science companies for the most part operate in sleek modern buildings in fifty one verdant industrial parks, which provide campus-like setting for research and manufacturing' (Bylinsky, 1974). One of the major bases of the most dynamic growth phase 'was associated with the birth of a revolutionary new industry, the semi-conductor industry' (Saxenian, 1985: 25), the first firm being set up in 1955. Electronic industries burgeoned based on a massive increase in educational establishments and on all the specialized inputs and services which were necessary. The population which had been 175,000 in 1940 had risen to 1,250,000 by 1980. But much of the attraction was not only that of employment in innovative industry, it was also the drawing power of a whole lifestyle. 'It's a particularly pleasant place to live and work – a beautiful landscape of hills and plains, a bounteous garden of nature where fruit trees and wild flowers bloom even in February.... Few places on earth so agreeably mix hedonistic delights with the excitement of urbanity' (Bylinsky, 1974). It is possible, however, that the significance of environmental attraction, at least in this physical meaning, is over-stressed. Castells and Hall write:

> the milieux of innovation are a goal in themselves. People do not live in them because of the quality of their life or the beauty of nature ... the attraction that Silicon Valley continues to exercise over the high-technology researchers and entrepreneurs of the entire world relies on the simple and fundamental fact of being the depositry of the most advanced knowledge in electronics and on its capacity to generate the next generation of such knowledge by processing the flow of information through its social networks and professional organizations.
>
> (Castells and Hall, 1994: 26)

Undoubtedly the best review of contemporary industrial developments is that by Castells and Hall (1994) called *Technopoles of the World* with a sub-title 'The making of 21st century industrial complexes'. In the book the authors present a typology which comprises five elements:

1. **Industrial complexes of high technology firms.** These are extensive complexes where there are both research and development and manufacturing which are linked together. The type example is Silicon Valley itself.
2. **Science cities.** These are strictly research complexes without the manufacturing role. The essential purpose is to create a high level of scientific excellence which will then

generate, by interaction within its own networks, its own momentum, creating that surge of scientific innovation which was at the base of the development of Silicon Valley. One of the examples most often quoted is Akademgorodok near Novosibirsk in Siberia, virtually an attempt by the former USSR to replicate Silicon Valley, although it was conceived as long ago as 1957. It was planned virtually as a university campus, with a zone for scientific institutes, a university area with student dormitories, and residential areas carefully segregated by 'academic' status. The other most familiar example is Japan's Tsukuba, built some 40 miles from Tokyo. 'It is a national research centre, funded totally by central government; its laboratories, engaged on basic research, are government laboratories' (Castells and Hall, 1994: 67).

3. **Technology parks.** These are created more with industrial growth as the target through the attraction of high technology firms. By definition, therefore, they are much more widely distributed, while the definition itself is somewhat elastic. Included would be the developments along the M4 corridor in Britain, as well as those in the vicinity of Cambridge. Also included would be Sophia-Antipolis near Nice in southern France. 'By 1989, Sophia-Antipolis housed some 400 companies employing some 9500 workers; about 70 percent could be regarded as high-tech' (Castells and Hall, 1994: 88) In the background is the concept of creating an extended sunbelt along the Mediterranean.

4. **Technopoles.** These are different in being part of a programme of regional planning in which peripheral and lagging regions are stimulated by 'multiple strategies: concentrating public and private research institutes, promoting hybrid technologies, upgrading local university laboratories, establishing technology centres, funding joint R&D projects, and providing R&D funding' (Castells and Hall, 1994: 112). Twenty-six such sites were identified. In their book Castells and Hall give extensive

accounts of four of the sites, such as Sendai, some 188 miles north of Tokyo, with its range of industrial and research parks, as well as residential areas. Into this category presumably would go the multifunctional polis (MFP) which was announced by the Japanese and Australian governments in 1987 to be built north of Adelaide (Haughton, 1994). That project also emphasizes the problems. Haughton summarizes the present status:

> for a prototype scheme based on supposedly new thinking about the relationships between economy, society and environment, the MFP proposal remains curiously devoid of a grand vision....The early documentation makes much of the genealogy of the MFP, from the Greek city state to Ebenezer Howard's Garden City and Australia's own Canberra. There is little sense however of a grand design, an underlying philosophy for the MFP concept; the case is not presented with the utopian conviction of a Howard or a Le Corbusier. Instead the MFP project has at times an almost post-modern abandonment of overarching principle or theory....The reality is that the MFP is a prototype city which is led by pragmatism and harsh market realities.
>
> (Haughton, 1994: 51–2)

5. **Regenerated older areas.** The older metropolises by their size and significance have succeeeded in attracting high technology industries. Perhaps Route 128, the major gathering ground of high-tech industries about Boston in the USA is best included in this category. Also included must be the development of a massive array of firms about Massy south of Paris.

It will be evident that much of the above account is derived from the book by Castells and Hall. It is an excellent account and reference to it will fill out much of the detail which has perforce had to be omitted from this account. But their end is somewhat different. It is to identify the bases of the success of 'technopoles'

and to distil those elements which can be used for future planning. The purpose here is to understand the impact which such developments will have on the spatial pattern of industry in the city, and on the form of the city itself.

In this typology of twenty-first century industrial complexes lie the locational factors of contemporary industry where new parks are established largely on the urban periphery. As a result the attempts to model and typify industrial locations in cities as set out above (p. 194) have become much less relevant, although relict elements remain. Rather, it is necessary to refer back to the 'galactic' city introduced in Chapter 2. Now industry is rather one of the floating elements, the fixture in space of which is as much related to the density of research and development networks and their emplacement as to any of the traditional explicands of location.

There are inevitable outcomes for the city. On the one hand there are the 'technopoles', for already companies like Mitsubishi are envisaging 'a knowledge oriented future for industrializing countries'. These are modular configurations for new technologically based settlements, restructuring and redeploying Silicon Valley in a formal, planned manner.

In direct contrast, there are the cities of the first Industrial Revolution with much of the space which once held heavy industry now derelict wasteland or seeking a future, often as theme park or heritage site. The old inner city areas, once the location of so much smaller-scale industry are now bereft of industrial employment and especially of unskilled jobs, and more and more they have become the abandoned no-go areas of the contemporary city. These problems will be considered later (Chapter 14) in relation to the residential areas of the city, but the basic problem derives from industrial change.

11.3 OFFICE LOCATION IN THE CITY

If industry and retailing have been dispersed from the city centre, then there is no doubt as to what has replaced them, for the great take-up of city centre land has been by office uses, employ-

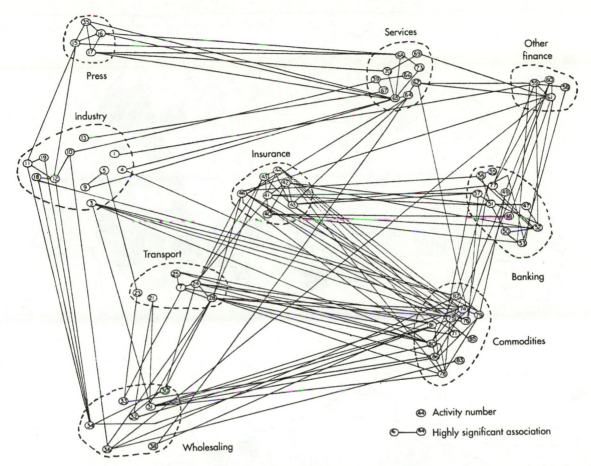

FIGURE 11.6 *The spatial linkages between offices in the city of London, mid-1960s: correlation bonds significant at 0.1 per cent level (after J. Goddard, 1968). The basic data were employment in 80 different types of offices or activities. Note: the numbers refer to the detailed categories used by Goddard. Owing to space they are not given here.*

ing that term in its broadest sense. Thus Coupland (1992: 30–1) demonstrates that between 1971 and 1978 some 49,405 manufacturing jobs were lost in Central London and although there was a gain of 4032 between 1978 and 1981, there was a further loss of 3842 between 1981 and 1984. In complete contrast whereas there were losses of 61,084 jobs in services between 1971 and 1978 and a further loss of 9108 between 1978 and 1981, the period 1981–4 saw a gain of 33,926 jobs. More pertinently still, jobs in business and financial

services increased by 13,580 between 1978 and 1981, and by 29,674 between 1981 and 1984.

These changes are a particular representation of the major switch in urban functions which has already been discussed in Chapter 6 but is here put in the context of its physical representation, or of land-use patterns and the built environment. 'The principal functions of major world cities today is to provide access to information users and providers engaged in the provision of advanced business services. As a result, technology intensive office buildings are

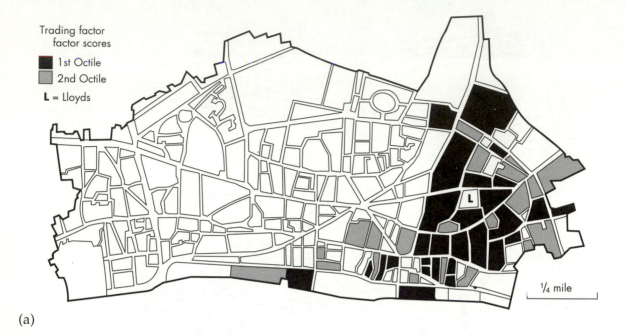

(a)

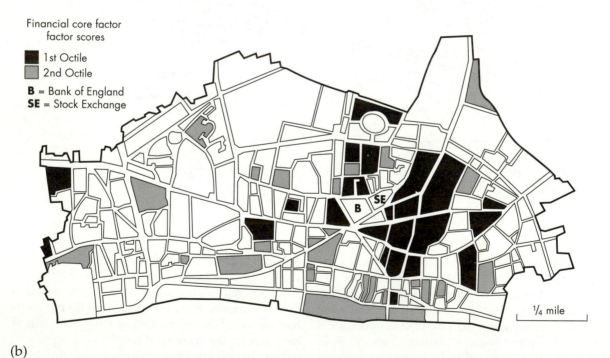

(b)

FIGURE 11.7 *The distribution of four factors identified from an analysis of activities in the city of London (after Goddard, 1968). A: The trading factor B: The financial core. Source: As Fig. 11.6.*

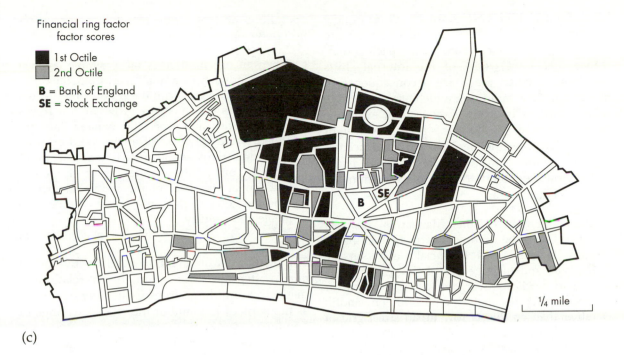

(c)

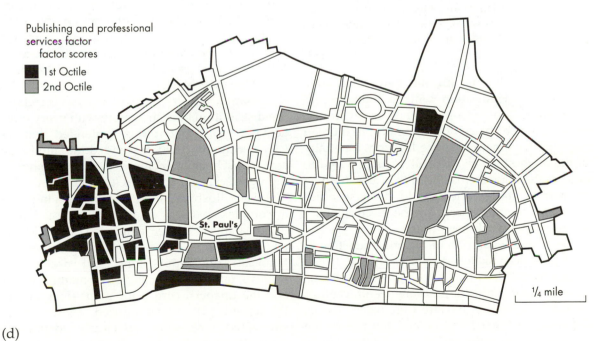

(d)

FIGURE 11.7 continued *The distribution of four factors identified from an analysis of activities in the city of London (after Goddard, 1968). C: The financial ring factor D: The publishing and professional services factor. Source: As Fig. 11.6.*

being built in high-cost centres – albeit in new locations – and advanced business services are continuing to concentrate near each other' (Moss, 1991: 184). Part of the internal shifts which have taken place are a consequence of the demand for buildings which are compatible with computer operations, especially extensive floor areas that can bear the necessary weight loads as well as the necessary ducted accommodation, for communication cables. These shifts, however, tend to take place within the general central area.

> The migration of the Manhattan financial district from the narrow caverns of Wall Street north to the World Financial Centre on the Hudson, and to large parcels on Water Street next to the East River, reflect the demand for mega structures that can be built on landfill rather than on the narrow urban street grid of the early twentieth century. The stock market is no longer defined by a place such as Wall Street – but is what Duffy and Henney (1989) calls 'an aggregate of activities' located in trading rooms linked to each other and to stock exchanges.
>
> (Moss, 1991: 183)

Or, in other words, 'in the new knowledge environment civilisation will be built around communities of people, and less around communities of place'. But the non-place urban realm in the sphere of office location, especially at the highest levels, is still a long way off. Indeed, the opposite has been the case, in direct contrast to earlier attempts to disperse office work and to the concept of the electronic city where people work from home.

The general evolution of the office function in towns has been sketched by Cowan *et al.* (1969) and Daniels (1979a and b). As Cowan develops the argument, until the mid-eighteenth century the main symbols of the city were palace, castle, cathedral and market place and the role of the office under such circumstances was very small. Even government was limited. Morgan (1961) estimated that in 1793 the British Treasury only employed 37 people! It was industrialism which brought the growth of the office with the crucial need for controlling the complexity of the new

operations and ensuring effective communication between their parts.

> When an organisation is fairly simple, as in a cottage industry or a small workshop, the office function is integral with the production process itself. But as the enterprise grows and increases in complexity, and especially as more links are added to the networks of internal and external communications, so the office function begins to separate off ... the separation of the office function at the level of the industrial firm is repeated on a city-wide scale during the nineteenth century.
>
> (Cowan *et al.*, 1969: 28)

The result was the creation of distinctive office areas, often growing about, or in relation to distinctive buildings from earlier periods, such as the Bank of England or the Stock Exchange in London.

The most effective analyses of traditional office areas were carried out by John Goddard (1968, 1970) on the city of London. An early study was based on local authority data and published in 1968. Employment in 80 categories for some 9000 establishments was aggregated into 216 street blocks, thus giving an 80×216 data matrix. The first stage of analysis correlated the distribution of each activity with every other activity. From that it was contended that the City's economy could be divided into nine basic sectors each being composed of a number of highly specialized activities (Goddard, 1968: 72). These are illustrated in Fig. 11.6, which indicates a complex pattern of interlocking activities. Factor analysis is then employed which identifies five factors called the trading factor, the financial 'ring', publishing and professional services, textile trading and other manufacturing and the financial core. Four maps (Figs 11.7(a), (b), (c) and (d)) are reproduced. These demonstrate the emergence of clearly identifiable concentrations. The first, the trading factor, is essentially eastern in its distribution as might be expected from its strong historical links with dockside. It focused on Lloyds as the predominant institution. The financial core, dominated by banks and insurance companies, formed a continuous zone around the Bank of England, while it itself was surrounded by the financial ring made up of stockbrokers and jobbers, accountants and finance houses. Publishing and professional services were essentialy western, grouped around the traditional focus of Fleet Street. Here, then, was the characteristic patterning of distinctive office areas, though with extensive interdigitation; there were no single office-type regions.

It is these long-standing locational patterns which have been greatly modified by the tremendous rise in transactional activities, as well as by pressures for more and more specialized space. In London this has been marked by eastward extension so that the old areas of dock and warehouse have become the sites of new developments. As Moss writes, 'in London the financial district – once defined by proximity to the Bank of England [note Goddard's financial core as defined above and in Fig. 11.7(b)] – has been extended to encompass new sites on the Dockland and over railroad land that can meet the spatial and technological requirements of the financial services industry' (Moss, 1991: 183; See Fig. 11.8). But not only that, other activities have shifted, above all, newspaper production from Fleet Street to Wapping already noted in industrial terms. In that case not only was it a locational break from a long-standing location, but it was also associated with a break of working practices which had also grown up over a long period. The main office movements are epitomized by Broadgate and Canary Wharf.

Broadgate has been called 'the showpiece of London development in the 1980s, pioneering and epitomizing the new locational shifts, design and development processes that emerged in the second half of the decade' (Punter, 1992: 78). It stands out especially as a mixed scheme with:

a restaurant, pub and shops on a public walkway through the scheme [and] the creation of a square that is generous and calm.... The first two phases of Broadgate complete the east and south sides of the square while the next three phases paralleled the western edge of Liverpool Street station, creating a link with both the new

station concourse and the Finsbury Avenue Square. A second new square doubles as an ice rink in Winter and a multi-purpose performance space in the Summer, fringed by a circular arcade of shops and two restaurants which reinforce public use.

(Punter, 1992: 78)

The location is shown on Fig. 11.8, which also stresses the way in which railway terminus land has played a significant role in commercial developments.

Canary Wharf, perhaps more noted for its problems in the recession of the early 1990s, is part of a larger scheme to regenerate London's docklands under the aegis of the London Docklands Development Corporation (LDDC) (Fig. 11.9). It fits in with the central area public/private development schemes which have already been discussed for city centre areas (Chapter 9). It is difficult to improve on Roger Lee's summary:

Located in the former West India Docks, Canary Wharf took its name from one of its former specialisms, the import of fruit from the Canary Islands. Today it is the scene of an audacious but flawed attempt to build a business centre in Docklands to rival the City of London and the West End... . Between 2 July and 17 July 1987 Olympia and York (O and Y, a private Canadian company and one of the largest property developers in the world) secured Canary Wharf...and signed an agreement with the LDDC to develop the site. A phased development programme to construct 10.9 million square feet of commercial, retail office and hotel space was begun. The scheme is so big that it has come to dominate, even to determine, the future prospects not only of Docklands but of the wider but diffuse and ill-coordinated attempt to shift the geography of growth in London away from the west towards the east Thames corridor.

(Lee, 1992: 7)

Ignoring the financial difficulties of O and Y, in the context of locational significance Canary Wharf epitomizes all those forces which influence locational change, not only national in the

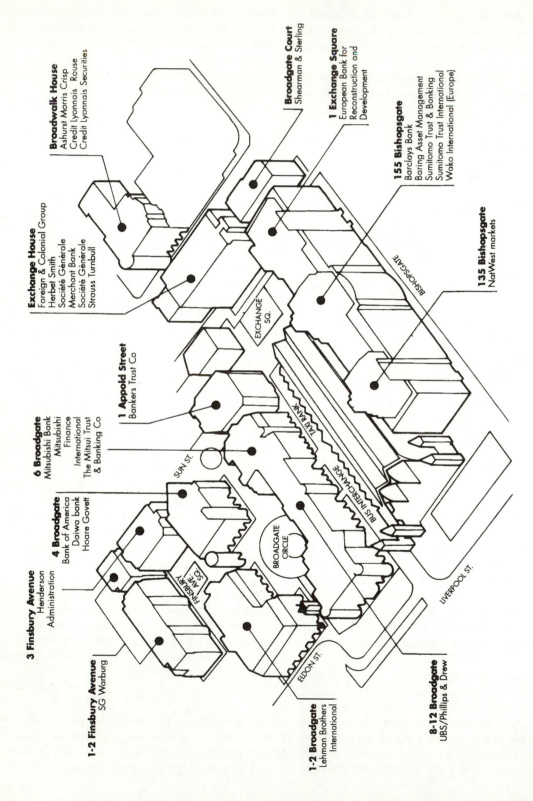

Broadgate Court
Shearman & Sterling

1 Exchange Square
European Bank for
Reconstruction and
Development

155 Bishopsgate
Barclays Bank
Baring Asset Management
Sumitomo Trust & Banking
Sumitomo Trust International
Wako International (Europe)

135 Bishopsgate
NatWest markets

Broadwalk House
Ashurst Morris Crisp
Credit Lyonnais Rouse
Credit Lyonnais Securities

Exchange House
Foreign & Colonial Group
Herbel Smith
Société Générale
Merchant Bank
Société Générale
Strauss Turnbull

6 Broadgate
Mitsubishi Bank
Mitsubishi
Finance
International
The Mitsui Trust
& Banking Co

1 Appold Street
Bankers Trust Co

4 Broadgate
Bank of America
Daiwa bank
Hoare Govett

3 Finsbury Avenue
Henderson
Administration

1-2 Finsbury Avenue
SG Warburg

1-2 Broadgate
Lehman Brothers
International

8-12 Broadgate
UBS/Phillips & Drew

SUN ST.

FINSBURY
AVE.
SQ.

BROADGATE
CIRCLE

TAXI RANK

BUS INTERCHANGE

EXCHANGE
SQ.

BISHOPSGATE

LIVERPOOL ST.

EDON ST.

FIGURE 11.8 *Who's Who in Broadgate. Source: Sunday Times. 27 Feb., 1994. Business 3–7.*

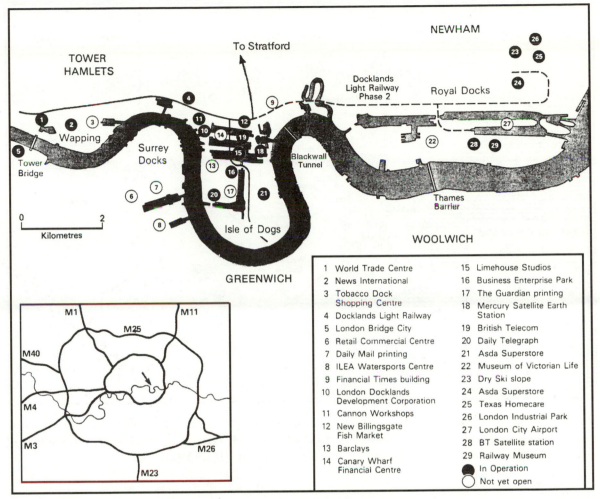

FIGURE 11.9 *London's Dockland. Development and Proposals in 1986.*

1	World Trade Centre	15	Limehouse Studios
2	News International	16	Business Enterprise Park
3	Tobacco Dock Shopping Centre	17	The Guardian printing
4	Docklands Light Railway	18	Mercury Satellite Earth Station
5	London Bridge City	19	British Telecom
6	Retail Commercial Centre	20	Daily Telegraph
7	Daily Mail printing	21	Asda Superstore
8	ILEA Watersports Centre	22	Museum of Victorian Life
9	Financial Times building	23	Dry Ski slope
10	London Docklands Development Corporation	24	Asda Superstore
11	Cannon Workshops	25	Texas Homecare
12	New Billingsgate Fish Market	26	London Industrial Park
13	Barclays	27	London City Airport
14	Canary Wharf Financial Centre	28	BT Satellite station
		29	Railway Museum

attempt at regeneration using all available means, the Isle of Dogs was declared an 'Enterprise Zone', but international as well, with the attempts to harvest profits to be derived from the restructuring being brought about in the transactional city.

Changes in Manhattan have already been noted (p. 204) but undoubtedly the most radical changes in American cities have been those in Los Angeles. Although there has been a major growth in 'high-tech' industries, Los Angeles has experienced the same decline as other large cities (Fig. 11.10). But alongside this there has been an enormous increase in financial transactions. Prior to 1960 the best known feature of central Los Angeles was the absence of commercial buildings. In that decade the nature of development was established as the Community Redevelopment Agency of Los Angeles began to clear land, not for housing but for business. 'The Bunker Hill project established the character of urban development in the next quarter century. By 1984, 60 municipalities in Los Angeles county had used redevelopment authorities to clear land at a cost of $3 billion in tax funds. On this redevel-

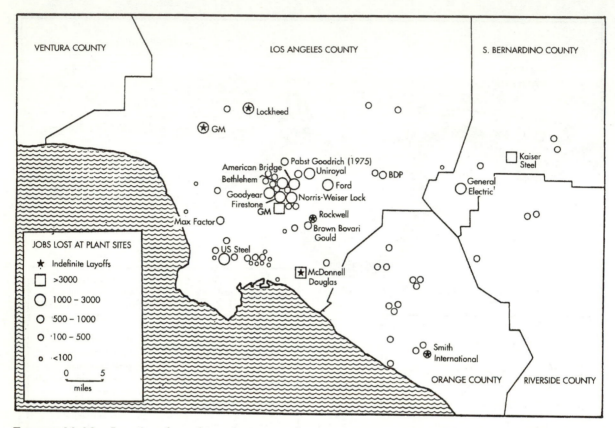

FIGURE 11.10 *Los Angeles: plant closures and major lay-offs, 1978–1982 (after Soja, Morales and Wolff, 1983). Source: E. Soja, R. Morales and G. Wolff, 1983: Urban restructuring: an analysis of social and spatial change in Los Angeles. Economic Geography 59, 216 (Fig. 3).*

oped land were erected 31 shopping malls and centres, at least 20 high-rise office buildings' (Light, 1988: 73) together with industrial parks, apartment complexes and hotels. 'In 1984, the City of Los Angeles announced an agreement between the Community Redevelopment Agency and Maguire Thomas Partners, real estate developers. This agreement authorized construction of three new office buildings with 3.2 million square feet of office space' (Light, 1988: 73). The driving force behind these developments was the role of Los Angeles as the financial centre for the western coast of the USA and also as the gateway to the Pacific mainlands, 'and moreover, to rival New York City as a management and control centre for

global capital' (Soja *et al.*, 1983: 112). As a result, the downtown area, once so anonymous as to be negligible, is now dominated by the office towers of financial concerns (Fig. 11.11). 'Today, downtown Los Angeles contains more than a third of the high-rise office space in the region, thirteen major corporate headquarters ... and the southern half of the Pacific Stock Exchange (including its computer headquarters)' (Soja *et al.*, 1983: 244).

The greatest contrast to Los Angeles is La Défense which has been established as an office quarter in Paris with the specific intention of abstracting from the central area, in an organized and formal way, those business uses which were likely to cause congestion. The point

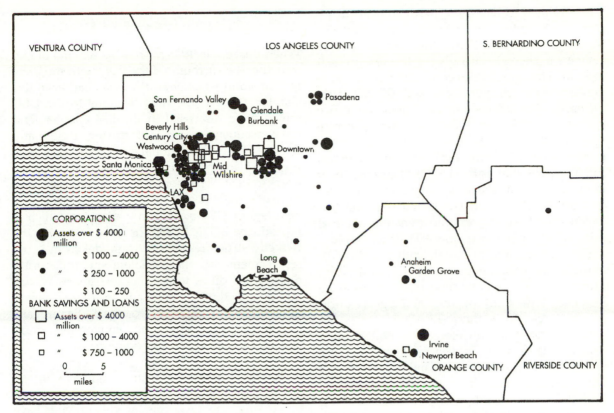

FIGURE 11.11   *Los Angeles: Corporate and banking headquarters in the Los Angeles region. Source: As Figure 11.10, p. 223 (Fig. 5).*

of origin must go back to the sixteenth century when Francis I built the royal palace, the Louvre. It was developed by Catherine de Medici, the wife of Henry II, when she created the Tuileries Palace and the Tuileries gardens. Thus came into being a westward extending axis (see Fig. 8.2(b)). It was transformed into one of the major features of Paris by Le Notre, under Louis XV, when he planned the Champs Elysées with the great Place de la Concorde at its junction with the Tuileries Gardens. At the western end of the Avenue is the Arc de Triomphe de L'Etoile and, still further, the Avenue de la Grande Armée culminates in the Porte Maillot. The line is extended in the Avenue Charles de Gaulle to the Pont de Neuilly and, beyond the Seine crossing was the rond-point La Défense.

In 1958 the French government set up the Etablissement Public d'Aménagement de la Région de la Défense (EPAD) with the role of organizing and rationalizing the somewhat untidy character of the area and developing a new core of development. Some 760 hectares were included about the rond-point of La Défense. EPAD does not itself construct buildings but sells building rights to investors. In 1964 approval was given for the construction of a series of office buildings which were to be 100 metres high and would contain 750,000 square metres of floor space. In 1970, in a second phase, the height was increased to 200 metres and the total office space increased to 1.5 million square metres.

The period 1974 to 1978 marked a hiatus, and indeed a period of considerable difficulty, partly related to recession, partly to over provision. But 'after four years of paralysis, Raymond Barre, the Prime Minister, decided to revive La Défense and give it the resources it needed. These measures coincided with a resurgence in the property market' (EPAD, Info Défense). La Défense is not solely an office quarter. In 1981 the 4-Temps shopping centre opened. It now has some 200 shops, three supermarkets, 20 restuarants and nine cinemas. EPAD claims that with 111,000 square metres it is one of the largest regional shopping centres in Europe. But, even so, EPAD describes La Défense as first and foremost an international business district with some 900 firms employing 100,000 people over half with executive or managerial jobs. Fourteen of the 20 leading French companies have their head offices at La Défense, whilst 13 of the 50 biggest world groups have offices there (EPAD).

But the fact that La Défense was on the major axis of the growth of central Paris was not ignored and President Mitterand ensured that it was included in his 'Great Architectural Projects'. The result was the *Grande Arche* designed by a Danish architect and appropriately continuing the monumental character of the westward extension. Nor is it finished there. EPAD has been charged with extending the axis a further 2 kilometres with residential, educational and office uses.

Here then is an office quarter of a metropolis of a very different nature and character. It is integrated into the whole planning history of the city, continuing a tradition that virtually starts with the building of the Louvre and represents a far-sighted scheme to meet the demand for office space where recent construction can provide all the necessary technical requirements and where other uses, including residence, are integrated. Perhaps one contrast with London can be noted. Canary Wharf and the associated developments are seen as attempting to counteract the westward growth of London by a deliberate shift to the east. In Paris, La Défense confirms a very long-standing direction of growth. But both are indicative of where the dynamic of the contemporary city is to be found.

11.4 CONCLUSION

When Burgess in 1925 set out his schema of city structure, he surrounded the CBD with the Zone in Transition which was discussed earlier in this chapter. Clearly, the modifications to the CBD which were discussed in the last chapter, and the consequences of manufacturing decline and the rise of the transactional city which have been reviewed in this, have made the concept an anachronism. The contrast between past and present has been neatly summarized by Beauregard (1989b: 255) in a diagram which is reproduced in Fig. 11.13. The main feature in the two city structures set out in model form is the replacement of the obsolete manufacturing districts by office towers, as in Canary Wharf, and by mixed office, hotel and residential development, as, in part, at Broadgate. Beauregard also includes the ubiquitous convention centre and its associated accommodation.

> In effect, mixed-use districts adjacent to the CBD, once labelled the zone of transition, and working- and lower-class neighbourhoods have been physically, socially and economically transformed. The zone has been replaced increasingly by high-priced residences, recreational spaces and new commercial corridors. It has become the zone of reinvestment. ... A central core surrounded by industrial, low-level commercial and transient residential uses no longer exists. Rather, substituting for that arrangement is a vibrant office core with a shrinking zone of transition and an expanding collar of relatively affluent areas.
>
> (Beauregard, 1989b: 255)

That conclusion is largely in line with what the last chapter on the CBD, and this on industrial and office uses have demonstrated. City centres have been transformed over the last quarter of a century. They are not only greatly different from the Chicago of 1925, which was Burgess's basis but also from the cities of the 1960s. In that process of change the decline of smoke-stack industry, the shift to out-of-town shopping centres and the great efflorescence of

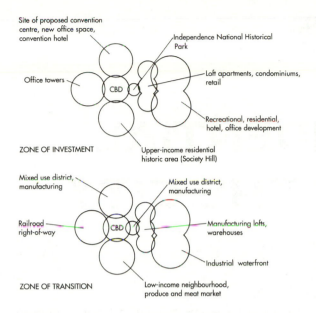

Figure 11.12 *The zone in transition and to the zone of reinvestment. Source: R. A. Beauregard 1989: Atop the urban hierarchy. Totowa, NJ., Rowman and Littlefield Publishers Inc. Figure 7.1, p. 255.*

office building have all played their part, along with schemes for central area regeneration. So, too, have all the 'players' which were set out in Chapter 8 on planning, including corruption. In writing of the office construction boom in Los Angeles, Light adds cryptically, 'instances of corruption surfaced' (Light, 1988: 73), although the comment is expanded in a footnote. It is a theme more explicitly considered in Mike Davis's *City of quartz* (1990).

At this point one reservation needs to be made. This discussion has concentrated on the 'mega-cities' which tend to dominate international discussion and analysis. Moreover, as Peter Hall argues

> agglomeration economies are undoubtedly even stronger for the information handling industries than they were for the old goods-handling industries. Because of the subtle relationship between face-to-face exchange and remote

exchange, and because the take-up of new technology is so much faster in information-rich techniques, a continuous process of synergistic innovation is set in these regions.

> (Hall, 1991: 20)

This is effectively a contemporary interpretation of circular and cumulative causation. Hall adds, 'in this process, the peripheral information-poor regions will be left ever farther behind' (Hall, 1991: 20). Translating that into an urban context, smaller and provincial cities will not experience the type of development sketched in Fig. 11.12. True, there will be parallels. There will even be the dispersal of offices where functions are predominantly routine. When the dispersal of offices was a dominant planning theme, classifications of meetings were proposed. Characteristic was the identification of orientation meetings undertaking broad reviews with a wide range of experts, planning meetings where high level officials were brought together, and programme meetings which were simply day-to-day reviews of work. Whereas the first two demanded accessible, central locations, the third could be located anywhere and therefore identified the routine procedures which could be dispersed. A British example is the Driver and Vehicle Licensing Centre which was relocated at Swansea in South Wales or the operation of National Savings from Durham. They carry out routine procedures. If these procedures were to be changed then the discussions would take place in government in London, not in Swansea or Durham. So routine office functions will be found in smaller towns and cities to a degree replicating the changes in the larger cities. But even so, the smaller the town and the farther it is from the driving forces of Western economies, the more like the towns of the 1960s it is likely to be.

It is apparent that the changes at, or near, the city centre which have been considered involve other changes such as gentrification. Moreover, the problems of the inner city, ignored in Beauregard's zone of reinvestment, are as acute as ever they were. But these are issues to be considered in the chapters which follow.

12

THE RESIDENTIAL AREAS OF THE CITY

12.1 INTRODUCTION: A BRIEF EVOLUTIONARY REVIEW

One of the prime functions of the city is to provide shelter for its citizens. There was a time when many surveys of settlements began with the assertion that man (at that time 'man' was universally used) was a gregarious animal and that association in space reflected some atavistic principle. But spatial agglomeration was always based on some need other than that of a psychological imperative. The Egyptian hieroglyph for the city, a cross within a circle, represented the two crucial demands of association about the cross-roads of trade under the protection of defensive walls. As was set out in the chapter on urban functions (Chapter 6) it is only through spatial association that economic advance via occupational specialization can be accomplished and maintained. But whether the city is conceived as a centre of trade or a defensive strongpoint, there still arises the notion of a hierarchical structure of its residents into those who control and order and those who respond and obey. And with that comes the implication of the spatial segregation of groups within the city based on status and prestige. Residence, therefore, is unlikely ever to have been a simple and uniform distribution.

Even so, it seems at times a tenet of faith in urban studies that if one goes back far enough in history, a stage is eventually reached when settlements were characterized by an undifferentiated accumulation of land uses. Within the residential category there was a corresponding incoherence. Subsequently as a response to growth, and especially the vigorous and unprecedented growth of the nineteenth century, land-uses became differentiated into specialized and discrete regions, while within the residential category segregation by social class or migrant origin created distinctive social areas. Certainly it can be argued that single class social areas were a product of greatly enhanced mobility and do not appear until well into the twentieth century.

But residential segregation in some form has been characteristic of cities at all times. It is of course true that the compression resulting from limited intra-urban mobility engendered a much closer association, a necessary nearness which might have supported a social order based more nearly on consensus rather than the conflict which derived from spatial distancing, but segregation there was. It was, however, on a more intimate scale. Thus in the nineteenth century the green baize door which separated 'upstairs–downstairs' rather than the 'tracks' of the 1930s, constituted the physical divide, a segregating boundary. Indeed it is possible to argue that all of the uses which became part of the differentiated modern city were at one time to be found within the single house (Carter, 1980: 183); craft, industry and selling, extended family, apprentices and labourers; but they were as effectively separated within the house as the slaves in the cities of the American South (Wade, 1964).

The reference point for consideration of the 'pre-industrial city' is Gideon Sjoberg's book which carries that title. His basic statement on spatial organization is well known: 'The pre-industrial city's central area is notable as the chief residence of the élite. Here are the luxurious dwellings, though these often face inward, presenting a blank wall to the street – a reflection of the need to minimize ostentation in a city teeming with the under-privileged' (Sjoberg, 1960: 97). Perhaps one should add also, providing maximum defence in a city where policing was minimal. 'The disadvantaged members of the city fan out towards the periphery, with the very poorest and the outcasts in the suburbs, the farthest removed from the centre' (Sjoberg, 1960: 98).

The explanation for this situation is twofold, for it was partly related to the need of the élite to control a power base, partly related to the technology of transport. Control was only feasible through having ready access to the headquarters of the governmental, religious and educational organizations. The highly valued residence, then, was where the fullest advantage could be taken of the city's strategic facilities; in turn these became tightly bunched for the convenience of the élite. Much more pertinent in Sjoberg's view was the impact of intra-urban transportation which was slow, uncomfortable and inconvenient and resulted in the minimization of the need to travel. Moreover, combining both aspects of the explanation, residence outside the defences meant exposure to attack, while the journey into the city was dangerous. The élite sought the security, comfort and control of the city centre and were willing to suffer its twin dangers of plague and fire, thus London's earliest building regulations sought to minimize the danger of fire. The lower orders were relegated to the urban periphery as were the noxious or dangerous occupations such as metal-working and tanning.

Ample testimony to this generalized pre-industrial pattern has been provided. In 1935, a quarter of a century before Sjoberg's book was published, Hoskins wrote in *Industry, trade and people in Exeter, 1688–1800*, using the evidence of the Hearth Tax of 1671–2:

> The seven richest parishes form a compact nucleus in the heart of the city, though mainly to the eastern side of the Carfax, while the five poorest parishes all lie against or outside the line of the medieval walls. In the rich heart of the city we have a population of merchants and craftsmen, with a flavouring of 'esquires' who have, for the most part, made their money in trade, and it was from here that the city was governed and had been governed from some unknown time in the past.
>
> (Hoskins, 1935: 116)

Subsequent studies by Langton on Newcastle (1975) and a review of Phythian-Adams's work on Coventry (1979) have confirmed the broad outline presented by Hoskins. But that confirmation is by no means simple and direct. Qualification is best introduced by a quotation from Susan Reynolds's book on English medieval towns:

> A society stratified by wealth is different from one stratified by groups defined by their occupations or their function in a market economy. Though the social stratification in medieval towns may well have been formed by an unconscious muddle of both principles, we shall not be helped to understand it by using merchant, craftsman and servant as synonyms of rich, middling and poor. Other factors, too, influenced contemporaries' views of the social hierarchy. Long settlement probably conferred prestige on a family and the influence of noble values helps lend weight on descent and land ownership. What exacerbated class divisions and class consciousness seems to have been less the direct conflict of the differing economic

interests of merchants and craftsmen than the control over local taxation and government which municipal independence gave to the dominant men in a town, whether they were craftsmen, merchants or landowners.

(Reynolds, 1977: 76)

Again, writing of medieval London, Reynolds maintains that 'at the end of the period, London had a prosperous elite, who were distinguished from their fellow citizens by wealth, municipal influence and less measurable elements of social prestige, rather than by their practice of any particular trade' (Reynolds, 1977: 78). In short, there was an urban patriciate derived from aristocratic, land-holding families who dominated urban governmental functions and who were associated with the elaborate ritual linked to them, Hence, the city centre location.

This can be illustrated most effectively by a centre far removed in space though not in time, Angkor Thom, part of the great ceremonial centre in Cambodia. Writing of the last city built on the site in the late thirteenth century, L. Briggs notes that the 'enclosure at Angkor Thom was simply a religious, administrative and aristocratic centre, where lived – clustered around the capital and principal temple (Bayon) – the civil and military functionaries, the priesthood, the rich families and the army... ' (Briggs, 1951: 219). This constituted an enormous patriciate which (according to a list for the temple of Ta Prohm at Angkor, built by Jayavarman VII) included 18 high priests, 2740 officials, 2202 servers and 625 dancing girls. To support these, tens of thousands of peasants were needed to produce food and craft articles. This is the specialization of occupation which is inherent in urbanism. But the markets and homes of the masses were in the suburbs along the Barays, or large artificial lakes to the east and west of the walled city, and along the banks of the Siemreap River even to the mouth (Briggs, 1951: 219). At this period the main markets were outside the southern and eastern gates: the central city square was used for festivals and parades. The markets were not characterized by permanent structures. According to Chou Ta-Kuan, a Chinese traveller who visited Angkor in 1296–7, 'most of the commerce was carried on by women... . They had no permanent shops in the ground. For a desirable place rent was paid to the authorities (Pelliot, 1951: 12–13).

The ceremonial centre, with economic activities relegated to less significant locations was characteristic of the early city.

To revert to the British example, however, it is interesting that over time the urban patriciate, exercising influence by way of descent and nobility, gave way to a more predominantly mercantile class which derived its significance from the wealth derived from trade. But to make such a distinction is difficult. As Christopher Brooke writes:

It has long been disputed whether the patricians who came to rule the cities of Europe in the central Middle Ages (to adapt Lady Bracknell's words) were born in the purple of commerce or had risen from the ranks of the aristocracy – or as the problem was more commonly phrased, were the sons of the feudal nobility or self made merchants. For two reasons, the problem thus stated is now seen to be quite unreal. First of all we know the origin of a tiny percentage of the patricians of the crucial period between 1050 and 1150. ... Secondly, it takes too little account of the pattern of the economic life of the age.

(Brooke, 1975: 77–8)

Even so, Phythian-Adams (1979) in his study entitled *Desolation of a city, Coventry and the urban crisis of the late Middle Ages*, is able to identify a point very much later in time when the old order gave way. He notes the last banquet of the Corpus Christi Guild in 1534, a meeting which symbolized the end of an era. A year later the Corpus Christi Guild was united into the Trinity Guild.

At a single stroke the social age compartmentalization at the upper end of society and the traditional sequence of senior civic office-holding were swept away. With this went the complete collapse of the ceremonial system as an ideal mirror of community. ... As outdoor ceremonies vanished, and the cathedral was demolished, so too did the ritual centre of the city lose its signficance.

(Phythian-Adams, 1979: 269–70)

This analysis, if its universality is accepted, portrays a period of urban change which had strong spatial implications. The abandonment of public ceremomy – the medieval mirror shattered – together with the transformation of the higher echelons of society, suggest that there must have been a falling off in the significance of the centre, a feature which Phythian-Adams explicitly maintains. Moreover, as culture became private and of the home rather than public and of the city streets, so the incentive for central residence falls away, certainly in relative terms. These changes might well mark a shift from a ritual centre dominated by an urban élite, a true patriciate, to a series of commercial sectors dominated by mercantile interests. These Phythian Adams identifies in Coventry as does Langton in Newcastle. Immobility still tied the rich merchant to the centre and control of city government and the operation of the markets was essential, but the home-centred privatism of the new élite was a critical change. It marked a significant inflection in urbanism in the Western world; the entry into the determination of city structure of the uninhibited, private ideal where the home and not the community became the controlling force and while the city was certainly to be under firm government, the prime function of that government was to provide an appropriate environment for private enterprise and private living.

It was against the dominance of such principles and within a mercantile prelude to industrial capitalism that the industrial period was to transform completely the structuring of the residential areas of Western cities as immensely powerful forces were brought to bear on the existing organization. It is a process that cannot be traced in detail here, but perhaps it can be resolved into the interplay of three sets of forces. The first was demand arising for housing; the second was the range of interests involved in

supplying this residential accommodation; the third was constituted by the constraints which mediated between demand and supply, above all the constraint resulting from public intervention. But this third and intervening force, weak but growing, was never more than a qualification of the driving force of the private ideal. Each of these can be considered in turn.

Demand for housing

It is possible to identify three separate though closely related factors.

1. When Eshrif Shevky and Wendell Bell set out the bases of what they called social area analysis, they proceeded from the assumption that the modern city has been shaped by the complex whole of society and that, therefore, the social forms of urban life are to be understood within the context of the changing character of the larger, containing society. Foremost among the changes was the variety of skills which developed among individuals and subsections of the population. Due to the processes of industrialization and urbanization itself skills became much more widely extended in their nature. That can be regarded as the classic division of labour of the capitalist state. This not only applied to industry itself, but to the great range of business, professional and administrative roles which developed during the nineteenth century. A highly complex and intricately stratified class system was created. But this was also a highly competitive situation where status was no longer ascriptive but achieved through the talents which the individual could deploy. Those who had attained a status defended it, particularly against dilution. Those lacking status sought to achieve it and since the basis was competition social mobility was characteristic. The term *nouveau riche* is central to the reality of such upward mobility. Each stratum of society sought to distance itself from that below and to copy the style of that above.

The élite of the upper classes were still drawn in part from the landed gentry, especially the younger sons, but in a class that was never so exclusive as to prevent their being joined by the very wealthy whatever their origins. 'The formation of a formal social life confined to *private* locations and rigidly defined by convention which was embraced by aristocratic and middle classes in both town and country made it possible for upwardly mobile individuals and parts of families to gain access to new groups if they had the necessary qualifications' (Davidoff, 1973: 27).

The italics in the above sentence are mine to throw emphasis on the critical condition. Whereas kinship had been the major criterion, sponsorship gradually developed by which a member of the aristocracy acted as a sponsor for the *nouveau riche* even to the extent of standing in as host or hostess at parties.

In order to formalize this aspect of the class system, an elaborate code of etiquette grew up from the second quarter of the nineteenth century. It performed the same role as initiation ceremonies and other rituals in trade unions. In a system where the aim was to keep those below at bay while gaining access to the next higher group, introductions were vitally important and subject to clear, prescribed custom. The leaving of cards also played an important social role.

The woman's role became a limited one. Gainful occupation for upper-class women was unheard of while large numbers of servants did all of the household work. An elaborate social routine was partly a response to these conditions and also to location, for as those highest in the social scale abandoned the city centre, the association of women with city government was broken. Charitable activities, the arts, and the social round replaced it, although the rearing of large families was the main occupier of time. Burnett has effectively presented the way in which, mediated by a bourgeois royal family, the ideals of the middle class dominated Victorian Britain.

This new class was the most family-conscious and home-centred generation to have emerged in English history. Originating perhaps in the 1820s a distinctively middle class family pattern grew up which was in full development by the middle decades of the century; although at first in sharp contrast to the family patterns of both the aristocracy and the working classes, it was one which, in time, the middle classes were successful in imposing on the orders above and below themselves. It had certain well defined characteristics, based on a set of values and beliefs which were imbued almost with religious authority – the belief in male superiority, from which it followed that man alone was capable of wielding economic, political and legal power, and that wives and children owed only obedience to this God-like creature; the belief that a "lady" did not work, that her vocation in life was a prudent marriage and the procreation of children. ... The home ... became almost a sacred institution.

(Burnett, 1980: 95)

Literature, of course, provides its illustrations of social forces. The epitome of this new class was surely the 'Man of Property' himself, Soames Forsyte. Indeed all of *Upstairs Downstairs* could be considered as written to illustrate the concept in the above paragraph. Soames's new house at Robin Hill (within 12 miles of Hyde Park corner be it noted) was due to the originality of Bosinney, his architect, but the Watney Lodge (visited by the Pooters, themselves archetypically lower middle class) was more in the general taste.

The Watney Lodge I seem to see:
Is gabled gothic hard and red
With here a monkey puzzle tree
And there a round geranium bed.

Each mansion, each new-planted pine
Each short and ostentatious drive
Mean Morning Prayer and beef and wine
And Queen Victoria alive.

(Betjeman, 1958: 259–60)

2. Given this situation, the place of residence (the house itself) became a critical element. It aped the aristocratic country estate; it was necessarily detached to give privacy; the grounds constituted the estate; the shrubbery the woodland. To achieve this it had to be suburban. But there were two other forces driving those economically best-off to the suburban rural fringe. First and foremost (and the second factor relating to demand) were the deteriorating conditions of the city centre as industry and overcrowding brought dirt, noise and above all, disease. Lewis Mumford sums up the movement in typical style;

> The early romantic suburb was a middle class effort to find a private solution for the depression and disorder of the defouled metropolis: an effusion of romantic taste but an evasion of civic responsibility and municipal foresight. The instincts which prompted this exodus were valid: caught in the new urban wreckage, the old cry 'women and children first' was a sound one. Life was actually in danger in this new urban milieu of industrialism and commercialism, and the merest counsel of prudence was to flee – flee with all one's goods, as Lot and his household had fled from the sultry hell of Sodom and Gomorrah.
>
> (Mumford, 1961: 492)

3. The third factor was related to demand in that the means were at hand and were to become increasingly so during the century. If, at first, movement depended on the carriage and the few good roads, by the 1860s the suburban railway was available. 'In Manchester the successful cotton magnates had by-passed the first ring of suburban development ... and escaped to the rural retreats of north Cheshire. In London the city businessmen could choose to ignore the developed inner suburbs such as Kensington and Hammersmith in the west, and travel swiftly to the distant countryside of the Thames Valley or the North Downs' (Gaskell, 1990: 160). In this way, perception

of class identity, accompanied by the glorification of the private ideal, resulted in physical distancing. Effective transport allowed the city to become segregated into homogeneous areas according to social status and because of distancing conflict became the mark of class relationships.

Suppliers of housing

If demand brought about the differentiation of social areas, so too did supply. Landowners, developers and builders were able to profit greatly by responding to demand, and by leading it. Perhaps in the supply of the vast areas of housing for those lowest in the social scale the greatest impact was made. The innumerable health reports of the nineteenth century are full of indications of the way in which profits could be made from the provision of housing for the so-called working class, or even for those who could obtain no footing in the social order. The standard exemplification is, of course, Engels's description of Manchester, but a good example is the Welsh town of Merthyr Tydfil (Carter and Wheatley, 1982). A guide of 1848 recorded:

> The best of the workmen's houses are, for the most part, those erected by the different iron companies for such as labour in connexion with their establishments. Some of these appear to have been unsold, especially at Dowlais. Speculators of various kinds seem to have built courts, alleys and rows of houses, wherever opportunities presented themselves, in order to meet the demand for the rapid increase of the town, entirely without regard to any order or system, and without any control as to lines, the form of the streets, or to arrangements for drainage.

The worst areas exercised the literary talents of those who visited them. A correspondent of the *Morning Chronicle* in 1850 wrote:

> There is a quarter of the town extending along a flat on the right bank of the Taff from the lowest point of High Street, towards Cyfarthfa

– the proper name of which is Pont Storehouse; but like the unhappy and lawless people who inhabit it, the place has an alias and is generally known by the name of 'China'. The houses are merely huts of stone – low, confined, ill-lighted and unventilated; they are built without pretensions to regularity and form a maze of courts and tortuous lanes, hardly passable in many places, for house refuse, rubbish and filth. In some parts they are considerably below the level of the road, and the descent is by ladders. Such houses are called 'the cellars'. Here it is that in a congenial atmosphere, the crime, disease and penury of Merthyr are for the most part located. Thieves, prostitutes, vagrants, the idle, the reckless and the dissolute, here live in miserable companionship. This neighbourhood formed the main scene of our enquiries; and what I, that day, saw of misery, degradation and suffering, I shall remember to the end of my life.

Government intervention in housing

The third force operative in the creation of social areas has been the constraints which directed demand and modified supply. As such, it can be interpreted more specifically as the interventions of central and local government, even the growth of municipal government itself. Again, an interpretation can be offered that the crises of capitalism beget the intervention of the state, To quote a definition: 'Planning (and by that I cover all the varieties of constraints) is a historically-specific and socially necessary response to the self-disorganizing tendencies of privatized capitalist social and property relations as these appear in urban space' (Dear and Scott, 1981: 13).

It is perhaps ironic that one of the major impacts of democratic government has been to promote dispersal and segregation. In early times the need to exercise local power, the pre-emptive need to control the mob, held the élite firmly at the centre. However, as municipal democracy grew and the riots of the mob against specific grievances became changed to the more coherent programmes of radical political parties, the need for absolute control of the

central physical institutions disappeared. Also, the need for physical presence diminished and the pressure to live at the centre was released. In short, the arrival of representative democracy transformed all of those aspects of power control which Sjoberg identified as inherent in the pre-industrial city and began the process by which the inner city could be virtually abandoned by the élite.

In more direct terms the growing complex of constraints was to become visible in Britain in the increasing web of public health and housing legislation which was the preface to town planning. Again this is a very lengthy and complicated process which cannot be traced here (but see Chapter 13). Essentially, there were two aspects: the control of building standards and the beginning of municipal housing. But the basic principle was an interventionist one – the provision of a series of safety nets to ensure minimal standards and hence, the health and stability of cities themselves (which might well have been the prime concern) and maybe, incidentally, to catch those who fell out in the competitive process.

In 1875 the Artisans and Labourers Dwellings Improvement Act was passed. It allowed local authorities to purchase and then clear areas of unfit housing. A subsequent improvement scheme drawn up by the authority could be carried out by the sale or leasing of land to builders. But it was enacted that the local authority itself could rehouse the inhabitants although the houses built had to be sold within ten years. The Act is usually referred to as the Cross Act since it was passed when Sir Richard Cross was Home Secretary. But here is Sir Richard, writing in 1882:

> I take it as a starting point that it is not the duty of the Government to provide any class of citizens with any of the necessaries of life, and among the necessaries of life we must of course include good and inhabitable dwellings. To provide such necessaries for any class is not the duty of the State, because if it did so, it would inevitably tend to make that class depend, not on themselves, but on what was done for them

elsewhere, and it would not be possible to teach a worse lesson than this – 'If you do not take care of yourselves, the State will take care of you.' Nor is it wise to encourage large bodies to provide the working classes with habitations at greatly lower rents than the market values paid elsewhere.'

The end result, therefore, of continuing legislation was certainly to raise standards, certainly to provide more and more municipal housing until it rose to some 31 per cent of British stock in 1981, but to make no modification of the basic process of segregation. Indeed the inter-war drive to suburbanization only exacerbated contrasts as those intermediate in the social scale followed the example of those above them and moved farther and farther out and were totally separated from those living in public housing. Perhaps, as David Harvey would have us believe, they were led by those seeking class monopoly rents (Harvey, 1974). But whether supply is the master or not, the result is the same.

All the evidence to this point has been derived from Britain, but perhaps the best summary is by an American. Sam Bass Warner, Jr, in his study, *The private city, Philadelphia in three periods of its growth*, wrote:

The quality which above all else characterizes our urban inheritance is privatism. By and large the productivity and social order of the metropolis flowed from private institutions and individual adjustments. So did its weaknesses. Privatism left the metropolis helpless to guarantee its citizens a satisfactory standard of living. Privatism encouraged the building of vast new sections of the city in a manner well below contemporary standards of good layout and construction. ... It was a private city and the public dimension of urban life suffered accordingly.

(S. B. Warner, 1968: 202)

Perhaps one can return to Phythian-Adams's comment on the sixteenth-century changes in Coventry. The culture which emerged, he wrote, 'was no longer a public, visual affair concerning all those who participated or cared to watch: but a private matter for the individual citizen in the furnishing or decoration of his house, his reading matter, and his worship'

(Phythian- Adams, 1979: 278)

This introduction to the chapter has tried to show that critical to the evolution of segregated social areas in the West have been the values associated with privatism, the fundamental belief in the city as the best arrangement to be devised to satisfy the competing individual interests of the inhabitants or of those inhabitants who could compete successfuly, rather than in a city as a community devised to promote communal ends. But there is an ideal. Even its consummation has been optimistically identified where individual wealth is so universally distributed that socio-economic status, or more critically wealth, is no longer a determinant of where people choose to live. Rather families move at the behest of family needs or notions of culture and territoriality. That is the ideal.

In a broader sense private property is seen as a guarantee, a necessary condition, of individual liberty against the coercion of the State. It is not by chance that a conservative government in Britain made the sale of municipal housing the foremost element in its housing policy. Nor is it surprising that it has been implacably opposed by socialism. Property owning necessarily sucks the individual into privatism and the upholding of the private ideal, or if it be preferred, of the capitalist system and engenders a segregated city.

Ultimately in Western democracies an uneasy balance is struck between the thrusting demands of privatism and the constraints exercised by democratic control. What these have produced is a city racked by a divisiveness of which the major outward and visible sign is a system of residential segregation of which the three extremes – affluent suburbs, deteriorating inner city and crumbling municipal estates – are the most visible symbols.

At this point, therefore, it is necessary to consider the fundamental bases on which residence in the contemporary city can be differentiated.

12.2 SOCIAL AREAS OF THE CITY: A GENERAL CONSIDERATION

An interpretation of residential distributions at a general level was offered when the structure of the whole city was examined in Chapter 7. The outline of this interpretation was most clearly sketched by Alonso in his comparison of historical and structural theories of urban form (Alonso, 1964b). At the heart of the generalized residential pattern lies the paradox that the richest people live on the cheapest land and vice versa or, in other words, that land values decrease with distance from the centre.

The historical explanation is that derived from the analyses of E. W. Burgess and in its simplest form maintains that as the city grows outward from the centre so new houses are built on the periphery. New fashions in housing appear at these points and such houses are desired, and can only be afforded, by the richer elements of the population and they, in consequence, move progressively outward. As this happens, the outmoded houses they abandon filter down the income scale. This produces concentric residential zones with increasing income, and by implication, increasing social status, from the centre out. These are conveniently named working or lower class, middle class and upper class zones.

Structural explanations on the other hand are based on the play-off between accessibility, the costs of commuting and the costs of land. If one assumes typical cultural values and tastes, which are a liking for ample land and a relative willingness to commute, it is clear that more distant and thus cheaper sites are more attractive to the wealthy than the poor. The result is a distribution of income groups which is similar to that of the historical type of explanation, for structural forces and historical process have been, over the last century, working in the same direction.

This simple picture can quite obviously be modified. Thus in structural terms it is possible to introduce the diminishing returns that occur with distance from the city centre. The greater the distance the larger is the city and hence the attractions of the facilities at the centre are very high. At the same time, the larger the city then the greater is the distance to be traversed to the central facilities and the more irksome the journey. It follows that the richest part of the population, the upper classes, will also be found in luxury houses or apartments adjacent to the centre where the nature of the buildings can shut out the environmental problems. Moreover, the very wealthy can maintain the town apartment and the country house, maximizing the advantages to be derived from both residential locations and minimizing travel difficulties by cutting them down to less frequent intervals than the daily journey, such as weekends only.

In terms of historical explanation, the introduction of a sectoral interpretation of city structure based on the maintenance of an upper class residential area at the centre and extending outwards, accomplished much the same sort of variation on the over-rigid concept of zones. In this way the early simplistic schemes were modified.

At this point it is necessary to consider precisely what are being employed as the defining criteria within the overall concept of the 'residential areas'. Undoubtedly the earliest geographical concern was with structures, that is with houses as part of the physically visible scene, rather than with areas defined in the more nebulous terms of income or social status. Again the problem of use and structures arises and once more the general trend has been to divorce the two in order to clarify procedures. This divorce is seldom complete, studies of social areas often include the physical condition and amenities of housing as a variable, while studies of housing as such usually employ terms such as 'working class houses'; but this sometimes reflects lack of clarity in research design as much as a desire to reconcile use (in this case interpreted as social class) with structure. In general the study of structures has become the domain of the historical geographer and has strong links, therefore, with historical interpretations of city patterns. The study of income, class and status as spatial variables has been pursued by social geographers and is linked more closely to structural theories of city patterns.

In view of the haphazard way in which references to contrasted city areas are couched, now in terms of a type of housing (inter-war semi-detached), now in terms of income (upper income bracket housing), now in terms of class (working class areas) it becomes vital to identify quite clearly which particular notion is being employed as the key criterion of definition.

The two bases of identifying areas, by type of building or by social category, can be separated for purposes of analysis, though once more it must be emphasized that this is only a convenient simplification, even if it is preferred to the confusion of terminology that has occasionally been employed, even to the extent of 'very large residence' combined with such terms as the 'middle wealthy'(Brookfield, 1952). In this chapter the built environment will be largely set aside and the characteristics of the population will be considered as the prime basis of differentiation of residential areas.

12.3 SOCIAL CHARACTERISTICS OF RESIDENTIAL AREAS

The zones and sectors which were identified by the Chicago ecologists were presented in social terms and it is, therefore, with such terms that the bulk of investigation of residential areas has been concerned. The first stage was the empirical testing of hypotheses which followed from the ecological models. These hypotheses were that populations are segregated residentially by social class and, since social distance can be translated into physical distance, observable patterns of residential segregation in spatial terms can be discerned within the city. The classical model implied that social class increased with distance from the city centre, but that was hardly tenable even within America.

In these hypotheses the concept of 'social class' is a little vague and for most operational

purposes it has been equated with 'occupation' which is the most easily accessible parameter:

> Occupation is only one of the criteria of social stratification, however. Its signification is clearly twofold because it relates to an economic relationship with the means of production but also defines a work-situation which will in turn have consequences on both the consciousness of class identity and also will help to define patterns of status estimation and attitude to the stratification system both within and outside the work situation.
>
> (Jackson, 1968: 3)

Even so it is far from being either a simple or a universally accepted basis (Coxon and Davies, 1986).

The major testing of the first of the above hypotheses, using occupation as a measure of social class, was that by O. D. and B. Duncan. They used two basic measures The first is the index of dissimilarity:

$$\sum_{i=1}^{n} \frac{(x_i/\Sigma x_i) - (y_i/\Sigma y_i)}{2} \cdot 100$$

where x_i represents one occupation and y_i represents another occupation of people residing in zone i. Essentially it measures the percentage of an occupational group (x) which would have to move to make its distribution identical with another occupation group (y). If this is computed between one occupation group and all others combined it gives the second measure, an index of segregation. Table 12.1 indicates the index of residential segregation for each major occupation group, for employed males in Chicago in 1950.

From this the Duncans argued that there was a U-shape pattern of indices of segregation. With high values at the top and bottom of the social scale, professional workers and managers and service workers and labourers, but with low values in the intermediate occupations where status is less clearly marked – 'residential segregation is greater for those occupational groups with clearly defined status than for those groups whose status is ambiguous'. From an analysis of

TABLE 12.1 *Index of residential segregation of each major occupation group, employed males, Chicago 1950*

Occupation group	By census tract	By zone sector segments
1 Professional, technical and kindred workers	30	21
2 Managers, officials and and proprietors, except farm	29	20
3 Sales workers	29	20
4 Clerical and kindred workers	13	9
5 Craftsmen, foremen and kindred workers	19	14
6 Operatives and kindred workers	22	16
7 Service workers, except private households	24	20
8 Labourers, except farm and mine	35	29

Source: After O. D. Duncan and B. Duncan, 1955.

the indices of dissimilarity in residential distribution for the eight groups in Table 12.1 the Duncans concluded that there was an 'essential correspondence of social and spatial distance among occupation groups'. The result of their study was to demonstrate that segregation in residential terms did exist and that the most segregated occupation groups are those at the extremes of the socio-economic scale.

There is a considerable danger that the multiplication of studies such as that of the Duncans provides no more than a series of exercises derived from and, more importantly, limited by, the context of the ecological models or the abstractions of the land economists. Use of the single variable of occupation means that they fail to take into account the great variability within residential areas. This failure was to be compensated in the attempt to increase the range of variables which are to be taken as diagnostic of distinctive social areas.

The immediate stimulus to derive acceptable social areas came from the need to provide 'community areas' for a study of Los Angeles, as well as from the awareness that one variable was an inadequate means of approaching a complex problem. It followed that the first departure was via the attempt to build up a more satisfactory and *ipso facto* a more complex means of classifying basic census units into homogeneous social areas. The early statement of Shevky and Williams in their study of Los Angeles in 1949 was followed by a volume by Shevky and Bell, *Social area analysis* in 1955, in which an attempt was made to put the study on a broader basis. The basic principle on which the key variables were related to the evolutionary changes already noted were contained in the sentence, 'we conceive of the city as a product of the complex whole of modern society; thus the social forms of urban life are to be understood within the context of the changing character of the larger containing society' (Shevky and Bell, 1955). From this three aspects were isolated epitomizing this changing character. These were:

TABLE 12.2 *Social area analysis: steps in construct formation and index construction*

Postulates concerning industrial society (aspects of increasing scale) (1)	Statistics of trends (2)	Changes in the structure of a given social system (3)	Constructs (4)	Sample statistics (related to the constructs) (5)	Derived measures (from col. 5) (6)	
Change in the range and intensity of relations	Changing distribution of skills: Lessening importance of manual productive operations – growing importance of clerical, supervisory, management operations	→ Changes in the arrangement of occupations based on function	→ Social Rank (economic status)	→ Years of schooling Employment status Class of worker Major occupation group Value of home Rent by dwelling unit Plumbing and repair Persons per room Heating and refrigeration	→ Occupation Schooling Rent	Index I
Differentiation of function	Changing structure of productive activity: Lessening importance of primary production – growing importance of relations centred in cities – lessening importance of the household as economic unit	→ Changes in the ways of living – movement of women into urban occupations – spread of alternative family patterns	→ Urbanization (family status)	→ Age and sex Owner or tenant House structure Persons in household	→ Fertility Women at work Single-family dwelling units	Index II
Complexity of organization	Changing composition of population: Increasing movement – alterations in age and sex distribution – increasing diversity	→ Redistribution in space – changes in the proportion of supporting and dependent population – isolation and segregation of groups	→ Segregation (ethnic status)	→ Race and nativity Country of birth Citizenship	→ Racial and national groups in relative isolation	Index III

Source: After Shevky and Bell, 1955.

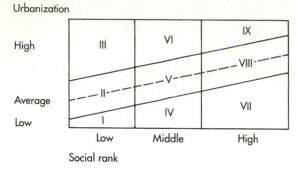

FIGURE 12.1 *The derivation of the classification of social areas in Los Angeles by Shevky and Williams (after Shevky and Williams, 1949). The criteria used are set out in the text. Urbanization is plotted against social rank which is divided into three. The regression of urbanization on social rank is plotted and divisions drawn two standard errors away. This procedure gives nine divisions. Two are shown in Fig. 12.2.*

1. Change in the range and intensity of relations.
2. Differentiation of functions.
3. Complexity of organization.

Via the arguments which are set out in Table 12.2 these were translated into three constructs which were identified as social rank (economic status), urbanization (family status) and segregation (ethnic status). These were measured for statistical purposes by the data indicated in Table 12.3 to provide the three indices.

Implicit in this whole process, as indeed the authors proposed, is the assumption that the selected indices related to the observed social differentiation between urban sub-populations, and at the same time they are unidimensional, that is, that they do not include similar aspects within them and therefore repeat the same measure three times. This point will be considered later. In order to classify areas the index of urbanization was plotted against that of social rank as a base divided equally into three. Urbanization was also divided into three intervals, 'with the middle or average interval determined by the space of two standard errors about

TABLE 12.3 *Indices used in social area analysis*

Index 1 Social Rank
A   Occupation ratio. Total number of craftsmen, operatives and labourers per 1000 employed persons.
B   Education ratio. Number of persons who have completed no more than grade school per 1000 persons 25 years old and over.
C   Rent – omitted from the list.

Index 2 Urbanization
A   Fertility ratio. Number of children under 5 years per 1000 females age 15 through 40.
B   Women in labour force. The number of females in the labour force per 1000 females 14 years and over.
C   Single family detached dwelling units ratio. The number of single-family dwelling units per 1000 dwelling units of all types.

Index 3 Segregation
The number of persons designated 'negro', 'other races' and 'foreign-born white', the last from south and east Europe. Shevky and Bell give a list of the countries. The sum of these is divided by the population of each tract and multiplied by 100 to give the index.

Source: After Shevky and Bell, 1955.

the regression line of urbanization related to social rank' (Shevky and Williams, 1949) (Fig. 12.1).

Segregation was added to this basic classification by high and low categories related to the mean figure.

Figure 12.2 shows a part illustration of the map of social areas of Los Angeles produced by Shevky and Williams using 1940 data. The wedge of high social rank and average or low urbanization picks out one of the best known areas of Los Angeles, swinging from the Pacific Ocean at Santa Monica along Sunset and Wilshire to Beverly Hills. To the east the suburban areas of Glendale and Pasadena also emerge quite clearly, although here a low index of urbanization reflects the suburban character of these parts.

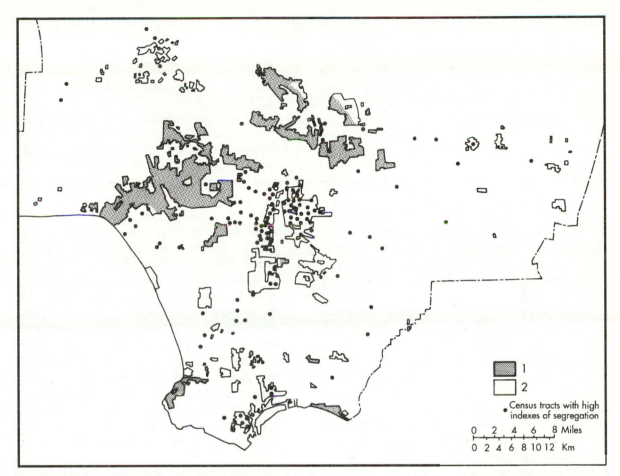

FIGURE 12.2   *The social areas of Los Angeles. 1. Census tracts with high social rank and average and low urbanization, that is categories VII and VIII on Fig. 12.1. 2. Census tracts with low social rank and average and high urbanization, that is categories II and III on Fig. 12.1. Note: This is a simplified extract from the original coloured map, but the major contrasts within the Los Angeles of the late 1940s are shown, especially the area of high quality residence extending from the coast at Santa Monica through Beverley Hills to Hollywood which contrasts with the inner city of low social status and of ethnic segregation.*

The work of Shevky and his associates has been the subject of a great deal of criticism. It has been urged that it has no theoretical background and is merely an attempt to delineate areas for their own sake, a view as already noted, not unfamiliar in regional geography or attempts to define the CBD. To some extent this limitation is not surprising, especially in the light of the many expressions of dissatisfaction with the existing ecological models and the need for empirical work upon which reformulation could be based. What theoretical background there is can be discerned if the way in which Shevky derived his constructs is examined in relation to the characteristics of the urbanization process. The dominance of Wirth's ideas becomes evident and the parameters which Shevky selected were intended to examine in small-scale areas the changes which were taking place in society at large as urbanization increased. There are immediate problems in this transfer of scales and to the extent that Wirth's ideas themselves are not accepted the whole conceptual basis of social area analysis in this context is undermined.

If the whole notion of the constructs is open to question so, too, is the selection of the statistical measures. For example, in most subsequent studies rental has been eliminated from the measure of social rank. 'In considering the whole range of possible parameters which might be selected as measures of aspects of urban social structure, one might, therefore, have even graver doubts as to the validity of isolating those few indices which Shevky suggested' (Robson, 1969: 52).

The third line of criticism relates to the unidimensional nature of the indices, which was noted earlier, that is, whether the three are discrete and unrelated to each other and not, in fact, overlapping measures of the same thing. Thus it can be shown that fertility is closely associated with occupation and education and hence has a significant linkage with social rank and does not stand outside it as a measure of that most nebulous construct which Shevky called 'urbanization'.

It will be apparent that a very familiar problem is now raised: how to collapse a large number of variables, which are somehow related to social character, into a smaller range of significantly associated variables denoting underlying components of social character. Factor analysis was clearly called for and hence the large number of studies which were subsequently published which attempted to reduce the large number of possible variables to a smaller number of uncorrelated factors by use of this statistical technique (Johnston, 1978).

At this point some difficulty arises over the use of principal component and factor analysis in geographical research (Davies, 1984). It is not possible here (it is far from the purpose of this book) to review statistical techniques and W. K. D. Davies (1984) presented an extended

discussion. It is only necessary to insist on three points:

1. Component analysis is only a taxonomic procedure not to be confused with a theoretical formulation.
2. The variables included will determine the components (factors) abstracted. If data on occupation and wealth, age and ethnic origin are put into an analysis, it will not be very surprising if socio-economic status, lifestyle and ethnicity emerge as the main components of variation. The critical decision is what variables are included and this is often dictated by what data are available in usable form. This has become critical in postmodern analyses which have moved away from these large-scale models and brought to the fore the city characterized by other social categories, such as single parents, homosexuals or the physically disadvantaged. Since they are not encompassed by the variables, they never appear in the outcome.
3. The labelling of the components (factors), although related to the loading of variables, is subjective and there is a strong incentive to 'find' the accepted components.

It follows from the above that in conceptual terms there is no great advance on Shevky's procedure: indeed, the theoretical basis is even more tenuous. The uncorrelated character of the components ensures, however, that the 'constructs' do not overlap and hence it does provide a real test of the nature of the three measures which Shevky proposed. At the same time, by use of grouping techniques, the individual areas can be associated into regions by the scores on each component so that the spatial arrangement of the social areas can be studied from an objective and uncommitted basis.

The general approach is best presented by means of one example chosen from the very many which have been published, at this stage restricting discussion to Western cities. W. K. D. Davies's analysis of social differentiation at the intra-urban scale which presents the social ecology of Cardiff in 1971 is a clear mainstream

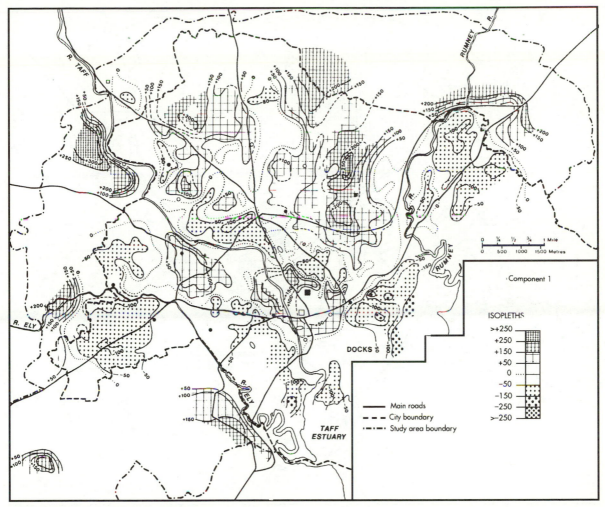

FIGURE 12.3   *Cardiff: distribution of components from a factorial ecological study (after W.K.D. Davies, 1983). Socio-economic status.*

example of much work undertaken at that time when factorial ecology was the height of fashion in urban geography (Davies, W. K. D., 1983).

Davies assembled a total of 27 variables all derived from the 1971 census, for 541 enumeration districts which covered not only the city of Cardiff, as it then was, but also a fringe of contiguous parishes outside the administrative boundary.

A two-fold procedure was used to produce the final set of factorial results that were inter-

preted, namely the use of R mode Principal Axes component and Image Analysis factoring methods followed by Direct Oblimin rotation ... . The use of a set of alternative factoring procedures does mean that if the results are shown to be invariant then a great deal of confidence can be placed in the conclusions.

(Davies, 1983: 80)

From the results a six axis solution was presented as being the most acceptable and where each axis accounted for more than 5 per

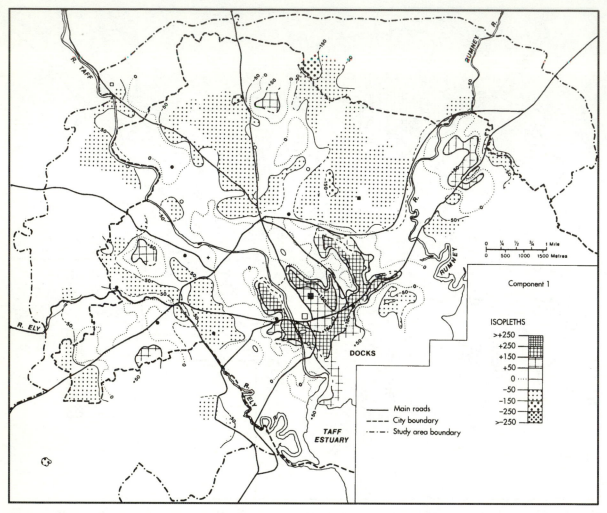

FIGURE 12.4 *Cardiff: distribution of components from a factorial ecological study (after W. K. D. Davies, 1983). Life cycle and tenure.*

cent of the variance. The six components were identified as: socio-economic status; young adult and non-family; life cycle and tenure; housing and ethnicity; late family and mobility; urban fringe. In order to avoid the representation of 541 small areas, the maps were drawn by the interpolation of isopleths. Figures 12.3, 12.4 and 12.5 reproduce the maps presented for socio-economic status, life cycle and tenure and housing and ethnicity, while Table 12.4 is Davies's comparison of his study of Cardiff with earlier versions and with his study with G.

Lewis of Leicester (Davies and Lewis, 1973, 1974).

From this study, and from Table 12.4, which has been introduced as a standard example, four brief conclusions can be drawn:

1. The simple, unidimensional constructs of social area analysis do not appear. Thus, for example, Davies writes, 'Social status is clearly indexed, but it is a more general axis than expected since it is associated with ethnicity (Welsh/English), economic and

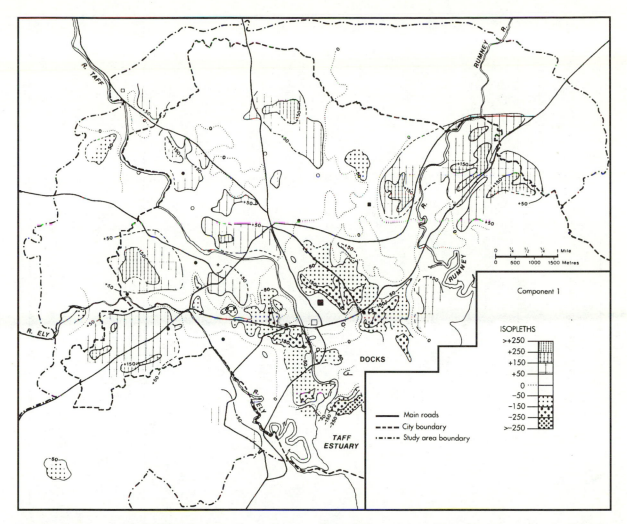

FIGURE 12.5 *Cardiff: distribution of components from a factorial ecological study (after W. K. D. Davies, 1983). Housing and ethnicity. Source: (Figs 12.3, 12.4 and 12.5) W. K. D. Davies, 1983: Urban social structure. a multivariate-structural analysis of Cardiff and its region. Cardiff: University of Wales Press. Social Science Monographs 8. Figures 15, 16 and 18.*

mobility variables.' The variable loadings on socio-economic status are shown in Table 12.4.

2. The patterning, as displayed in Figs 12.3, 12.4 and 12.5, is generally interpreted with socio-economic status being sectoral, the life cycle components being zonal and ethnicity characterizing inner city areas. Even so, none is as clear as theory demands. Thus, the high status sector is discontinuous and irregular.

3. There are other aspects which break across the simplicity of generalized interpretations. Thus in Britain, the tenurial contrast between private and public housing, at least in 1971, clearly interposes both in the character of components and their distribution.

4. By extending the analysis across the city boundaries an 'urban fringe' component is generated, largely distinguished by commuters, two-car households and incomers to the city.

TABLE 12.4   A comparison of factorial ecological studies of Leicester and Cardiff

	LEICESTER 1966 (283,260)	CARDIFF 1971 (278,221) a. Results. b. Postulated Axes	CARDIFF 1966 (253,000)	CARDIFF 1961 (256,682)
City Date (Pop.) Author	(Davies and Lewis, 1973)	(Davies, 1980)	(Evans, 1973; Herbert and Evans, 1973)	(Herbert, 1970)

LEICESTER 1966

Higher Order
I SOCIAL STATUS
- −61
- 4. Substandardness
- +76 — 8. Urban Fringe
- +69 — 1. Socio-Economic Status
- +74 — 3. Life Cycle/Tenure

II FAMILY STATUS
- −64 — 7. Economic Participation and Late Family Socio-Economic Status
- +54 — 5. Young Adult (Mobile)

III ETHNICITY-MIGRANTS
- +50 — ...
- +87 — 6. Ethnicity
- +66 — 2. Mobility (Females)

CARDIFF 1971

Higher Order
I NON-URBAN
- −87 — N.F. (part of 4)
- 5. Urban Fringe
- 37 — 3. Life Cycle and Tenure
- −59 — 3. Life Cycle and Tenure

II FAMILY AND SOCIAL STATUS
- +62 — 5. Late Family Mobility
- +71 — 1. Socio-economic Status

III ETHNICITY-NON FAMILY
- 60 — 2. Young Adult/Non Family
- −77 — 4. Housing and Ethnicity
- −37 — ?N.F. (Part of 7.) (Life Cycle and Tenure)

CARDIFF 1966

First Order
1. Substandardness — 4. Tenure and Life Cycle (Residential Quality)*
2. Urban Fringe — N.F. —
3. Socio-Economic Status
4. Life Cycle — 2. Tenure and Life Cycle
5. Late Family — N.F.
6. Young Adult/Pre-Family — 1. Social Status
7. Ethnicity — 5. Urbanization (Young Adult)
8. Mobility — 3. Residential Quality (Ethnic, Housing)*
9. Females — —
10. Tenure
11. Economy

CARDIFF 1961

First Order
2. (Ethnic, Residential Quality)*
N.F. —
1. (Tenure/Age)**
N.F.
?N.F.
4. (Non Family/Children)*
3. (Ethnic)*
5. Not shown

*Titles in brackets represent either a re-naming of axes, on the basis of the loadings, or the addition of titles when they were absent in original study.

	LEICESTER 1966	CARDIFF 1971	CARDIFF 1966	CARDIFF 1961
Model:	Component Principal Axes	Component and Factor Principal Axes	Component Principal Axes, Image	Component Principal Axes
Techniques: Rotation:	Oblique; Varimax	Oblique, Varimax	Varimax	None
Variables:	56	27	40	26
Areas:	143	541	119	334
Axes:	9	6	5	5
Variance:	72.2%	67.0%	70.6%	73.0%

Source: After W. K. D. Davies, 1983.

At this point, and against the background of Davies's example, some attempt to summarize the conclusions from factorial ecologies can be made.

1. **The components of social variation.** Murdie (1969) produced a table of the components which had been identified from a number of analyses, mainly of American cities. In nearly all the cases three were outstanding, and usually made up the first three extracted. They are economic status, family status and ethnic status. There is a fairly clear element of agreement in all these studies, although British evidence emphasized housing conditions rather than economic and social status. The components which emerge are, therefore, closely allied to those constructs proposed by Shevky and Bell. The rather vague 'urbanization' construct is, however, reinterpreted as a measure of 'family status' or family characteristics. This gives three basic components which play a part in urban social variation. Here, however, a further and major reservation has to be made. The life cycle dimension is very much the product of factorial ecologies of the 1960s and 1970s. It is significant that it is referred to as the family cycle and based on standardized stages of family development. But the 1980s and 1990s have seen crucial changes in the contribution of the conventional family of married parents with children, some would even propose its dissolution. Gober *et al.* (1991) in a study of household instability, residential mobility and neighbourhood change in Phoenix, Arizona, conclude that:

> while the family life cycle has proven to be a rich and productive theme around which to organize research in residential mobility, it is increasingly unable to capture a great deal of contemporary intra-urban population change. Not all changes in living arrangements occur as a result of life cycle changes, nor are all changes in residence life cycle induced. People divorce and cohabitate (*sic*), move into and out of households

as economic need and social circumstances dictate, and their residential mobility may or may not occur in conjunction with these shifts.

> (Gober *et al.*, 1991: 81)

They suggest in consequence that a notion called 'life style' (Clausen, 1986) must now replace life cycle as an organizing framework, and that has clear implications for ecological studies.

This discussion links back to the comments made on p. 226 that by its very selection of census data as input variables, factorial ecologies might well miss the crucial changes which are taking place in the formation of social areas in cities. Certainly the life cycle dimension is likely to remain as identifiable, but two reservations have to be made:

(a) As long ago as 1976, B. S. Morgan cast doubt on the occurrence of the life cycle dimension in British cities. In a study of Exeter he wrote that 'residential immobility among households who enter new housing early in the family life is a more important influence on the residential patterns of family status groups than changing housing needs generated by life cycle changes' (Morgan, 1976: 105). In brief, economic resources are needed to respond to changing family composition. But if those resources are not available, or inertia intervenes, there will be no adapting by movement to a new property and the emergence of a life cycle dimension will be inhibited.

(b) Since the structure of the family, or the way of living, has been transformed in the last decade, and as marriage is conceived as not for life but for a temporary span to be determined, and as other fashions of living, such as the family with a single parent develop, then the concept of a life cycle may not be the best basis for the comprehension of social areas. Lifestyle must surely be

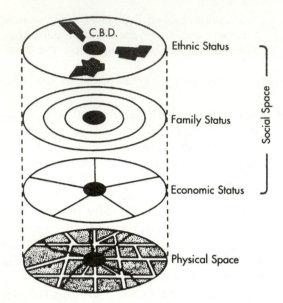

FIGURE 12.6 *A model of the residential structure of the city (after R.A. Murdie, 1969).*

more appropriate – and the impoverished family locked into an inadequate home is little more than the reflection and indication of a lifestyle rather than of a family cycle.

2. **The spatial articulation of the components of social variation.** Once again it is true to contend that most studies which have been concerned with the spatial characteristics of identified components have not propounded any new theory of city structure, but have examined the results in relation to the earlier ecological models. To a large extent these studies have been placed in the context of zonal as against sectoral characteristics of distribution. The terms 'zonal' and 'sectoral' are to some extent pejorative and the real crux is whether the locational character of the component is controlled by distance or directional forces in relation to the city centre. The clear conclusions are set out in Murdie's set of three hypotheses and in the diagram he presented (Fig. 12.6).
   (a) Economic status tends to be associated with measures of income, occupation

and education and tends to be distributed sectorally.
   (b) Family status tends to be associated with fertility, type of household, and labour force participation by women, and tends to be distributed concentrically.
   (c) Ethnic status tends to form 'groupings' which can be superimposed upon the cellular structure created by combinations of sectoral and concentric patterns.

On the basis of his work on Cardiff and other cities (1983) Davies extended the three dimensions and argued for six as follows:

1. **Migratory status or mobility**. An axis which differentiates between high and low mobility and migration. Three spatial patterns are suggested:
   (a) The new growth of single family dwellings on the edge of the city.
   (b) Transient or bed-sitter areas of the inner city.
   (c) Low migration, settled middle zones of the city with late family characteristics.
2. **Residential quality, substandardness, skid row.** This marks out the poorest city areas with the worst housing stock and often ageing population. In behavioural terms it is linked to a 'down and out' population and to alcoholism and presumably in the 1990s to drugs. 'What does seem clear is an inner city location, either concentrated in one area or partially surrounding the inner city core in a U shape' (Davies, 1983: 324).
3. **Pre-family (young adult) or working women.** This is an inner city area characterized by footloose young adults before marriage, or in semi-permanent partnerships. Consequently there is a high proportion of working women. Again this is characteristic of inner city areas for accommodation demanded is greatly contrasted to that of the 'settled' family.
4. **Late or established family, or postgenitive or high economic participation.** This component is characterized by middle-aged,

late family groups where there is a high participation in work since mothers are returning to work and late teenage children are starting work. It characterizes the middle zones of the city although seldom in a neat and continuous pattern.

5. **Tenure.** This is generated by the significant, though now declining, role of public housing in Britain. Its extended characteristics can be determined by the allocation policies of local authorities. Spatially, it appears as a series of clusters.

6. **Urban fringe.** This appears partly at least as a consequence of extending the area surveyed beyond the city boundary. Agriculture appears as an associated variable along with those already noted, commuting two-car households and recent immigrants.

Davies notes that there are other dimensions which have been identified, usually variations on social and family status. Thus one notable addition is a family dissolution or family disorganization axis (Hadden and Borgatta, 1964; Timms, 1971). It is possible to speculate that single parent families would appear in the 1990s. Davies further comments that the 'increased use of social welfare indicators or mental illness variables will clarify the pressure of such an axis' (Davies, 1983: 325). But such a possibility takes one back to the initial proviso (p. 126) that results must depend upon the input variables.

In the recent work there has been much less attention devoted to factorial ecology, largely on the basis that given that it needs census data to be practically viable, then there is little more that can be revealed. In addition, post-modernist reaction to these large-scale models has generated an awareness that the infinitely complex mosaic of the city means that other dimensions, not identifiable by the variables fed into ecological studies, characterize its space. The contemporary trend, therefore, has been away from these aggregate statements of structure in two directions. The first is to an examination of the aspatial political and economic forces which

generate the conditions under which spatial differentiation occurs. The second is to consider the way in which a much larger range of population groups experience and use city space. These will be considered later but before doing so it is necessary to change the scale of review and consider how the behaviour of individuals plays its part in creating the conventional, broad patterns which this chapter has considered.

12.4 RESIDENTIAL PATTERNS: THE PROBLEM OF DISAGGREGATION AND RESIDENTIAL CHOICE

An approach to the contrasted disaggregated scale can be made by considering further the life cycle or family status dimension of factorial ecology. As has been noted the dimension gives rise to a zonal pattern since the inner parts of the city are characterized by single, unmarried people who are generally younger in age, whereas the middle life, married with children population is found in suburbia. The explanation clearly lies in the different needs of the household in terms of its collective lifestyle and activities.

This shifts the analysis to an examination of the manner in which choices of where to live are made. In the making of these household activity systems are crucial. Household activities include a vast range extending from work to social and recreational activities to shopping. There are few studies of this sort and results from investigation are not very fruitful, but one study from Detroit (Smith, R. V. et al., 1967) illustrates the principle. For each of five major sectors of metropolitan Detroit, the centres of white and black populations had been calculated in a previous study (Mayer and Hoult, 1962) and joined to form a 'centre line' of white and negro population (Fig. 12.7). A similar technique was now used to define a centre line based on a stratified sample of population questioned as to 'the location of meeting places of all formal associations they belonged to, as well as the residence location of all friends,

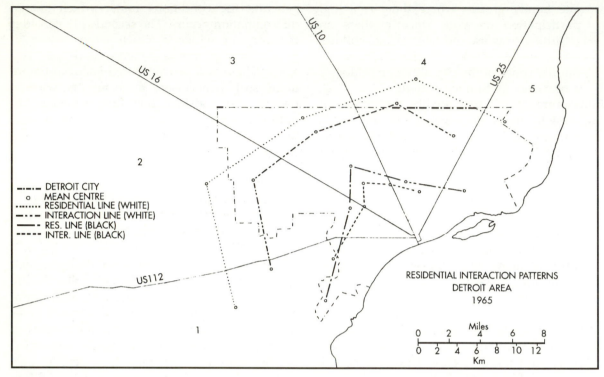

FIGURE 12.7 *Residential reaction patterns in the Detroit area (after R.V. Smith et al., 1967). The method of construction of the residential and interaction lines is given in the text.*

neighbours, relatives and co-workers with whom they interacted. In this manner point locations were derived for all formal and informal associational activities of each respondent' (Smith, R. V. *et al.*, 1967: 13). From this it is clear that residential segregation is associated with segregated interaction. The activity systems of these two populations, in the study context, are related to different areas.

To some extent it can be suggested that residential choice – particularly in intra-urban relocation which plays a large part in determining the total pattern – is the function of the total of these activity systems. But there are two sorts of activity which could control residential choice. The one is directed towards employment and place of work and the other to social and leisure activities.

Moriarty examined the relative stress to be put on these two controls considering them as

competing hypotheses (Moriarty, 1970). The first he called the 'economic competition' hypothesis in which the ability to pay is the basic factor and in consequence journey-to-work and site costs are the major influences in the residential decision. 'Differences in the locational behaviour of residential decision makers are due to differences in their budget costs and income resources, and it is this difference that determines the spatial distribution of socio-economic groups in urban space' (Moriarty, 1970: 98). Moriarty maintained that such a proposition is nowhere supported by empirical evidence and set out in contrast the second 'social choice' hypothesis which argues that residential decisions stem from conscious, or indeed subconscious, social choice related to the differing values, needs and desires of people.

Most empirical studies reject the 'economic competition' interpretation and find no relation

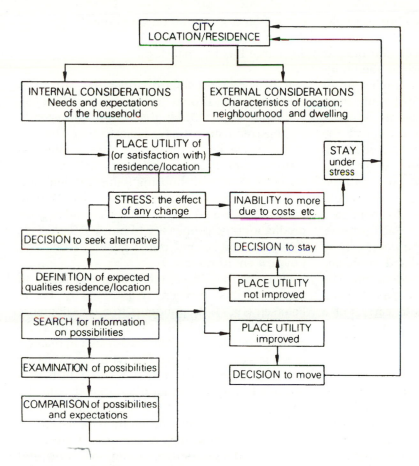

FIGURE 12.8 *A residential location decision-making model (modified from D. T. Herbert, 1972). Note: Place utility is a complex notion and will be determined by an elaborate series of 'trade-offs', of which the most obvious is a desired house in an undesirable location against an inadequate house in a desirable area.*

between social class and distance to work. For example, Halvorson reported 'the major conclusion to be drawn from the data summarized is a negative one ... the journey to work or work access constitutes a rather minor factor in the residential location decision' (Halvorson, 1970: 178). Halvorson went on to positive conclusions which related residential location decision to the quality of housing and the quality of the housing environment. Similar results have been widely reported and have led to an emphasis on decision making and environmental perception as main areas of current investigation

These areas of subsequent investigation are best presented in a residential location decision model which sets out the reasons operative in the selection of a particular location. This has been done in diagrammatic form by Herbert and Fig. 12.8 is derived from his model (Herbert,

1972: 246–56). The key concepts are to be found in the work of Wolpert (1965) and of Brown and Moore (1970). The individual household can be considered to be under the influence of two sets of forces. One is internal, in that it is generated by the household itself and defined in terms of its own needs and expectations, the other is external and defined by the characteristics of the locale. These two sets interact in creating 'place utility', which essentially measures an individual's level of satisfaction or dissatisfaction with a given location. 'If the place utility of the present residential site diverges sufficiently from his immediate needs, the individual will consider a new location' (Brown and Moore, 1970: 1).

The divergence between present site and possible new locations is set up by the operation of 'stressors', or by any change which alters, or threatens to alter, the *status quo*. These could be related either to internal factors – for example, an increase in family size by birth of children or the coming of in-laws to live with the nuclear family – or to external forces such as the building of a nearby motorway or the buying up of proximate houses by people of different ethnic origin. They need not be as tangible as these examples, for instance, a change in the way the family regards its social status could give rise to stresses. More directly related to life cycle are the changing demands for space and location with ageing. Thus an over-large garden could become a stressor as the capacity to manage it diminishes. From the impact of such stresses comes the decision to consider the possibility of an alternative location. Two related problems face the household in their consideration. The first is to define the qualities to be expected at a new location and the second is to carry out a search in order to find such a location.

Definition of house quality

A number of studies have attempted to list and compare desirable qualities. One was by Butler and Chapin (1969: 2). From a nation-wide survey in the United States they reported that metropolitan households prefer the following:

1. Better neighbourhood quality with either a less desirable housing unit or less accessible location over a less desirable neighbourhood with either a better housing unit or better accessibility. (Overwhelmingly – approximately 70 per cent to 27 per cent.)
2. A house that is desirable inside but appears less so outside to one where an attractive exterior belies the interior. (Overwhelmingly – 80.4 per cent to 14.2 per cent.)
3. Access to better than average schools in an area with higher taxes to lower taxes and inferior schools. (Overwhelmingly – 78.3 per cent to 15.2 per cent.)
4. A conflicting combination of a new or fairly new house together with a well-established neighbourhood.
5. Modern architectural style to traditional. (But barely – 45.5 per cent to 37.6 per cent.

6. A housing unit all on one floor.
7. Few children in the neighbourhood.
8. Large lots to small lots.

A listing of possible qualities has also been presented by Brown and Moore (1970: 5). Attempts to identify the most consistently emphasized features have isolated two. The first is the house itself, that is its various physical features, especially the space available. Clark *et al.* in a study of housing consumption and residential mobility concluded that 'square metres per person was a consistently good predictor of the propensity to move' (Clark *et al.*, 1984: 42). The second is the general social and physical environment.

One of the most interesting attempts to approach this problem was that by G. L. Petersen (1967). Respondents to a questionnaire were asked to detail their judgements on photographs of residential areas rated in relation to ten selected variables. The replies were factor analysed to produce a first factor which was essentially physical quality associated with the age of structures, and the second an environmental quality named 'harmony with nature' as reflected in such things as amount of greenery, privacy and open space. Again structures and environment are paramount, although in reality those are always the obvious criteria.

Search process for housing

After the definition of the desired qualities in a house the next problem to be faced is the search for information on the possibilities available. A diagram constructed by Silk (1971, 1972) is reproduced as Fig. 12.9 to indicate some aspects of the search procedure. It is a truism that one limitation on the individual will be the extent of knowledge; choice will be restricted to a known area which can be termed the 'awareness space', modern jargon for the *terra cognita* of the ancients. This is the area with which the person is familiar through varied activities, such as the journey to work, shop or to visit friends. It is virtually the same as 'action space' although that term is best reserved for the area familiar through direct

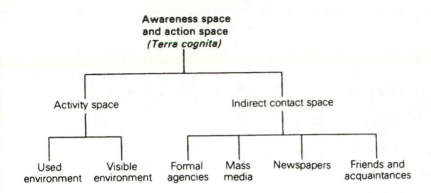

personal activity as opposed to indirect contact space which would include familiarity derived from the media and reports from relatives and friends. To that must be added the information derived from real estate agents.

Having considered the desired qualities of a new location and where these are, the household will proceed to review all the possibilities and compare them with the existing location. From this a decision will be made to move or to stay in relation to the measuring rod of place utility. Again there seems to be clear evidence that 'the evaluation [of locations] occurs in two stages: residential areas are evaluated and then vacancies in acceptable areas are considered' (Preston and Taylor, 1981: 447). Interestingly, studies of location decision making abstract the same two criteria of structure and environment as studies of residential mobility, but reverse them in significance.

At this point it is worth recalling the studies of consumer behaviour introduced in Chapter 5 (pp. 79–83) for buying a house is an aspect of consumer behaviour.

Social networks and housing choice

A further resolution of intra-urban mobility at an individual scale remains. The intense mobility associated with modern urbanism makes Webber's non-place urban realms (Webber, 1964: 310) all the more significant. If people structure their lives within these realms then the problem arises of identifying that structuring and assessing its impact on spatial behaviour and the

choice of where to live and if to stay there. This introduces the field of social networks in urban situations which has not received a great deal of attention from urban geographers. J. C. Mitchell, although concerned primarily with African towns, developed basic notions which can be generally applied. In order to give some indication of the flavour of Mitchell's work it is worthwhile outlining very briefly his identification of the morphological and interactional characteristics of social networks (Mitchell, 1969). The morphological characteristics are:

1. **Anchorage.** This defines the point of origin of the network. It has no absolute meaning but is defined in the context of an investigation as the individual on whom the network is fixed or anchored.
2. **Reachability.** This defines the degree to which an individual can be contacted via the network or the extent to which others can be contacted. The ease with which others can be reached is of obvious importance to any network.
3. **Density.** This is a different concept from reachability for it measures the interaction between all the members of the network.
4. **Range.** This assesses the number of people in regular and direct contact with an individual.

The interactional criteria are:

5. **Content.** This defines the basis on which contacts within the network are made. It

LEVEL IN SPATIAL HIERARCHY	LEVEL OF MEASUREMENT	
	NOMINAL	HIGHER LEVEL
LOW The dwelling plot	Presence/absence of Garage Central heating Garden Structural character, e.g. building material style	Price Age Size of house Size of plot
MEDIUM The immediate neighbourhood	Presence/absence of Made-up roads Tree-lined avenues Other house types in the area	Housing density
HIGH The larger neighbourhood The region shops	In country By river	Aggregate cost of journey to work Nearness to services

FIGURE 12.10 *Decision making in house purchase in relation to scale (after J. A. Silk, 1972).*

could be, for example, kinship, friendship, religion or work.

6. **Directness.** This is a measure of reciprocity, of the extent to which contacts are one way or not.

7. **Durability.** Some networks come into being for a very specific purpose, perhaps to advocate a local public policy of some sort, and then disappear. Not all networks are permanent.

8. **Intensity.** Individuals will vary in the degree to which they feel themselves committed to the network, ranging from the complete commitment that its members may feel to an organization like the Mafia, to a more casual involvement in minor causes. It is worth observing that in common everyday terms the word 'mafia' is used to denote a network.

9. **Frequency.** The frequency of contact between the members will vary and provides a measurable feature of interaction.

These nine measures can be applied to any interactive situation whether it be a complex of business links or a pattern of golf club membership. It is implicit in the ideas of Wirth and of Webber, that in an urban context these various reticules become discrete as urbanites segregate the roles they play. In all this the geographer would seem to have but a tenuous interest but, although the networks described in Mitchell's terms are intrinsically aspatial, implicit within them are relative locations and inevitably spatial manifestations. As Connell has written, 'the content of these spatial links remain critical and morphological networks alone are inadequate; social networks, by definition, are only marginally spatial' (Connell, 1973). But they do have spatial consequences and those consequences can be seen in their impact upon residential decisions. Much discussion has been centred upon environment and house, but the relationship with actual or potential social networks cannot be ignored.

The end result

The end result of these studies of the individual or household's residential location decision should be to build up from this microscopic level to the explanation of the location and characteristics of residential areas. This supposedly avoids the impasse of inadequate explanation to which macroscaled studies are tied, just as consumer behaviour studies supposedly provide new directions from the unsatisfactory aggregate model of central place theory. But reference to central place will reveal that here again the problem of scale is fundamental and that contrasted conclusions can be derived at different scales.

The priority of influences in residential decision making will be related to the scale at which the decision is being made (Fig. 12.10). Unless that scale is very carefully specified then confusion will arise. At the crudest aggregate scale of the whole city, decisions will relate to the ease of journey to work and *cost*, for to suggest that price is not a highly pertinent factor is to fly in the face of reason. At this level and appropriately the various aggregate models are pitched, such as those derived by Wilson (1977), from the Lowry formulation. At a somewhat larger scale the constraints of workplace and of crude cost will be set aside and, probably, as assumed, general neighbourhood characteristics might well become pertinent as 'residential neighbourhoods', sieved out at the first scale of the search, are compared. Finally, at the largest scale, and with the area sieved out, the qualities of the house itself become paramount. It is certainly not easy to disaggregate into a logical system progress across the various scales of decision for that presupposes that individuals knowingly structure complex decisions rather than act on impulse or without logic. The argument here presented assumes an unreal hierarchical ordering in decision. Perhaps the most interesting point is to relate the standard dimensions of factorial ecologies to the stressors in the location decision model. The family cycle is certainly one of the major precipitators of household moves: marriage, the production of children and the schooling of children are obvious stressors. Clark et al. (1984: 42) write of their study, 'for owners as well as renters the addition of a child always had a significant influence on the housing adjustment'. Socio-economic status is a complex notion. At its crudest it is simply the capital available, but it certainly subsumes the household's own view of its social standing, which it might wish its house to display, and also the money it can deploy to externalize that view in bricks and mortar or stone. Finally, ethnicity can be related broadly to the notion of territoriality which might define those parts of the city acceptable to the household. To equate territoriality with action space is perhaps open to criticism but the basic notions are not far apart. Thus the interaction study of Detroit presented earlier as one of an aggregated scale of study is not as far apart as is sometimes suggested from behavioural studies at the micro-scale.

Future progress seems to lie in the attempts to specify scale in order to analyse the way in which the pay-offs between scales are operated. But like studies in consumer behaviour the ultimate reduction of this approach is to the individual level where the psychologist should be most competent. Even the individual has to be specified, for many studies in the past referred to 'his' decision, but it is the household which moves and which partner predominates in the process is relevant.

12.5 CONCLUSION

In conclusion it is again worth observing that studies of decision making in relation to residential choice have become less frequent in the literature of urban geography than they were at one time. This perhaps is due to the fact that behavioural studies too often viewed decision making as a process of free choice whereas in reality it was greatly constrained. To a family in public housing it was well-nigh meaningless as the crucial decisions were made by a public authority housing committee. As a consequence the consideration of the constraints has become more significant than the way a choice is made and of those, the housing market, a major influence in that problem, must be considered.

RESIDENTIAL AREAS OF THE CITY: THE HOUSING MARKET AND INSTITUTIONAL INFLUENCES

13.1 INTRODUCTION

A critical and unjustified assumption has been made in regard to all the material considered in the last chapter. That assumption is that socially distinct residential areas are the direct consequence of free choice, or in more general terms, of demand. Because people want properties of certain types, located in particular areas to suit their perceived needs, then the market responds by providing for those wants. The assumption, it is true, has been hedged around with limitations, but these appear as asides rather than as direct statements. If socio-economic status, or the ability to pay, is a fundamental control of residential segregration, then only those who can deploy the greatest resources can choose freely, and only very few of those without some limitation. At the lower end of the scale it has already been noted that municipal or public housing can greatly modify the patterns set out in the standard models of urban structure. But the location of that housing is in no way the free choice of the residents, neither is the acquisition of an individual property which is the consequence of an allocation process. Again, in dealing with disaggregated residential choice, it was indicated that although many households would find themselves subject to 'stresses', no action could be taken in the form of a search for a different house or a different location because of the problem of cost. Such households remained where they were under stress or the household itself was restructured. Part of the restructuring process is the creation of homelessness.

All this implies that the basic creators of urban residential patterns are not those demanding, for income limits 'free' choice, but those supplying. There obviously is a demand for specific products as the varied types of property imply, but it is only met in so far as the supplier can see a potential profit. The minimal return to building cheap housing, and the cheapness is constrained in most Western countries by an insistence on minimum standards, means that a minimum quantity is built, unless there is some form of subsidy. The form of the city is, therefore, a product of the combined interests of the developer and the builder operating within a nexus of institutional stimuli and the constraints exercised by the policies of government. Such a simple reversal of the earlier approach is too stark a reaction, but the implication arises that in order to understand the aggregate pattern of residential areas, which factorial ecologies only describe, the geographer must turn aside to consider the housing market. Patterns alone are insufficient, processes must be pursued. Residential location decisions are not free and unconstrained for they are made in relation to a stock of existing buildings and a complex market situation. Thus a house bought with a large mortgage at a time of boom in the housing market can in a time of slump fall to a price below the mortgage incurred. The consequent negative equity locks the buyer into a location and inhibits the free mobility which is supposed to operate in all the aggregate models. But even further, in the first instance the whole nature of the stock depends on the constraints and stimuli offered by governmental policies and so political attitudes have to be brought into even a geographical review because ultimately the whole process of the provision of shelter becomes an overt political issue. The range of views can be indicated under four groupings.

1. **The far left.** Housing should, like education and health, be a social service provided wholly by the state and in which no advantage should accrue to accumulated wealth. There would be no private housing but each individual or family or partnership would be allocated housing related to their needs. Since housing would be regarded as a state service there would be no rents since they would be covered by that part of the gross national product appropriated by the state. Only by these procedures can social justice in the proper provision of shelter be achieved. The price is total state control, and the enormous power given to the various gatekeepers.

2. **The moderate left.** Housing is again regarded as a social service like health care and education, but the attitude to private provision of such services is ambivalent and owner occupation is accepted. This is usually justified by the argument that owner-occupiers acquire property for use rather than for accumulation in the Marxist sense and therefore, have more in common with tenants than with landowners and the exploiters of property.

> While ownership of housing may involve an accumulative potential, it does not constitute ownership of a productive resource, and thus does not form the basis of class formations of any kind. Land-owners, developers, and others involved in the supply of housing have a monopoly over productive capital in the city which is not in any way shared with those who buy their own houses. Thus in terms of Marxist political economy, owner occupiers are not a distinctive class since although they accumulate wealth, they do so through consumption rather than production, their property does not reproduce itself, and their 'profit' (if such it is) does not represent the extraction of surplus value.
>
> (Saunders, 1977)

Private profits from renting accommodation are of a different nature. The state must, therefore, be the active, preferably the sole, provider of properties for rent. The rents charged by the state should, because housing is a service, be nominal rather than economic, if they are exacted at all. There should be no means test so that the subsidy is based on the house and not on the individual or family.

3. **The far right.** Housing is a commodity like any other good and should be bought and sold on the open market through the bidding process without any state intervention. The individual provides his or her own housing, as health care and education, and has the choice of how much of disposable income is allocated to it as against other desirable goods and services. The ultimate goal is the direct opposite of the extreme left. A rising standard of living would enable everyone to make a successful bid for some housing and so the archetypal 'property owning democracy' would be achieved. In the meantime a Darwinian 'law of the jungle' operates where the successful and wealthy become the owners of the most expensive property and housing filters down the social hierarchy. The poorest people, and the worst housing drop out at the bottom. However, the rich can obtain profit by renting to the poorest sections of the population so that all are housed. Choice is maximized, although only for those who have the money to ensure it, for it becomes zero for the destitute. Only by such a system can individual freedom be assured, although the price is a vast inequality.

4. **The moderate right.** Housing is essential for all, like health care and education, but the amount to be consumed is the individual's choice. However, there must be a basic provision below which no one falls. This is best achieved in a mixed economy where the most desirable state is to be an independent owner-occupier. Where that is not possible due to lack of means, or where it is not desired, accommodation for rent is made available both by the State and by private individuals or companies acting as landlords. Since the form of tenure and the sums spent constitute a choice in family resource allocation, rents cannot be subsidized and a full economic rent must be charged. In order to ensure the basic provision, however, subsidies or rebates are paid to individuals or families who can show that they have not the means to meet the charges. Subsidies go to means-tested people and not to the housing as such. The State provides a safety-net to ensure that minimum standards of shelter are available for all citizens.

Each political group will, no doubt, feel that its views have been oversimplified or misrepresented in the foregoing paragraphs. An attempt has been made, however, to ascribe laudable ends to each group. But the crucial implication for the geographer is that the implementation of policies based on the different principles will produce considerable, even fundamental, differences in the distribution of residences within the city and in consequence social areas are, at least in part, the product of housing philosophies and the resultant housing policies.

13.2 PRIVATE AND PUBLIC HOUSING IN BRITAIN

To some extent most of the argument in this book has assumed a degree of universality. Central place systems or concentrically zoned cities are conceived as common to the Western world. That universality is derived from the conventional freedom of Western capitalist democracies where price is the arbiter and where, in consequence, a similar suite of locational forces will operate. From the working out of those forces over standardized space, generalizations as to patterns can be established. However, as one moves into the housing market then the intervention of government becomes more apparent and generality much more difficult to sustain. Acccordingly the consideration in this chapter will be largely, though not exclusively, concerned with Britain. But it must be emphasized that Britain is only being used as an example of how the supply side imposes on distribution patterns; the purpose is not to describe a unique situation.

The total housing stock of Great Britain in 1991 was 23.1 million (Dept. of Environment, 1993). It

TABLE 13.1   *Categories of housing tenure and stock of dwellings in Great Britain 1914–91*

Date	Rented from local authority or new town		Privately rented		Owner occupied		Total
	Number	%	Number	%	Number	%	
1914	0.1	1.8	7.5	88.2	0.9	10.0	8.5
1944	1.6	12.4	8.0	62.0	3.3	25.6	12.9
1951	2.5	17.9	7.3	52.5	4.1	29.6	13.9
1961	4.4	26.8	5.0	30.5	7.0	42.7	16.4
1971	5.8	30.6	3.6	18.9	9.6	50.5	19.0
1981	6.6	31.0	2.7	12.6	11.9	56.4	21.2
1991	5.7	24.8	1.7	7.4	15.6	67.7	23.1

Source: Minford et al., 1987: 30.
DoE Housing and construction statistics, 1993
Note: Numbers are in millions. Owing to rounding there are some minor discrepancies in totals.

can be considered from a number of points of view. 'Fitness', for example, would be one criterion. But from the viewpoint of housing provision the critical classification is by tenure, for that identifies the providers of accommodation. Table 13.1 indicates the categories of housing tenure in Great Britain between 1914 and 1991.

There are three clear conclusions from the data. The first is that two types of tenure – owner-occupier and the public sector (mainly but not exclusively municipal housing) dominate. The second is that it has come about mainly as a result of a large and rapid decline in the privately rented sector in the second half of the twentieth century. Thus the proportion of such accommodation has fallen from 52.5 per cent in 1951 to only 7.4 per cent in 1991. There have been two causes of the decline. The first was the rapid increase in municipal housing and the second the decreased availability of rented property [particularly after the Rent Acts of 1965 and 1974]. There are reciprocal elements in the changing pattern. After the Acts the letting of property by private landlords, especially at the bottom end of the market, became much less attractive, largely due to measures which controlled rents, laid down minimum standards and, above all, gave security of tenure to tenants. Britain has thus moved from

an extreme where private renting was dominant, but the exploitation of tenants was widespread, to one where renting has been so restrained as to be hardly worthwhile and where, in consequence, the supply has greatly diminished. Recent legislation seeking to redress the balance will be considered later. Also to be reviewed later is the third conclusion from the housing data. In the 1980s both the number of council houses and the percentage decreased to give a clear dominance to owner occupation.

13.3   MARKETS IN HOUSING

The existence of markets in housing presupposes a non-socialist economy, for if all housing is owned by the State no market will exist and the critical process will be allocation. Given a mixed economy, however, as in the United Kingdom, the varieties of tenure and the demand for finance will generate a number of associated markets. These have been set out in a simple diagram form by Susan Charles (1977) and this is reproduced as Fig. 13.1. There is little point in describing the diagram which is self-explanatory. But its relevance will emerge as the discussion proceeds.

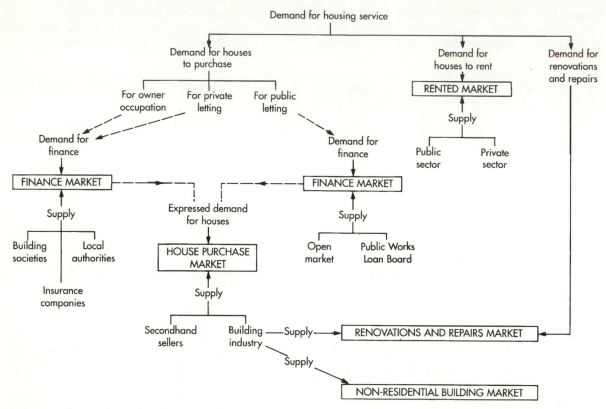

FIGURE 13.1 *The housing market (after S. Charles, 1977). Note: This must be related to the date of its construction and some subsequent modifications would need to be made. Thus, for example, the banks have entered the 'finance market' for owner occupation to a much greater degree.*

If, at this stage, the small privately rented sector is ignored, then the supply side can be considered under the two dominant tenures of public housing and owner-occupied housing. But there are two aspects common to both forms of provision. The first is the actual construction of new houses and the second is the obtaining by, or the allocation to, individuals of the houses. These have to be considered separately in relation to each tenurial type.

13.4 MUNICIPAL OR COUNCIL HOUSING

The building of houses

The British Government's consultative document on housing policy set out the 1977 situation. 'Public sector authorities provide houses for roughly a third of all households in England and Wales. Their stock has grown from 2.2 million houses in 1951 to 5.5 million in 1976. In this period 3.2 million new houses have been completed by local authorities, 0.2 million by new towns and 0.5 million by housing associations' (Dept. of Environment, 1977: 75). This form of provision has a long history extending back to the legislation of the nineteenth century and the Housing of the Working Classes Act (Merrett, 1979: 307–19). 'Local authorities had been given powers to build working class houses by Lord Shaftesbury's Act of 1851, but the amount of new building for which they were responsible was quite insignificant .... Municipal building remained unimportant because it was regarded as undesirable and unnecessary'

(Ashworth, 1954: 91–2). It was not until 1949 that the words 'working classes' were dropped and subsequent measures became simply 'Housing Acts'. There seems little purpose here in discussing the not irrelevant but complex problem of the capacity of the building industry, but rather it is more useful to concentrate on the critical provision of finance, which is at the heart of the municipal housing issue.

Each local government authority operates a statutory Housing Revenue Account (HRA) relating to the accommodation it rents. In 1975–6 the proportion of the HRA costs met by rents was 45 per cent (DoE, 1977) so that in order to operate the system the remainder, which involves both operating costs and the financing of new building, has to be obtained from elsewhere. There are three sources. The first is the local rates or community charge (taxes) which make an essential but limited contribution. In 1975–6 the Rate Fund accounted for but £244 million of a total investment in the public sector of £2798 million. The second source is capital borrowed on the open market at current rates, and the third is government subsidy.

In 1975–6 general subsidy accounted for £966 million and rent rebates added a further £292 million. It is the nature of this subsidy which has become a political issue. In 1967 the then Labour Government, in an attempt to meet electoral promises in relation to the number of houses to be built, introduced a system by which local authorities were able to borrow on the open market at going rates of interest but government subsidy maintained that interest rate for the authority at a standard 4 per cent. The conditions allowed central government control of standards but established an open-ended financial commitment which inevitably caused problems. The response of the succeeding Conservative Government in 1972 was the Housing Finance Act which attempted to shut off the open commitment and introduced the controversial notion of 'fair rents'. Although there were provisions made for phasing, the basic intention was to introduce economic rents which covered the costs of building and expenditure on maintenance, and ensured a balanced

HRA. In order to cushion the less well off, a rent rebate scheme was introduced by which those who could show need were given rebates on the economic or fair rents paid. In 1975 the Labour Government, which came into power in 1974, repealed the 1972 Act and substituted the Housing Rents and Subsidies Act, although it was only intended as an interim measure. By this the notion of a fair or economic rent was abolished and subsidies restored (for example 66 per cent of the annual loan charges on approved capital works), though the rent rebate principle was retained.

A major change came with the Conservative victory at the 1979 election for under Mrs Thatcher for the first time radical right-wing policies were put into operation. They are effectively epitomized in the White Paper 'Housing, the Government's Proposals' (HMSO, 1987) which was published in 1987. 'The Government will encourage local authorities to change and develop their housing role. Provision of housing by local authorities as landlords should gradually be diminished and tenant choice should increase' (HMSO, 1987: 3). There were as many as 10 Acts during the 1980s dealing with housing and associated issues (Malpass and Murie, 1990: 96) but the most important were the Housing Act 1980, the Housing Act 1988 and the Local Government and Housing Act of 1989.

The most significant measure in the 1980 Act was the right of council tenants to buy their houses. Part 1 of the Act is headed 'Public Sector Tenants' and Chapter 1 'The Right to Buy'. It is the implementation of that right which has led to the substantial decline in the number of public authority housing and its decreasing proportion of the national stock. The peak of 200,393 sales of properties was reached in 1982 immediately after the Act came into force. But there was a secondary peak of 185,975 in 1989 (HMSO, 1991). In addition, there were major modifications to the process of subsidizing municipal house building from central authority. 'The most fundamental difference between the 1975 subsidy system and the housing subsidy system is that producer subsidies are no longer guaranteed. They can now be confiscated when there are reductions in the deficit of the "notional" Housing Revenue Account due to an excess of the change in "reckonable expenditure"' (Gibson, 1981: 1). Moreover, the Secretary of State indicated a guideline for rent increases, with a further indication for the future so that 'it is likely that in a few years all housing authorities will be receiving zero subsidy' (Gibson, 1981: 1). Given that a significant part of most local authorities' revenue expenditure arises from debt charges from borrowing, some 27 per cent in 1991–2 (Malpass and Warburton, 1993: 12), it follows that the finances available to build new houses were severely limited. In brief, the diminution in the numbers of houses by sale could not be offset by house building. Thus in the first year of operation, 1981–2, council rents rose by 48 per cent on average and in three years the real value of the subsidy fell by over 80 per cent (Malpass and Warburton, 1993: 94). All this is symptomatic of the basic policy of the residualization of public housing, that is, it becomes no more than a residual element in the housing stock.

The major feature which distinguishes housing in Britain from that in other members of the European Union and from the USA is the very limited role played by private rentals (Table 13.2). The private rental sector provides especially for those who do not qualify for the limited amount of social rental, council housing mainly in Britain, but who cannot afford to buy. That category especially includes the young and much of the problem of homelessness is a consequence. The Housing Act 1988 sought to solve the problem by deregulation but, as the data (Tables 13.1 and 13.2) indicate, with little success. A further measure in the 1988 Act allowed local authority tenants to opt for a new landlord and established Housing Action Trusts (HATs) as a way of meeting the most pressing problems in the public sector, problems which were seen as the product of poor management and inadequate maintenance. HATs 'were to be corporations appointed by the Secretary of State to take over large, run-down local authority housing estates. Ownership would be transferred to new private or voluntary hands' (Karn,

TABLE 13.2 *Housing tenure in selected countries 1988–90. Percentages in each category*

	UK	USA	West Germany	France	The Netherlands
Owner-occupied	67	64	40	51	44
Private rented	7	33	45	26	13
Social rented	26	3	15	23	43

Source: Hallett, G. (ed.) 1993: *The new housing shortage. Housing affordability in Europe and the USA.* London: Routledge.

1993: 74). Effectively more housing would be removed from the public sector.

The 1989 Local Government and Housing Act again modified the financial regime with the effect of raising rents and cutting capital expenditure. 'The essence of the system is very simple. The HRA is now ring fenced, so that authorities cannot make discretionary payments to or from the general fund, and subsidy is the difference between notional income and expenditure' (Malpass and Warburton, 1993: 96). Moreover, the variety of subsidies has been reduced to one and the 'rate' fund contribution eliminated. 'The effect of these moves is both to restore government leverage on rents across the country and to give it complete control of subsidy income to local HRAs (Malpass and Warburton, 1993: 96).

The impact of all these measures can best be demonstrated by the numbers of starts on new housing. In 1982 there were 192,400 starts divided into 140,100 in the private sector and 52,300 by local authorities. In 1992 there were 156,000 starts (the lower number reflecting depression in the housing market) but whereas there were 120,200 in the private sector, the local authority contribution was only 2600 (the figures are to the nearest 100 and are taken from the DoE Housing and Construction statistics, March Quarter, Pt. 2, 1993).

The legislation on housing and the financial arrangements are much more complex than this brief summary has implied. Special studies are available in Malpass and Means (1993), Minford *et al.* (1987), Cooper (1985), Balchin (1985). In itself the legislation is of more concern to students of housing economics and local government finance. But the construction of housing of a particular type in specific localities is a major factor in the creation of social areas. In so far as council house building is restricted and the continued placement of new estates inhibited, the shape and character of the city is constrained. Further, and it is a point which will be taken up later, the right to buy council housing has operated differentially. Thus Sewel *et al.* (1984) have demonstrated that in Aberdeen sales of council houses have been higher on estates with a higher socio-economic profile and disproportionately located in the popular estates, with the less popular estates and diffi-cult-to-let areas generating far fewer sales. Such a situation will increasingly lead to differentia-tion between estates which will be reflected not only in the character of social areas but in the general social stability of the less favoured estates . It is a contribution towards the creation of 'sink' estates (see Chapter 15, pp. 299–301). There is therefore a geographical or locational product of housing legislation.

Allocation of houses

The allocation of public housing is in itself a matter of considerable interest since it is indica-tive of the system which might operate in a non-market, non-capitalist situation. There are two stages in the process. The first is the sieving out of the eligible and the ordering of families into degrees of priority. Eligibility might not seem to apply where there is only a state source of housing but even under such circumstances it could have a regional or locational reference. That is, in a totalitarian situation migration of popula-tion is contained and controlled by assigning eligibility for housing in a regional context. Within contemporary Britain there is no uniform system of establishing eligibility and priority and each local housing authority operates its own system. There are, however, some overriding

principles in the priority to be given to the homeless after the Housing (Homeless Persons) Act of 1977, and to single parents by virtue of need and homelessness, although this has been modified after public reaction.

Two bases of priority can be identified. The first, which is more concerned with eligibility, and is the more controversial, is length of residence or of working in the area. Because eligibility through a residential qualification has resulted in immigrant homelessness it has become an issue of public concern. The result was a declaration against it by Government, at least in terms of eligibility if not explicitly in terms of priority. To some extent this is a response to situations easily exploited by the media; the disadvantages and the injustices of abandoning qualification by residence are hidden and less exploitable. The second basis is need, and a wide variety of criteria is used in its establishment. Included are measures such as size of family; separation of family; accommo-dation available, especially in number of bedrooms related to the sex of children; disabil-ity; ill-health; the sharing of facilities such as kitchen or bathroom. Each authority can score those families on its housing list by such criteria and allocate accordingly.

Ultimately all schemes give rise to ill-feeling for none can be totally objective, even if need could be defined. Moreover, incompatible crite-ria have to be played off against each other, such as length of residence against need, the charac-teristic dilemma of long-standing claims of those resident and wishing to marry as against the single parent. At best, a whole complex of influ-ences is brought to bear – personal, social and political – at worst, powers over people are placed in the hands of gate-keepers, that is those who within the bureaucratic system control access to housing. The powers of such gate-keepers are every bit as alarming, perhaps more alarming because they are absolute, as the exploitative ability of capitalist landlords. The role of the manager or gate-keeper in urban society, although central to this whole theme, cannot be considered at length in this brief chapter. An excellent case study is that by

TABLE 13.3 *Status of tenant and type of property, Hull, 1972*

Age of dwelling	Local authority grading of tenants		
	Poor/fair (%)	Fair/good (%)	Very good/ Excellent (%)
1919–44	62.5	13.9	10.7
1945–65	29.2	38.0	28.6
Post-1965 used dwelling	8.3	23.4	17.9
New dwelling	0.0	24.8	42.0

Source: After Gray, 1976.

Henderson and Karn (1987), *Race, class and state housing. inequality and the allocation of public housing in Britain.*

In summary, there are three problems related to eligibility and priority. The first is the difficult assessment of the significance of residence in a geographically mobile society. The second relates to problems engendered by the requirement to show need, for it offers no encouragement to improve existing conditions; improvement of conditions lowers need and reduces priority. The third is the power given to gatekeepers and, to revert to an element introduced in Chapter 8, corruption in the process of allocation if not widespread is extremely well attested and widely reported. However idealized a society, there are unlikely to be all the houses required, of the appropriate type, in the right location, and as long as there is a shortage these problems will remain.

The second stage in the allocation process is that of matching individual families with particular houses. This might seem to be an actuarial problem in relating houses built to family sizes, and a managerial problem in detail. A characteristic difficulty arises however, in relation to life cycle changes. A large house allocated at the peak period of family formation and child rearing becomes inappropriate at a later stage after child launching when an elderly couple or a single person is left. But the direction of families, parallel to the free market pattern of life cycle movements, is socially unacceptable. People cannot, in a reasonably free and caring society, be moved around like so many pieces on a chequer board.

A much more contentious procedure arises, however, where the managers attempt to concentrate problem families on one estate. This practice is both condemned and denied, but Gray in a study of Hull was able to show that it is operated (Gray, 1976). He established that all new tenants were, in 1972, interviewed by a 'housing investigator' who, not only recorded details, but also assessed households in general terms as 'good' or 'bad' tenants. Table 13.3 shows the results.

Gray's summary of these results is significant. In terms of the class structure of residential areas, the effect of local authority policies both in Hull and elsewhere is to accentuate the social and spatial contrasts between classes: the worst-off groups in the city are increasingly segregated in the poorest council house estates and the contracting private rented sector, while the most privileged (and generally higher class) council tenants tend to be concentrated in the new high-status estates elsewhere in the city (Gray, 1976: 44). More recent studies by Skellington (1981) and that already quoted by Henderson and Karn (1987) provide additional evidence of the process. Here, then, is a detailed illustration of the process by which socially contrasted city areas are created, not by demand in the market but by the manipulation of supply and allocation by managerial decisions. Whether this is a 'proper' procedure is another matter and a subjective one. It can be argued that it gives an element of providing compatible neighbours which those in the owner-occupied sector can and do value highly in their purchasing strategy.

13.5 OWNER-OCCUPIED HOUSING

Building of houses

A complex of interests is involved in the provision of houses for sale and it is useful to follow the order set out by Robson who considers in turn the landowner, the developer and the planner (Robson, 1975). To his three interest groups it is worth adding the builder. It is not, however, easy to keep these apart since they both overlap and clash in the business of building houses.

1. **The landowner** at one time held a predominant control over residential development and it is one of the standard exercises in historical urban geography to show how land ownership patterns, and the willingness or otherwise of owners to release land, were determinants in the nature of town extension (Slater, 1976). Such absolute control has, however, been whittled away during this century. Even some nineteenth-century legislation modified the extreme *laissez-faire* situation and from the first Town Planning Act of 1909 constraints have been successfully built up. The 1947 Town and Country Planning Act marked a significant step, for the initiation of compulsory development plans and development control, together with the powers of compulsory purchase, meant that influence on the internal structure and external extension of towns was placed more firmly in the hands of planning authorities.

   Land-use was by this time effectively nationalized and state controlled via local government representation. What remained was the critical issue of betterment. Once agricultural land is scheduled for residential development, or when such scheduling seems likely, its value vastly appreciates and the question arises as to whom this increased value should go. Again this is a political matter. The left maintains that it is the community which creates this increase which therefore should revert to the community; that is, that all land should change hands at current-use value. Beyond this is the view that all land should be nationalized in any case, or should come into public ownership at development. It would then be expropriated at current-use values. The far-right argues that it is a basic guarantee of freedom that the individual should hold property and should be free to sell it where and when he or she wishes for as much as the free market will offer. If an entrepreneur takes the risk and is enterprising enough to buy land and hold it in anticipation of development he should reap the reward of his risk taking and initiative.

   In 1974 Britain moved substantially (though far from completely) to the left with the Community Land Act. In briefest summary, this gave local authorities the right to purchase land needed for 'relevant' development in the next 10 years from 1976 at its undeveloped cost. In order to achieve this each authority had to set up Land Acquisition Management Schemes (LAMS). To cover the interim period of change, during which land could be sold privately, a Development Land Tax on profits at 66.6 per cent was levied on such transactions. Eventually all development land was to be acquired by the local authorities. Behind all this also lies the critical issue of the supply of land for residential building. Prior to the Act it certainly paid the large development companies to build up large land banks where both inflation and betterment on release for building provided rich profits.

   After the general election of 1979 a Conservative Government came to power and the leftward trend of the mid- and late

1970s was reversed. The Community Land Act was repealed by the Local Government, Planning and Land Act of 1980. To ensure an adequate supply of land, a public sector land register was compiled in order to identify land which could be released for development though it has made little contribution to the areas of great need. Thus the see-saw effect of party politics continues to exert a critical influence on the land market.

2. **The developer.** Already in considering the supply of building land both developer and planner have been introduced. The word 'developer' now has pejorative overtones. A developer's interest is profit and in order to achieve this the main aim is to upgrade use, that is to change land from a low revenue to a high revenue use (Community Development Project, 1976). This can be done by raising rents, intensifying use by such means as adding floors to an existing building, or by upgrading use, that is, for example by replacing housing with offices.

It can be argued, therefore, that the developer has had three major influences on British cities. The first is that concentration on the construction of large residential estates on greenfield sites at the city margins has dominated, thus strongly reinforcing if not creating suburbanization. This involves a simple and profitable process of upgrading land in agricultural use. The second is that having no interest in building low cost housing, especially on high value city centre land, the developer has been active in increasing returns by the construction of office blocks on decaying residential areas. An excellent example of this was given by Ambrose and Colenutt in their book *The property machine* (1975). A chapter was devoted to 'The property development system in Brighton'. It considered in detail the redevelopment of the twenty-five acre Brighton Station site. The alternatives were either a new station, an hotel, conference facilities, a hypermarket, a vast car park, a telecommunications centre, a recreation centre, general offices and a small amount of housing, or restoration of the Victorian station and council housing. There is no room here to trace the details of this conflict but the situation neatly points the opposition between maximizing returns in either economic or social contexts – and there are jusifications for both views.

The third influence of the developer operates via that of the **builder**. The builder, like the developer and landowner is in the market to make a profit and like anyone in such a situation responds to demand, not need. Moreover:

> the price of houses for sale is not determined by what they cost to build but by what buyers can afford. And what buyers can afford depends on the levels of wages, credit and subsidies. But if wages rise, house prices rise. If government subsidy increases, house prices rise. If credit gets cheaper and easier to get, house prices rise. As soon as decent housing comes theoretically within reach of the poorest workers, house prices rise and takes (*sic*) it out of reach again.
>
> (Ambrose and Colenutt, 1975)

The undefined adjective 'decent' in the quotation makes the statement that much weaker but it does set out the injustices the market creates.

3. **The planner**, too, as has been indicated, operates a major influence on residential character. In a situation where land-uses are designated by planning authorities, decisions can have critical effects on the value of land, either blighting it or raising it to a level which, without compulsory powers of purchase at current-use value, the local authority cannot meet. The planner, as bureaucrat, has to hold the ring between individual and community interest and so is usually the most easily accessible for condemnation. But the planner can only operate within the framework of national legislation and some indication of the nature of this has already been given.

Obtaining housing

In terms of owner occupation the fundamental control of who gets which house is the ability to pay and since few people have sufficient capital to buy outright, then the control is access to loan finance, usually through a building society. It was estimated in the 1970s that between 80 and 95 per cent of all advances for house purchase came from the building societies (DoE, 1977: 50), with small contributions from local authorities, insurance companies and banks. These other contributions have become greater in the 1980s. Thus whereas in 1973 some 71 per cent of the net advances for house purchase came from building societies, the figure had fallen to 58 per cent in 1982, with the banking sector's share having risen from 11 per cent to 36 per cent (Balchin, 1985: 231). Building societies in Britain are non-profit-making and also, unlike other financial institutions, they borrow short and lend long. Because of this, and the fact that they depend on investment money, the societies maintain that they have an absolute duty to ensure security for their loans. This they do by using two restrictions on advances, the one related to the individual concerned; the other to the property. The personal restriction is a measure of creditworthiness. Most mortgages are limited to some 80–90 per cent of the purchase or valuation price, whichever is the smaller. However, 100 per cent mortgages became more common in the 1980s although subsequently producing many problems and a spate of repossessions by building societies. The criteria employed are usually that monthly payments should not exceed weekly income or that the amount advanced should not be more than two or three times the annual income. Mortgages are usually for terms of 25 years and any building society therefore looks for stable career prospects and secure earnings. A salary from a professional career which increases with age and experience is far more attractive than the wage of a manual worker in heavy industry which might peak with physical capability and then decline. The government publication 'Housing Policy' concluded that 'there are

probably a lot of people who want to become home owners, but whose needs do not quite fit into the conventional pattern of mortgage facilities and lending criteria applied across the building society counters'.

The second restriction on mortgage relates to the property. Since this is the security on which the loan is made it has to be judged as a sound investment. Certain rules of thumb seem to be applied though there are no firm fixed guidelines. The property should be characteristic of the district in which it is located; public transport should be reasonably available; the district should not be showing evidence of deterioration. In general the more 'standard' the type of property the easier it is to obtain a mortgage. Thus in Britain, the best regarded property is the three-bedroomed semi-detached house.

Building societies dislike new, idiosyncratic styles of housing but it is their attitude to older property which gives rise to most criticism. Property which is old, in poor condition and in deteriorating areas does not offer good security and the building societies, therefore, refuse to advance loans. Taken to its furthest extent this involves the controversial process of red-lining, that is of identifying whole areas, usually of the deteriorated inner city, where mortgages are refused. Lambert, in a study of Birmingham, concluded that:

> the building societies deliberately steered clear of certain of the older areas... or exercised extreme caution, although they usually emphasized that there was no written policy which precluded all houses in those areas. However two or three managers did say that they had marked off areas on a map (in one case we were shown this) and rarely looked at any property falling within these boundaries.
>
> (Lambert, 1976: 33–4)

Where such red-lining had occurred it was often said to have been done at the instigation of surveyors who, it was argued, knew the city well and all the trends and markets involved. After a review of the evidence Peter Williams writes, 'Although our conclusions must be tempered by the lack of any comprehensive

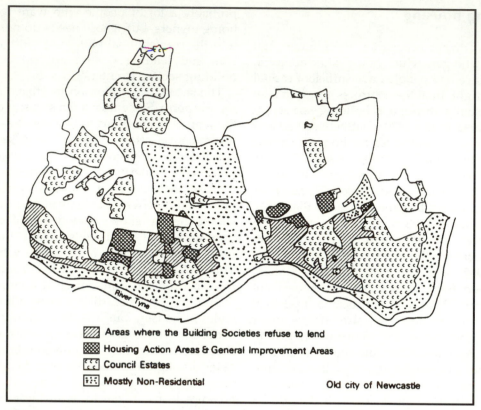

FIGURE 13.2 *The 'redlined' areas of Newcastle in 1976 after Community Development's 'profits against houses'.*

analysis, there can be no doubt that red-lining exists and that building societies, at least, believe themselves to be justified in this practice' (Williams, 1978: 32).

The real problem remains to produce mapped evidence. The authors of *Profits against houses* (Community Development Project, 1976) did include a map of Newcastle (Fig. 13.2) but it is not discussed in the text and no indication of the evidence upon which it was based is given. Presumably, it is only an impression derived from experience rather than a map actually used by one society or drawn from a meticulous examination of all mortgages offered in Newcastle over a set period. Under social pressure during the 1980s, and with a boom in house purchase these limitations became less significant. However, the immediate conclusion

is that the building societies, through the mortgage system, exert a considerable influence upon city structure. At a more general level, the taxation relief given to mortgages represents a huge subsidy to private home ownership. In 1982–3 the tax relief on mortgages amounted to £2225 millions, compared with a general subsidy to municipal housing of £987 million (Balchin, 1985: 216). By 1994 the tax relief had risen to £4300 million, whereas the total value of output in public housing ran at just over £400 million per quarter.

At this point it is useful to introduce parallel evidence from the United States. Owing to increasing concern with regard to declining residential areas the Home Mortgage Disclosure Act was passed in 1975. By this all depository institutions were required to disclose the number and value of all mortgage loans issued in all SMSAs. Such data are amenable to spatial analysis and they have been used by Dennis Dingemans in a study of red-lining and mortgage lending in Sacramento. He contends that although his investigation does not enable him to determine the exact role of mortgage lenders as causal agents of neighbourhood change, it does contribute a basic understanding of the geography of mortgage lending in a relatively untroubled metropolitan area (Dingemans, 1979). Dingemans's conclusions are that mortgages and home improvements are strongly associated with several of the traditional measures of the socio-economic status of neighbourhoods. He adds, 'many of the pioneering students of intra-urban lending patterns have found these patterns elsewhere and have been quick to reach the conclusion that lenders were red-lining – discriminating against mortgage loan applications from certain areas regardless of their personal qualifications for the loan sought'. Properly, however, he contends that specific investigation of the processes and behaviours underlying these patterns is essential rather than the making of direct assumptions from aggregate data.

There is a further group which can operate to influence social areas within cities. These are the estate agents. The traditional role of the realtor in the USA, for example, was either to preserve or tip, that is to change rapidly, the racial character of areas (see Chapter 14. p. 274). This rather crude impact upon social space has to a great degree been modified by legislation against racial discrimination both in the USA and Britain. Even so the influence of the estate agent can still be significant for he or she is not a passive and neutral gatekeeper, but brings both unconscious and conscious influences to bear. Palm, writing of the influence of realtors in the USA in the late 1960s, identified three control techniques (Palm, 1976a and b). The first is the automatic matching of a customer to an area where no specific area is specified by the client; the second is by warnings of unhappiness, indications that a household will not fit in easily to an established ethos, while the third is

the simple distortion of facts, for example, describing a house as sold if the client seems inappropriate.

The general argument is that the agent feels a 'commitment' to residential areas and that a successful business is run by avoiding mismatches and mistakes. Even if such blatant management does not occur, it can be shown that estate agents cover only limited portions of the housing market and will recommend in relation to perceived social categories (Palm, 1976a and b). Peter Williams's summary of the influence is, 'thus through their role as advisers to landlords, building societies and property companies, as well as having personal control of all the different organizations, estate agents are in an important position with regard to the generation and manipulation of change as well as being crucial intermediaries in the process whereby households obtain accommodation' (Williams, 1976: 61). There is the danger, however, of assigning predetermined and, indeed, malevolent intentions to estate agents. They themselves are only operating within a social context, carrying out what the community around them wishes, otherwise they would soon go out of business. They have no brief to be agents of social change in directions perceived as 'good' or 'bad'. The notion of 'manipulation' and still further of the 'generation' of change needs some qualification. It can be argued that the change derives from large-scale social processes and the realtor is literally merely the 'agent'.

13.6 THREE EXAMPLES OF HOUSING DEVELOPMENTS

This review of the impact of the supply side on the structuring of the city, like many such studies of process, has had some difficulty in firmly linking the matters discussed to patterns on the ground. In order to demonstrate the link three specific topics which have been in the forefront of discussions of housing can be considered.

The high-rise block in public housing

One of the distinctive features of nearly all large British cities during the 1960s and the early 1970s was the building of high-rise tower blocks to accommodate council tenants. The result was a very distinctive contribution to the townscape as well as to the pattern of social areas. The construction of these blocks shows a very distinctive peaking towards the end of the 1960s and then a marked tailing off in the 1970s. This is clearly illustrated in Table 13.4. The 1980s and 1990s have been largely concerned with their destruction.

Two points can be derived from the table. The first is the growth in the contribution in high-rise blocks to total accommodation provided, to reach a level of just under a quarter between 1964 and 1967. The second is that the percentage of total costs in each year is greater, that is, that high-rise development was a more expensive form of provision. This had been clearly demonstrated by 1959, the year before the data in Table 13.4 began. P. A. Stone (1959) had indicated that the net construction costs per three-bedroomed dwelling with a standard floor space were just under twice as much in a 12-storey block as in the traditional two-floored house. These special costs were met by additional subsidy from central government to the local authority. But the question must be put as to why this form of development occurred, for there was certainly no demand for it from the population involved. For an explanation one has to turn to consider the influence of two of the controllers of supply.

The local authorities and their professional representatives, the planners, faced a major difficulty in the post-war period. On the one hand there was the need for a massive slum clearance and redevelopment programme, especially in relation to very high density working class housing dating back to the nineteenth century. On the other hand the urban sprawl which had characterized the inter-war period had been the source of much criticism. Ian Nairn's book *Outrage* of 1955 was a characteristic blast:

TABLE 13.4 *Local authority high-rise construction: tenders approved 1960 to 1975*

Year	Percentage of all dwellings in five or more storeys	
	By numbers	By cost
1960	11.91	17.27
1961	10.92	16.15
1962	13.35	19.43
1963	19.99	26.93
1964	29.08	28.30
1965	19.49	24.99
1966	21.81	27.25
1967	21.38	26.67
1968	15.78	19.84
1969	9.93	12.66
1970	6.80	8.65
1971	5.16	6.61
1972	2.80	3.76
1973	2.63	3.79
1974	1.58	2.37
1975	0.042	0.48

Source: CDP, 1976: 29.

The city is today not so much a growing as a spreading thing, fanning out over the land surface in the shape of suburban sprawl. ... This thing of terror, which will get you sweating at night when you begin to realize its true proportions, we have called ... Subtopia. It consists of the universal suburbanization not merely of the country, or of the town, but of the town-and-country – the whole land surface.

(Nairn, 1955: 365)

Moreover, some of the most distinguished urban theorists urged the same point. The city, argued Le Corbusier, should be compact, lively and concentrated (Le Corbusier, 1947). The way forward, therefore, seemed evident. High-rise, high density development would solve critical problems. In a foreword to a Ministry planning bulletin of 1962 called 'Residential Areas: Higher Densities', the then Minister of Housing and

Local Government wrote, 'over the next 20 years we are going to need at least six million more houses. This is a formidable problem with two inescapable consequences: we need to allocate more land for housing and we need higher densities especially in pressure areas. We need not one or the other, but both – more land and higher densities' (Ministry of Housing and Local Government, 1962: 3). There seemed no reason why such a policy should not succeed. Some of the most expensive and exclusive of luxury accommodation was in high-rise apartments. In many European cities life in high density apartments was a commonplace, even a tradition, and caused no apparent problems.

The second influence involved was that of the big construction companies and their architects. The authors of *Profits against houses* accuse them of having a clear interest. 'In the 1950s many big civil engineering firms wanted to move into council housing; they wanted to develop new technology and use mass production methods – to make a profit. High-rise flats fitted the bill' (CDP, 1976: 29). The argument continues that a massive persuading programme, together with the corruption of local officials which has since come to light, enabled them to achieve their ends, including large public subsidies for this expensive form of provision which went directly to the builders. Having obtained their profits, and acquired an expertise, these firms moved out and on to the lucrative field of high-rise office block construction.

Much of the criticism of high-rise building is *ex post facto* in character. There is no doubt that at the time the tower block was seen as a solution to a range of problems and a means of keeping the city compact and lively. That it went disastrously wrong is now widely agreed, although the reasons are perhaps not quite as obvious and are worthy of a brief review.

The first reason put forward is that such accommodation is completely alien to the British tradition (apart from the tenements of Glasgow), and, therefore, greatly disliked by those allocated to it. That this is so is unchallengable. Surveys both by government and by other groups have revealed a consistent dislike of

high-rise flatted dwellings. A review of an Opinion Research Centre report in 1967 began, 'One in six of the homes currently being built by local authorities in Britain are (*sic*) flats in tall blocks of ten stories (*sic*) and upwards. Yet only one person in a hundred wants to live in a block that high' (*The Sunday Times*, 1967). Mere dislike is, perhaps, too weak a reason to account for the disastrous physical decline of many of these blocks, but it significantly shows that the general pattern was determined by the supplier and not by demand which was overwhelmingly for detached or semi-detached houses (79 per cent in the survey quoted and 71 per cent of a government survey of families living at high density) (Ministry of Housing and Local Government, 1970: 33).

The explanation for the rapid deterioration of these high-rise blocks must, in part at least, be an environmental one. But as Michelson has pointed out, it is necessary to make a distinction between the properties inherent in high-rise building and properties that appear connected to them in so far as the buildings are constructed in a particular way:

A good number of the social implications of high-rise apartments are, in fact, implications of these buildings *as they are conventionally constructed*. It takes no great stretch of the imagination to contemplate buildings with total sound proofing, with cavernous living space, with service and recreational facilities near every suite, and so on. We may not expect such buildings under current economic conditions or under current methods of finance, but the limitations of the status quo are not necessarily limitations to high-rise buildings per se.

(Michelson, 1977: 47)

The two major problems which derive from high-rise living are related to density and access. High density generates noise in particular, but a whole range of other less easily identifiable conditions which are often said to be associated: physical ill-health, 'heightened aggression, family breakdown, inadequate child care and personal disorder' (Michelson, 1977: 49). But this is the complex field of environmental psychology and little has been convincingly demonstrated. There are in Michelson's words many frightening insinuations, but few firm conclusions. More immediate is the second problem of access. There is the direct problem of playing space for children within the parents' direct view, but indirectly there is the problem of access via lifts, which are either inadequate or break down and remain out of order, and via concrete walkways. The ill-lit corridors and balconies and the graffiti covered walls become the ideal locations for petty crime, which soon escalates into major crimes of violence. Once this happens the whole situation degenerates into an uncontrollable sequence of vandalism and destruction until the block, with windows boarded up and every fitment broken and wall defaced, becomes uninhabitable for all but the criminal. The outbreak of violence on British housing estates during the mid 1980s in places such as Handsworth, Brixton and Tottenham can in part be ascribed to the environmental conditions on high-rise housing estates, where repairs have not been carried out due to limited housing budgets.

The classic case of a well-intentioned project designed by a specifically commissioned architect which moved from inauguration in 1954 to demolition in 1974 is, in fact, not British but the Pruitt-Igoe complex in St Louis, Missouri. Pruitt-Igoe consisted of 33 blocks each intended to house 12,000 residents. It was conceived as a revolutionary project to free the blacks of St Louis from 'the vicious circle of poor housing, unemployment, crime, disease and eternal poverty which had haunted them for so long'. It was intended to be the best possible low-rent housing (Rainwater, 1970; Winchester, 1974: 16). It was a Utopian endeavour which utterly collapsed. The history of the débâcle is too long to narrate here but has been ascribed to complete lack of provision of shops, nurseries, play areas, day-care centres and general services. As one commentator recorded, 'The fact that so many people were crammed into so small a space, and that so many of the people came with problems far in excess of their ability to cope, added mercilessly to the dilemma' (Winchester, 1974: 16).

This discussion can be taken a little further into the field of environmental psychology which has already been opened up. Oscar Newman, in his book *Defensible space* (Newman, 1972) was much concerned with the sorts of minor crimes and senseless violence and vandalism which characterize housing projects like Pruitt-Igoe. Newman starts from the isolation that characterizes high-rise living where access to neighbours 'across the garden fence' is not possible. This, he argues, leads to a withdrawal from community life and hence there is no consensus community control of antisocial behaviour. All space is public space and designed and made to look so. Nobody owns it, so nobody defends it. From this comes Newman's notion of defensible space, whereby such anonymous areas are fenced and attached and made subject to surveillance. In short, a 'territory' is established which can be clearly identified and defended. 'Another feature of many low-rent housing projects ... that exacerbates crime statistics is that they tend to be designed so that they stand out from the surrounding context by virtue of their starkness and lack of architectural embellishment.' It is almost as if they were being set up as a target for attack. The interiors, too, tend typically to be 'institutional' in character, with tiled walls, uncarpeted floors and 'vandal proof' fitments. The message is defiance, a physical challenge to those who do not wish to be contained to show that they are superior to their symbolic prison walls. Newman argues that 'such environments make the residents devalue themselves, lose their self-esteem and self-respect and so create a vicious circle of damage and repair and greater damage' (Mercer, 1975: 93). One, thus, returns to Pruitt-Igoe where a situation was reached where, because of the danger of attack, workmen were unable to make any repairs and by 1968 copper sheeting was being torn off roofs. The result was flooding and in the winter of 1971 hundreds of flats were inches deep in ice.

This brief foray into the field of environmental psychology has been made to demonstrate the direction in which process studies lead.

From aggregate social areas as the consequence of demand, consideration shifted to the need to consider supply. The details of supply of municipal housing led, in turn, to a review of the characteristics of one type of provision and some of the social and physical responses which followed. A position has now been reached where at least the groundwork has been established for a study of the process by which such an aggregate social area as 'The Piggeries' in Liverpool can be undertaken. But as in behavioural studies of individual choice, a substantial departure has been made from conventional geographical analyses of spatial patterns into studies of the impact of the built environment on social and community character.

Here, too, the assessment of the failure of high-rise accommodation will be related to political viewpoints. Those to the left, starting from the romantic egalitarian assumptions that all people are alike and inherently good, will ascribe these disasters to the economic system and the social and physical environment it necessarily generates. Those to the right, starting from the realistic assumption that all people are not alike and are inherently selfish and competitive and that there is only varied redemption from something approaching original sin, will relate the degeneration of such buildings to the fact that they become the social sinks into which the indolent and the criminal descend. In the words of nineteenth-century morality, 'those who wish to be clean, clean they will be' or in the more modest phrase quoted about Pruitt-Igoe, 'so many of the people came with problems far in excess of their ability to cope'. Neither extreme is justified. It is probably unfortunate that those families most unable to cope and seeking public help are those given accommodation where there are severe environmental stresses. But as Michelson comments, it is not high-rise living of itself which creates those stresses but rather the way in which it is established when costs have to be kept to the minimum. It is not very logical to rail against the luxury of penthouse suites on the one hand, and against high-rise blocks for council tenants on the other, without looking a little more carefully at the total context. A more balanced treatment of tower blocks is beginning to become apparent (Glendinning and Muthesius, 1994).

The suburban estate

It could be maintained that few wish to live on a suburban housing estate and that given an unconstrained choice, most would probably opt for a large house in its own grounds but with immediate accessibility to all desired services. The decision to buy a house on an estate is, therefore, a compromise accepted as being a feasible solution. From this rather elementary point of view it is evident that the creator of the settlement form, indeed the house design itself, is not the demanding customer but the supplying developer. In a more complex way this can be seen as the product of the market system.

These arguments have been most cogently advanced by David Harvey with empirical exemplification from Baltimore (Harvey, 1974):

> Residential differentiation in urban areas has long been explained in terms of social ecological processes, consumer preferences, utility maximizing behaviour on the part of individuals and the like. The Baltimore evidence suggests that financial and governmental institutions play an active role in shaping residential differentiation and that the active agent in the process is an investor seeking to realize a class monopoly rent.
>
> (Harvey, 1974: 249)

Harvey defines such a rent as a form of absolute rent which arises ' because there exists a class of owners of "resource units" – the land and the relatively permanent improvements incorporated in it – who are willing to release the units under their command only if they receive a positive return above some arbitrary level' (Harvey, 1974: 241). It follows that the class-monopoly rents obtained will depend upon the outcome of a conflict of interest between the owners and the consumers. The characteristic middle class seeker of housing has, to use Harvey's terms for the process, been subject to a 'blow-out' from the inner city which has been subject to an influx of low-income populations and a rapid decline in the services of all sorts.

Here then is a well-defined market which the landowner can exploit. The population concerned can be convinced of the advantages, and especially of the relative prestige, of living in a particular area which may lead to house purchase and the creation of a positive return to the owner. The whole operation is conducted within the institutional framework of national and local government but the home buyer is led, he does not do the 'demanding'. The residential structure of the city is, therefore, shaped by the derivation of profit returns for the owners of the monopoly rents and the financial institutions which support them:

> This is not to say that considerations of race and ethnicity, social status and prestige, life style aspirations, community and neighbourhood solidarity and the like are irrelevant to understanding residential differentiation, ironically, all of these features *increase* the potential for realizing class monopoly rent because they help to maintain the island-like structure, to create the absolute space of the parochially minded community.
>
> (Harvey, 1974: 245–50)

Indeed, as Harvey goes on to maintain, it is possible to argue that consumer preferences are produced systematically from this process rather than generated spontaneously from the community. They are certainly as liable to manipulation as are the preferences for consumer goods, and effective manipulation within the sub-market can increase returns.

The argument advanced above is one slanted from a particular angle and the simple notion of house seeking populations led by the nose to a situation which they are manipulated to believe they want, is far from completely credible. But the view of residential segregation by socio-economic status presented by Harvey is sufficiently impressive to emphasize that the middle class suburban estate can equally be regarded as the creation of the realizers of class-monopoly rents, as the spontaneous generation of an

environment out of the wishes of the families who live on it.

Location of council housing

For the urban geographer the location of council housing within the city must surely be the core of interest, but it is surprising how few studies deal with it. An exception is *Housing and residential structure* by Basset and Short (1980) from which Fig. 13.3(a), (b) and (c) is taken.

Location is related to the nature and quality of housing, which, in turn, depends on the conditions, largely the legislation, under which it is built. It is possible to suggest an outline classification (Jennings, 1971):

1. **Housing built before the First World War.** This was mainly a response to the Housing of the Workng Classes Act of 1890 which provided for local authorities to erect, convert, alter, enlarge, repair, fit up, furnish and supply lodging houses for the working classes, the term was defined to include separate houses or cottages (Merrett, 1979: 309). There was no great take-up and the properties built were mainly of the cottage, garden suburb type at low density.

2. **Inter-war housing.** This was related to three Acts. The Housing and Town Planning, etc., Act of 1919 (the Addison Act), the Housing Act of 1923 (the Chamberlain Act) and the Housing (Financial Provisions) Act of 1924 (the Wheatly Act). These Acts resulted in the building of extensive and monotonous estates of terrace housing, usually of a mediocre quality. The process was accelerated by the Housing Act of 1930 (the Greenwood Act) which dealt with slum clearance.

3. **Post-war housing.** Perhaps the most significant legislation was the Housing Act of 1949 which dropped the pejorative 'working class' limitation and allowed local authorities to provide accommodation for any member of the community. Setting aside the temporary accommodation in the immediate post-war period, building was once again characterized by large estates provided with

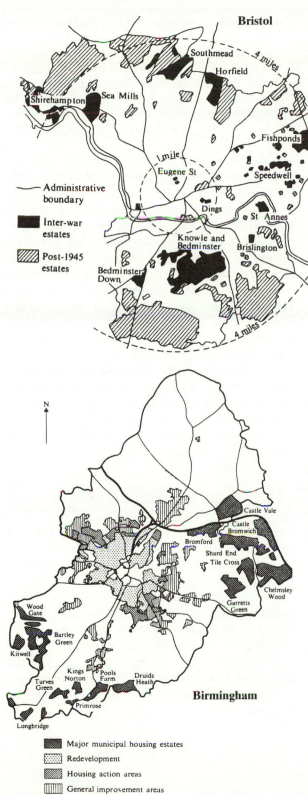

Bristol

Southmead
Horfield
4 miles
Shirehampton
Sea Mills
Fishponds
1 mile
Eugene St
Speedwell
— Administrative boundary
Inter-war estates
Post-1945 estates
Dings
St Annes
Knowle and Bedminster
Brislington
Bedminster Down
4 miles

Nottingham

N

Bestwood Park
Edwards Lane
Highbury Vale
Sherwood
Stockhill Lane
Broxstowe Strelley
Beechdale
Bilborough
Wollaton Park
Lenton Abbey
Clifton

0 1 2
Miles

Estates built 1919-39   — Major roads
Estates built 1945-66   ----- City boundary
Some small groups of houses built 1919-66

N

Castle Vale
Castle Bromwich
Bromford
Shard End
Tile Cross
Wood Gate
Chelmsley Wood
Bartley Green
Garretts Green
Kitwell
Kings Norton
Pools Farm
Druids Heath
Turves Green
Primrose
Birmingham
Longbridge

Major municipal housing estates
Redevelopment
Housing action areas
General improvement areas

FIGURE 13.3 *Public housing areas in the British Cities (after K. Basset and J. Short, 1980) of Bristol, Nottingham and Birmingham in the 1970s. Source: K. Basset and J. Short, 1980: Housing and residential structure. Alternative approaches. London: Routledge and Kegan Paul. Figures 7.4 and 7.5.*

only a minimum of services. The space standards had fallen from pre-war levels in the effort to maximize house construction. The reaction led to the report of a committee under Sir Parker Morris called 'Homes for Today and Tomorrow'. These Parker Morris standards became a basic measure of quality but it was not until 1969 that they became required (Merrett, 1979: 109).

4. In the late 1950s and the 1960s a phase of inner city redevelopment and the removal of slums led to the phase of high-rise building which has already been considered. It was an attempt to maintain densities in areas where space was necessarily restricted but it produced an inner city ring of tower blocks.

5. After the disaster of the high-rise era, building has returned to low-rise solutions, often in the form of infill. But the decline in council house construction has meant that the physical impact on the city has been limited.

Against the above background the location of council housing can be considered. Figures 13.3(a), (b) and (c) indicates that there is a distinct ring of estates which correspond to the suburban ring of Burgess's zonal scheme. But it is also clearly not continuous but rather in the form of a series of sectors. These fill in the gaps left by private development. That social status is sectoral in character in British towns, therefore, is largely a response to the location of municipal estates. There is no major locational contrast between pre-war and post-war estates. Indeed, the latter were sometimes extensions of the former. The main departure from the simple scheme of a discontinuous fringing zone is the sporadic developments nearer to the city centre and related to more recent clearance. In Birmingham especially, these form an inner city high-rise ring (Fig. 13.3(c)). These were a result of the use by the city of the regulations of the 1944 Town and Country Planning Act and the comprehensive redevelopment it allowed through land acquisition. This meant that the older residential/industrial area could be cleared out and five 'new towns', Ladywood, Newtown, Neckells Green, Highgate and Leebank created. They now form a distinctive inner ring.

The differential allocation of council housing has already been discussed and there is clear evidence, widely attested, that council estates are not socially similar. Thus Twine and Williams have clearly demonstrated that in Aberdeen:

> a definite and regular pattern of segregation was discovered. In particular a concentration of households in the semi-skilled, personal service and unskilled categories was found in estates built in the 1930s under slum clearance and overcrowding legislation. An attempt was also made to measure the distribution of multiple deprivation... and this was also found to be concentrated in estates built during the 1930s.
> (Twine and Williams, 1983: 253)

The authors explain these patterns by the constraints on letting. Thus they note the 1930s policy of dividing the public sector housing stock into higher rented non-rebated properties and lower rented rebated properties. 'The pressure of these financial constraints fell upon the poorer existing and prospective tenants who were less able to agree to higher non-rebated rents for the more desirable houses' (Twine and Williams, 1983: 265). After 1971 they suggest three possible mechanisms for the perpetuation of the pre-war social segregation – differential rents, the grading of tenants and patterns of selective refusal of the less desirable estates. Be that as it may, the final outcome is clear, that council estates demonstrate social segregation and that is to a degree related to age, and to the nature of the housing offered (see Chapter 15, p. 299).

13.7 CONCLUSION

This chapter has attempted to show that the residential patterning of the city cannot be completely understood either by inferring large-scale ecological processes from aggregate patterns or from seeking to identify individual behaviour in relation to residential decision making. Over-arching both of these is the housing market operated by public authorities and private interests and only within the context of that market can the complex mosaic of residential areas be properly comprehended.

14

PROBLEM AREAS OF THE CITY: INNER CITY

14.1 INTRODUCTION

The heading of this chapter is as inappropriate as identifying one part of the city as a zone in transition. All parts of the city are in transition, just as all have problems in some form. But there are very specific areas where problems are acute and demand separate discussion. One forms the concern of this chapter. Most problems are locationally associated with the inner city, but in accepting that, two essential points need to be made. The first is that it would be an error to regard inner city problems as features recent in date, or as first identified in Burgess's 1925 zone in transition. In the sense that it generated specific social problems then the inner city dates to that time in the nineteenth century, or indeed earlier, when the gentry and the urban bourgeoisie began to abandon their centrally located town houses for suburban residence, and when the processes of invasion and succession began to operate. Thus, for example, the immigration of the Irish replicated in the mid-nineteenth century all the aspects of the ethnicity factor derived in contemporary ecological studies. The second point is that the greatest social problems do not necessarily occur in the inner city. Peripheral public housing estates present severe social disruption against a background of multiple deprivation. Again, at a different level, the tendency for the city to spread into greenbelt land in the rural–urban fringe creates another, if very different, set of problems. But it is apparent that all the terms used are locational or geographical – inner city, peripheral estates, rural–urban fringe – and crucial themes in urban geography.

14.2 ETHNIC SEGREGATION OR THE URBAN GHETTO

In the dictionary of political correctness maybe the ghetto is an improper term, but the concentration, segregation and isolation of population groups within the city has been characteristic from earliest times. From the outset, cities were meeting places and, by virtue of that role, assembled people of different cultural backgrounds. Once lodged in the city such populations, distinctive either by their cultural practices or by their physical appearance, and frequently by both, tended to keep together, for spatial concentration maximized security in an alien environment, as well as providing the most effective way of retaining identity and maintaining traditional customs, some of which demanded buildings to which access had to be immediate. But there was always a darker side in that indigenous populations, fearing incomers from both economic and cultural motivations, sought to isolate immigrant populations and, either by custom or by law, to contain them within prescribed limits within the city.

This isolation of culturally alien people always generated ambivalent reactions for it has never been clear whether those discriminated against sought to retain their separate identity or wished to merge as rapidly as possible with the host population. It is a dilemma which faces all such groups. Thus the Black in America is caught between regarding the ghetto, or the ethnic segregation which exists, as a temporary phase in an historic process which will eventually result in diffusion throughout and integration with American society, or as a means by which a different way of life can be preserved, one based on a value-system different from that of White Anglo-Saxon protestant America (WASP). It is epitomized in the demand for the use of the term Afro-American, which stresses cultural rather than physical difference. But if such contrasts are to become pivotal, then it is difficult to envisage how they can operate on a non-spatial basis since different cultures use space in contrasted ways. The ghetto, therefore, becomes a permanent feature, even if by a fairer and more equal allocation of resources its nature and its name change.

There is, of course, a third possibility: out of interaction a new cultural identity is derived. This is already partly true of the USA where popular music, for example, owes more to Black cultural traditions than to any WASP inheritance. But that may be no more than the assimilation process by which alien elements are taken in by the majority culture. More directly out of the discussion the fact emerges that the urban ghetto can play two quite opposite roles.

1. **The temporary ghetto.** This is a segregated area through which populations become adjusted to new ways of life. In particular, immigrants to a country find immediate refuge with their own kind until they have become adjusted to urban living in a new country and, with a rise in their socio-economic status and acquisition of native mores, they become diffused throughout the population. An excellent example of this process has been provided by Jakle and Wheeler (1969) in a study of Dutch immigration into the city of Kalamazoo. In this case the immigrants were less concerned with retaining their identity than in becoming true Americans. A series of maps shows the Dutch population, identified by Dutch surnames, between 1910 and 1965. The major Dutch immigration into Kalamazoo occurred in the late nineteenth century and by 1910 clear concentrations were identifiable and they remained so until 1929. By that time two distinctive Dutch clusters had appeared, one to the north and the other to the south of the city centre. Here were what can loosely be called Dutch ghettos, and they were closely associated with churches, the crucial buildings about which identity was preserved. But by 1965 complete dispersal had occurred, although there was one relict concentration to the north of the CBD. The Dutch had been completely assimilated and were American in all ways and were only identifiable by their names: the ghetto has performed its role.

2. **The permanent ghetto.** This is better contrasted by degree than by nature, but it is the means by which a cultural group can actively resist being weakened and lost in the larger community of which it is a part. In this case the ghetto does not aid assimilation by being a stage in the process, but quite the reverse. It is a means by which assimilation can be resisted and identity preserved.

Two brief case studies, one partly historical and considering Jewish populations, the other contemporary at least in its manifestation, can be introduced to illustrate the varied nature of ethnic segregation, although both are based on religion as the basis for separation.

That the pre-industrial city was characterized by distinctive ethnic enclaves is indisputable, for it was in such a condition that the ghetto itself came into being. The segregation of the Jews can, in part, be considered not as a direct result of external pressures against a 'peculiar' people, but due to needs arising from their own religious customs, particularly ways of preparing food, the demands of attendance at a synagogue and the need to take part in various aspects of communal life. Their migration after the diaspora can be compared to that of the Irish after the famine or that of the Blacks in twentieth-century America; and the formation of distinctive areas within the city was the result of much the same process. But the Jews became subject to much harsher, legalized conditions which institutionalized the ghetto, in the same way as apartheid legalized racial segregation in South Africa. The origin of the word 'ghetto' is itself disputed although there now seem two possibilities. The first is derived from the relations between the Jews and the Venetian state at the beginning of the sixteenth century. Venice was prepared to tolerate the Jews in return for the financial advantages they brought through taxation. But constant friction and periodical attacks occurred and the solution seemed to lie in creating a designated and protected area:

> The practice of establishing Jewish quarters separated from Christian dwellings by actual physical barriers was not without medieval precedent. ... But it fell to the Venetians in 1516 to contribute the word ghetto to the vocabulary of persecution... . Zaccaria Dolfin, a member of the Collegie, referred on 26 March 1516 to Franciscan warnings about the 'corruption of the state' ... which would inevitably ensue if Jews were still entertained in the city. He suggested that they be enclosed in the area which was already – before Jewish occupation – known as the Ghetto Nuovo, the new foundry in Venetian dialect. This was 'like a fortress', with a single entrance that could be guarded at night ... It could therefore perform the double function of protecting the Jews from violence and plunder, and of enabling an effective curfew to be imposed upon them.
>
> (Pullan, 1971: 486–7)

If this is the origin of the term 'ghetto' then it appropriately stands for the distinctive nature of ethnic areas in pre-industrial cities. But the second possibility must be added that the word could also be derived from the Hebrew word 'get' meaning divorce or separation (Encyclopaedia Judaica, 1971, 7: 542).

The degree of segregation varied greatly and in many cities the Jewish presence was too limited to provide distinctive areas. In British cities the concentration is still preserved in the word Jewry, as in Old Jewry in the City of London. In Europe, however, the areas were larger and more clearly marked. Perhaps the best known of the ghettos was that at Frankfurt-on-Main. The Jews had early settled in that significant commercial centre, initially under the protection of the Emperor. They were subject, however, to periodic pogroms, the worst in 1349, when, blamed for the Black Death, the whole community was massacred. Eventually in 1462, partly as a result of demands from the Emperor, and even the Pope, in order to ensure their security, they were transferred to a specially constructed street, the Judengasse. That was situated in a sparsely inhabited portion of the city, removed from the rest of the inhabitants, on the border between the old and the new city, on a part of the dried-up moat which ran along the wall of the old city (Wirth, 1928: 42). The Judengasse was enclosed by walls and entered by gates so that it formed a clearly defined and discrete part of the city. There were 110 registered inhabitants in 1463. The number increased to 900 by 1569 and 3000 by 1610. But since the ghetto was enclosed by walls there was no opportunity for extension, with the result that as the population grew, storeys and back premises were added to houses already subdivided. At the peak of its prosperity this small area not only produced some of the most famous financial families, the Rothschilds for example, but became a centre of Jewish learning. It suffered attack and devastation in 1614–16, and was destroyed by fire in 1711. But it was not until 1798 that the prohibition which maintained the ghetto was lifted.

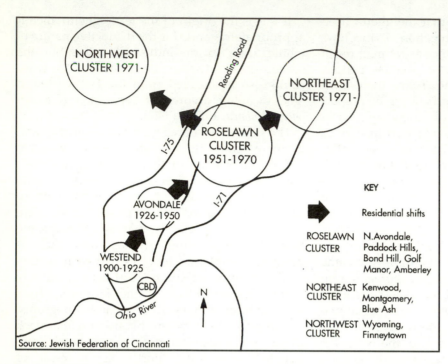

Source: Jewish Federation of Cincinnati

FIGURE 14.1
Residential shifts of Jewish population in Cincinnati, Ohio (after D. P. Varady et al., 1981). Source: D.P. Varady et al., 1981: Suburbanization and dispersion: a case study of Cincinnati's Jewish population. Geographical Research Forum 3, May 1981, 9.

'The typical ghetto of the sixteenth century', wrote Wirth, 'is a densely populated, walled-in area usually found near the arteries of commerce or in the vicinity of a market' (Wirth, 1928: 51). The location characteristic is epitomized in the ghettos of Venice and Frankfurt. Away from the main residential cores of the settlements, they were symbolically set apart but still retained an adjacence to the centres of commercial activity.

Industrialization in the nineteenth century, allied to the movement of Jews from East Europe, created ghettos in many large European cities (Carter and Lewis, 1990: 136–7), but the development in the contemporary city can be considered in relation to the United States where the Black ghetto, with which it can be compared, has become so crucial a feature of race relations.

The changing location of the Jewish population in Cincinnati, Ohio, has been the subject of a study by Varady *et al.* (1981) which centres on the crucial issue of suburbanization and consequent dispersal. 'Jewish communal leaders perceive this increased spatial dispersal as a problem. Unless a critical mass of Jews remain in contiguous areas of the city and/or suburbs it is difficult and expensive to maintain viable communal and religious facilities' (Varady *et al.*, 1981: 5). And, as already emphasized, Judaism is a religious system which affects the whole life. 'While everyone had conventional food laws, Jews had divine commandments governing food. The most striking point about Jewish law is that it brings the entirety of life, including civil and domestic practices, under the authority of God. Jews were not free to assimilate' (Sanders, 1993: 37).

The study uses lists of donors to the Jewish Welfare Fund of Cincinnati to trace changes of residence. These are summarized in Fig. 14.1, where four stages are apparent. The first dates to the 1920s when, with rising incomes, the first shift took place from the characteristic immigrant inner city location, like the one Wirth had described for European Jewry for much earlier times, to Avondale, an inner suburb. Between 1930 and 1950 Avondale was a clear Jewish enclave with synagogues, religious

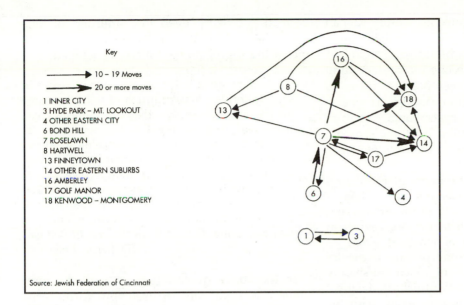

FIGURE 14.2 *Internal moves of Cincinnati's Jewish population 1976–9. Source: As Fig. 14.1, p. 11.*

schools, kosher butchers and all the associates of Jewish cultural identity. But a change began in the late 1940s and by 1960 Avondale was predominantly Black. The Jewish response, in classic invasion and succession sequence, was a reclustering in a third location, in suburban communities which the authors call the Roselawn cluster (Fig. 14.2). By 1970 Varady *et al.* report that some 62 per cent of Jewish families in Cincinnati lived within the cluster. 'Furthermore, the Roselawn community, Cincinnati's third generation Jewish ghetto, alone contained more than one fifth (21 per cent) of all Jewish families in the Cincinnati area. Roselawn's population was approximately one half Jewish' (Varady *et al.*, 1981: 9). In brief, because of the demand for propinquity and contiguity, the ghetto had been reformed.

During the 1970s suburbanization increased. The Roselawn cluster decreased to include only 45 per cent of Jewish families and two new clusters came into being. Again, Black pressure on Roselawn was partly responsible, although the changes were due to the formation of new families rather than actual migration. But the result was a thinning. 'If Roselawn could be called a ghetto in the 60s, certainly the north-eastern and northwestern clusters which had

been created by the latest moves [Fig. 14.2] could not be termed ghettos in the 80s' (Varady *et al.*, 1981: 10). Thus in the area of the north-eastern cluster only 10 per cent of the population is Jewish, and about 15 per cent in the northwestern.

Here, in the case of Cincinnati's Jews, lies the dilemma of an urban ethnic minority. Some older families still remain in the city, others in the inner suburbs. How can a way of life, an identity dependent upon communal practices, be sustained as suburbanization and dispersal become dominant? The existence of urban ghettos might well be a problem, but their disso-lution brings problems too. A parallel British study can be found by Newman (1985, 1987) with comment by Waterman and Kosmin (1987) where the problems of valid data are also discussed.

The second example is the contemporary segregation of Protestants and Catholics in Belfast. As Emrys Jones (1960) demonstrated, before the nineteenth century there was no evidence of conscious segregation between Protestants and Catholics, but with the large-scale immigration of Catholics in the first half of the century tensions were generated and a clear separation of the two population groups developed. In the mid-eighteenth century Roman Catholics formed only 6.5 per cent of the population, but by 1861 the proportion had risen to 34.1 per cent. This was the economic and cultural threat which has already been noted as the basis of reaction. Out of tension rioting arose and spatial separation was exacerbated; security being again a crucial element in ethnic segregation. Jones quotes a report on the riots of 1886:

> The extremity to which party and religious feeling had grown in Belfast is shown strikingly by the fact that the people of the artisan and labouring classes ... dwell to a large extent in separate quarters, each of which is given up almost entirely to persons of one particular faith, and the boundaries of which are sharply defined. In the district of West Belfast, the great thoroughfare of the Shankhill Road, with the network of streets running into it, and the side streets connecting those lateral branches, is an almost entirely protestant district... the great catholic quarter is due south of the Shankhill district, and consists of the thoroughfare known as the Falls Road.
>
> (Jones, 1960: 191)

This situation has remained virtually unchanged to the present. Indeed, the 'troubles' of the last twenty years, and the threat posed to those not living among people of their own persuasion, has greatly increased segregation as populations have moved after being attacked or threatened with attack. The result is a segregated city (Figs 14.3 and 14.4) where now the boundaries are marked by physical barriers, the so-called 'peace line', designed to keep the two opposed groups apart. Boal's studies of Belfast in the late 1960s (Boal, 1978, 1982; Boal and Pook, 1973; Boal and Royle, 1986) showed that even then some 99 per cent of Catholic households were located in areas where over 90 per cent were Catholic, while some 69 per cent of Protestants lived in streets where over 91 per cent were Protestant. More recently, a study by Doherty (1989) using 1971 and 1981 census data calculated indices of dissimilarity for nine subdivisions of the Belfast area. The data presented problems due to a campaign of non-cooperation

with the census, even the murder of one enumerator, but all the evidence pointed to increasing segregation. Doherty concluded:

> the present study has demonstrated increases in segregation levels over much of the urban area in the inter-censal period. In Belfast West, particularly high segregation has persisted over time in an area which contains almost 40 per cent of the Catholics in the urban area. The size of the area, the level of residential segregation, coupled with the activity segregation identified by Boal (1969) and the nationalist orientation of the people, all combine to produce a ghetto area which is quite distinct from the rest of Belfast.
>
> (Doherty, 1989: 157)

In the above quotation the use of the term 'ghetto' is significant in itself. But further, the distribution of Catholics is not greatly different from that of a standard ethnic minority, as identified in Chapter 12. Of the six virtually exclusive Roman Catholic areas, four are located in separate enclaves adjacent to the city centre, with one of them (the Falls Road) having grown outward as a distinctive sector to the city boundary and beyond. The remaining two are the Ardoyn in the north-west and an area to the east of it.

These two examples are illustrative of the main features which both cause and characterize ghettos. The central theme is the presence of a minority population which differs from the host population in some significant cultural attribute. But there are specific factors which prevent the sort of assimilation which occurred with the Dutch in Kalamazoo. In the case of the Jews generally, and the Belfast Roman Catholics, there was present a strong determination to preserve identity and to react against assimilation. That reaction in both cases has given rise to violence which, in turn, accentuates spatial concentration.

The most widely quoted example of the ghetto in contemporary cities is that of the Black people in the USA. There the ghetto, instead of being the temporary feature it has been for almost all other immigrant groups, at least historically as for the Dutch in Kalamazoo, has

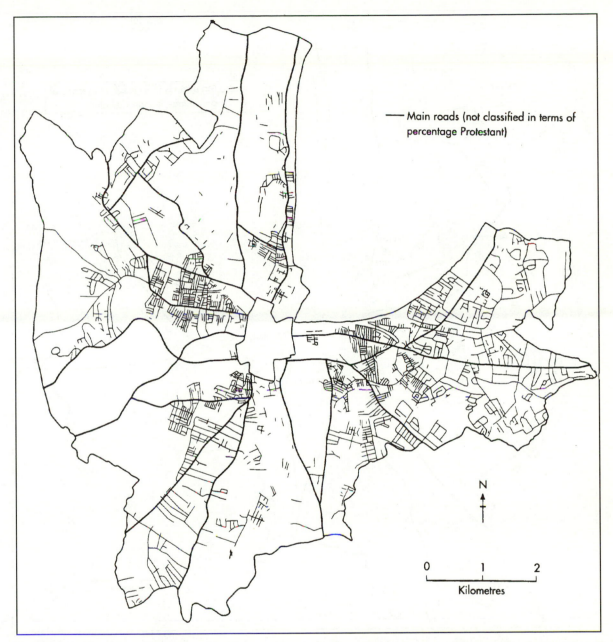

Main roads (not classified in terms of percentage Protestant)

FIGURE 14.3 *The distribution of streets in Belfast with under 9.5 per cent Roman Catholic (after F. W. Boal, 1973).*

become permanent whether the Blacks wanted it that way or not. 'The early pattern of negro settlement within each metropolitan area followed that of previous immigrant groups. Migrants converged on the older sections of the central city because the lower cost, and often subdivided, housing was there, friends and relatives were likely to be there, and the older neighbourhoods then often had good public transport' (Kerner, 1968: 243–4). But unlike other

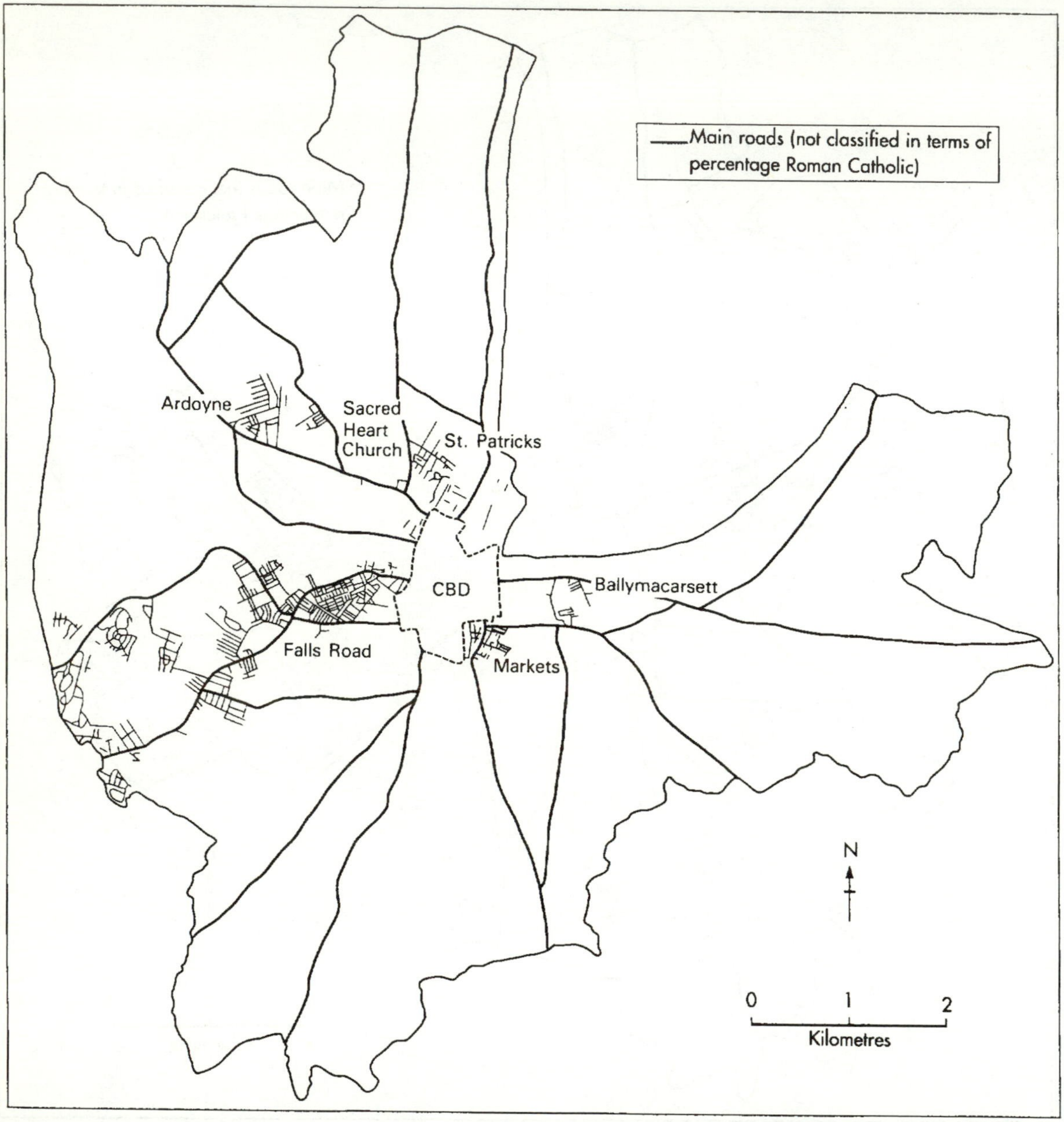

FIGURE 14.4 *The distribution of streets in Belfast with over 90.5 per cent Roman Catholic (after F. W. Boal). Source: Figs 14.3 and 4. F. W. Boal and M. A. Poole, 1973: Religious residential segregation in Belfast in mid-1969: a multi-level analysis. In B. D. Clark and M. G. Gleave ed. Social patterns in cities. Institute of British Geographers Special Publication No. 5.*

incoming groups a subsequent pattern of dispersal has not happened. Morrill (1965b) suggested four reasons for this:

1. **Prejudice and discrimination.** There is no need to elaborate on these basic emotional responses which now go under the general

term 'racism'. But it is necessary to observe that there is no simple response to ethnic contrasts of custom and lifestyle, but a more complex one where, although way of life differences can exist, the major basis of discrimination is skin colour.

2. **Characteristics of the Blacks.** In this Morrill included all those difficulties faced by members of a minority group when they leave the shelter of the ghetto. Cultural traits can be disguised, even abandoned, but physical differences remain evident. There is, of course, a history of attempts 'to pass', that is to modify skin colour or hair character in order to pass as a member of the dominant group.

3. **Real estate and associated financial institutions.** Apart from any ideological commitment or prejudice, the interests of these people is to avoid any change in an area which would result in prices falling. Once they do begin to fall then the greatest profit is to be derived from the volume of sales and hence from encouraging rapid selling.

4. **Legal and governmental barriers.** These were nominally eliminated by anti-discrimination legislation. But could still be manipulated (see pp. 274–5).

In summary, the concentration of blacks in central cities results from a combination of forces. Some of these forces, such as migration and initial settlement patterns in older neighbourhoods, are similar to those which affected previous minorities. Others – particularly discrimination in employment and segregation in housing and schools – are a result of White attitudes based on race and colour. These forces continue to shape the future of the central city (Kerner, 1968).

Morrill, in the same paper quoted above, identified the expansion of the ghetto as a process of spatial diffusion. The mechanism is related to the gradual spread of Black occupancy as marginal blocks are penetrated and then rapidly change in character (Fig. 14.5) so that any form of racial integration is most difficult to sustain. There can be identified for most blocks

or streets what is called a 'tipping point'. 'Once the proportion of non-whites exceeds the limits of the neighbourhood's tolerance for inter-racial living (this is the tip point) the whites move out' (Grodzins, 1958: 6). Perhaps one should add that a fear of declining property prices is as significant as prejudice. A figure of 30 per cent Black occupancy has been suggested as the balancing fulcrum, though it will vary in relation to a wide range of factors (Wolf, 1963).

The result of these forces is the emergence in large cities of clearly marked Black ghettos. The location of these areas is implicit in their history. As the exodus of White population from the city centre to the suburbs has proceeded, so the ghetto has been intensified as a portion of the inner city. If one accepts a process of spatial diffusion for ghetto expansion then this will take place where White resistance is least effective in an invasion–succession sequence. Since areas high in socio-economic status are sectoral, often retaining a link with the centre, the ghetto will be squeezed into a sector-like arrangement, made up of a series of blobs around the inner city. This is the situation which Murdie proposed for ethnic areas on his model (Fig. 12.6) set out in Chapter 12 (p. 232).

A comparison can be made with the areas in British cities where immigrants have clustered. In his study of Birmingham between 1961 and 1971 P. N. Jones argued that there were no ghettos in the strict sense of the term since there were no areas showing exclusive domination by one cultural group, rather there were mixed areas (Jones, 1967, 1970, 1976). In the clusters of immigrant areas he identified in Birmingham the immigrant population from the new Commonwealth only formed some 14.6 per cent of the population, with a maximum of 16.8 per cent. That compares with figures from Cleveland, Ohio, where 80 per cent of the Black population lived in areas which were 75 per cent Black in composition.

Nevertheless, in Birmingham, examination of change between 1961 and 1971 showed that 'intensification has remained the dominant spatial process rather than dispersal. Movement to areas of post-1920 housing, whether private

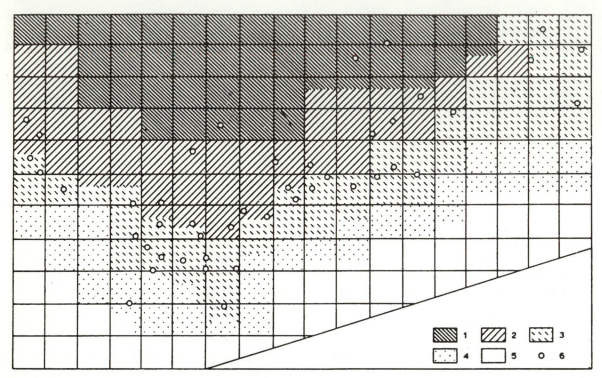

FIGURE 14.5 *Black purchases on the edge of the ghetto in one year in the mid-1950s (after R. L. Morrill). The five shaded boxes refer to the five identified areas within which the proportion of purchases in 1955 was as follows:*

Area	No. of White Purchases	% of Total Purchases	% Area Black
1	8	3.9	32
2	26	4.3	16
3	67	40.6	5
4	72	98.7	1
5	112	100.0	1
6	*New streets with sales to blacks in 1955.*		

or municipal, is clearly not enough to make any significant impact in comparison with the concentrations of the middle ring of the city' (Jones, 1970: 217). Figure 14.6 shows 11 clusters of immigrants which Jones could identify in 1960. They take up a distribution which is both zonal and sectoral. There is a clear zone of occupation which avoids the inner city and is closely linked to a high-density middle zone. That is a consequence partly of structural, partly of institutional factors. The inner zone is an area of late nineteenth-century slums, particularly of small houses or cottages built around courts. These are inherently ill-suited to immigrant needs because of their very small size. Moreover, at the time, large tracts were municipally owned and scheduled for comprehensive redevelopment.

The immigrant clusters were, therefore, pushed outwards and took over the high-

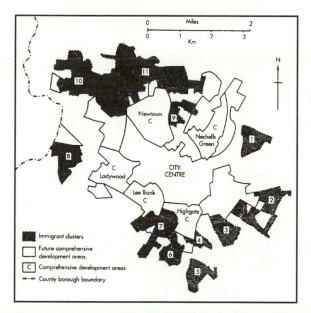

FIGURE 14.6 *Immigrant clusters in Birmingham (after P. N. Jones, 1967). Immigrant clusters were identified by including all contiguous enumeration districts which belonged to the upper quartile (10.1 per cent immigrant population). Completely surrounded districts which fell below the critical value were excluded. The clusters are: 1. Saltley; 2. Small Heath; 3. Sparkbrook; 4. Highgate; 5. Balsall Heath - North Moseley; 6. Cannon Hill; 7. Calthorpe Park; 8. Summerfield Park; 9. Newtown Aston; 10. Handsworth; 11. Aston.*

density, later bye-law terraces and even larger but contemporary houses and villas. The latter had often leases with but a short time to run which made normal mortgages difficult to obtain so that they became available for a quick cash sale. They were large enough to house the often extended immigrant families. The 'invasion' was made all the easier since the houses lacked modern amenities, not only internally but externally as well, as for example the absence of garages or garage space. The result was that 'the clusters are not typically situated in the slum-ridden heart of the city, but in the tree lined, often attractive townscapes of the middle ring' (Jones, 1967: 22).

Two spatial influences created this situation in Birmingham. The first was the physical and institutional restraints operating in the city centre which prevented the takeover of the oldest, poorest properties. The second was the resistance from the suburbs, together with the particular conditions required by the immigrants, the economic resources they had, and their need to associate with people of a like cultural background. The result was a zone about the city centre which was almost complete. Indeed, if other immigrant groups, such as the Irish, were added the zonal picture became even clearer. That situation conformed to the notion of a black collar around the CBD which was used in the Kerner report, even if it was pushed out further from the actual core. In detail, however, the ring is rarely complete and changes as the processes of invasion and succession proceed. It is more consistently reliable to think of 'ghettos' as combining sectoral and zonal properties.

A further example can be added from the British West Midlands. Wolverhampton presents a classic case of the formation of an inner ring of immigrant occupation about the city centre (Evans, 1984). Again, however, local circumstances intervene and there is a clear extension into a southeastern sector, significantly away from the more affluent suburbs which are to the west (Fig. 14.7). Evans in his study of Wolverhampton considers change since 1961. In that year the Index of Dissimilarity was 64.4. (It will be remembered that on the Index 100 represents complete dissimilarity in distribution from the population as a whole, while 0 represents total equivalence.) In 1966, at the only mid-decadal census the Index was 74.5, but in 1971 it was 60.9 and in 1981 it was 54.8 (Evans, 1984: 8). The figures seem to represent a general trend in the UK with early high and increasing concentration succeeded by some dispersal. Even so, the immigrant populations remain in well-defined areas.

This lowering of the Index of Dissimilarity, and the possibility of dispersal it implies, raises the crucial question of the extent to which the ghetto is being broken up, or ethnicity as a

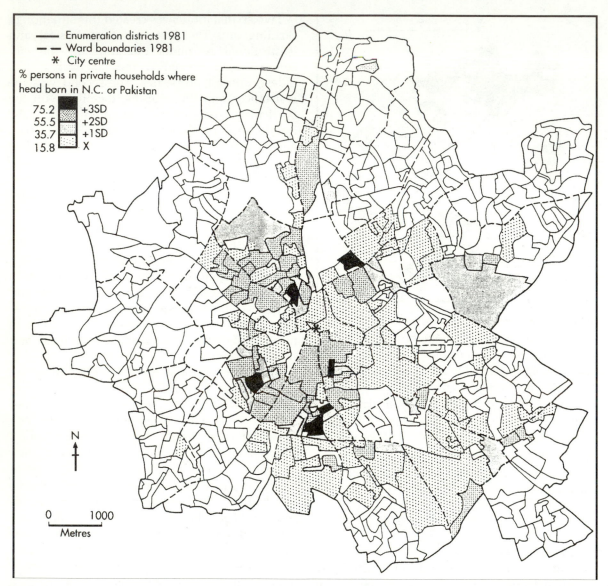

FIGURE 14.7 *Wolverhampton: the distribution of population born in the new Commonwealth and Pakistan 1981 (after D. J. Evans, 1984). The shaded areas relate to standard deviations (SD) above a mean of 15.8 per cent. Source: D. J. Evans, 1984: The segregation of the New Commonwealth population in Wolverhampton Municipal Borough, 1961–1981. Department of Geography and Recreational Studies, North Staffs Polytechnic (now Staffordshire University), Occasional Papers in Geography 3. p. 6, Fig. 4.*

significant differentiator of urban space is decreasing. This has become a particular issue in the United States where it has been argued that Black suburbanization, for which there is direct statistical evidence, demonstrates that a process of assimilation is under way. Thus, for example, the topic was considered in a book by Thomas A. Clark (1979) called *Blacks in suburbs*. Clark summarized the changes by noting that:

the average annual rate of central city black increase has fallen from 2.97 per cent during 1960–1970, to 1.64 per cent from 1970 to 1974 and subsequently to an annual loss of 0.79 per cent (–0.79) during 1974–77. Suburban rates of increase have varied more irregularly in recent years. The average rate of black suburban increase rose from 2.53 per cent during 1960–70, to 4.55 per cent during 1970–74, and more recently it has fallen to just 3.87 per cent between 1974 and 1977. The emerging picture is one of absolute annual black decline in central cities, primarily due to accelerating rates of net out-migration from the largest central cities, while both suburban areas and non-metropolitan areas experience comparatively high rates of black increase.

(Clark, 1979: 24)

In brief, Blacks are becoming more suburban. So much must be accepted; but a further question arises as to what that suburbanization means, and can it be interpreted as signifying the break-up of the ghetto and the decline of segregation.

There are two contrasted reponses to that question. The first is based on changing socio-economic status. It is contended that incomes of Black families have risen, the median Black income in suburbs rose by 12 per cent between 1970 and 1977, as the traditional ghetto mechanism is at work transferring the upwardly mobile professional Blacks into integrated suburbs – communities where the main motivation is not to exclude the Blacks but to exclude the poor. Behavioural surveys purport to show a declining prejudice towards Black professional entrants into suburban housing. This theme has been most cogently advanced by W. J. Wilson (1978), a Black sociologist, in his book *The declining significance of race*. He accepts the fact that there is a large Black underclass, but maintains that it is based on class rather than race. 'The argument is not that particular groups have failed to utilize opportunities or advancement but that previous experience of social exclusion, coupled with the new spatial division of labour, creates conditions in which an "underclass" emerges which in time does indeed come to possess characteristics of a pathological kind' (Cross, 1992: 113). Social mobility, as a response

to declining discrimination, exacerbates the problem by abstracting to the suburbs those who would become community leaders. But to reiterate, the problem lies not in skin colour or race, but in the way the city has developed, and especially the redistribution of employment away from the city centre which has been examined in earlier chapters. 'A shift away from the convenient focus on "racism" would probably result in a greater understanding and appreciation of the complex factors associated with recent increases in the rates of social dislocation among the urban underclass' (Wilson, 1985: 159).

It is not surprising that these ideas, downgrading the politically correct explanation of racism, have provoked quite violent reactions. The most polemical was that of Alphonso Pinkney (1984) in a book the title of which is self-explanatory, *The myth of black progress*. But in geographical terms the closest analyses have been those of Joe T. Darden (1973, 1985, 1987). In an examination of the Kansas City SMSA at the 1980 census Darden showed that high levels of segregation, as measured by the Index of Dissimilarity, were characteristic. He tabulated 17 income levels and demonstrated that high indices were recorded across them all. Thus the highest figure was 95.6 for incomes over $75,000; but even for the lowest incomes the index remained high, the $2500 to $4999 level returning 80.8. When the basis was changed to occupation the same situation was apparent. The lowest index figure was that for 'professional speciality', but it was still 72.0. Segregation between Whites and Blacks with the same level of education again produced the same result. The mean for elementary education was 83.4, for high school 81.6 and for College 74.1. Darden concludes, therefore, 'neighbourhood segregation between blacks and whites does not occur merely because blacks are poor, less educated, or in lower-status jobs. The "nature of the beast" is race not class. Therefore, black socioeconomic mobility does not guarantee spacial mobility' (Darden, 1987: 17).

In a substantial investigation of Detroit in 1980, Darden set out the general distribution of Black and White as in Table 14.1. These figures

TABLE 14.1 *The racial composition of the population of the Detroit SMSA 1980*

Area	White (%)	Black (%)
Detroit City	34.3	63.0
Lapeer	98.1	0.2
Livingstone County	98.5	0.4
Macomb County	97.1	1.3
Oakland County	93.1	4.7
St Clair County	96.6	2.0
Wayne County	62.3	35.9

Source: After Darden, 1987.
Note: the percentages do not sum to 100 since other small groups, such as Eskimo, Asian and Pacific Islanders, have been omitted.

indicate the total imbalance of the Black population distribution. But even with these figures, there was an increase of over 1000 per cent of Blacks in 11 suburbs located in the three counties, Wayne, Oakland and Macomb. However, that increase was from a very low base indeed, indicated by the fact that only 4.2 per cent of the suburban population was Black.

In considering trends Darden noted that changes in the definition of the SMSA created problems, but, using the 1970 base, the mean level of Black segregation was 58.9 per cent in 1970 and 58.4 per cent in 1980. He wrote:

> segregation increased in nine or 37 percent of the suburbs. Thus by 1980, after more than a decade of the Fair Housing Act, blacks in eight of the 24 suburban municipalities were more segregated than blacks in the city of Detroit.
> It was revealed that most blacks in the suburbs are restricted to only a few neighbourhoods... Obviously the blacks living in Detroit's suburbs have not escaped ghetto-like conditions despite suburbanization.... If this trend continues, the ghetto pattern which characterized the city of Detroit at the turn of the century will emerge in the suburbs. Little suburban ghettos ... are already occurring.

> (Darden, 1987, 15: 18)

The explanation for the manifest suburbanization of Blacks but without a significant diminution in segregation, lies in what is usually termed overspill and leapfrogging. They are illustrated in Fig. 14.8. The ghetto grows and extends. In so doing, it develops beyond the administrative limit of the city, that is it moves into suburbia. Also, leapfrogging leads to the seeding of new segregated enclaves beyond the city bounds and in the suburban ring. Thus suburbanization in formal statistical terms does occur. But segregation is unchanged: suburbanization does not mean desegregation. Perhaps the situation has been presented here on too absolute a manner. Even Darden's data for Kansas City show lower, though still high, Indices of Dissimilarity for those with professional occupations and more education. Kain's (1987) review of Chicago demonstrated the same processes operative as in Detroit:

> The growth in black population in 9 established black communities accounted for more than half of the 1970–1980 growth in black households in Chicago's suburbs. Most of the remaining black suburban growth occurred in suburban communities on the periphery of the central city ghetto and represented a continued peripheral growth of the ghetto across the boundaries of the central city.

> (Kain, 1987: 84)

Even so he writes that, 'the former nearly absolute barriers to black occupancy of many suburban areas may have been breached' (Kain, 1987: 84). But if 'black progress' is a 'myth', and in terms of residential city space the ghetto remains, then the question of why this is so must be posed, especially in the light of the Fair Housing legislation which was noted above.

The means by which, prior to any control, segregation was maintained by prejudice operating through estate agents, real estate agents or realtors in the US is well attested and has already been discussed in Chapter 13 where three main procedures were outlined. The first was the direct refusal to sell houses to Blacks in White areas. The other two were less overt. The second was to maintain two lists, one to be

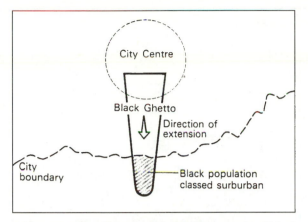

FIGURE 14.8 *The relation of the growth of black ghettos in the USA to administrative boundaries.*

presented to White clients, the other to Black. The third was racial steering, that is using persuasion to urge buyers into areas deemed appropriate by suggestion, such as the buyer would not be happy in a certain location. All this was made illegal by the Fair Housing legislation which was contained in Title VII of the Civil Rights Act of 1968. That legislation explicitly prohibited discriminatory practices, including racially motivated refusals to sell or rent, any statement of racial preference or exclusion in advertising, any denial that a property is available when it is so and any form of racial steering. This immediately re-poses the question as to why against such a background segregation remains.

The answer lies in the fact that the Federal Government adopted a somewhat muted approach to the enforcement of an Act which is in itself limited in application. If an individual considered that discrimination had taken place, then a complaint could be made to The Department of Housing and Urban Development's (HUD's) Office of Fair Housing and Equal Opportunity. If conciliation had not produced a settlement in 30 days then civil action could be taken in a Federal Court which would issue a restraining order and impose a fine of $1000. Clearly the case-by-case approach which was implicit assumed the market was intrinsically

fair and only exceptional cases would arise. Moreover, the damages were insignificant and there was no authority to enforce the conciliator's decision. Some attempt was made to repair these loopholes in the Fair Housing Amendment Act of 1980 but that failed. As a result, the crucial controls devolved down to localities and at that level the use of zoning, which has already been discussed (Chapter 8, pp. 149–51) became paramount. The problem now resolves itself into a conflict between localities and the Supreme Court with a complex of decisions at the core of the issue of Black housing and desegregation. It is too lengthy a task to review these decisions here but, fortunately, there is a quite excellent study by R. J. Johnston (1984) *Residential segregation, the state and constitutional conflict in American urban areas.* The complexity is attested by the 125 entries in the 'Index of Cases Sited'. Perhaps it is sufficient to quote Johnston's conclusion:

> As long as they do not explicitly and/or intentionally practice (*sic*) racial discrimination, suburban municipalities remain free to pursue the narrow sectional interests of their residents – which in the case of the higher income areas is to exclude those who might depreciate the social character of the milieu, its tax rates, and its schools. ... Zoning is a powerful tool for the protection of suburban interests.
>
> (Johnston, 1984: 104)

The segregation of ethnic or racial minorities characterizes urbanism almost universally, as it always has. It remains one of the basic problems. Morally it can be posed in terms of equity, but physically it presents a substantial threat to stability. To revert to Wilson, 'the communities of the underclass are plagued by massive joblessness, flagrant and open lawlessness, and low-achieving schools and therefore tend to be avoided by outsiders. Consequently, the residents of those areas whether women or childen of welfare families or aggressive street criminals, have become increasingly socially isolated from the mainstream pattern of behaviour' (Wilson, 1978: 58).

14.3 INSTITUTIONALIZED ETHNIC SEGREGATION: THE APARTHEID CITY

Ethnic segregation is generally the outcome of unconstrained forces operating within the city and based on a complex of social, economic and plainly racial motives. But in one case, that of South Africa, until 1993, it was the policy of the State, so that cities were formally organized into racial enclaves. The South African legislation was complex but was centred on the Group Areas Act initially passed in 1950, although it was amended in detail. The central principle of the Act was that in defined areas of the city, land and property could only legally be owned and occupied by people of a specified racial group. In South Africa the main groups were the White, mainly Afrikaner (Dutch) or British by origin; the Coloured, produced by interbreeding between the White population and indigenous Bushmen and Hottentots, now called the Khoi-San; the Black or Bantu; the Indian, mainly brought into the sugar plantations of Natal or attracted as traders during the period of initial gold and diamond exploitation. All of these groups were allocated consolidated areas which ideally were separated by marked physical barriers, such as rivers, or by buffer zones of unused or non-residential land. These were designed not only to act as boundaries but to prevent contact and mixing at the margins. Travel of one group through the area of another was also minimized. Taking these principles into account a sectoral form of city was inevitably to be predicated. Figure 14.9 is an idealized diagram or model of an apartheid city prepared by Professor Ron Davies of the University of Cape Town (Davies, 1981). The Central Business District is the focus although within it Blacks could not hold property, the ownership of businesses was reserved for Whites only. The Indian trading tradition was, however, acknowledged by a separate Indian CBD, the best known being in Durban. White higher social class areas occupied distinctive sectors although the innermost parts were in lower class White occupation. The Blacks and

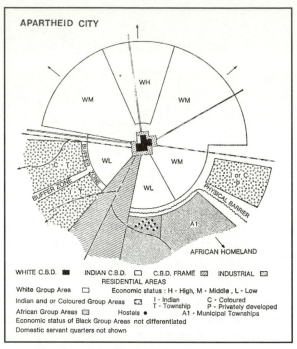

FIGURE 14.9 *A model of the apartheid city in South Africa (after R. Davies, 1981). Source: R. Davies, 1981: The spatial formation of the South African city. GeoJournal, Supplementary Issue 2.*

Coloureds were disposed in sectors clearly separated from those of the Whites. There were clearly marked buffer zones formed by physical barriers, industrial zones or communication lines. Until 1986 legislation, it must be remembered that no Black was a citizen of the Republic of South Africa (RSA) for they were regarded as belonging to one of the Homelands, the nominally independent states such as Bophuthaswana, set up within the RSA territory. Blacks were, therefore, regarded as only temporary residents renting temporary accommodation in the townships. Many of the Black workers were indeed migrants from countries outside South Africa and were housed in hostels shown on Fig. 14.9. Ideally under the apartheid system the Black areas themselves should be divided according to ethnic origin so that there should be separate sectors for the Xhosa, Zulu

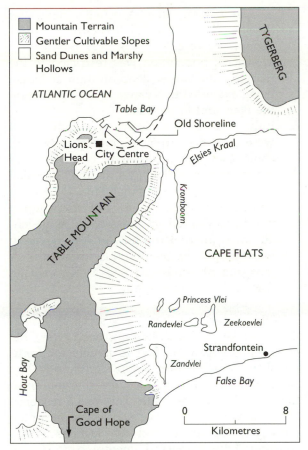

FIGURE 14.10 *The site of Cape Town (after J. Western, 1981).*

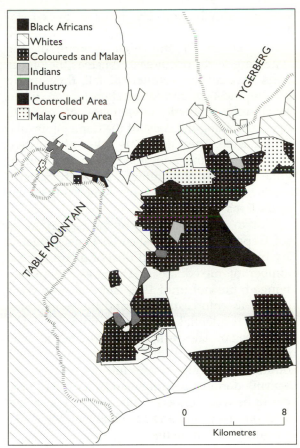

FIGURE 14.11 *Cape Town under the Group Areas Act 1979 (after J. Western, 1981). Source: of Figs 14.10 and 11. J. Western, 1981: Outcast Cape Town. London: Allen and Unwin.*

or Sotho peoples. To a degree, there was an attempt to implement such a scheme in the large townships, such as Soweto outside Johannesburg. Coloured and Indian populations were also located in distinct sectors but unlike the Blacks they were allowed to own property and hence the division in the figure between municipally built townships and privately developed areas for the Coloureds.

Figure 14.9 is a highly idealized diagram. It does not include squatter settlements which were made up of illegal immigrants who constructed shacks on derelict or unused land. Crossroads outside Cape Town is the best known example. Also, when the Group Areas Act was passed towns were already segregated

in highly localized ways. New plans had to be drawn up and approved to conform to the Act but the resultant schemes were quite removed from the ideal. One example can be considered (Western, 1981).

The site of Cape Town (Fig. 14.10) must surely be one of the best known and most spectacular in the world. It is dominated by three elements. Table Mountain is the first with its characteristic flat top and steep sides. The lower slopes are gentler and have provided cultivable land. To the east are the Cape Flats, forming the second element, an area of calcareous sand covered by blown sand. Across this low plain the rivers meander slowly and are blocked by coastal dunes, forming extensive lagoons or *vleis*. The third element is the high ground of the Tygerberg to the north-east. Whereas the higher lands have a heavy winter rainfall, they are by far the most preferred locations, even when the purely aesthetic advantages of outlook, especially along the coast of the Cape peninsula, are set aside. The Cape Flats suffer from extensive waterlogging and flooding in the winter, while in the dry summers the southeasterly winds whip up the sands into unpleasant sandstorms.

Prior to the application of the Group Areas Act, Cape Town was most certainly a segregated city but there was a good deal of mixing and a number of mixed areas. This was completely changed, but not until 1960, since the city itself refused to enforce the Act and the reorganization of residential areas was imposed by a Commission. The principles were the obvious ones: that the White population should command the higher ground for reasons of security, dominance and amenity; and that there should be well-defined buffer zones. It is easiest to set out the group areas as shown on Fig. 14.11 in a series of comments:

1. The whole of Table Mountain and its lower slopes were allocated to the Whites. It is the area with major environmental attractions for it has elevation, outlook and is well drained and flood free. In addition, the White area surrounded the CBD and the industrial zone about Table Bay.
2. The coastal areas (mostly to the west of the area on Fig. 14.10) with their spectacular scenery and high quality suburbs were also to be White.
3. The high lands of the Tygerberg were to be White giving access directly to the mainly industrial sector extending east from the city centre.
4. The Cape Flats were largely allocated to the Coloured population, with the Kromboom River together with a suburban railway and an industrial zone forming the buffer with the White area to the west. A new settlement

of over 250,000, called Mitchell's Plain, was developed on the Flats with a major shopping centre.
5. The Black population was restricted to the margins, the earlier concentration at Langa being extended to Nyanga and Guguletu on the Flats and to the east.
6. Influx control, the means by which Blacks were prevented from migrating to the city, could not be completely enforced and movement especially from the Transkei produced squatter settlements. These in general were cleared away but international publicity resulted in Crossroads being left, although internal troubles in 1986 severely undermined its continuous existence. A government sponsored site at Khayelitsha was planned to replace it.

Within the scheme there was a series of anomalies. A Malay area was left adjacent to the CBD even though it overlooked it, largely for traditional and tourist reasons. More significantly, Fig. 14.11 shows a substantial White extension on to the Cape Flats in the south. This was a response to the attractions of the coastal beaches, one of the classic surfing areas, and the lagoons which constituted a well-marked amenity.

This brief survey demonstrates how application in reality deviated from the model though retaining its essentials. It also indicates how the most attractive locations were retained for the controlling group. Here, then, is the ultimate stage of ethnic or cultural segregation where it is enforced by law or institutionalized. It then becomes not a means of integrating immigrant populations into the community at large but of keeping them permanently separate, as was noted on p. 262. It now seems that tragically the cities of the former Yugoslavia, including Sarajevo, will revert to some form of institutionalized segregation as the warring ethnic groups will refuse to settle together, partly because of memories of the past, partly because of fears of the present and future. Apartheid is not something confined to the South African past, it is ever present where ethnic conflict is sufficiently sharp.

14.4 INTRA-URBAN INEQUALITIES: MULTIPLE DEPRIVATION AND THE INNER CITY

If the views of William Wilson recorded in the quotation above (Section 14.2) are accepted, in so far as they relate to Blacks in Western cities, then that section is largely an irrelevance, for the problem is not racial prejudice but rather the emergence of a deprived section of the population, an underclass largely, though by no means exclusively (see Chapter 15) concentrated in the inner city. This immediately poses the question as to why this underclass, located as it is, has been created.

The answer to that question is in essence simple, even if all the associated aspects are greatly complex. One of the recurrent themes of this book has been the predominance of peripheral expansion and the growth of the metropolitan fringes. The consequence has been the siphoning off of those elements which constitute the good life – whether they be employment, education, leisure pursuits or virtually anything else. Those with the ability or the opportunity have moved with them, leaving behind the disadvantaged and those who are discriminated against. In line with the traditional role of the ghetto as the means by which newcomers to the city could establish themselves before moving out in physical terms and possibly upward in social terms, so too slums have been categorized into 'slums of hope and slums of despair'. While the latter were the receptors of society's failures and its inadequates, the former were regarded as a stepping stone to higher things, a temporary condition only. The real problem with the Black ghetto is that the traditional mechanism has seemingly jammed. It now appears that the same has happened with the slum: the contrast between the affluent and the underclass has become so great and the disadvantages of the underclass so all-pervading that no transference is taking place, so that the dispossessed become further alienated into a different world 'isolated from the main stream patterns of behaviour'. This is illustrated in Fig. 14.12 where the

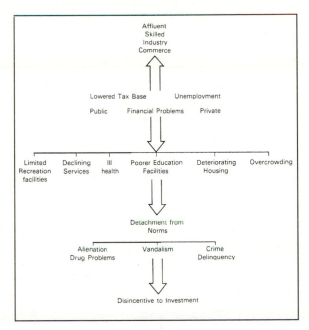

FIGURE 14.12 *The process of decline in the inner city.*

additional problem of the lowered tax base in an area of poverty is included. The finances available to treat the severest urban problems are diminished, although government usually attempts, by a variety of measures, to compensate.

At the expense of some repetition, it is worth while briefly to outline the causes of isolation:

Ageing and obsolescence

The inner part of the city is of necessity the earliest built and the oldest. Moreover, in most Western cities these were the areas built during the nineteenth century when building regulations were minimal so that the quality of housing is low. New building has taken place progressively farther away from the centre where decay adds to the problem of poor quality building. But given the cycle of rebuilding of the central business district, then some extra reasons need to be invoked to explain the lack of redevelopment in the adjoining inner city.

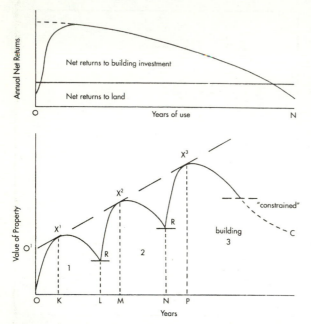

FIGURE 14.13 *Upper diagram. The expected net returns from investment in a new building. Lower diagram. Idealized replacement cycle for individual buildings. (after L. S. Bourne, 1967). The upper diagram is self explanatory. In the lower diagram O represents the time of original construction whilst K, M and P indicate the time of the theoretical optimum net returns on the original investment. R is the level at which replacement occurs and L and N redevelopment in the form of a new building. The line x^1 -x^2 - x^3 represents the theoretical increase in total property value with replacement by more intensive uses.*

The prime reason is that buildings are long-term investments and modification is costly. They are not easily detachable from the area in which they are sited. The result is that with age and use the value of a building declines in terms of the net return on investment. A point can be reached when redevelopment would give a greater return: that is, it would compensate for the cost of removing the building, the cost of a new one and the loss of returns during the process. Rising land values imply an increasingly intensive use (Fig. 14.13). This can be achieved

by the subdivision of houses thus creating the situation where slum rents per unit area and return on capital are among the highest. If this does not happen a wide range of problems faces the potential developer, among them the difficulties of widespread ownership of small units and of acquisition. Moreover, any large area will include property in a variety of conditions – many planning authorities have been accused of the demolition of houses of good quality in a policy of comprehensive redevelopment.

The result is that, faced with all these difficulties, capital can be better devoted to the commercial property of the centre or the residential suburb. It is not economically viable for private capital to provide housing for low income groups, and thus the inner areas must either deteriorate into slums or be redeveloped by private capital as luxury apartment blocks or by public capital for housing. As they deteriorate they provide the milieu where the newcomer to the city can find refuge, together with the dropout. In relation to the urban society at large those on their way to integration within it and those on their way to rejection of it, find common ground in an environment of decay and dilapidation.

The British Department of the Environment commissioned studies of the inner areas of Birmingham, Liverpool and Lambeth (London) and the consequent reports provided a mass of detailed information on housing conditions. The Birmingham report, *Housing policy for the inner city* (Llewelyn-Davies *et al.*, 1976), demonstrated the range of problems which include not only physical decay of fabric and the lack of standard amenities, but also paradoxically both overcrowding and under-occupation. The former is self-explanatory, but for the elderly widow, characteristic of the sorts of population left in the inner city areas, the problem of coping with a house which is too large, including the physical demands of cleaning and the financial demands of effective heating, can become a major one.

Suburban and peripheral growth

This has been so much a theme of previous chapters that nothing needs to be added here.

Town and regional planning

In many countries the most immediate problems in the second half of the twentieth century have been congestion at the centre of cities and the lagging performance of peripheral regions. The result was that governments adopted policies which were designed to relieve congestion and promote regional development. In Britain, the classic example of the former was the New Towns policy which was advocated in Sir Patrick Abercrombie's *Greater London Plan* of 1944 and formulated in the New Towns Act of 1946. Writing in a volume published for the Town and Country Planning Association, Peter Self (1972) suggested three purposes of the new towns:

1. As answers to 'megalopolises', or the excessive concentration of people and activities within great cities or conurbations.
2. As ways of organizing the vast volume of new development which will occur anyhow, along more beneficial lines.
3. The creation of 'balanced' communities.

The first two were partly achieved, the third was not. Self had to admit, 'in some respects the new towns are noticeably "unbalanced"'. That is they did not present a British average array of ages and skills, for they abstracted from the city the young and the skilled. This is in no way a criticism of a very successful planning achievemant, but its consequences for the inner city were never assessed. The more able and the younger and more active were decanted out of the conurbation centres. The same effect was brought about by the use of Industrial Development Certificates to spread industry to lagging regions. The transference of the Royal Mint from London's east end to South Wales, to Llantrisant, which was nominated as a new town, is a classic example of government policy. There is little doubt, therefore, that inner city blight was, at least in some measure, brought about by planning policies which were dominated by considerations of central congestion and regional equality. The impact upon the inner city was conceived as beneficial.

Exploitation

Most socialist interpreters would identify the inner city's problems as the inevitable product of the capitalist system and the predominance of the search for profit over considerations of humanity. The implications of the contrast between affluent suburbs and decaying slums has best been illustrated by Bunge's scenario for Detroit (Bunge, 1971). Bunge sets up an equation

$$R = A(P - C) - 30\ ATD$$

where R is the rental profit in dollars per month per square mile; A is the number of dwellings per square mile; P is the rental in dollars per month; C is replacement and maintenance costs in dollars per month; T is cost of transport in dollars per mile per day (multiplied by 30 to give the monthly total); and D is distance from the city centre in miles.

Reference to Chapter 7 will confirm that this uses the simplest of measures of land rents related to distance from the city centre (compare with Fig. 7.5, p.138). Bunge proceeds to make the substitutions indicated on Fig. 14.14(A) and hence arrives at the conclusion that at one mile from the city centre there are profits of $190,000 per square mile per 30-day month or $2,311,345 per annum being generated, i.e.

$$R = 5000\ (80 - 30) - 5000\ (0.2)\ 2\ (30) = \$190,000$$

Bunge's second diagram (Fig. 14.14(B)) indicates the flow of these rents, together with other money transfers of a variety of natures. The implications are now apparent; those in affluent suburbs are deriving their standard of living in part from the exploitation of the inner city.

This exercise is, of course, highly oversimplified. It is designed to illuminate an issue rather than as an empirical investigation. One has to accept Bunge's data for they are presented without reference to any sources of confirmation and are nothing more than subjective assessments. This is even more dangerous in Fig. 14.14(B) where flows are indicated of various dimensions but without any values being

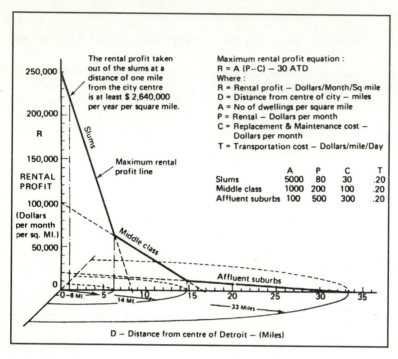

The rental profit taken out of the slums at a distance of one mile from the city centre is at least $ 2,640,000 per year per square mile.

Maximum rental profit equation :
R = A (P–C) – 30 ATD
Where :
R = Rental profit – Dollars/Month/Sq mile
D = Distance from centre of city – miles
A = No of dwellings per square mile
P = Rental – Dollars per month
C = Replacement & Maintenance cost – Dollars per month
T = Transportation cost – Dollars/mile/Day

	A	P	C	T
Slums	5000	80	30	.20
Middle class	1000	200	100	.20
Affluent suburbs	100	500	300	.20

FIGURE 14.14A *The urban geography of Detroit's rental profits (after W. Bunge, 1971). 14.14B Direction of money transfers in metropolitan Detroit (after W. Bunge, 1971).*

(a)

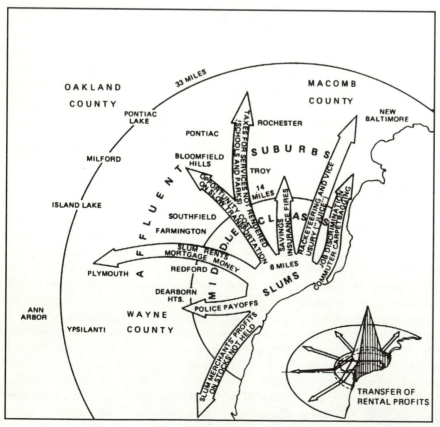

(b)

assigned. Moreover, none of the reciprocal inflow of money from taxes to the centre is included. The various welfare services generate a substantial return flow which would have to be evaluated. Deliberately to omit those inflows from the calculation raises the suspicion over the whole probity of the exercise. But Bunge's presentation makes the point which it was intended to do – there is more than simple inequality, there is exploitation.

Positivist statements of the implacability of geometry are becoming unfashionable, but even so an axiom can be propounded that space predicates inequality and necessitates relative deprivation. Every point on the earth's surface cannot be provided with equal access to goods and services, nor can a total population be located on a single point. It has been demonstrated in Chapter 3 that one of the consequences of the time–cost expended to traverse distance is an urban hierarchy, or at least a size ranking of some form; in Chapter 7 it was noted how the friction of distance structures intra-urban land uses. In any urban system there will be differential access to the highest order services, and that, for a sector of the population with least access, constitutes deprivation. No economy has yet developed, or will in any likelihood in the foreseeable future develop, in a spatially uniform way. Both regional and hierarchical contrasts are the inevitable outcome of the tyranny of distance.

Since the contrast between geometrical and political interpretations of urban inequalities has been introduced it is appropriate at this point to pursue radical and Marxist critiques of the causes of urban deprivation. Writing on geographic inequality under socialism Fuchs and Demko concluded that 'the discussion of issues such as spatial inequalities is hardly constructive when it is done in order to condemn a specific ideological system. Since East and West share the problems of spatial inequalities, and express a commitment to their elimination or reduction, this should be recognized as a common problem to be removed from the area of political debate' (Fuchs and Demko, 1979: 318). This ascribes the basic problem to

geometry, but in spite of the decline of Marxism as the controlling ideology of states it is still a fundamental misreading of the critique continued to be offered by radical geographers. Most exponents of the Marxist approach have little in common with those who emphasize welfare concerns; spatial inequalities, they assert, derive from the fundamental economic order and anything but revolution in that, a thing centrally political, is merely the rearrangement of a superstructure to make an outward appearance less damning and more defensible, while the underlying driving mechanism of the capitalist system, the infrastructure, is left untouched.

Unfortunately there is no immediate exposition of a 'complete' Marxist urban geography to which to turn to derive the social make-up, in spatial terms, of the city. Most radical writing has been concerned with urging the need for such a venture rather than undertaking it, with condemnation of the capitalist city rather than the presentation of the socialist city in all its detail. David Harvey has, however, attempted a sketch of the urban process under capitalism in order to provide at least a framework for analysis (Harvey, 1978). He confesses that 'to broach the whole question of the urban process under capitalism in a short article appears a foolish endeavour'; to summarize his sketch must appear doubly foolish. Yet the brief diversion (or this transfer to the central *thema*, according to view) is worth while even at the risk of travesty, to present some indication of the nature of the approach.

The starting point is the main motivation of the capitalist system, the making of profit. Two basic themes follow, accumulation and the class struggle. Profit, or the derivation of surplus value, accrues to the capitalists who dominate, organize and exploit the work process. Labour can only sell its work power in the market and hence conflict arises as the capitalists ensure the securing of surplus value produced by labour, at the expense of labour. Accumulation and class conflict are in this way part of the same process. To maintain domination and speed accumulation the capitalist class behaves with increasing violence to the labouring class; to

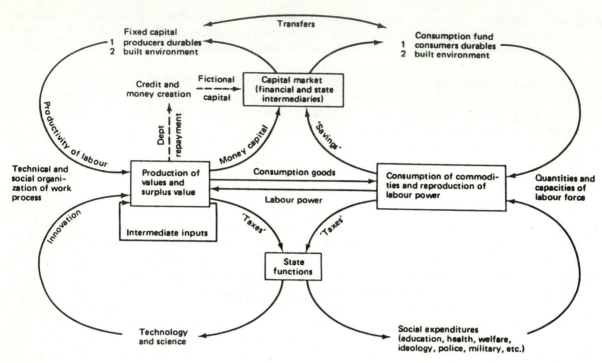

FIGURE 14.15   *The structure of relations between the primary, secondary and tertiary circuits of capital (after D. Harvey, 1978).*

battle for its own interests the working class unites. 'Through competition, however, the inherent laws of capitalist production are asserted as external coercive laws having power over every capitalist. A world of individuality and freedom on the surface conceals a world of conformity and coercion underneath' (Harvey, 1978: 102). Freedom to choose in the capitalist world is, therefore, a myth; it is part of the superstructural arrangements to conceal infra-structural impositions.

At this stage the direct question as to how all this relates to the urban geographer immediately arises. But it is still necessary to follow Harvey through his analysis of the 'laws of accumulation'. He sketches what he terms three circuits. In the primary circuit he demonstrates the tendency for the system to move towards over-accumulation, for its prime *raison d'être* is to expropriate surplus value, even as an end in itself. This surplus, however, can be used by

being switched to the secondary circuit by progressive investment in fixed capital, either to aid the production process or to support those external features which support production and which are called the built environment: housing is an example. Finally there is a further tertiary circuit which absorbs surplus through investment in science and technology further to enhance production or in social expenditure to maintain the efficiency of labour. A health service is not concerned primarily with the total well-being of the individual but with ensuring that the labour input can be maintained. These three circuits are shown in Fig. 14.15.

It is against such a structure that spatial processes must be viewed. Two examples can be sketched. The first is derived from the fact that the underlying contradictions within the capitalist process of accumulation generate crises. The potential of capitalism for balanced growth is never achieved. Thus over-production, increasing the quantity available in the market, precipitates a decrease in value and, as the potential for further productive investment is exhausted, a crisis follows. Harvey identifies:

1. **Partial crises.** These occur within an economic sector or geographical region and are met by local adjustment.
2. **Switching crises.** These follow on the redirection of investment within
   (a) sectors of the economy
   (b) geographical regions
   although the two are so interrelated as to be difficult to separate.
3. **Global crises.** These affect the whole capitalist world as in the growing crisis of the late 1970s and the early 1980s.

To demonstrate the direct relevance of this analysis it can be argued that the inner city problem, which has already been considered in this chapter from a conventional cartographically based viewpoint, is a switching crisis. Investment has shifted from non-profitable to more profitable sectors/regions to maximize accumulation. The crisis in this case becomes deeper because of the spatial immobility of investment in the built environment. But capital simply abandons the 'region' and leaves it to deteriorate with the result that costs are eventually shifted to the working class by the necessity for State intervention via comprehensive redevelopment or one of the many forms of rehabilitation. Alternatively, residual profits can be made by the attraction of the poorest into these deteriorating parts of the city. Slum and ghetto are the inevitable products of the system. In the end, the community (the working class) absorbs the burden of the contradictions of capitalism either in a direct financial way or through a devalued environment.

Here, then, are the 'real' reasons, the causes of inner city decay. Mapping measures of social deprivation, devising 'enterprise regions', these are simply containing, at community expense, the costs of the contradictions of capitalism, in a cosmetic exercise.

The second example is not unrelated to the first and here it is useful to revert to a point already discussed – the explosive development of residential surburbia in the USA after 1945 (see Chapter 13, p. 257). At that time, Harvey argues, potential over-accumulation faced the country. This dilemma was in part resolved by the creation of a whole new lifestyle through the rapid proliferation of the suburbanization process. Furthermore, the social unrest of the 1930s in that country pushed the bourgeoisie to adopt a policy of individual home ownership for the more affluent workers as a means of ensuring social stability. This solution has the added advantage of opening up the housing sector as a means for rapid accumulation through commodity production. So successful was this solution, epitomized by Levittown, that the housing sector became a Keynesian 'contra-cyclical' regulator for the accumulation process as a whole.

It will now be apparent that these two examples of inner city decay and suburbanization are not only related to each other but also have a common genesis from the operations of the capitalist system. It will also be clear why Marxist geographers argue that the initial point of departure for urban geography should be neither measurement of a real objective world, nor the uncovering of the patterns of the perceived world, but the 'first cause' of the politico-economic system.

One final point must be made. All this is a critique of the city in the existing and real capitalist world, it goes no way whatsoever to specifying an alternative. So one returns to the authors whose views were set aside at the beginning of this section:

> It should be apparent to all but those wilfully ignorant or already ideologically committed that the imposition of a Marxist political socialist system is in itelf no panacea leading to removal of spatial inequalities; decision makers under state socialism appear to hold values of growth, productivity, efficiency and rewards similar to those of capitalists, leading to similar problems of spatial inequality.
>
> (Fuchs and Demko, 1979: 318)

14.5 SOLUTIONS TO INNER CITY PROBLEMS

It would be very misleading to leave a discussion of inequality and multiple deprivation without a brief indication of the action which has been taken to offset it. Though according to the views developed above it is no more than 'a historically-specific and socially necessary response to the self disorganizing tendencies of privatized capitalist social and property relations as these appear in urban space' (Dear and Scott, 1981: 13). Indeed, in that quotation probably lies the inherent weakness of most actions, for they have seldom been based on a thorough understanding of the underlying causes, but rather on pragmatic attempts to solve specific issues.

Since specific actions are central then a case study is necessary. Here the example is Britain. As with so many topics in this book, it is wrong to pretend that attempts to solve the decline of the inner city constitute anything but an issue of great complexity. It has been dealt with admirably by Atkinson and Moon in their book *Urban policy in Britain* (1994). What follows here is a brief outline which cannot hope to include all the detail of a large number of government reports and much legislation, itself changing with the political party in power. The three main targets for improvement have been housing and the environment, employment opportunities and personal and social conditions.

Housing and the environment

Grants of money have been available for the private renovation of houses since 1949, and various Housing Acts since then have changed the sums available and the detailed conditions. Since 1964 local authorities have had powers to undertake more general improvements to tracts of land rather than to individual buildings. After 1969 these were called General Improvement Areas (GIAs) and the scope was broadened to include the provision of amenities and the general upgrading of the environment. In 1974 Housing Action Areas (HAAs) were introduced for special and more difficult parts where the amount of rented property was high for example, or where significant environmental problems were leading to houses being abandoned. In these areas the local authority had greater powers of compulsion and there was a higher financial provision. There was a third class of area called Priority Neighbourhoods which were usually adjacent to the other two and designed to prepare them for improvement.

Industry and employment

Employment has been one of the major difficulties of the inner city and its significance as a causal factor has been noted (p. 273). In the 1970s the Government of the day had sponsored a massive series of inner area studies and in 1977 a White Paper called 'Policy for the Inner Cities' was published. Out of the discussions came the general view that the central problem was the one of unemployment derived from the collapsed economic base of the inner city. When the Government moved to legislation, therefore, in the Inner Urban Areas Act of 1978, its main financial provisions were directed to sponsoring partnerships between central and local governments and industry, as in the Inner City Partnership Programmes. There were eventually seven Partnerships, 15 Programme Authorities and 19 Designated Areas. Their main purpose was to generate jobs. This was done by financial incentives to firms as well as to projects directed towards improvement of the environment in order to make it more attractive to industrialists. Subsequently, under Mrs Thatcher a much greater role for the private sector was created, leading in 1989 to the Urban Programme Review. City Challenge in 1991 introduced competitive bidding by authorities for finance. Three other initiatives were Enterprise Zones, which were devised to offer investors freedom from the constraints which surrounded most entrepreneurs, Free Ports and perhaps most important, Urban Development Corporations. By 1992 there were some 13 such corporations, but undoubtedly the most significant was the

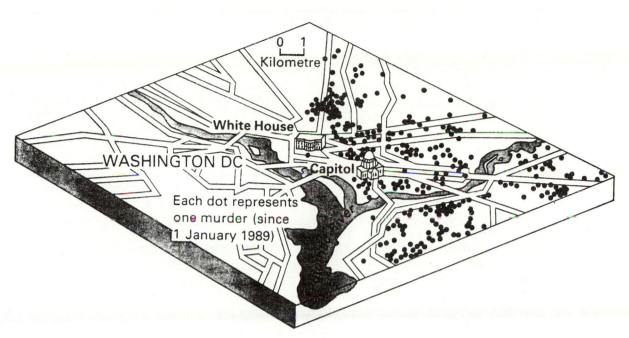

FIGURE 14.16 *Distribution of murders in Washington DC between January 1st and mid-March 1989. Source: Times Newspapers.*

London Docks Development Corporation which became the flagship of government promotion and which has already been discussed in Chapter 11. These sorts of initiatives continued in a series of publications – *Action for cities* (HMSO, 1988), *Progress in cities* (HMSO, 1989) and *People in cities* (HMSO, 1990). In 1992 an Urban Regeneration Agency was established primarily to deal with derelict land in cities. All these varying and overlapping attempts are well assessed by Atkinson and Moon, 'We see urban initiatives as symbolic; set up to mitigate the worst excesses of urban decline and demonstrate governmental concern without raising fundamental questions or grasping the nettle of consequences. The economic costs of intervention and the social costs of non-intervention were more than governments were prepared to risk' (Atkinson and Moon, 1994: 265).

Personal and social conditions

The Inner Areas Studies generally agreed that personal poverty itself was not specifically a problem of the inner cities for it was as prevalent on peripheral municipal housing estates. Even so, such initiatives as the Urban Programme of 1968 were designed to offset the special disadvantages of these areas. The general social services are intended to meet major needs and the objectives of the Urban Programme were to provide immediate help with such things as community centres, family advice centres and nursery schools. In general, after Policy for the Inner Cities of 1979 attention swung to the more general causation of inner city decay and away from action on a limited and more personal directed basis.

It has to be admitted that all the measures which have been taken have had only partial success as is evidenced by the continuing problems of unrest. The apparent failure of wide-ranging measures is due to the fact that they were in opposition to two very strong forces. The first was the massive stress on suburban growth and the substantial economic forces ranged behind it. The second was that at the time the measures were becoming operative the general problem of economic recession arose and widespread national unemployment. Under such circumstances the claim of the inner cities became only one among many others.

The inner city must be the most studied section of the whole urban area. But in spite of an immense expenditure on academic and policy studies, as well as the schemes of improvement and the programmes of central area redevelopment which have been discussed (Chapter 11. p. 205 *et seq.*) large inner city areas remain as the often burnt-out and degenerate environment of an alienated underclass. The young have particular problems. Without training or qualifications, without employment or the prospect of employment, young people are drawn into vandalism and petty crime as a way of life. From that some 'relief', the kicks which make life tolerable, can be achieved through drugs. The path to addiction and to hard drug use is then open. And drugs can only be paid for by escalating crime. That has become the major problem of the 1980s and 1990s. Control of the drug supply, and the immense profits to be gained, has brought in gang warfare and murder. Figure 14.16 shows the distribution of murders in a short period – 1 January to mid March 1989 in Washington DC. The figures include 97.8 per cent black as against 2.2 per cent white, in what, of course, is a predominantly black city. But there is a most noticeable absence in the white gentrified area of Georgetown. These are the conditions which have made inner cities into no-go areas where residents keep off the streets and barricade themselves in at night. The good life which the city is supposed to promise is very remote in these parts.

14.6 INTRA-URBAN PATHOLOGIES

All these problems are clearly illustrated in analysis of intra-urban pathologies. Studies which fall within this group are essentially within the positivist empirical tradition of urban geography and urban ecology. *The city*, that seminal work of the Chicago School of Human

Ecology, which was edited by Park and Burgess and published in 1925, contained the following paragraph:

> A recent study by Neils Anderson of what he calls 'Hobohemia', an area in Chicago just outside the 'Loop', that is to say, the down-town business area, which is almost wholly inhabited by homeless men, is a study of... a human junk heap. In fact, the slum areas that invariably grow up just on the edge of the business area of great cities, areas of deteriorated houses, of poverty, vice and crime, are areas of social junk.
> (Park and Burgess, 1925: 109)

In the acerbic terms of the 1920s, such is the identification of a pathological condition.

Two primary aims can be identified in the context of intra-urban pathology:

1. The identification of those parts of the city which are characterized by elements of social disorganization, as indicated in the quotation from R. E. Park above. Crime, in all its manifestations, and disease are the two most obvious symptoms, but others, such as the break-up of families, or the failure to meet civic obligations in the inability to pay rates or taxes, can be included.
2. The identification of the socio-economic, demographic and ethnic correlates of social disorganization.

It must be stressed that the correlates are not interpreted as either the cause or the consequence of the condition, but rather that they add depth to the characterization of distinctive city areas.

One of the best examples of investigation under the heading of intra-urban pathology is that of J. Giggs into the distribution of schizophrenia in Nottingham (Giggs, 1973: 1983). His later study (1983) is the more complex and developed but the earlier one is presented here. Schizophrenia is an extreme form of the set of mental disorders which are termed 'psychoses' which involve the disintegration of the personality to the extent where an individual becomes incapable of dealing effectively

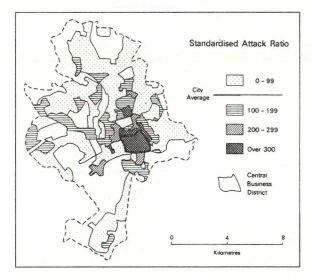

FIGURE 14.17 *The distribution of schizophrenia in Nottingham (after J. A. Giggs, 1973).*

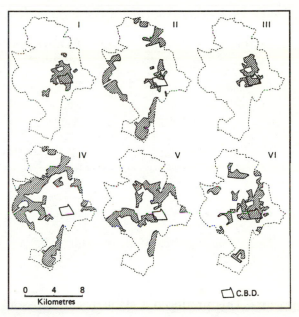

FIGURE 14.18 *First order promax factors derived from factor analysis of 41 variables for Nottingham (after J. A. Giggs, 1973). Shaded areas are enumeration districts exceeding 0.5 over the standard deviation. For identification of factors see Table 14.2.*

either with himself or his environment. Giggs mapped the distribution of this illness in Nottingham and the general pattern is shown in Fig. 14.17 in terms of Standard Attack Ratios (SARs). These ratios adjust the crude figures of incidence per enumeration district to take in variations in population density and population at risk by age bands. The SAR is expressed in terms of the average attack rate for the city so that a ratio of 300 means that the SAR for that district is three times that for the city as a whole. Setting aside minor differences betwen male and female distributions, the simple conclusion from Fig. 14.17 is that schizophrenia demonstrates a well-defined locational pattern and is closely associated with the inner city centre. Having established the primary distribution of the condition, Giggs proceeds to undertake a factor analysis in which 41 variables for each of the 140 enumeration districts of Nottingham were employed. Of these variables 12 were related to the schizophrenia population itself, 11 were demographic in nature, four were socio-economic, nine related to housing tenure and eight to housing and household conditions. The remaining one was a measure of distance from the city centre. An oblique (promax) rotated

solution was developed and six factors abstracted. These are shown in Tables 14. 2 and 14.3.

The distribution of these factors by standard factor scores on the enumeration districts is shown in Fig. 14.18, where only those districts with a positive standard deviation of greater than 0.5 have been isolated. The relation of the first schizophrenia factor to the distribution of SARs in Fig. 14.17 is apparent, as would be expected. The distribution of the remaining factors can be convincingly related to the interpretation of city social space already presented. Simple subjective pattern matching suggests the association of the illness with rooming house areas of single people locationally at or marginal to the city centre. But since an oblique rotation was employed and orthogonality not retained, these six factors were themselves subject to further analysis in order to derive higher order

TABLE 14.2 *Schizophrenia in Nottingham: factors and loading variables*

Factor	Positive loading variables	Negative loading variables
I. Schizophrenia	All 12 schizophrenia variables	
II. Life cycle	Young; mobile married	Elderly; small households
III. Urbanism–Familism	Rooming house	Large households; large dwellings; married
IV. Rented housing– housing amenities	LA housing; young; suburban	Private rented; unfurnished dwellings; sharing amenities.
V. Socio-economic	High social class (1 and 2)	Low social class (5)
VI. Marital status	Single	Married

Source: After Giggs, 1973.

factors. This produced a stable result across all the factoring methods which were tried. The primary factor method, without rotation, gave a first factor explaining some 86.8 per cent of the variance. This second order factor, as it is called, can be interpreted in relation to the loadings of the six first order factors as shown in Table 14.3 where a complex of factors is revealed associated with schizophrenia. The negative correlates confirm the suggestion that 'such conditions as having a stable marriage, being gainfully employed, being educated, and being intelligent may all summate to a condition called security and freedom from anxiety. Having all the "advantages" may be ... critical ... in determining how easily and successfully one copes with the tasks of living' (McNeil, 1970: 27). The real problem of causation remains, however, for such an argument could well be circular. One still does not know whether people are schizophrenic because they cannot cope, or cannot cope because they are schizophrenic.

The geographical relevance of Giggs's study is that 'it has ... been demonstrated that there are close correlations between the incidence of schizophrenia and a large set of social and environmental factors' (Giggs, 1973: 71). The distribution of that set is coincident with the inner city. 'Importantly, these unfavourable social and economic traits collectively assume

TABLE 14.3 *Schizophrenia in Nottingham: second order factors*

Strong positive		*Negative*
III	Urbanism	Familism
IV	Rented housing pre-1914	Rented housing post-1920
	Unfurnished	Local authority
I	Schizophrenia	No schizophrenia
Moderate positive		*Moderate negative*
II	Persons over 45	Persons 15 to 44
IV	Single persons	Married persons

Source: After Giggs, 1973.

their greatest intensity in the inner, slum areas of the city' (Giggs, 1973: 71). It is but a short step back from this conclusion to that earlier quoted from R. E. Park written in 1925, that here are the city's 'junk heaps'. Park's statement now seems unsympathetic, even brutal, in the way it was expressed, but it continues to pose the same problem of causation. Do these sorts of people gravitate towards the inner city because having a specific physiological or mental condition, they are unable to withstand the pressures generated by living in an intensely competitive urban environment, or has the urban environment by its very nature created such stressful conditions that psychoses are generated? At the aggregate and descriptive geographical level the question is immaterial, for what is clear is that a section of the city can be isolated where such pathologies are located and where they have identifiable characteristics and correlates.

Process investigations which seek to resolve some of the above problems, characteristically move to the behavioural level, but comparatively little has been done in the field of mental illness. An exception was a study by Dear of the movements and distribution of psychiatric patients discharged from Hamilton (Ontario) Psychiatric Hospital (Dear, 1977). Follow-up presents obvious problems of tracking down and locating the discharged persons and, as Dear noted, it has ethical implications as well as detection problems. His study showed two clear concentrations of the sample he succeeded in tracing who had been discharged with no institutional arrangements. One was located in the downtown area (the inner city), the other in the census tracts nearest to the hospital. 'The former concentration may reflect the distribution of the population in need of care; it may also suggest the ghettoization of expatients. The latter concentration supports the general distance decay hypothesis that patients in need of care will reside close to the source of that care.' Here, then, is some evidence of process, although evidence of pre-hospitalization is not given. Discharged patients, without family or other support, filter through urban space to an inner city location:

> One recurrent phenomenon of increasing prominence is the tendency for psychiatric patients discharged from mental hospitals to congregate in transient areas of the inner city core. In many ways, the discharged patients resemble other minority groups who have gravitated towards areas of cheap rental accommodation in order to 'establish' themselves. The mentally disabled, however, usually lack the skills to spiral upwards along with the other self-improving groups.
>
> (Dear, 1977: 588)

In this way the inner city takes in the mentally ill as it does any disadvantaged or outcast section of the community. This has been exacerbated in Britain by the introduction of the principle of 'care in the community' together with the diminution of the number of hospital beds for the mentally ill. Without the resources for effective community care, the mentally ill are all the more liable to become footloose and form a distinctive element of those sleeping rough in the city centre, the group with which this discussion of intra-urban pathologies began.

A similar conclusion can be derived from studies of urban crime. Again, to argue simplistically that the causes of crime can be derived from the analysis of distributional evidence is naive, but the socio-economic correlates can be sought and characteristic areas isolated, and that is at least one step in a complex research problem. The identification of particular parts of the city which show a high crime rate is long established. The Chicago social ecologists necessarily became involved in the consideration of criminality as an aspect of ecological processes and, as part of that tradition, C. R. Shaw in 1929 published his book, *Delinquent areas; a study of the geographical distribution of school truants, juvenile delinquents and adult offenders in Chicago.* In Britain, Terence Morris took up the same theme after the Second World War. His book, *The criminal area: a study in social ecology,* was published in 1957. Both these books had specific territorial connotations in their titles, one coining the phrase 'the delinquent area', the other using the term 'the criminal area'.

But only comparatively recently have urban geographers undertaken the linkage of crime with the standard analyses of social areas. There are reasons for that which can be briefly reviewed. The first and most obvious relates to data. Most ecological studies rely on census data but the figures for crimes committed are not always easy to obtain and are most certainly partial for not all crimes are reported. Moreover if the criminal and not the crime is the basis of analysis then one has to cope with the fact that only a varying proportion of those who have committed crimes are caught. In addition, there

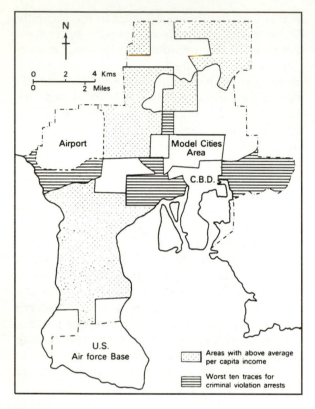

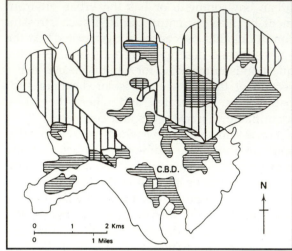

|||| High soci-economic status factor. Areas with positive scores

≡ Above average delinquency rates by residence, per 1000+ at risk

FIGURE 14.20 *Delinquency rates and social status in Cardiff (after D. T. Herbert, 1976).*

FIGURE 14.19 *Aspects of crime and social status in Tampa, Fla. (after D. M. Smith, 1974).*

is the problem of whether to analyse crime by the location where it is committed or by the place of residence of the offender, for in studies explicitly spatial it is a decision of fundamental importance. Finally, definition presents difficulties, especially at a time when the view is held, if not widely, that laws with which the individual does not agree, or which appear to be unreasonable to any dissenting group, can avoid what can loosely be called moral censure.

The associated, but legitimate, point is also argued that most studies of crime are selective and concentrate on lower class criminality rather than that of the upper classes. Thus tax evasion, fraud and corruption are not as frequently studied in this ecological mode as robbery and rape.

D. M. Smith has carried out investigations into crime in Tampa, Florida, and Fig. 14.19 is based on his work (Smith, 1974). It shows the distribution of

the ten census tracts which returned the highest rates for all criminal violations by place of occurrence in the city during 1971. At a superficial level the distribution was as expected. The central area of the city was surrounded by a ring of tracts characterized by high crime rates. The only major break was to the north where an area under the Model Cities Programme showed some amelioration, although that might have been due in part to the under-reporting of crime in a conscious element in the developing of better community relations. Again, the actual areas defined may reflect police activity rather than crime itself. All these 10 tracts were located outside the areas with above average income. Smith attempted to correlate crime rates with a variety of measures of deprivation or socio-economic disadvantages but only demonstrated what he called a 'limited correspondence' rather than a clear association. Even so, the aggregate pattern is a very familiar one.

In Britain the major geographical contribution has been made by D. T. Herbert, particularly in the study of delinquency, that is crime committed by adolescents with the usual age limit set at 20 (1976a and b, 1977). Figure 14.20 reproduces material from his work on Cardiff with two distributions superimposed (Herbert, 1976a and b). The first is the first factor derived from an analysis using 40 variables and categorized as a social class factor, since the high positive loading variables included the percentage of the population in social classes 1 and 2, having two or more cars, travelling by car to work and having exclusive use of all the household amenities recorded by the census. The second superimposed distribution is that of above average delinquency rates by residence per 1000 of the population at risk. A clear relationship can be seen. Once again there is the characteristic surround of the CBD, marking the inner city with lower social class and high delinquency levels.

It has been shown in Chapter 7 that socio-economic status is essentially sectoral in its intra-urban disposition and on Fig. 14.20 the sectoral extensions of the lower quality areas and of delinquency are well matched. Again the apparent simplicity of a pattern matching exercise should not be over-stressed for there are anomalies. There is certainly a difference between what are conventionally called 'rough' and 'respectable' working class areas, while municipal housing tends to emphasize the sectoral extensions of the delinquency distribution (see Chapter 13, p. 258). But it can be argued that, in like manner to those suffering from mental illness, those who are recidivist will gradually lapse into the inner city sink; those who are not will eventually take their places within standard working class areas, while those who make crime pay will graduate to the affluent suburbs. To maintain that there is an exclusive association of crime with the inner city is certainly mistaken, although it is correct to argue that the inner city is associated with crime. And as was indicated earlier, the link with the sale of drugs has made the inner city a great deal more lawless than ever.

14.7 THE INNER CITY: THE REALITY

The academic analysis of the inner city can be far removed from the reality of living within it.

The following is a brief extract from an article by two journalists, Russell Miller and Peter Warren, entitled 'A cry from the streets'. It relates to the Moss Side area of Manchester, England.

Shootings are not in the least uncommon in Moss Side: no fewer than 100 incidents were reported in the last six months of 1992. The beleaguered residents talk of often hearing gunfire at night as rival gangs battle for turf and control of the drugs trade – the area's most thriving business. ... Local dealers use children with mountain bikes as couriers to deliver packages and collect money from junkies who come from far and wide in the certain knowledge they can score in Moss Side. [Since the police removed some of the biggest dealers.] Two leading gangs have been struggling ever since to move into the vacuum thus created. The 'Goochies' (named after Gooch Close, a Moss Side Street) control the west side of Alexandra Park Estate, the 'Doddies' (named after Dodington Close) control the east. Members of the Gooch gang wear red bandanas, the Dodington gang wear blue. Machetes were once their favoured weapons, but the increasing use of firearms indicates that both gangs are now equipped with a formidable armoury of 'shooters'. Walking through Moss Side Leisure and Shopping Centre is like walking through a film set depicting a peculiarly chilling urban apocalypse. Most of the shops are shuttered, having gone out of business long ago, and the only people hanging around within the barren precincts are predatory groups of youths with mountain bikes and menacing bad-ass stares, who glare malevolently at strangers and make curiously threatening clicking noises with their tongues. Some wear baseball caps and bandanas – the uniform of drug couriers. At the last count 53 per cent of teenagers between 16 and 19 in Moss Side were unemployed.

... Moss Side is an alien pocket of poverty, urban deprivation and lawlessness stitched into the cityscape, yet a world apart. A mile to the north of Moss Side is Manchester city centre ... a couple of miles to the south is leafy Didsbury, with Edwardian villas of mellow red brick and glistening white paint. You can buy a fine pair of jodhpurs in Didsbury; you can buy a nice Uzi sub-machine gun for £300 in Moss Side.

(*The Sunday Times*, 7 March 1993)

An equivalent report on Detroit has been presented under the heading 'An urban nightmare becomes a reality'.

Whole swathes of Detroit are now burnt-out ghetto land. It is a city of broken glass and abandoned buildings. The wind whistles through gutted blocks of flats, scrubby weeds grow in empty lots and seagulls feed on the debris of a dying city. Blackened stumps of rotting buildings jut out on the horizon ... . Barbed wire is a permanent feature of many streets, which remind visitors more of TV pictures of Beirut than an important city in the capitalist world. It is like a nightmare vision of a post-atomic future as groups of down-and-out men stare aggressively at police-escorted VIP motorcades. ... Detroit is a city that has sunk to a state of devastating self-destruction.... Detroit has moved from boom to bust and is now often referred to as Murder City. Only Washington DC beats it for the number of murders each year ... a total of 595 [in 1982]. The worst evidence of decay is along the Cass Corridor, a mile-long stretch of burnt-out or boarded-up tenements which cuts through the centre of the city. It is interspersed with empty lots where once grandiose Victorian mansions have been razed to the ground.

(*The Sunday Times*, 20 March 1994)

14.8 GENTRIFICATION

The area about the city centre is manifestly one of great complexity. A review of ethnicity and the use of terms such as 'a black collar', as well as the consideration of areas of deprivation, and of urban pathology, may give the impression that the zone in transition as described by Burgess in 1925 is still extant. But in Chapter 11, a diagram was reproduced (Fig. 11.12) which depicted a 'zone reinvestment' and displayed a whole series of new developments which surround the soi-disant CBD. Among the uses

Beauregard includes in that diagram are loft apartments and condominiums, and he comments that the zone in transition has been 'replaced increasingly by high-priced residences' (Beauregard, 1989b: 253–4). The result is that within the broadest regionalization of the city which ecological studies portray, there are within the inner city the sharpest of transitions between high-quality improved housing and slum-quality dwellings. Thus Hall writes 'central London especially is remarkably heterogeneous in its social make-up, with dilapidated properties sitting next to those with the builders' skips outside in the gentrified "burys" of Islington (Barnsbury, Canonbury and Highbury) and public housing alongside the private' (Hall, 1990: 8).

Gentrification is presented as a problem in that it impinges on the accommodation available for the less well off. Moreover, it can be seen as part of that social polarization which characterizes the post-industrial, transactional city. While low level jobs in cleaning and catering remain, they are a response to the highly paid employment of those who carry out the transactions of the transactional city. Middle range jobs are scarce and hence the sharpness of the boundaries between the groups who now occupy inner city space.

The best definition of gentrification is that by Hamnett, on whose work much of this section relies. It is:

simultaneously a physical, economic, social and cultural phenomenon. Gentrification commonly involves the invasion by middle-class or higher-income groups of previously working-class neighbourhoods or multi-occupied 'twilight areas' and the replacement or displacement of many of the original inhabitants. It involves the physical renovation or rehabilitation of what was frequently a highly deteriorated housing stock and its upgrading to meet the requirements of its new owners.

(Hamnett, 1984: 284)

Hamnett goes on to argue that closely associated with such a process is not only a rise in the price of property but also a change in tenure from rental to owner occupation. The crucial point of the definition is that gentrification is considerably more than the 'tarting up' of older housing. It involves fundamental changes in tenure and value, in habitat, economy and society. It transforms ways of living and it accentuates polarization.

There are two quite different interpretations of gentrification. The first emphasizes the consumer, that is, it sees the process as demand driven. Those who work in the city react against the time-consuming daily journey to place of employment and also, maybe, against what is seen as the sterility and gaucheness of suburban living. A location in the inner city gives access to the substantial attractions which still characterize the centres of large cities – the museums, art galleries, theatres and restaurants – and a quasi-Bohemian life, as against a lengthy journey to an outer suburb. This is especially true of the young, the upwardly mobile professionals who were given the acronym 'yuppy' in the 1980s. They were, and are, predominantly single or living with partners and without children. A whole way of life came into being, no more so than in those who adapted the industrial lofts of Manhattan to give the 'loft living' of Beauregard's diagram. All this is summarized by Ley:

job growth, [in] the white collar complex of downtown head offices, product services, and indirectly, [in] public institutions and agencies in ... nodal centres ... leads to the 'production' of professionals, managers and other quaternary employees working downtown, who then provide the demand for housing reinvestment in the inner city ... this population, as it gives political and economic expression to its own predilection to urban amenity, will restructure the built environment and accelerate the gentrification process.

(Ley, 1986: 532)

In contrast to this concept of a demand-led transformation of parts of the inner city is the view that gentrification is part of the structural adaptation of the capitalist city to the post-

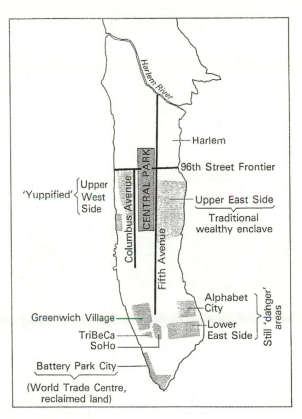

FIGURE 14.21 *High cost residential areas in lower Manhattan. Source: Times Newspapers, Sunday Times, 16 Nov. 1988.*

industrial world. This view is best summarized by Smith:

The so-called urban renaissance has been stimulated more by economic rather than cultural forces. In the decision to rehabilitate inner city structure, one consumer preference tends to stand out above the others – the preference for profit, or more accurately, a sound financial investment. ... A theory of gentrification must ... explain why some neighbourhoods are profitable to develop while others are not. What are the conditions of profitability? Consumer sovereignty explanations took for granted the availability of areas ripe for gentrification when this was precisely what had to be explained.

(Smith, 1979: 540–1)

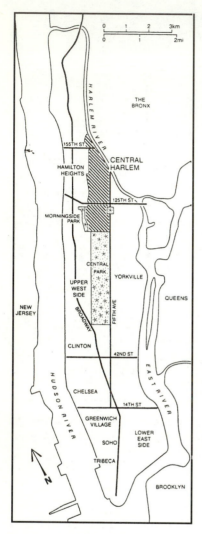

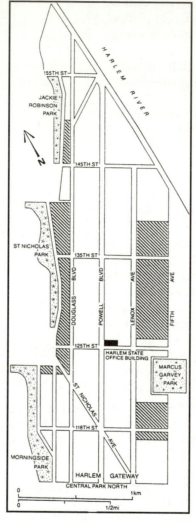

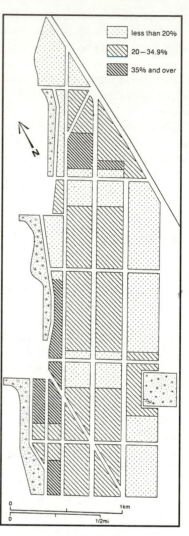

FIGURE 14.22A *Manhattan and the location of Harlem.*

FIGURE 14.22B *Harlem: above average increases in per capita income, 1970–1980 (after R. Schaffer and N. Smith, 1986). Source: R.Schaffer and N. Smith 1986: The gentrification of Harlem? Annals of the Association of American Geographers, 76(3), 354. Figure 2.*

FIGURE 14.22C *Harlem: private residential turn-over rates, 1980– 84. Source: As Fig 14.22b, p. 356. Figure 4.*

The resultant explanation is usually called the rent gap theory. It can be closely linked to the diagram of the circuits of capital set up by Harvey (Fig. 14.15). Capital will always seek areas of maximum returns and after the exten-sive development of suburbia, they were seen to be turning to the possibility of the redevelop-ment of the inner city. Those poor and uninflu-ential could easily be displaced. Nowhere is that made more apparent than in Zukin's book, *Loft living*. She comments on the conversion of the Manhattan lofts as interpreted by succeeding demands:

> the presence or supply of underused loft buildings supposedly inspired an inventive adaptation. Demand for lofts emerged among worthy, though unworldly artists and performers. They settled bravely in the urban tundra and carved neighbourhoods out of the wilderness. Just when they had succeeded in taming their cast-iron environment, a band of new arrivals – who were interested in domesticating an industrial aesthetic – moved in on their territory. Recognizing neither claims nor conventions, this wave of loft tenants bid up property values, started boutiques and crowded the original settlers with their purely residential ethos. They outlawed mixed use. They had no sense of mission. But that is mythology not urban history.
> (Zukin, 1988: 174–5)

Zukin goes on to assert that the real controllers were the corporations engaged in driving out low value uses. 'What is really at stake on this terrain is the heart of the city: the reconquest of the downtown for high-class users and high-rent uses' (Zukin, 1988: 175). Corporations in the quaternary sector, dominant in the post-industrial city, are concerned to push up the cost of space to its highest level and all those who cannot meet the new costs are simply impediments.

As Hamnett (1991) points out, these views are not mutually exclusive and the gentrification process is a product of both supply and demand, but both are closely related to the changes in the nature of the city which have taken place and which have been discussed in this book.

New York is one of the prime examples of the impact of gentrification (Fig. 14.21). As early as the First World War the white middle class had moved out from Manhattan and the former mixed middle and working class district of Harlem developed as a black ghetto. The process was complete by the 1920s. Away from the main axis of business the remainder of Manhattan followed the same process of social disorganization and of physical deterioration, the one exception was the Upper East Side, overlooking Fifth Avenue and Central Park, which was the traditional location of the very wealthy and has remained a vastly expensive and luxury area of the city.

Increasingly since the late 1960s a process of reclamation by the better-off Whites has been taking place, alongside new commercial developments. The Upper West Side has become the home for the upwardly mobile professional people, with Greenwich Village having its distinctive artistic aspect. However, gentrification has spread to other parts. The recent development at Battery Park City, near the World Trade Center, is built on reclaimed land. 'Soho... is no longer an eerie no-man's land of warehouses. It is now bustling with chic boutiques, art galleries and $600,000 luxury lofts... the area directly south of Soho known as TriBeCa (Triangle below Canal) is now almost as expensive' (*The Sunday Times*, 18 November 1988). But perhaps the most interesting extension is from the Upper West Side towards the north, for the borders of the redevelopment edge into Harlem, 'most of which is still a no-go area, though the property developers are eyeing the burnt-out brown stones hungrily, "breaking the 96th Street frontier" as *The New York Times* calls it'.

One of the most intriguing processes of gentrification in Manhattan is, therefore, that of Harlem. Schaffer and Smith (1986) in a paper with an appropriate question mark – 'The gentrification of Harlem?' – have discussed in detail the evidence of the 1980 census. They conclude that there is evidence for the process, especially on the western edge, in what they call the western corridor (Fig. 14.22), 'in the western corridor there is firmer evidence of the beginnings of gentrification' (Schaffer and Smith, 1986: 355). This is attested by above average increases in income and rents and, especially above 126th Street, an area-wide increase in high-income households.

But the gentrifiers are primarily Black, although the eventual supply of high earning Blacks is limited and although gentrification by

Whites of predominantly black areas (Harlem is 96.1 per cent Black) is rare, it is nevertheless identifiable, as in Capital Hill in Washington DC. Schaffer and Smith suggest that 'the constraints on gentrification in Central Harlem are, then, considerable but not insurmountable' (Schaffer and Smith, 1986: 360). All this has broader implications especially in relation to Black suburbanization which was discussed earlier in this chapter. Schaffer and Smith suggest 'that if the evolving spatial pattern of gentrification in the central city continues, then not only will the suburban ghettos burgeon, but the inner city ones will shrink at the hands of the white middle-class migrants' (Schaffer and Smith, 1986: 362). That is a long way off, but it is a significant prognostication of what the city of the future might be, with a White core and very scattered and dispersed outer suburbs, and an inner Black suburban ring.

PROBLEM AREAS OF THE CITY: URBAN PERIPHERY AND SUBURBIA

15.1 INTRODUCTION

The inner city dominates popular imagination as the area generative of urban problems. But, as was indicated at the beginning of the last chapter, all areas of the city have their problems and the difficulties which characterize the periphery are every bit as significant as those of the city centre. In most general terms these problems of the periphery arise from two sources. The first is the isolation, by cost if not by time and distance, from the city centre and all the facilities and services which it does still provide. The second is the tension which is created by the impact of city extension upon the open countryside. This second source is much more acute in small countries where space is limited, such as those of Western Europe, than in the large continental states such as the USA, Canada and Russia. Even the stable middle class suburbs have their problems, traditionally of isolation and the consequent boredom of the housewife left on her own throughout the day,

husband at work, children at school and household tasks diminished by machinery. That was the condition which brought about the situation greatly loved by fiction writers and film makers where the lonely wife takes to prostitution or other sexual adventures as an escape to a concealed, but more colourful lifestyle. But feminism has made the term 'housewife' politically incorrect, while the rise of female employment opportunities, together with escape routes provided by second cars and the suburbanization of city centre facilities, have all contributed to change the former condition.

This chapter, therefore, deals with three selected aspects of the periphery; the municipal housing estate; the defended private suburban estate; and the rural–urban fringe itself.

15.2 THE MUNICIPAL HOUSING ESTATE

The municipal estate as such is not, of course, a generator of problems. But there are estates,

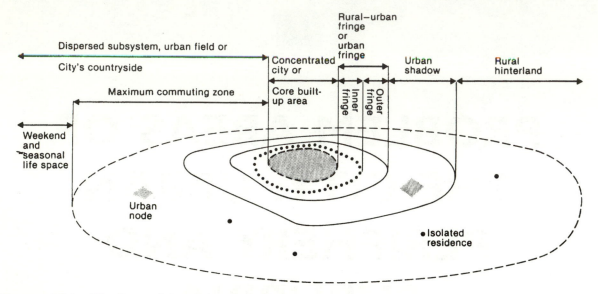

FIGURE 15.1 *The form of the regional city (after C. R. Bryant et al., 1982). Source: C. R. Bryant, L. H. Russwurm and A. G. McLellan, 1982: The city's countryside. Land and its management in the rural–urban fringe. London: Longman; p. 12, Fig. 1.2.*

which are usually well and widely known locally, where social disruption is characteristic. David Herbert used a questionnaire survey to identify those areas of Cardiff to which people least wanted to move. He mapped residential areas on the basis of low residential attraction. There are two elements among the high scoring, least desired areas. The first is the expected inner city ring, but the second is the peripheral housing estates, with Ely, to the west standing out. Herbert quotes a local newspaper as commenting 'Ely, one of the largest housing estates in Cardiff, does not have much going for it. It has poor social facilities, traffic problems and poor shopping. Among key problems facing the district are truancy, vandalism and delinquency' (Herbert, 1979: 132). This is reflected in ambitions and hopes. Whereas in a sample from another estate some 70 per cent hoped their children would go on to higher education, the figure for Ely was as low as 15 per cent (Herbert, 1979: 134). In the summer of 1993 there was an outbreak of communal violence and rioting parallel to that in other British housing estates.

The causes of such situations are manifold. Certainly prime among them is the policy of municipalities of creating 'sinks', that is concentrations of the most inadequate families, through their policies of allocation. That was discussed briefly in Chapter 13. The best study of these estates is that by Frances Reynolds (1986) called *The problem housing estate*. Her data are derived from a survey of a peripheral estate in a midland city which is not identified but called 'Omega'. It was compared with another estate within the city bounds called 'Chalkbury'. She writes,

an information sheet provided to an applicant for housing by the city gave facts and figures on each of the estates administered by the city. Each estate was categorised as having 'little or no turnover', 'very few vacancies', 'occasional vacancies', 'reasonable turnover' or 'fairly regular turnover'. This was a code which told prospective tenants that if they wanted an offer in a reasonable period of time then they had better put their names down for somewhere in one of the last two categories. Chalkbury was in the first category. The three large peripheral problem estates, of which Omega was one, were in the last category.

(Reynolds, 1986: 175)

It is significant that 'many [applicants] . . . wrote on the form that they would under no circumstances consider a tenancy on one of the unpopular three estates' (Reynolds, 1986: 175). In spite of urgings to adopt contrary policies, it is inevitable that the awkward, quarrelsome, noisy and antisocial families, as well as the homeless single mothers who had to be housed as a priority until 1994, were pushed into these 'sink' estates.

In addition to allocation procedures, there were other physical and social circumstances which exacerbated the condition. The physical element is that the estates were too large to develop that community feeling which creates restraints much more effective than those of the law. Moreover, they have been under-provided with community facilities, ranging from inadequate shopping to an absence of leisure and recreation centres and youth clubs and meeting places. Again, another physical problem matches that of the high-rise building which was discussed in Chapter 13 (p. 225). Terraces of housing without adequate soundproofing and defined gardens have meant that here too 'defensible space' does not exist. Of the social circumstances one stands out above all others and that is unemployment. Reynolds quotes 15 per cent for 1981. 'Two or three years later it was estimated locally at closer to twenty per cent for Omega as a whole and over fifty per cent for those aged sixteen to nineteen years old on the estate.' By the very nature of housing priority families, there is a predominance of young people on these estates and they are crushed from three directions. The location of the estates often removes them from locations where unskilled jobs are available. The change in the nature of the city, which this book has set out, means that there is limited opportunity for those without training while the mechanized, even robotized, operations of contemporary industry have done away with much manual labour. Finally, economic recession bites early and

deeply in these areas. On top of all that, as Herbert's work showed, these are the parts of the city where training and education are less regarded. They are also areas, for the names are well known, which employers will regard with suspicion when they appear as addresses on job applications.

The result of a combination of the least adequate and the most troublesome elements of the population, with high unemployment and lack of community facilities, is wanton vandalism and crime. Car crime has been especially prevalent with its provision of thrills to relieve boredom and as a basis for the creation of prestige, esteem and hierarchical ordering. Another relief from boredom is, or can be, even more disastrous. Starting with glue sniffing, drug addiction can grow with the usual consequent petty crime to pay for acquired habits increasingly likely to escalate.

Reynolds identifies five factors which had played a role in making Omega the most unpopular estate in the city. 'These were, its size and segregation on the city periphery; the composition of the population; the cumulative effects of the housing allocation system; the conflict between types of tenant; and the design and structure of the dwellings together with the layout of the estate as a whole' (Reynolds, 1986: 183). All of these have been included in this discussion, except perhaps, explicitly, the conflict between types of tenants, which was that between those elements of the working class usually called 'respectable' and 'rough'. The end result has often been a degeneration into riot and lawlessness which the police can barely contain and which serves only to blacken the reputation of the estates and to increase discrimination against their inhabitants from all aspects of city life.

15.3 DEFENDED SUBURBS

The violence of the inner city, the riots and mayhem of lower class housing estates, and the tendency of both to spread from their points of origin, together with the escalation of urban crime, have produced a strong feeling of insecurity among those higher up the socio-economic scale. One response in the USA has been an increase in gun ownership and a much greater willingness to use weapons in self-defence. But in spatial, or geographical, terms the reaction has been to create walled and defended suburbs. These residential estates, fenced or walled and patrolled by guards, will be familiar to any reader of the standard crime novels located in Los Angeles. Here is Lew Archer:

> It was only a couple of miles from downtown Malibu. The gate was ten feet high, topped with barbed wire. On either side of it, a heavy wire fence plastered with 'No Trespassing' signs stretched off into the hills as far as I could see. The man who was waiting for me at the gate. . . made no effort to conceal the heavy revolver in the belt holster under his jacket.
>
> (Ross Macdonald, 1968: 39).

Since the late 1960s when that was written, the situation has become more extreme. The best discussion is in a chapter called 'Fortress L.A.' in Mike Davis's book *City of quartz*. He begins the chapter:

> The carefully manicured lawns of Los Angeles's Westside sprout forests of ominous little signs warning 'Armed Response'! Even richer neighbourhoods in the canyons and hillsides isolate themselves behind walls guarded by gun-toting private police and state-of-the-art electronic surveillance. ... In the once wide-open tractlands of the San Fernando Valley, where there were virtually no walled-off communities a decade ago, the 'trend' has assumed the frenzied dimensions of a residential arms race as ordinary suburbanites demand the kind of social insulation once enjoyed only by the rich. Brian Weinstock, a leading valley contractor boasts of more than one hundred newly gated neighbourhoods, with an insatiable demand for more security. 'The first question out of their [the buyers'] mouths is whether there is a gated community. The demand is there on a 3 to 1 basis for a gated community than not living in a gated community.
>
> (Davis, 1990: 233, 246)

At an early point Davis calls the building of walls 'the architectural policing of social boundaries' (Davis, 1990: 223). But it is certainly not a new phenomenon, nor one exclusive to the United States. In Britain, the most publicized *cause célèbre* dates back to the 1930s. In 1934 the so-called Cuttlestone Walls were built in Oxford to separate slum-clearance tenants of the Cuttlestone Estate from private development by the Urban Housing Company. 'On September 22nd [1934] the Company wrote to the Town Clerk and announced that it had decided to keep the roads on its estate as private roads and, in addition, to enclose the whole estate by building walls' (Collison, 1963: 52). The walls remained until under post-war planning legislation the City Council were able to demolish them in 1959. It is possible to contend that the tradition of ensuring tranquility by the erection of walls is one of the longest associated with cities. It is no coincidence that in the pre-industrial city in the West those high in the social scale lived within the walls, those at the lowest level without (Carter, 1983: 171–5).

That might well be a specious comparison and the 'carceral city' as Davis calls it, one where people are imprisoned within clearly defined parts, even buildings, is a very different matter. 'Anyone who has tried to take a stroll at dusk through a strange neighbourhood patrolled by armed security guards and signposted with death threats quickly realises how merely notional, if not utterly obsolete, is the old idea of the "freedom of the city"' (Davis, 1990: 250). And lest this should be thought of as a solely Californian, or American, feature, it is instructive to consider an advertisement by John Mowlem Homes in *The Sunday Times* for 17 May 1987:

> Step inside Tadworth Park and you'll think you're dreaming. Never before has there been such a superbly designed development of substantial detached properties set back from leafy avenues, or nestling among lawns and courtyards, The houses are among the finest we've ever built. ... We've even installed security systems and gated the main access

roads so that your personal property will be as safe as your investment.

This is some way from gun-toting security guards, but the trend is in the same direction.

The impact of such developments is to deny the very freedom of movement and interaction which the city originally provided. Rather than being a meeting place where different ideas are reviewed and new notions developed, it can become a segmented system where the individual moves in the enclosed box of a car between highly segregated pieces of the city each conforming to his or her lifestyle. The galactic city may well be a complex of shifting locations, but the spaces between can be as isolating as space itself. To return to the crime novel with which this section began, most of John Ross Macdonald's novels, as those of Raymond Chandler, are pivoted about the narrow-minded limitations of the illiberal residents isolated in their mansions or their secluded suburbs and beset either by problems of the past or by the nonconformity of their children. Isolation from external threat apparently generates its internal tensions and the promise of tranquility given by protective walls is illusory. After riot, fire and earthquake perhaps Los Angeles is 'the junkyard of dreams'. There are problems even in the most affluent suburbs.

15.4 RURAL–URBAN FRINGE

The space into which the town extends as the process of dispersion operates has created the concept of a rural–urban fringe, an area with distinctive characteristics which is only partly assimilated into the growing urban complex, which is still partly rural and where many of the residents live in the country but are not socially and economically *of* it (Herington, 1984). The foregoing introductory sentence reveals two somewhat different terms of reference. The first refers directly to the physical characteristics of an area, the second to the social characteristics of the occupants and it is as well that these are kept carefully apart. Three aspects can therefore be considered:

1. The notion of the fringe as a distinctive physical area or region of the city, primarily designated by characteristic land-use associations.
2. The notion of the fringe as that area where urbanization impinges on rurality and where conflict between ways of life is generated.
3. The impact of urban expansion on agricultural land.

It has been suggested that together these three aspects produced a complex pattern of transition which can be seen as having three constituent parts (Fig. 15.1). There is an inner zone where land is in an advanced stage of transformation 'from rural to urban uses – land where there is little doubt over much of its area about its urban-oriented function and ultimate conversion to urban uses' (Bryant *et al.*, 1982: 13). The outer fringe is an area where although primarily rural, there are clear urban elements present indicating that city influences have begun to infiltrate. Beyond lies the urban shadow where there is only a sporadic and scattered representation of the city in some non-farm residences and estates with commuting patterns to the city.

The fringe as physical area

A large amount of literature concerned with the physical delimitation and the defining features of the rural–urban fringe, appeared during the period from the mid-1940s to the beginning of the 1950s. In land-use terms Wissink (1962) called it an area of 'great differentiation', while Golledge (1960) used the term 'a geographical no-man's land'. Both these expressions are derived from the wide variety of uses found in an area which has been only partly brought into the urban complex. The city does not grow outwards in well-defined advancing rings of rapidly completed development. It extends haphazardly, making rapid advances at one point and hardly moving at all at another. It is this process which occasions the incoherent land pattern which is taken as representative of the fringe. This is not something associated particularly and uniquely

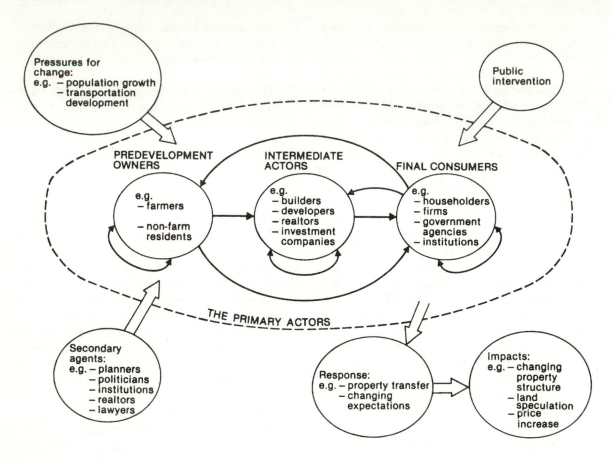

FIGURE 15.2 *An interpretation of the land market in the urban fringe (after C. A. Bryant et al., 1982). Source: Fig. 4.1, p. 53.*

with present metropolitan growth. The fringe belts which Conzen and Whitehand have recognized as significant determinants of the morphology of urban settlements (see Chapter 19) represent the continuing influence of former fringes long after they have been encompassed by the advancing urban front. Schnore and Knights (1969) demonstrated how early a feature metropolitan expansion was in the United States.

To some extent research on the rural–urban fringe became less popular in the 1980s, largely because the extensive development of megalopolis brought comparable but somewhat different issues to the fore. As cities changed from mononuclear settlements into large-scale regional organizations, the specific character of

an encircling fringe lost a good deal of its meaning. Even so, it is worth pursuing since the impact of urban extension on the countryside still generates a set of distinctive problems.

At the city margins, therefore, in the context of the fringe, a wide mix of land-uses is characteristic, ranging from a variety of commercial developments, including out-of-town shopping centres, to the city services and industries which are conveniently located at the margins. Wehrwein (1942) described the fringe in the USA as an 'institutional desert' because of the uncontrolled location there of unpleasant and noxious establishments such as slaughterhouses, junkyards and wholesale oil storage, and of utilities such as sewage plants and cemeteries.

In a contemporary context the various land-uses, older villages, newer residential extensions, commerce, industry, city services and the underlying farming, are not neatly sorted out into homogeneous areas but are intermingled in random fashion and it is this which gives its distinctive quality to the land-use pattern of the rural–urban fringe.

An attempt has been made by Bryant *et al.* (1982) to summarize the forces operative and their outcomes in the rural–urban fringe in a diagram which is reproduced in Fig. 15.2. The responses indicate the dynamic impact of urbanization upon the fringe and the variety of conditions consequent. These reflect the nature of the fringe and demonstrate not only the push of young, mobile, middle class populations, but also the adaptation of farming, the use of fringe land for leisure pursuits, as well as the many demanders of extensive land holdings or objectionable uses which find a place at the city margins.

More recently other aspects of the fringe have become conspicuous. In the UK two of the most notable land-use modifications are the growth of a number of nurseries and garden centres selling plants and equipment to the newcomers, and the development of horse-riding schools and stables, catering especially for the young female members of the incoming middle and upper class families. A further use is as recreational land, for instance where the fathers of the horse-riding daughters can play golf! Similar developments can be identified both in the United States and Australia but perhaps the major feature in the USA has been the out-of-town shopping centre, which has already been discussed in Chapter 10.

Impact of urbanization

R. E. Pahl (1965) attempted to summarize the characteristics of the fringe under four main headings:

1. **Segregation**. The ability to pay for the new housing of the fringe results in a pattern of segregation appearing. Giggs in a study of Nottingham showed that not only are there basic differences between the north and west, which was built up and industrialized in the nineteenth century and is suburb rather than fringe, and the south and east, which is still underdeveloped and can be considered part of the fringe, but that within the south and east clear differences can be identified. A cluster analysis of 78 parishes using 15 variables was carried out, the variables representing population growth, age structure, housing characteristics, socio-economic status, employment, workplace and mode of travel, mobility and dependency. Significant differences between the parishes were revealed and four sets were identified. Giggs recognizes these as: Set 1, modern mining settlements; Set 2, large residential suburbs; Set 3, small residential suburbs; Set 4, small villages, i.e. unmodified villages.

He concludes, 'the major residential tracts of the suburbs – old villages, private housing, council estates – all are clearly segregated, with the private estates generally located closer to the old village cores. Large planned tracts are given to commercial, educational and institutional uses' (Giggs, 1970). A characteristic variety of settlement form is accompanied by segregation of population by 'class'. To such an extent is this developed that many parts of the fringe become status symbols in a residential context, they are the places in which to live.

2. **Selective immigration**. The rural fringe will attract in particular 'mobile middle-class commuters who tend to live and work in distinct and separate social and economic worlds from the established populations' (Pahl, 1965). Pahl argued the case for the 'metropolitan village'. This is, in other words, the commuter village, where the mobile middle class builds a highly dispersed pattern of activities based not on a place, but on the region. To a large extent their choice of residence is associated with class and life-cycle stage. But Pahl adds two

other features which influence the way of life. These are the necessary associations with other lifestyles which occur in small and heterogeneous settlements, and the relationships developed in the general social organization of the village, an involvement and an interaction. 'The sociologically most significant feature of this settlement type is the interaction of status groups which have been determined nationally – by the educational system, the industrial system and so on – in a small scale situation' (Pahl, 1968: 276).

3. **Commuting**. This follows from the previous point and needs little comment except to note that it is not confined to the more wealthy, but the availability and cost of transport necessarily confine the less well-off.

4. **Collapse of geographical and social hierarchies**. This is one of the most interesting of Pahl's conclusions and advances the concept of a distinctive fringe. With the populations partly directed towards other parts of the city for certain services, then the service content of fringe settlements becomes modified. They do not need to carry an array of goods and services commensurate with the population they serve, but can become specialized in particular directions. It is possible that conventional central place ideas would not apply in direct fashion and that something akin to the dispersed city is appearing in the fringe. Instead of rounded bundles of functions at particular hierarchical levels collected at appropriate nodes, the various functions are being dispersed in several nodes in specialized or segregated bundles, the whole process being precipitated by the mobility of the population. This is the basis of the out-of-town shopping centres noted above. Likewise the segregation of incoming groups, with their links back to the city, undermines the traditional social hierarchies of the rural areas.

Pahl summarizes the above points by maintaining that a new population is invading local communities bringing in national values and class consciousness at the same time that a new type of community, associated with dispersed living is emerging. Here, however, the argument has moved far away from the concept of a fringe as a physical area to one associated with particular social processes. One view considers the rural–urban fringe as identified by static features, a mix of land-uses brought about by the incomplete extension of the city as well as the demands which it makes on its marginal areas. The other view sees the fringe as showing distinction in the nature of the communities which occupy it, brought about by the migration of mobile middle class families orientated to the city and dominated by urban lifestyles.

The contrast between what is 'local' and what is 'national' may well be a better way of looking at the contrasts between what are usually termed 'rural' and 'urban' although the impact of the modern mass media, particularly television, make these substitute terms hardly more appropriate. It is the direction of orientation that matters and one feels that terms such as 'inward orientation' and 'outward orientation' are preferable to the rather meaningless 'local' and 'national' dichotomy.

Indeed the most useful concept to introduce is that of Melvin Webber's 'non-place urban realm' – 'an urban realm is neither urban settlement nor territory, but heterogeneous groups of people communicating with each other through space' (Webber, 1964). Everyone participates in different realms and shifts from one to another. 'This is, of course, especially true of the highly specialized man who may turn from a transatlantic phone call to arbitrate an intra-office personal problem, then read his mail from customers in various places, and then join other motorists in the peak-hour rush before reassuming his role as parent, newspaper reader, and member of a friendship circle.' Webber goes on to assert that no urban settlement is a unitary place but a part of an array of shifting and interpenetrating realm spaces. In this interpretation the conflict in the rural–urban fringe is between the limited realm participation of the rural or local population and the large range of realms in which the immigrant participates. Nevertheless the intensely 'local' or 'inward orientation' or limited realm participation can be found lodged in the city centre, in the urban village or in the mining village, 'the "national" outlook or outward orientation' and above all, wide-ranging realm participation can be found in the semi-rurality of the city fringe. Even so this does not destroy the concept of the fringe areas as those where the conflict of orientation is particularly acute.

Impact on agricultural land

Agriculture in the fringe becomes inevitably influenced by urban penetration, although the most controversial issue in smaller, highly industrialized countries is related to the loss of productive farmland. The purchase of land for development leads to both smaller units and fragmentation of holdings. Where 'betterment' is not exacted for the benefit of the community, speculation by the purchaser of fringe land in anticipation of development can create a deterioration in agricultural standards. Also where the 'urban fence' is ill-defined or not well defended, vandalism from the encroaching housing estates can create problems, for such antisocial activities are not restricted to the inner city. Again the advance of the actual urban frontier spreads the wave of 'urban retreaters' and the rather large scale 'hobby farmers' who seize opportunities to live in a rural environment and 'work' on the land.

The impact upon agriculture itself is of less significance in an urban geography. The most extensive study is by Munton in a book devoted to London's green belt, referred to as the Metropolitan Green Belt or MGB. He concludes that because of the differences in the farming pattern within the MGB, 'there is as a result no clearly defined urban fringe type of farming' (Munton, 1983: 77). He quotes Thompson's (1981) conclusion in support:

For while he felt able to conclude that farms close to the urban edge were small, have a low proportion of cropped land, a high proportion

of permanent grass, rough grazings, land in non-agricultural uses, and low stocking rates – probably the effect of the census failing to record the number of horses kept on farms – all of which might be seen as 'urban' farmng features, he went on to note the variations in land quality in determining the spatial pattern of farming activity around London.

(Munton, 1983: 77)

In his own analysis Munton confirms the presence of substantial tracts of poorly maintained, if not derelict, farmland. He ascribes this to the short-term letting of land by companies, principally developers and gravel companies, hobby farmers and occupiers who keep horses in substantial numbers although reliant on farming for the greater part of income. An urban fringe type of farming there might well not be, environmental variations engender variation in farming, but there is no doubt that fringe conditions do substantially modify agriculture.

Munton's (1983) book, referred to above, is an intensive study of London's green belt. Its review of both conforming and non-conforming land-uses provides an admirable account of the tension between the demands to maintain the MGB primarily as agricultural and open land, and the pressures to initiate development.

15.5 CONCLUSION

Writing as long ago as 1967, Harold Mayer identifies problems of competition for land and preservation of open space as the two most important areas of research interest in the rural–urban fringe (Mayer, 1967). Ten years later a conference on research on the urban fringe organized by the British Countryside Commission (Phillips and Veal, 1979) revolved around five topics:

1. Agriculture and the urban fringe.
2. Land management and development pressures.
3. Recreation in the urban fringe.

4. Land-use relationships and conflicts.
5. Interaction between policies in the urban fringe.

All these are more detailed presentations of the basic two problems identified by Mayer. Agriculture through the fragmentation of holdings and an unwillingness to undertake long-term investment takes on a distinctive character. Speculative buying of land in antici-pation of development creates a complex land market, difficult to monitor. Recreational demands increase, not only from the affluent but from the urban poor and deprived. All these interact, creating conflicts in which some sort of mediating influences are sought. The manner in which a city expands at its fringes, the process by which land is taken into the city, the mecha-nisms of the various decision processes all these are critical in the interpretation of the urban fringe which, indeed, has a claim to be the real transition zone of the city.

IMAGES OF THE CITY: CITY LIVING AND USE OF CITY SPACE

16.1 INTRODUCTION: APPROACHES TO PERCEPTION STUDIES

Over most of this book it has been implicitly accepted that the city can be treated as a natural object, a phenomenon in space which is perfectly perceived and perfectly comprehended by all those who form part of it or establish relations with it. This view has been modified at two points. The first was in the consideration of consumer behaviour in relation to the central place system where it became apparent that, although on an aggregate scale such a system was discernible and could be interpreted as the product of the activities of shoppers and entrepreneurs over time, on a disaggregated scale consumer behaviour was by no means in accord with the principles derived from the aggregated analyses. In the resolution of this apparent conflict a more sensitive approach was found to be essential. Likewise in looking at residential patterns it was evident that, although distinctive social areas could be identified, any family's decision-making process on where to live was extremely complex and derived from the way in which the members assessed their needs in relation to their knowledge of the city.

Thus it can be asserted that for the citizen the objective city does not exist. Every citizen has most certainly a partial, and most probably an idiosyncratic view of the urban environment in which he or she lives. Moreover, gender as represented by the 'he' or 'she' will produce different reactions to city space. Again, the view from the drawing board, or the view from the aeroplane, is not the view of the citizen on the ground and in the street, for there the prospect is limited and the scene partial.

A strongly growing element of urban geography has focused on the establishing and analysis of images of the city and in the geographer's particular line of interest, the development of mental (cognitive) maps which translate the images into a spatial framework. These mental maps, or images, of the urban environment can then be examined in relation to the characteristics of the people concerned. There still remains, of course, the problem that these characteristics are ultimately associated with psychological or personality types. With Golledge it is possible to view consumers as Marshallians who behave as economic man, as Pavlovians who develop repetitive patterns of behaviour via a learning process, as Freudians who are fantasizers and react capriciously to stimuli such as advertisements, and as Veblenians who follow the example set by their peers (Golledge, 1970). But in spite of the difficulties introduced by personality characteristics it is still possible to consider images of the city in relation to the standard variables which have been consistently used in social geography – socio-economic status, sex and age for example.

Downs has proposed three types of approach characteristic of work on geographic space perception as follows:

1. The structural approach which is concerned with the way in which the array of information about a place is perceived. It is evident that all the sense perceptions, all the impinging data, about an environment cannot be remembered. There is, therefore, a process of selection and ordering, a structuring which has to take place. On this basis it is postulated that mental maps of a city are constructed in which useless detail is discarded and the data necessary for the purposes of the individual, such as finding the way from A to B, are retained.
2. The evaluation approach goes somewhat further in that it is not only concerned with the way the environment is structured but also how it is evaluated in relation to decisions to be made and subsequent action to be taken.
3. The preference approach is directed towards the way in which preferences are developed among a set of objects distributed in space.

Evaluation and the identification of preferences are somewhat similar and they have been in part considered in relation to consumer behaviour in Chapter 5 and to location decisions as to residence in Chapter 12. This chapter will, therefore, concentrate initially on the structural approach and concern itself mainly with the physical and visible character of the townscape.

16.2 A TYPOLOGY OF PERCEPTION

Before considering more precisely how images of the townscape can be put into some structured array it is useful to consider a typology of urban perception which identifies the way such images are derived. Appleyard (1973) proposed a threefold classification.

Operational perception

People become aware of many elements in the city because they use them as reference points in their everyday life in getting about in the city or simply getting to work. Bus stops or traffic intersections, key buildings or distinctive physical features are noted, remembered and probably in turn exaggerated in importance, in the process of creating a working mental structure of the city. This is undoubtedly the most universal of the types of perception, derived as it is from repetitive activity. At the same time in this process some parts of the city will remain virtually unseen, certainly unremembered, and perceptually invisible as the individual journeys from distinctive feature to distinctive feature. For example, only parts of buildings may be remembered. A shop may often be recalled in terms of its ground-floor identity and the window it displays to the public while the architectural character of the whole building may remain completely unknown.

Responsive perception

Awareness of city images will be closely related to distinctive or unusual features which generate an immediate response: 'bright, isolated, singular and distinctive elements intrude on the operational search patterns of the traveller or catch the eye of a gazing passenger'. The role of the advertising industry is to elicit this sort of response by the use of distinctive designs or colours, by employing sexual imagery or establishing a conditioned reflex to easily remembered, catch phrases.

Inferential perception

This awareness relies on past experience and is realized by inference from parallel situations. An unknown city is likely to conform in much of its general character to that of a large number of known cities. From experience it is possible to find one's way about in a strange city by using accumulated knowledge of the structuring of urban areas in general.

Given the *types* of perception the major academic problem was the devising of some analytical system which reasonably represented the way in which city images, the perceptions themselves, are structured. This was solved, at least partially, by Kevin Lynch in his book published in 1960 and appropriately entitled, *The image of the city*. Lynch proposed five elements through which the physical structure of the city was visualized.

1. **Paths** are the channels along which people move within the city and, as a result, they tend to predominate in urban imagery since movement usually promotes observation. The characteristic problem of anyone in a city is how to get from A to B and hence the path becomes the dominant remembered feature.
2. **Edges** are linear elements which represent distinctive physical breaks within the city. They can be natural features, such as breaks of slope, sea or lake shorelines, or river fronts. They can also be man-made features, such as the distinctive breaks brought about by railways or urban motorways.
3. **Districts** are sections of most cities immediately identifiable to the inhabitants and usually have local names. The most universally known are associated with distinctive

activities, cultures, or groups such as Soho in London, Montmartre in Paris or Harlem in New York.

4. **Nodes**: certain points in the city, most often road junctions, stand out as nodes or foci. The citizen can enter into or pass through these nodes and they represent easily identified stages in movement within the city. Often they are clearly demarcated physical elements, as in the city square: Piccadilly Circus, Times Square, Red Square, the Etoile are all nodes.
5. **Landmarks** differ from nodes in that they are features which can be observed but not normally entered or passed through. The name itself indicates the role these have played to the navigator at sea in its earliest meaning and to the stranger in the city at the present. Distinctive buildings or features stand out as being easily identifiable and easily retained as images to be used in structuring the mental map of the city. The landmark can be used to epitomize the whole city. When a tired film or TV producer wants to indicate that his action is taking place in Paris then the opening sequence of shots is the Eiffel Tower, if it is in London then Big Ben appears and if it is in New York it is either the Statue of Liberty or the Manhattan skyline.

It is evident that few individuals consciously use these five elements as such, and that they constitute together an arbitrary system for analysing the image of the city held by people or groups of people. The elements will also have different meanings for different people, or even for the same people at different times: an urban motorway is a path to a motorist but an edge to a pedestrian. Lynch compares two maps, one showing an outline map of Boston, and the other 'the visual form of Boston as seen in the field' (Lynch, 1960: 18–19) but although derived from field reconnaissance this must be described as an urban analyst's view of Boston for in this sort of study there are no absolutes, only images. Lynch's major contribution was to provide the simple but basic tools by which the way people structured their image of the city could be realized and the perceptions of different groups compared.

Lynch proceeded by means of office and street interviews to build up a number of Boston images. But perhaps the critical concept is the notion of 'different groups' and the contrasted way in which they perceive city space. The problem of the nature of such groups immediately follows and it would seem sensible to relate them to the dimensions which in factorial ecological terms differentiate the urban population. A large number of individual and particularist views can be derived, but will add little to the purpose of this investigation. The psychologist might be interested in how the city appears to the sufferer from claustrophobia but such abnormalities are rarely the geographer's immediate concern (see Section 16.6). The dimensions of ecological studies imply that the image of the city will vary according to socio-economic status, life cycle stage and ethnic origins.

16.3 CITY IMAGE AND SOCIO-ECONOMIC STATUS

Francescato and Mebane have reported in a preliminary way their attempt to consider images of Rome and Milan. Their respondents were broken down into groups by status, age, sex and whether or not they were natives of the respective cities (Francescato and Mebane, 1973). Figure 16.1 reproduces the images of Rome of the middle class (Fig. 16.1(A)) and lower class (Fig. 16.1(B)) subjects using Lynch's analytical techniques. Social class was determined by using occupation and education as indices: 33 respondents were ascribed to the middle class and 27 to the lower class. The contrasts are immediate and obvious.

The middle class group identifies a much larger number of elements over a much wider spatial extent. The reasons seem straightforward; greater wealth provides greater mobility, a wider range of cosmopolitan interests and a greater propensity to use all the widespread resources of the city. Poorer people are less

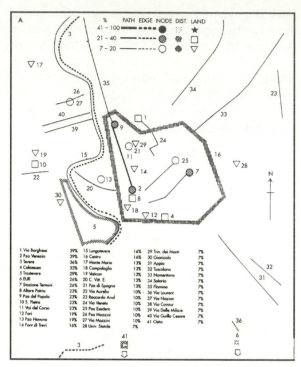

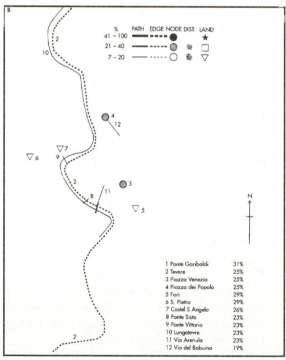

FIGURE 16.1 *A: Images of Rome: the middle class image. B: Images of Rome: the lower class image (after Francescato and Mebane, 1975).*

mobile, are more likely to have a shorter journey to work, will not use the range of city resources and so will be less exposed to city space. The authors suggest an alternative explanation. They suggest that the tendency for the lower class to produce maps covering only very small areas but indicating a good deal of local detail, reflects the home orientation which is characteristic of poor Roman families. This is undoubtedly a relevant feature, but such localism is in itself an aspect of poverty, although certainly a stress on home life and the neighbourhood exacerbates the limitations following directly from lack of wealth.

16.4 CITY IMAGE AND ETHNICITY

One of the most frequently quoted examples of an imagery of the urban environment is the pilot study by the Advance Planning Section of the Los Angeles City Planning Commission (1971). In this 25 respondents were asked to draw maps of Los Angeles, from which a general representation for each of the groups was produced. Three of these composite city image maps are reproduced here (Fig. 16.2). Figure 16.2(A) is the city image of the sample from Avalon, a delightfully wry name for a predominantly black area in the south of the metropolitan area and a little to the north of the better known Watts. Figure 16.2(B) is that of the sample from Boyle Heights, an almost exclusively Spanish-speaking tract adjacent to the central industrial and commercial areas. The third image of Los Angeles is shown in Fig. 16.2(C) and represents the sample from Westwood, a white upper class neighbourhood, located on the southern fringes of the northern hills between Beverley Hills and Santa Monica.

These three maps require but little exegesis. The white upper class image is extensive in reach and detailed in content; that of the Avalon residents is limited in reach and has less city-wide detail: only a low proportion of the respondents are even aware of the universally known residential areas such as Hollywood. It is also noteworthy that the white image is dominated by the east–west grain along the well-known boulevards, such as Wilshire and Santa Monica, which skirt the Hollywood Hills and Santa Monica Mountain. The black Avalon image, in contrast, is dominated by the north–south grid of the plain and related to access to the CBD. Finally the Boyle Heights composite image is most severely restricted and is limited to what is little more than the immediate neighbourhood.

Although these three images are related to ethnic and segregated groups it is not possible to sustain this argument without qualification. Socio-economic status, or access to mobility via wealth, must be invoked also. Nevertheless, these results lead on to notions of ethnically based territoriality, and most certainly participation in the whole city-wide range of activity is related to assimilation into the larger urban community. The ghetto has been considered in its spatial context in an earlier chapter but now it can be clearly seen as a constricting feature.

Gerald Suttles (1968) in his book *The social order of the slum* subtitled 'Ethnicity and territory in the inner city', develops the way in which restriction grows:

For persons in the Addams area [his name for that part of Chicago with which the book is concerned] only the adjacent neighbourhoods are well defined. Beyond this, their notions of established boundaries become vague and uncertain. ... Each little section [of the city] is taken to be a self-sufficient world where residents carry on almost all their legitimate pursuits. A person who leaves his own area, then, is suspect so long as he has no visible and justifiable reason for straying from his home grounds.

... Individuals in the Addams area achieve a positive association with co-residents of the same age, sex and ethnicity, primarily because conflict with other persons forces them together into small face-to-face groupings ... so positive a role for conflict cannot be appreciated unless it is placed in a developmental sequence. At the outset parents ... do not prescribe a definite set of persons ... [but] voice a variety of proscriptions: 'Don't go out of the neighbourhood'. ...

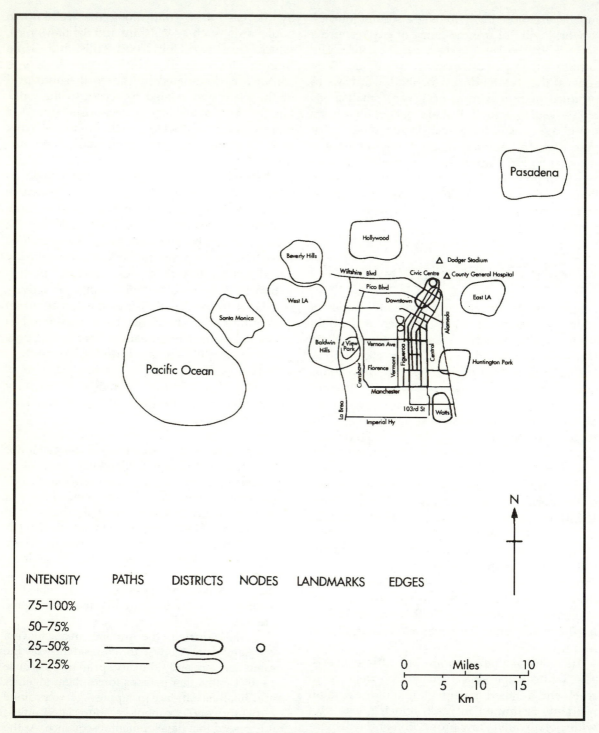

FIGURE 16.2A   *Composite images of Los Angeles (after Los Angeles Department of City Planning, 1971). A: Avalon.*

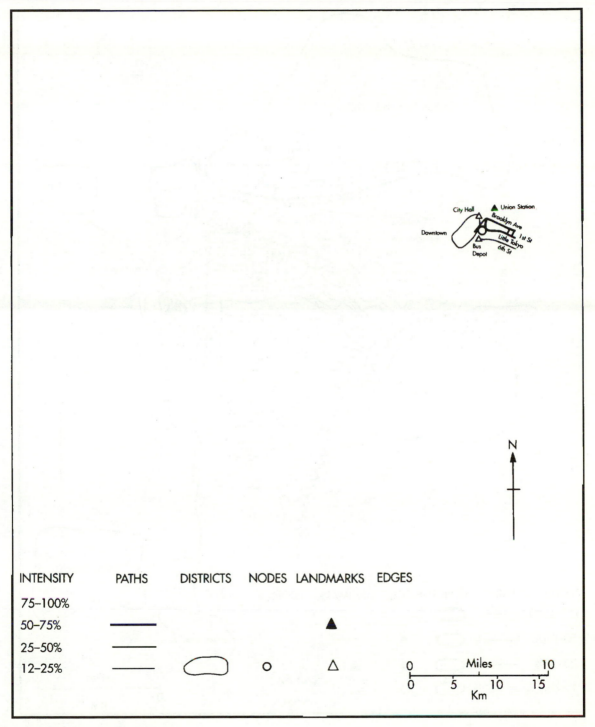

FIGURE 16.2B *Boyle Heights.*

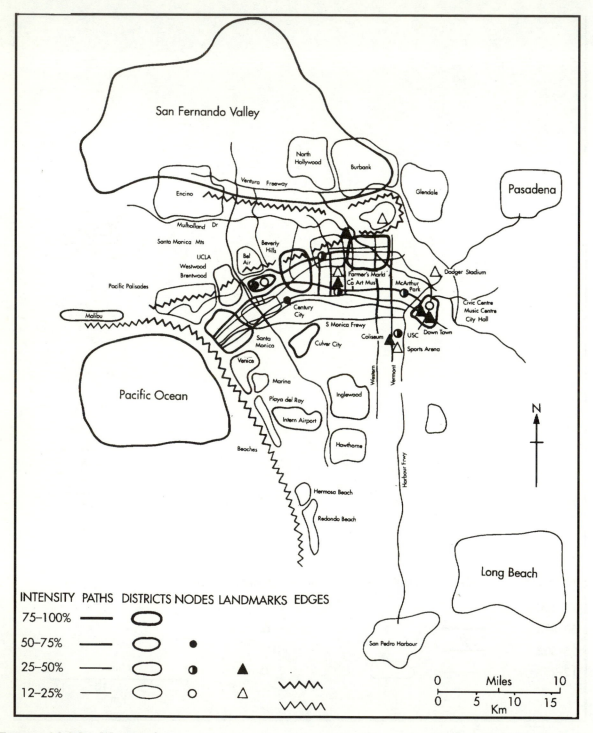

FIGURE 16.2C *Westwood.*

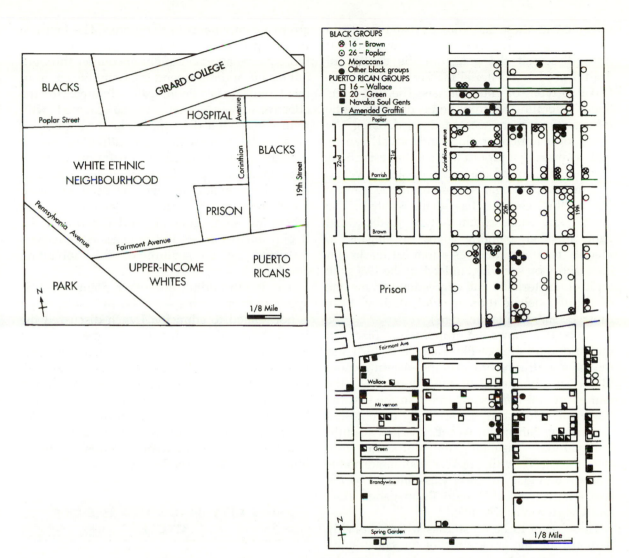

FIGURE 16.3   *A: Philadelphia, Social groups in Fairmount and vicinity. B: Gang graffiti in the eastern part of Fairmount and vicinity. The street names, and the locale of the prison, can be used to relate the areas covered in the two maps (after D. Ley and R. Cybriwsky, 1974). Source: D. Ley and R. Cybriwsky, 1974: Urban graffiti as territorial markers. Annals of the Association of American Geographers, 64(4), p. 497, Figures 5 and 6.*

'Don't you get off the block'.... Injunctions of this sort do not initially produce positive associations but only territorial aggregates.

(Suttles, 1968: 15, 228)

In this detailed study Suttles outlines the processes by which territorial restrictions emerge and it is these which are eventually revealed in the severely restricted images of the total city which many ethnic minorities display.

The clearest manifestation of a restricted image of the city, or perception of city space, is associated with the defence of the territories known as the 'turfs' of street gangs. Some reference to these has already been made in Chapter 14 (p. 293). Ley and Cybriwsky (1974) have discussed the gangs in the context of graffiti used as territorial markers. They identified two types of graffiti artists. The first is the 'loner' operating far from home, often on exotic targets. One even succeeded in daubing the wings of a jumbo jet. Such individuals can be seen as making a protest against the restrictions which hedge round the ghetto or slum dweller, a manifest sally against the constrained image which this chapter has demonstrated. The second and more common type is the street gang member painting logos or slogans to identify territory, much as a tom cat marks out its territory by spraying. Indeed, in the 1960s the topic of territoriality and aggression became one of popular debate (Lorenz, 1961; Ardrey, 1966; Morris, 1967; Montague, 1968).

Figure 16.3 shows the proportion of Blacks in the Fairmount area of Philadelphia in 1970 and the gang graffiti recorded in the eastern section. The clear dominance of the 'Moroccans' (the names are not real) is apparent. 'In the east the powerful Moroccan gang had uncontested supremacy; Fairmount Avenue marked a southern boundary insulating the Moroccans from close contact with 16-W. The amorphous western boundary of 16-W was confused by the two relatively weak Puerto Rican gangs' (Ley and Cybriwsky, 1974: 499).

There seem to be no studies which explicitly demonstrate the association of the constrained image of the city on the one hand with the manifest identification of a similar territory by street gangs. Even so, the implicit linkage is a significant indicator of deprivation through a limited capability of exploiting the whole resources of the city.

The notion of the restriction of the individual's use of city space introduces the ideas contained in Hagerstand's 'time geography' model of society (Hagerstrand, 1970). Most human activities are space consuming but 'while the space packing of urbanization allows time-savings for both individual and society it also creates inter-regional and intra-regional social and economic inequalities ...' (Pred, 1973). There

are two components in the model. The first is a life-perspective environment which takes in all those places a person might encounter throughout his life and the second is a daily-life environment which is theoretically limited to the effective distance that can be covered with return, in a single day, rather like the bounds set on a medieval market. Every individual can be thought of as contained by constraints within a time–space envelope. Some of these constraints are directly biological, like the need for sleep which restricts the reach of daily travel, but others are socio-economic and determined by deployable resources. In this sense city-image is restricted by the nature of this inhibiting envelope.

It is worth adding one other point which is apparent from the maps in Fig. 16.2, although it is not related to ethnicity. Lynch discussed the imageability of the city, that is, the extent to which it threw up clear and distinctive images. Examination of the Los Angeles maps will indicate that they are dominated by paths and districts and that there is a lack of nodes and landmarks. This tends to confirm the popular impression of the city as dominated by the automobile, as a mass of suburbs (districts) linked by freeways (paths).

16.5 CITY IMAGE AND THE LIFE CYCLE

Two aspects of this relation arise:

1. **The development of spatial cognition.** This sort of study is concerned with the way in which spatial cognition develops in the child from birth. This book is most certainly not the proper place to include a review of work on this topic for it is a highly specialized field. It is included here, as a heading, since it is a part of the life-cycle relation to images. Perhaps the dismissal of studies of the development of spatial cognition as a specialized field too easily eliminates studies of the city image among children, 'a group whose very numbers and unique

characteristics cry out for more attention than geographers have given them in the past' (Hill and Michelson, 1981: 224). Even so, in spite of the development of the theme by Hill and Michelson (1981) and by Piche (1981) there has been but little development of the field within urban geography. It is perhaps noteworthy that none of the quoted authors is a professional geographer.

2. **City image and age groups.** Different age groups, different life-cycle stages will be concerned with different aspects of the city. One of the best known studies in Britain was carried out by Brian Goodey *et al.* in Birmingham (1971). A rather unusual means of collecting data was used: readers of the *Birmingham Post* were asked to send in spontaneously drawn maps of the central area of the city. Goodey reported that 'we found evidence for youth, housewives' and old people's maps of Birmingham and for the two latter the maze of recently opened city underpasses emerged as very severe blockages to mobility and urban navigation' (Goodey *et al.*, 1971: 70)

It would be wrong to give the impression that the three dimensions which have been discussed are the only determinants of city images, or indeed that only the visible elements that can be translated into Lynch's terminology are involved. In Chapter 9 the central business district was considered as an objective area of the city and appropriate methods of delimitation sought. Even so it was noted that the earliest attempts were based upon local opinion; thus through image studies the wheel turns full circle and geographers are once more concerned with local views, for if consumers behave in a way not in keeping with the dictates of economic man then it is likely that views of the CBD will also vary.

This was the subject of an investigation by Klein who attempted to establish varying definitions of the town centre by contrasted groups (Klein, 1967). A random sample of subjects, 1118 in all, was selected from a number of clearly defined residential areas. These respondents were then given a set of 24 photographs of the city centre and were asked to divide them into three categories, 'town centre', 'not town centre', 'unknown'. Analysis of the results was then undertaken in relation to a number of variables characterizing respondents, including socio-economic status, age, sex, length of residence and location of residence. This was done by calculating, for each of the photographs, the percentage of the total sample which judged it to be of the 'town centre' and then calculating the deviation from the average for each of the subgroups determined by the variables. Two of the subgroup analyses can be considered briefly.

Figure 16.4(A) represents a west–east profile across the city identified by five of the photographs. The responses of three groups are examined in relation to these five points, those who lived to the west of the town centre, those who lived to the east and those who lived in the inner city. These groups are represented by graphs which indicate the deviation from the average in percentages. This figure demonstrates that those who live in the west exaggerate the town centre by extending it in a westerly direction, while those who live in the east do the same in an easterly direction. Those who live in the inner city overestimate the central point and underestimate the extensions both to east and west and, although there are no statistical tests to match place of residence against the other variables, it seems that this locational bias is independent of them.

Klein suggests that this result may be due to the fact that people tend to use that part of the central area nearest to them, or at least to be more familiar with it. But he also adds that it may be wishful thinking and a determination to believe that the city centre is nearer than it actually is, thus introducing another problem in this field, the notion of subjective distance.

The second example from Klein's study is shown in Fig. 16.4(B) where seven points along the north–south profile are related to length of residence in the city. Those who have been in the city the shortest time, like the younger age group, have a much more restricted view of the town centre. Especially remarkable is the

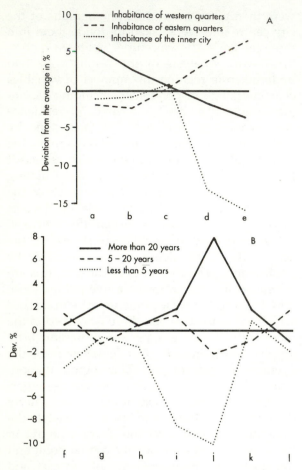

FIGURE 16.4 *Karlsruhe: the town centre as identified by residents (after Klein, 1967). A: West–East profile in relation to place of residence. B: North–South profile in relation to length of residence. In these two cross-sections the points labelled a to l represent the points from which photographs were used in the study. a to e are five places ranged from west to east, f to l are seven places ranged from north to south. Note that in the period before the study the city centre of Karlsruhe had tended to shift westward. This is clearly reflected in the responses of the inner city inhabitants to graph A, and by the newer residents in graph B.*

disagreement over point 'j', the Festplatz. This is a point on a southern extension of the centre dominated by cultural elements and a traditionally distinctive feature of the city. Those who

have lived in the city most of their lives insist that it is part of the town centre whereas the newcomers have no such sentiment. Clearly there is great complexity in the image the citizen builds up, for not only is it related to standard social variables, but to location, length of residence and to all those personal quirks which belong to the individual. Beyond the objective city is the city which people use, controlled by the images they hold for 'behaviour depends on the image'.

16.6 GENDER AND THE CITY

The phrase 'the city which people use' immediately triggers a reaction to the limited bases of socio-economic status, ethnic identity and age which, derived from standard factorial ecologies, have been, up to this point, the bases for the analysis of city image and perceptions of city space. But it is characteristic of post-modernism and its eclecticism that the derivatives from traditional aggregate studies are no longer sufficient. An awareness of a whole series of other groups has arisen.

There is a problem, of course, as to how far geographical investigation should be spread over the infinity of different groups within the population. Thus, to the blind the city presents a very different environment from that of the sighted. That this is a crucial matter of investigation for those concerned with the training of the blind is without question. But the extent to which standard urban geography can follow such an issue is another matter. It has certainly been a topic for research and for publication in geographical journals. However, of an extensive reference list in a paper by Golledge and collaborators (1993) only one-third are to geographical papers and half of those are by Golledge and collaborators themselves. Others of a geographical nature are to general deprivation and the remainder are to specialized journals. Equally important is the perception and use of the city by children, particularly the identification of the way in which spatial cognition develops which was introduced in Section 16.2. But while in no way diminishing the status and significance of such studies, they are perhaps better regarded as specialist issues which any urban study will recognize but which are better left as specialisms to be followed by those who can acquire the necessary background in environmental and behavioural psychology. The major study is of the groups themselves, the city is but one of a large number of relevant environments.

There is one section of the population, indeed forming half of it, however, which has not appeared in any direct way in this book. At the outset of this chapter it was stated that it was possible, indeed necessary, to consider images of the city in relation to the standard variables of social ecologies – socio-economic status, sex and age were given as examples. But gender has been ignored. To develop a more specific feminist theme, deconstruction of this book would reveal it to have been written by a politically centralist male. A Marxist female would have produced a very different volume. But, significantly, although an inordinate amount of 'theory' and generality has been generated, a feminine orientated urban geography, to the present author's knowledge, has yet to appear. This author clearly cannot attempt such a volume, but only speculate how different it might be. There are, of course, volumes which deal with women and the city, such as *Women in the city* (Little *et al.*, 1988). But although that book deals with 'gender and the urban environment' it is a series of separate studies rather than a coherent urban geography.

Aside from such consideration of a 'total' urban geography, however, it is manifest that the female view of the city will be very different. It is, perhaps, insulting to treat the female and the city as a subsection in a chapter on the city's image, but again this author can do little else. If all the chapters to this point, and after this point, have assumed a male or androgynous perspective, one can only recognize and acknowledge it. The real problem is that no great corpus of work exists on which to draw: it is 'questionable whether the fundamental challenge posed by feminist urban research has yet had any effect on the dominant mode of thought' (McDowell, 1985: 10).

That the female view of city space should be different must depend on two factors. The first is role, specifically where that role as housewife and mother, rightly or wrongly, is tangibly different from that of the male. The second is directly biological and is the difference in physical strength, so that, for women, city violence assumes a quite distinctive aspect.

The first of these two aspects, that of role difference, has declined over recent decades, although the decline may have brought more problems. The old stereotype of man the breadwinner and woman the homemaker has become outmoded. The creator of that older situation, setting aside the remote past where women were part of the husband's goods and chattels, was the divorce of place of work from place of residence at a time when a combination of improving medical skills and the absence of effective means of contraception produced large families. Before that time women were by propinquity bound up with the daily process of earning a living and with urban ceremony. Again, the factory system took 'making' out of the home, and the male away from it, so that, even if for the working class the divorce of living from working was short in distance, the effect was the same as for the urban bourgeoisie. The impact of nineteenth-century industrialism, therefore, in different ways according to class, was to maroon women in suburbia or isolate them in working class terraces. The response was to create new lifestyles. In the suburbs in the late nineteenth century among the upper classes an elaborate ritual of card leaving and entertaining developed. In less pretentious times, with a loss of servants, the coming of domestic labour-saving appliances and of smaller families, other rituals were developed. Gardens were tended with care, charities were supported and a variety of societies established, mainly concerned with domestic matters. Shopping had become a leisure pursuit as much as a necessary chore. The pre-industrial system where luxury goods were brought to the house and everyday goods were bought in the market by servants disappeared. The new shopping centres were attractions in themselves so that awareness of city space included the CBD, but it was an awareness of only one aspect of it.

The rise of female employment, changes in family structure, and the feminist movement, have transformed the situation. To some extent it would be easier to have written a feminist geography of the city for the period between the two world wars when male and female roles were so contrasted. It would have been possible to present two different city images. Now, few, if any, occupations are not open to women, although the occupational structures of male and female remain contrasted. Even so, in principle, if roles are no longer accepted as different, then neither is perception and use of the city. Mobility, especially with the coming of the two car family, is no longer a limitation. But such conclusions are still unreal. Much female employment is part time and low paid and the reconciliation of wage earning with child rearing and home making is extremely difficult. It certainly engenders different patterns of movement which are neglected in most studies of transport in the city. There remains also an undoubted residual isolation, for the need to look after very young children inhibits mobility. And work patterns and limited movement are both reflected in a contrasted image of the city and use of its space.

If the contrasted roles of female and male have been somewhat diminished in recent times, then physical disadvantage has moved in the opposite direction. It is well expressed in Valentine's paper, *The geography of women's fear*. 'It is well established in the sociology and criminology literature of Western Europe that women are the gender more fearful of crime and that this is related to women's sense of physical vulnerability to men, particularly to rape and sexual murder' (Valentine, 1989: 385). The geographic aspect of that fear is the relation of public space and its nature to safety. 'Women develop individual mental maps of places where they fear assault as a product of their past experience of space and secondary information. In particular girls are socialized into a restricted use of public space...' (Valentine, 1989: 386).

Perhaps it must be added that there is also a space–time aspect; the safety and security of place is not an absolute, but will vary with the time of day, those accessible by day are not to be used at night.

Those areas of the city to be avoided are generally, though not always, self-evident. The term 'street wise', although it has other connotations, is an everyday representation of an aspect of city image. The two types of area which are most threatening by virtue of the past crime record and self-evident exposure to risk, are the less frequented wide open spaces and those where isolation is accompanied by the absence of an escape route, such as alleyways or even multistorey car parks. But space has a social character too. White women would tend to avoid predominantly black areas, while middle class white women would probably avoid local authority housing estates with a bad reputation:

> There is a vicious circle in operation. The majority of women still adopt a traditional gender role, and as a consequence are pressurised into a temporally segregated use of space. The subsequent control by men of public space in the evening means that despite the career success and independence gained by women in the past decade ... the fear of male violence deters the majority of women from being independent.
>
> (Valentine, 1989: 389)

Confinement to the home, if now not solely an aspect of role differentiation, is the product of crime and the fear of crime. The crucial geographical point is not the feminist one of loss of independence and of discrimination by sex as issues for action, but of the differing use of city space which they produce. The ultimate situation is the elderly (age) woman (sex) living by herself (the feminist anathema of male protection) who is afraid to move out at all, even in daylight when marauding unemployed youths are about (see p. 293). All the vast resources of the great city are reduced to a single room and to that person discussion of those resources and their locational controls must seem the ultimate irrelevance. To the women in the street, or at home, the 'objective' city of the textbook is an abstraction of no meaning because its space is not available. It is also a mistake to dwell at too great a length on the problem of physical violence alone for there is a deeper and more significant differentiation. More fundamentally 'cities are environments built by men and especially the public areas, are perceived as male spaces' (Holcomb, 1986: 449).

16.7 PRACTICAL AND ACADEMIC RELEVANCE

At this stage it is likely that the question will be put as to what use there is in these sorts of studies, for are they not just another passing geographical fad? It is possible, however, that these represent a most important area for at least a departure is made from the academic's city to that of the people who live and work in it. Four uses of city image studies can be suggested:

1. Attention is drawn to the imageability of the city. It is easy from the geographer's viewpoint to forget that the city is a built form that should give visual pleasure, 'voluptas' as the Renaissance theorists called the quality. Aesthetic pleasure must be related to the quality of the images the city creates and hence city image studies call attention to qualities too easily ignored in social science and passed over to the architect's care.

2. On a large scale the efficiency of a city depends on its imageability for the ease with which people can get about, with which motorists can find their way, is closely related to its legibility, that is, the ease with which distinctive images produce a sequence that can be followed. On a small scale these qualities become related to any single building, for social processes depend upon environmental organization:

The concepts of sociopetal space and behaviour watching may be applied to explicitly social areas as well as to lobbies or hallways. Every natural traffic pathway is a potential supplier of encounters which can become interactions. Therefore, the social space located at the far end of a hall, or even half way down the hall, from the entrance has part of its therapeutic function diluted.

(Lawton, 1974)

Again it is quite evident that though the geographer will have an interest in these studies since they are concerned with the organization of space, the prime mover in them will tend to be the architect or the psychologist.

3. The third way in which these studies are of value relates to the fact that they reveal the images of the city of the everyday user and therefore also indicate the existence of major problems. Reference has already been made to Goodey's (1971) Birmingham study, where the problem of housewives having to navigate baby carriages through an underpass system was revealed. City image therefore directs attention to the diminished possibility of the use of the total city in the West by all but White, well-off males. It is at this point that disadvantage is most obviously recognized.

4. The last area of relevance is one of critical importance to the cities of the twentieth century. Deprivation, and its identification by means of socio-economic indicators, is a fashionable academic study. To a great degree the emphasis here is structural and related to systems of social stratification. But it should also be locational and geographical and nowhere is this more clearly revealed than in the maps drawn by the poor and the ethnic minorities. It is true that social and economic disadvantage and locational disadvantage are part of each other, though not exclusively so.

These four uses of studies of the image of the city have practical applications in the field of planning. There remains the significance of such studies within academic geography. Here the major problem is one of circularity for image and behaviour interact and to varying degrees condition each other. Even so the key issue is not the construction of images as an exercise in itself, but the interpretation of behaviour through the images held by sets of people acting in reference to sets of opportunities. To a large extent these opportunity sets will relate to particular decisions – where to shop, where to live, where to work – and these can be considered in the appropriate contexts of consumer behaviour and residential location behaviour.

In the explanation of patterns and movements in the phenomenal world, the observed objective city, the geographer must look to those characteristics of a citizen's behaviour which play their part in structuring it. Perhaps the most significant point of all to be derived from the discussions in this chapter is that the cities are environments built by men, with the implication that there is a very different urban geography which still needs to be written (Warmsley, 1988).

17

THE SOCIALIST CITY

17.1 INTRODUCTION

The whole of the book, to this point, has been based on the principle of a free market. Within the city, individuals, corporations or institutions bid for land and the highest bidder succeeds. The land can be put to any use the successful bidder wishes but the imperatives of the market operate and to retain viability profits have to be maximized. Otherwise predators will take over, and the process will be repeated until the most profitable or 'perfect' use is established. That is the basis on which most models of urban land-use depend. Likewise, the same principle underpins central place theory. Competition between centres results in the fittest surviving in a process of locational selection; a hierarchy is the product, a class system.

It should be noted, however, that the competition takes place within the context of the set of rules which make up the legal system. There is a further extreme whereby land is acquired outside that legal system by force or by fraud. The forcing of families off land with some unrealized potential, for mineral extraction, water supply or for development of some sort, has long been a favourite theme for film makers. It was one of the stereotype 'western' plots.

Between the employment of violence and strictly legal procedures, lies a whole range of operations where fraud and corruption, that is ways around the legal constraints, can operate. These have been noted in Chapter 8. But setting aside these procedures without the law, the basic processes so far invoked are those of the free market with such limitations which are needed to constrain them, in the general interests of the community, which have been built into the legal system.

In contrast to such conditions can be set a socialist political system where profit and its maximization is made illegal, that is it is subject to such further constraints which put it outside the allowed procedures. The 'community' allocates the disposition of all resources: in pre-feminist terms – 'to each according to his needs, from each according to his ability'. The basic concept, therefore, in understanding the spatial results, is the process of allocation. At this point three problems arise:

1. Much of the very considerable criticism of the Western city which is represented in an enormous, and it must be said a repetitive, literature (Pahl, 1970; Harvey, 1973, 1985a and b; Castells, 1977, 1978; Harloe, 1977; Smith, 1980; Ravetz, 1980; Mingione, 1981;

Cox and Johnston, 1982; Badcock, 1984) was based on a critique of the reality of the Western, capitalist city with an implied contrast with an unspecified ideal which was socialist. But the ideal was never given any very clear form or structure. Thus the question as to the nature of a socialist system of cities, and how it would differ from those of the capitalist world, was barely asked, let alone answered with clarity and precision. Thus whereas it is possible to build models of a capitalist city on a free market basis, it apparently cannot be done for a socialist city based on socialist principles of allocation which structure space.

2. Where there was the possibility of critical analyses in the cities and city systems of the former USSR and Eastern Europe, the collapse of the political structures has made it outmoded. Even Havana is being influenced by market forces.

3. Even an examination of what has been created in socialist regimes cannot be carried out with the same detail as in capitalist systems for the data do not exist in the public domain. Further, it is never easy within authoritarian regimes to distinguish between what is reality and what is propaganda.

Because of these three problems it has been difficult to write effectively about the socialist city. There is neither an idealized model, nor examples in the real world which can be critically examined in detail. Even so, some attempt can be made.

17.2 SOCIALIST CITY SYSTEMS

It is more intellectually fashionable to write of the inherent contradictions of capitalism than those of socialism. But there is a fundamental contradiction which undermines the concept of a truly socialist city system. The basic point of departure for a socialist system must be that all goods, services and facilities, necessarily provided by the state, must be equally available and accessible to all. Thus either everyone must be within the same time and cost (convenience, since cost should not come in to the equation) range of the Bolshoi Ballet, or establishments equal in status to the Bolshoi Ballet should be so distributed across the country as to make accessibility equal. But given the friction of distance that is manifestly impossible. It is certainly possible, though hardly feasible in practicable terms, to envisage a system of equally sized cities equally spaced. But the two imperatives of threshold and range make it inevitable that equal access to higher level goods and services is impossible. In brief, a truly and properly socialist city system is ruled out by the tyranny of distance.

The way forwards, therefore, is a more effective control of the city system, and the way it develops, by central government. F. E. Ian Hamilton (1979) writing on Russian cities before the dissolution of the USSR noted four ways in which such control can be achieved, and a fifth can be added:

1. **Direct designation of administrative status**. At a time when administrative bureaucracy is rapidly increasing, and when other transactional functions are attracted to centres of government, then considerable control can be exercised by the location of administrative functions with a consequent significant impact on growth.

2. **Central decisions on economic policy and the provision of infrastructures**. The direction of industrial and other development, together with the necessary infrastructures, especially in road and rail provision, will have a marked effect on town growth. Even decisions on infrastructure, without formal specification of development, will have crucial consequences.

3. **Identification of tributary areas**. This will be primarily in the administrative field as centres are designated for specific functions and the areas to be controlled defined. But it will have clear consequences in the economic field.

4. **Laying down of norms by central ministries**. Any such norms, such as living

space per person, will have an impact on city growth.

5. **Control of population movement**. City populations are to a degree brought about by population migration, especially a rural to urban drift. In a centralized state this can be controlled and manipulated by formal 'licensing' of permission to move. In the Soviet Union movement to towns was controlled by such means:

> Without an internal passport or equivalent documentation, movement to cities, or between them, is precluded. All urban inhabitants over 15 years of age are required to have a passport in their possession... But residence in any city is not assured simply by virtue of possession of a passport. Each urban inhabitant must have a propiska, a residence permit the equivalent of a visa, which is entered in the passport and which gives the legal recognition of the right to live in a specific city. ... In theory one cannot live in a city without such documentation. ... In summary there is an elaborate bureaucratic procedure for acquiring permission to reside permanently in a Soviet city.
>
> (Bater, 1980: 57–8)

The reciprocal of powers preventing migration to the cities is the forced dispersal of population from cities. There is an inherent antipathy between pure socialist regimes and urbanization for cities are centres of conflict and competition which offend dogma. Moreover, they are the points at which dogma is challenged. It is not surprising, therefore, that socialist regimes have sought to disperse city populations, especially those elements, such as the intellectuals, which are likely to create problems. 'In China, urban unemployment has been limited by controls on rural–urban migration and by "sending down" campaigns, most notably the rustication of middle-school leavers that took around 17 million youths to the countryside during the Cultural Revolution' (Gugler, 1988: 57–8).

The main comment on these methods of moulding the character of the city system must be that, with the exception of the last, they will be very familiar to any student of capitalist regimes for they have been deployed from time to time as part of regional policy in most Western countries. Thus W. K. D. Davies in discussing urban system development policies proposes a four stage degree of intervention – 'persuasion' to 'inducement' to 'control' and finally direct government 'involvement'. Of this last stage he writes:

> The most obvious example is in direct building of new towns. This "complete settlement" initiative may be contrasted with (sic) the dispersal of offices from London – either by persuasion in the private sector through the Location of Offices Bureau, or by direct re-location of government facilities. Such an economic generator initiative is designed ... to share out the benefits of employment to other cities – frequently those in depressed areas.
>
> (Davies, 1991: 228–9)

In such a context, studies of city systems in socialist countries, such as those in the volume edited by Bourne *et al.* (1984), are not greatly different from those in capitalist countries, despite being allocated to a separate section headed 'settlement systems in centrally-planned economies'. The conclusion by Bater is worth quoting:

> In a Soviet economy it is theoretically possible to manipulate socio-economic processes so as to conform with stated objectives or principles. The planned development of a unified settlement system has been part of Soviet theoretical formulation almost from the outset. But aside from a few general locational guidelines there has been a notable absence of specific policies and programmes to achieve such an objective. It was not until the late 1970s that ... 'the Central Urban Planning Institute in Moscow formulated the task of shaping the process of settlement with the ultimate aim of fostering the interconnected development of places on a national scale' (Kochetkov and Listengurt, 1977: 661).

Not only is a programme of action late to be developed, but detailed analysis of the evolving urban system itself is notably lacking.

(Bater, 1980: 80)

One possible exception to these comments is a study by Grimm of the city system of the then German Democratic Republic (GDR). It is worth considering just a section of his argument to demonstrate his attempt to conform to the socialist imperative.

He begins by dividing the city system into two. The first is the B settlement system (where B is *Bedurfnisse*) which aims at the fulfilment of social goals. The second is the E or economic system which is derived mainly from industrial activity. This is simply a division between the central place functions common to all towns (B) and the specialized functions (E) which characterize only some towns (see Chapter 6. p. 99). Grimm writes 'the B settlement system reflects the immediate realization of the requirements of human life and work. Its function is to realize the social needs of a nation: labour, housing, retail trade, education, health service, culture, etc.' (Grimm, 1984: 381). Within this system the crucial centres are those defined by daily commuting ('within reach of everyday traffic circulation'). These correspond to the centres of Kreises or administrative districts. 'The spatial pattern of centres [Fig. 17.1] shows the comparatively even distribution of centres of everyday social interaction (Kreis centres) and the less equal distribution of centres of higher rank. It illustrates the success of the government policy stressing social harmonization in all parts of the country' (Grimm, 1984: 380). It must be added that here is a proper sentiment of socialist ideology; but one smacking of idealism rather than reality.

Grimm proceeds to describe the Kreis towns, most with fewer than 20,000 inhabitants. Many of the functions are linked with their administrative status, such as offices of political parties, trade unions and of political, social and cultural organizations – 'Kreis centres of education with a secondary school and adult college; the editorial office of the special newspaper for the Kreis;

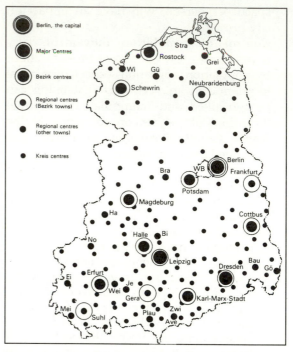

FIGURE 17.1 *The former German Democratic Republic (East Germany). A classification of settlements according to their importance to their hinterlands (after F. Grimm, 1984). Source: F. Grimm, 1984, The settlement system of the German Democratic Republic; its structure and development. In L. S. Bourne et al. ed. Urbanization and settlement systems. International perspectives. Oxford: Oxford University Press. p. 382, Fig. 18.1.*

Kreis centre of economy and trade with insurance offices, offices of banks and saving banks. As a rule about 50 different functions are located in every Kreis town of the GDR' (Grimm, 1984: 381). Grimm goes on to stress the way in which because of the diversity of jobs offered and the effective public transport system then the Kreis town is a universal and effective hierarchical level. He concludes 'the present administrative division of the GDR has led to a stable, well balanced integration of settlement systems within the Kreis framework. The even distribution of towns promotes the equalization of working and living conditions in accordance with the political and social aims of socialist society' (Grimm, 1984: 381). All this sounds more like an exercise in an appropriate form of political correctness than an in-depth analysis of a city system. Certainly there would be little difference between the concept advanced by Grimm and the fully-fledged town in England and Wales set out by Smailes fifty years ago. Moreover, once he moves to consider the E system the notion of equality falls away 'because of the diversity of industrial production E settlement systems in industrialized districts are less homogeneous than the rural systems, and there is little coincidence between B and E systems' (Grimm, 1984: 387).

It is difficult to outline the whole of Grimm's quite effective analysis, but the end is hardly in contrast with that of a capitalist system with the same complex interaction of central place and specialized functions. There is a gloss of socialist equality, but it is that and little more. Certainly the forms of intervention as outlined above are much more widely deployed under socialism as attempts are made to constrain the spatial inequalities which necessarily follow from unhindered free market operations. But the differences which result are of degree, not of kind. City systems in socialist countries are essentially similar to those of the capitalist world. Both are in part based on forms of regional planning to offset unequal development.

17.3 INTERNAL STRUCTURE OF SOCIALIST CITIES

The study of the internal structure of socialist cities raises two dominant issues: one operational, one substantive. The first is the quality of the data on which any judgement can be made, for there is nothing like the extensive small area statistics which have been so widely used in the factorial ecological analyses of Western cities. The second is the substantive issue of whether real progress has been made towards creating a city where equality is such that there are no

socio-economic or ethnic variations. In ideal terms, just as there should be no differential access to any urban facility within the city system based on location, so too, there should be no variation within the city. Whereas that is relatively straightforward as regards economic status and ethnic character, it is not clear in relation to the stage in the life cycle. Under the principle of to each according to needs, those needs will vary according to life-cycle stage and any system of housing allocation responding to that would be likely to be reflected in residential patterns within the city. But the ability effectively to examine the substantive issue is, in part, undermined by the relative paucity of the necessary data. There is a third difficult issue to which allusion has already been made in the previous section. Most socialist regimes have been authoritarian and the free expression of views certainly discouraged if not prohibited. The urban literature of the West is totally dominated by critical studies, usually from a left of centre viewpoint (see above, p. 325), but there is no equivalent in any socialist country to which reference can be made. The result is that it is difficult to differentiate between studies which are no more than propaganda, and those which are detached, which is perhaps a more appropriate word than objective.

The first detailed examination of the socialist city in English was that by F. E. Ian Hamilton (1979). He constructed a model of such a city identifying eight concentric zones:

1. The historic medieval or renaissance core.
2. Inner commercial, housing and industrial features from the capitalist period.
3. A zone of socialist transition or renewal.
4. Socialist housing of the 1950s.
5. Integrated socialist neighbourhoods and residential districts of the 1960s and 1970s – continued into the 1980s.
6. Open or planted isolating (green) belts.
7. Industrial or related zones.
8. Open countryside.

A number of comments can be made on this zonal structure which is shown in Fig. 17.2.

Where possible, the historic cores have been preserved as symbols of traditional culture. Indeed, where they were destroyed they have often been rebuilt to replicate the former situation, as in Warsaw. Cultural identity outweighs socialist ideals! The inner area is an amalgam of all the uses and buildings of the pre-socialist city. Among those uses are the old inner high quality residential areas. Outward from these inner parts are progressive zones of housing. The earliest were hurriedly built blocks of apartments, but succeeding them are organized residential neighbourhoods where the necessary services such as health care and education are provided along with essential shops. The predominant theme was to avoid the excessive commuting which was the consequence of central area employment and shopping. The new city centre was to be a monument to socialist control, so it was dominated by buildings such as palaces of culture and sports stadiums, monuments to people's uprisings and communications and media centres; shopping although clearly present was shifted out to the regional and sub-regional centres. On the same bases industry was made peripheral in its location and carefully segregated from residential areas although within easy access. The impact of the tendencies apparent in the model can be examined in the light of two studies – one of Warsaw, one of Kazan.

Weclawowicz's study of Warsaw in 1979 provides considerable insight into a city in the process of transition both from its capitalist past and its wartime destruction. Two maps are reproduced. The one is the distribution of the population with higher education (Fig. 17.3), the other (Fig. 17.4) a summary of a component analysis which uses the scores on components to identify four types, each subdivided. The three components are identified as socio-economic position, housing and social situation, and economic position. Thus type A has high positive scores on all components and is characterized by high positive indices of socio-occupational position, housing conditions and economic position. In contrast Type 4 shows negative scores on all the measures, although 4b returns a positive economic score. Weclawowicz finds some difficulty in interpreting these patterns and describes the result as a mosaic which is the

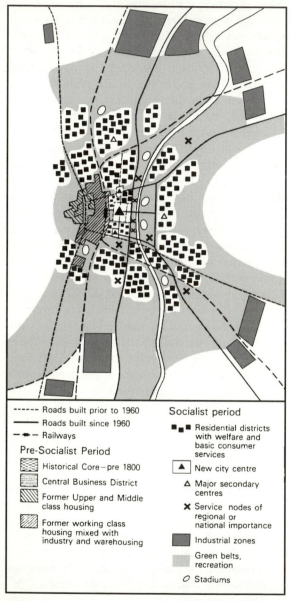

FIGURE 17.2 *A model of the growth of an East European city (after F. E. I. Hamilton, 1979). Source: F. E. I. Hamilton, 1979: Spatial structure in East European cities. In R. A. French and F. E. I. Hamilton ed. The socialist city. Chichester: John Wiley, p. 195, Fig. 9.3.*

Legend:

- - - - - Roads built prior to 1960
——— Roads built since 1960
–●– Railways

Pre-Socialist Period
▦ Historical Core – pre 1800
▤ Central Business District
▨ Former Upper and Middle class housing
▧ Former working class housing mixed with industry and warehousing

Socialist period
■■ Residential districts with welfare and basic consumer services
▲ New city centre
△ Major secondary centres
✕ Service nodes of regional or national importance
▬ Industrial zones
▒ Green belts, recreation
○ Stadiums

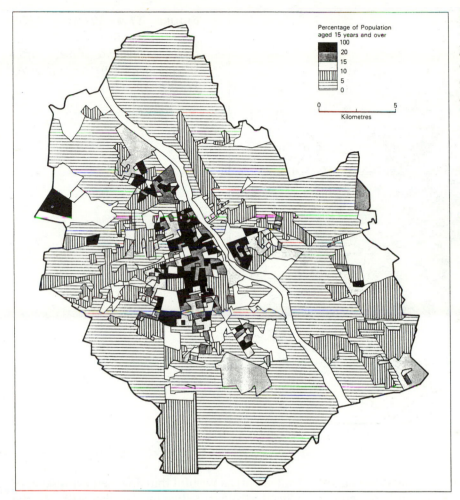

FIGURE 17.3   *Warsaw, 1970. The distribution of population with higher education as a percentage of the population aged 15 and over (after Weclawowicz, 1979). Source: G. Weclawowicz, The structure of socio-economic space in Warsaw 1931 and 1970: a study in factorial ecology. As Fig 17.2, p. 413.*

consequence of the process of transition. Nevertheless one dominant theme does stand out in both maps. Those of higher socio-economic status are centrally located, while the periphery is predominantly of lower status.

The study of Kazan is taken from what is now the best summary of socialist urbanization (D. M. Smith, 1989). Again two maps were simplified from a study by Rukavishnikov (1978) (Figs 17.5 and 17.6). The main point made by Rukavishnikov is that ethnic segregation of Tartar as against Russian was not significant. He wrote:

no rigid relationship between an individual's status in society and his place of residence is to

be found. There are no social groups living solely in the centre of the town or on its outskirts, nor has any social group taken over the ecologically comfortable zones of the city. We were unable to draw boundaries around territorial localizations of individual social strata of citizens or smaller socio-occupational groups. ... All that can be seen ... are large regions (in the old downtown districts, for example) where there is a small plurality of white collar and professional people or of workers, but there are no sharp lines of demarcation between such regions, and the social composition within tracts in each region is quite heterogeneous.

(Rukavishnikov, 1978: 68)

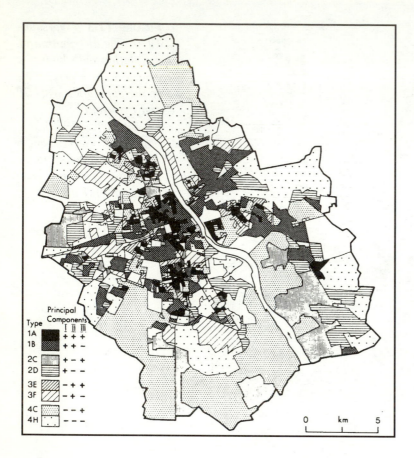

FIGURE 17.4 *Warsaw, 1970. The spatial distribution of the three component types (after Weclawowicz, 1979). Source: As Fig 17.2, p. 420.*

But Smith comments that the maps do suggest 'a peripheral dominance of workers, and a concentration of professionals in central parts of the city'; and 'while high social status and Russian origin are much less closely associated than before the Revolution, there must be some relationship between ethnic group and living conditions in Kazan, because Tartars predominate in the original (and poorer) Tartar parts of the city as well as in the incorporated villages and in the urban fringe' (Smith, 1989: 25).

Smith concludes his review of the Soviet city quoting a Soviet authority (Yanitsky, 1986: 273) who claimed that the city in the USSR was free from all the internal spatial contrasts of Western cities. Smith then comments, 'There is little if any doubt that, objectively, the Soviet city lacks the stark inequalities of its counterpart under capitalism. But the evidence ... clearly reveals Yanitsky's [statement] as wishful thinking – an expression of official ideology rather than of the reality of the Soviet socialist city.' Smith then presents a broad typology of socio-economic and environmental differentiation in large Soviet cities.

1. Inner, high status areas of good housing, occupied largely by the professional groups.
2. Inner, low status areas, of old and deteriorating property waiting for renewal.
3. Outer areas of relatively high status (more or less distant from the centre, depending on the size and growth pattern of the city) with a relatively high proportion of cooperative flats. Local employment may be predominantly white collar.
4. Outer areas of lower status, with a predominance of state housing; manual employment predominates.

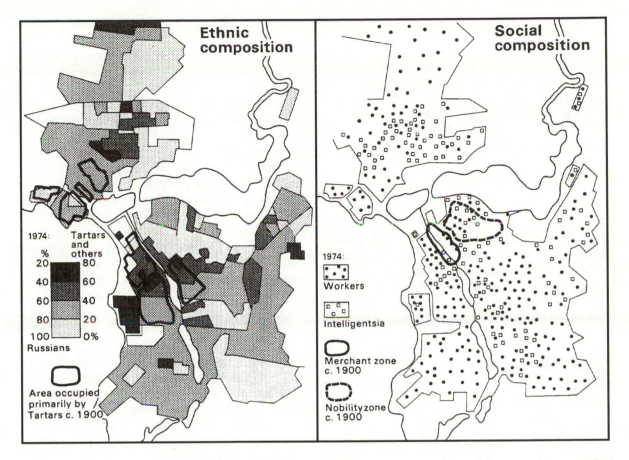

FIGURE 17.5 *Kazan: the location of ethnic groups c. 1900 and 1970 (after D. M. Smith, 1989).*

FIGURE 17.6 *Kazan: social composition c. 1900 and 1970 (after D. M. Smith, 1989). Source: Figures 14.5 and 14.6. D. M. Smith, 1989: Urban inequality under socialism. Cambridge: CUP. p. 24, Fig. 3.2a and b.*

5. Peri-urban areas and suburban enclaves of private housing of very poor quality much of it occupied by recent migrants from the countryside.
6. Quarters occupied by distinctive ethnic groups, possibly but not necessarily in lower status occupations.

From this very general review it is possible to add some further general comments. One of the contrasts between Western capitalist and socialist cities is the nature of the city centre. In the West the buildings of commerce predominate,

the skyline is the creation of private enterprise. But the central areas of socialist cities are marked by public buildings generally of a twofold character. The first group is made up of structures emphasizing political control and the monuments which go along with it. The second group symbolizes cultural identity, mainly through exhibition halls, museums and theatres.

Bater (1980), writing of the 1935 general plan for Moscow, notes that earlier planning notions had been against any distinguishable centre but these were modified:

By means of unified and uniform architectural ensembles, thoroughfares and squares, the city centre was to cater for massive demonstrations. ...The objective was to have the resultant ensembles reflect the glory of the state.... As the cultural and political uses of the centrality were emphasized, the customary central city functions were downgraded. The result was the conscious decentralization of administrative and distributive services into fully developed secondary centres.

(Bater, 1980: 30)

Because of its nature and the basic need to be in close contact with the manipulation of government which is centralized, the élite seek location near the centre, where advantage can also be taken of the cultural life. The élite, therefore, are often still found at the centre of cities much as they were in the pre-industrial West. One of the key changes which led to suburbanization in the West was the creation of local democracy (Carter and Wheatley, 1982: 115) for once oligarchic power is effectively devalued, and palace revolutions outmoded, then central location loses one of its main attractions.

The other general comment is on the nature of the residential districts created in the period following the earlier and immediate post-war reconstruction. The provision of an effective range of services for the community stands in sharp contrast to the extensive housing estates without services which certainly in Britain have generated so much social unrest. And there is some validity in the suggestion of a mosaic pattern of status rather than the segregated regional one of Western cities.

17.4 CONCLUSION

This has been a brief chapter and, in the light of recent events in the former USSR and Eastern Europe, perhaps it has been an unnecessary one, or a theme for historical rather than contemporary geography. The consequences of the relaxation of central authoritarian control have brought crime on a much greater scale to cities

such as Moscow and maybe a suburbanization process akin to that in the West will follow. The break-up of states along ethnic lines casts doubt over the notion that ethnicity had become irrelevant under socialism and of no significance as a dimension of intra-urban difference. The towns of the former Yugoslavia will most likely show greater ethnic division than ever. But in reviewing urbanism in the socialist states perhaps the greatest sadness is that no new concept of the city emerged.

There was very considerable discussion in the earlier days of the Soviet Union. Bater (1980: 22–7) suggests that by the late 1920s it revolved around two schools. The first, the urbanist school, proposed a system of self-contained urban centres with multistorey blocks designed for communal living, for, in theory, the family would disappear in favour of a totally communal way of life. These centres would have a population of some 50,000. As Bater comments, there was a clear mixture of Ebenezer Howard and Le Corbusier in these proposals. More significantly, they attempted to realize the notion of equally sized centres, equally spaced. In contrast, the de-urbanist school saw the proposals as no more than a small town policy. It was not Marxist.

They wanted an essentially townless, socialist society in which the age-old contradiction between town and country would be abolished. ... They proposed dispersing the population over the whole of the habitable parts of the state. Settlement would take the form of ribbon developments. Individual dwellings located in natural surroundings would be allocated as housing, but within easy access would be communal centres for dining, recreation and so forth. In other words, some privacy would be provided in terms of living quarters, but the life style would be essentially communal.

(Bater, 1980: 23)

The linear city, like the garden city and the 'unités d'habitation', had a clear Western tradition. But neither of these ideal schema was attempted as the exigencies of industrialization were paramount and the simple spatial controls of threshold and range exerted their influence. Ironically, perhaps the outsiders in urban societies (Sibley, 1981) in the West, the new age travellers, have come nearer to some of the socialist ideals. But it is a rejection of the city not its apotheosis. Perhaps it is apposite to question whether a socialist city as such is conceivable since cities are based on competition and are the product of mercantile and capitalist systems. The most appropriate socialist reaction was that in Kampuchea in 1975 when Phnom Penh and other large cities were evacuated in the determination to establish 'an equitable, self-sufficient and socialist country' (Gilbert, 1982: 164).

David Harvey at the end of what has been the most influential book in urban studies in the second half of the twentieth century – *Social justice and the city* – called for a genuinely humanizing urbanism and the establishment of a path from an urbanism based in exploitation to an urbanism appropriate for the human species. 'It remains', he wrote, 'for revolutionary practice to accomplish such a transformation' (Harvey, 1973: 314). That practice in socialist systems went little way towards the transformation he demanded.

THE CITY IN THE DEVELOPING WORLD

18.1 INTRODUCTION

As in the last chapter, most of this book has assumed a free market, so too most of it has been concerned with the developed West. Three presumptions have, therefore, been inherent from the start. These are: an industrialized or developed economy; a capitalist system; a Western cultural environment. An ideal procedure would be to relax these assumed conditions in turn and then to consider the consequences of such modification on cities, both in locational and functional character and in internal structure. In the last chapter a switch was made to consider the consequences of a socialist economic system. In this, therefore, it is necessary to turn to consider the city in undeveloped or developing countries and in non-Western cultures. On this topic a great deal has been published in recent decades and it demands review, even if analysis in any depth is beyond the reach of a single chapter.

In relation to the material already presented in this book two themes need to be investigated. The first relates to the process of urbanization and inter-urban relations. The second concerns the internal structure of towns.

18.2 CITY SYSTEMS IN DEVELOPING COUNTRIES

Three aspects related to urbanization need to be considered.

- The colonial origin of towns
- Over-urbanization and primacy
- Systems of periodic markets.

The colonial origin of towns

The colonial origin of towns is certainly not a unique feature of the developing world for many European towns, and all those of the United States and Australia were colonial in origin. Again, the great colonizing movements in medieval Europe were largely accomplished through a process of town founding; the bastide towns were by their nature colonial towns and carried many colonial characteristics. It is indeed possible to trace the way the towns were reclaimed by the indigenous populations (Carter, 1983). The great contrast with the cities established in the developing world is, however, the major cultural contrast between the predominantly European founders and the indigenous populations. In the USA and Australia that problem was offset by the effective complete dispossession of the native populations who, in any case, were not town dwellers; there were no towns, although that was not the case in South America. Elsewhere the native population remained in a condition of exploitation. The critical feature, therefore, of the cities founded in the tropical areas was the aim not of settlement but of exploitation. There have been attempts (Horvath, 1972) to distinguish between colonialism where a significant number of settlers migrate permanently to a country, and imperialism, where there are few such migrant settlers. But it is a distinction difficult to maintain (King, 1976: 17).

Because cities were centres of exploitation they tended to have a limited range of functions, mainly related to the export of primary raw materials and the military and administrative bases sufficient to make that role viable. Where the Church went as well to convert the inhabitants to Christianity it largely relied on the administrative system, but it could create its own centres as it did in California where the mission stations were the forerunners of many towns now bearing their names (Carter, 1983: 52).

Essentially, though not completely, these cities were parasitic rather than generative of an economic development which trickled down to the surrounding areas. For example:

the development of the urban system in colonial Latin America gave a primacy to the political and fiscal needs of the empire. The urban hierarchy of city and village, often imposed upon pre-existing native jurisdictions and urban institutions, preserved and enforced status distinctions between categories of population. The categories of Indian, slave, mestizo (of Spanish and Indian parents), mulatto (of Spanish and black parents) and Spaniard had different sets of rights and obligations in terms of tribute, forced labour, taxes, military service, trading and debt contraction.

Glade (1969)

Glade contrasts this form of colonization with Turner's characterization of the North American frontier as a source of economic and political autonomy and innovation. . . . 'In this situation, towns in Spanish America . . . did not serve to generate economic progress in a "backward" countryside' (Roberts, 1978: 38). Again Gugler and Flanagan writing of West Africa stress that urbanization in that area owes much to the colonial past. 'This is true not only of the atmosphere and location of towns but also with regard to the nature of the role these towns have played and are continuing to play in the economic development of the independent states' (Gugler and Flanagan, 1978: 26). Three significant consequences follow from this inheritance:

1. Because of their colonial nature, cities are characterized by cultural pluralism. They were mostly planned and developed to ensure and ingrain residential segregation. In post-colonial times this role has been completely reversed and cities have to be the leaders in integration (King, 1976: 23). The difficulty of the transformation is epitomized by the role of the city in post-apartheid South Africa.

2. The large cities often have highly skewed employment structures. Development as export and administrative centres has resulted in the basic range of industrial employment remaining undeveloped. This is particularly critical at a period of high urban in-migration.

3. Because of a location initially orientated towards external linkages, many capital cities are eccentrically located in the new national territories:

As a consequence of their location on the geographical periphery of the new states, the young capitals of West Africa have acquired a political character that is more regional than national. Because of major ethnic and economic differences between the capital region and the rest of the nation, the geographically marginal capital comes to stand as a symbol of faction rather than national unity.

(Gugler and Flanagan, 1978: 31–2)

Over-urbanization and primacy

The nature of urban growth in the developing world has already been dealt with in Chapter 2 when the process of urbanization was considered. Also at that point an explanation of the massive nature and rapidity of that urbanization was suggested. Certainly over-simple explanations must not be accepted. At the root is the process of demographic transition characterized by a fall in the death rate and consequent rapid population growth. Milton Santos (Santos, 1971: 26) shows that whereas in Sweden it took 100 years for the death rate to fall from 21 to 14 per 1000, in Costa Rica it fell from 23 to 7.9 per 1000 in 42 years (1920–62), in Sri Lanka it fell from 20 to 9 in 15 years (1946–61) and in Jamaica from 28 to 13 per 1000 in 29 years (1921–50). This population 'explosion' has taken place against two critical changes. The first is the nature of agriculture itself, which has become increasingly mechanized and commercialized and hence much less labour intensive. The second is the undermining of the traditional bazaar peasant economy by the import of manufactured goods from the developed world. In this context notions of 'dependency' can be put forward or notions of the evolution of a world system using the core–periphery concept. Thus Wallerstein (Wallerstein, 1974) for the early phase of European development distinguished 'between the core industrial regions such as England, semi-peripheral regions specializing in commerce, such as Spain and peripheral regions such as Latin America and Eastern Europe, which were organized to provide primary products for core countries' (Roberts, 1978: 15).

The expanding, modern world system has pushed the developing countries into the role of peripheral providers of raw materials, and the divergence which has characterized core and periphery on the national scale has been echoed at the world scale. A situation comes about, therefore, where not only is population growing but the bonds which held that population to the countryside are loosened and a dominant urban migration precipitated. Moreover, the attractions of the urban areas are themselves evident and much more clearly realized through kin links, as well as the growth of the media. A whole genre of studies carried out multiple regression analyses of migration flows in underdeveloped areas and the variables which emerged as significant, apart from those of distance and size, relate to the attractive pull of the cities in terms of wages, education, health and welfare (Masser and Gould, 1975).

This transfer of population into the urban areas is, as indicated above, partly related to the attraction of size and consequently there has been a trend for the largest city to grow most rapidly. This has been exacerbated by the investment policies of many countries which have diverted a disproportionate amount of funds to what is usually the national capital where political, administrative and cultural authority is centred. Gugler and Flanagan, in a section appropriately headed 'the hubris of capital cities', present a revealing account of such a trend in West Africa. 'Anyone who ventures beyond [the national capitals] to the vast expanses of West Africa is stunned by the disparity between the concentration of resources in the capital cities and the neglect that is the fate of much of their hinterlands' (Gugler and Flanagan, 1978: 40–1).

They go on to exemplify the process of concentration from the work of M. A. Cohen (1974) on the Ivory Coast. There 499 out of the country's 617 industrial enterprises were located in Abidjan and 60 per cent of the jobs in the modern sector of the economy. 'In 1968, the eight storey, 500 bed Centre Hospitalier Universitaire, one of the largest and most modern hospitals in Africa, was built in the luxurious Cocody quarter, home of the high government officials, but the fund given by

Table 18.1 *A comparison of employment structure in developed and underdeveloped countries*

Country	Date	Primary (%)	Secondary (%)	Tertiary (%)
Argentina	1960	22	21	57
Chile	1960	25	17	58
Venezuela	1960	32	12	56
France	1954	28	37	35
USA	1900	38	27	35
Germany	1929	30	41	29

Source: After M. Santos, 1971.

France was originally intended for twelve regional hospitals' (Gugler and Flanagan, 1978: 42). This sort of bias has resulted in a population of some 5000 in 1920 and of about 36,000 in 1940, becoming 242,000 by the mid-1960s and 555,000 in 1970 with an astonishing average growth of 10 per cent per annum from the 1950s to the 1970s. In 1975 the total urban agglomeration was returned as 1,423,323 out of a population of 9.3 million (UN, 1990). In reviewing Asia's probable urban future Dwyer has properly used the word 'frightening' (Dwyer, 1975: 15). The United Nations' Department of Economic and Social Analysis estimates that even using an urban definition of 20,000 there will be 993 million urban dwellers in Asia by the year 2000 compared with 265 million in 1960 when even then the cities were described as in a state of crisis.

As a result of the growth rates, which have been discussed, and the concentration of population into the capital cities, two consequences follow. The first is that of over-urbanization, a very nebulous concept, which has certainly not been without its critics, but it does emphasize an increasing urban proportion in countries whose economic bases are too feeble to sustain it. It is partly responsible for widespread unemployment. The mass of urban migrants move into a range of unproductive jobs, especially in the retail and service occupa-

tions so that the diagnostic feature of such situations is a high proportion of employment in the tertiary sector. Santos (1971: 330–1) has compared the state of the employment sector in three Latin American countries with that of three industrialized countries at appropriate dates in their development. It is not that there are great differences at the present between the tertiary sectors of developed and developing countries, but that, as Table 18.1 reveals, the countries of the so-called Third World have not undergone the prior and controlling process of industrialization. 'In developed countries technological progress, bringing urbanization, has transformed the occupational structure, whereas in underdeveloped countries it has been demographic evolution which has most greatly influenced the employment structure' (Santos, 1971: 329). To put it in rather a different way, though with the same implications, Gugler writes:

Third World cities are characterized by an excess of labour with limited skills. Open unemployment constitutes only one facet of urban surplus labour. A second element is underemployment, that is, the tasks at hand could be satisfactorily carried out by fewer persons. Finally substantial numbers, while perhaps fully employed, produce goods or provide services that can be judged to contribute little to social welfare; such persons may be labelled 'misemployed'.

(Gugler, 1988: 75)

This is the traditional view of urbanization without industrialization.

The second consequence of urban growth rates and especially the concentration of growth into capital cities, has been the characteristic of 'primacy'. The Law of the Primate City has already been discussed (Chapter 4) and the relation of primacy to underdevelopment considered. Linsky's (1965) conclusion that the concept is related to a particular process of historical change and that there is a need to relate primacy to social, economic and geopolitical conditions, must be recalled at this stage, for it would seem that the sorts of transformations discussed in this section and the dominance obtained at the core by virtue of its political and cultural authority, as well as by virtue of its economic and social supremacy, is the crucial factor in the emergence of the primate city.

Systems of periodic markets

These have already been briefly considered in Chapter 5 (p. 63). A corollary of the vast population increases in the large cities, is the lack of development in the areas removed from metropolitan influence. In such conditions the lack of demand for urban services results in a time–space substitution on the same lines as that which characterized medieval Europe. In a continuation of the search for a central place geometry most investigations have attempted to indicate how markets were organized so that they could be visited in sequence by itinerant traders. The classic and most quoted study, referred to in Chapter 5, was that by G. W. Skinner (1964) relating to markets in rural China, although as early as 1961 Hodder had described rural periodic day markets in part of Yorubaland. There he (Hodder) pointed out that such markets were generally evenly spaced out and independent of the village or hamlet distribution. 'Some markets ... have been simply established in clearings in the bush at the

junction of paths: others lie athwart a motorable road, sometimes near to, or in a village. In some cases the market has in fact created a village' (Hodder, 1961). Hodder describes the complex system as follows. Rural periodic day markets in Yorubaland operate on a ring system, each ring being composed of a complete and integrated sequence of markets taking place over four-day or multiples of four-day periods. Markets may thus occur on Monday, Friday, Tuesday, Saturday – four-day markets; or on Monday and Tuesday of the following week, Wednesday of the succeeding week – eight-day markets. This is a common phenomenon in West Africa and appears to be related to time factors operating prior to the adoption of the seven-day week towards the latter part of the nineteenth century as European control extended inland from the coast.

According to P. Talbot (1926), 'the original week appears always to have been composed of four days. ... The subdivisions into weeks in all likelihood originated from the necessity of differentiating between the days on which the various markets were held.' As Fig. 18.1 indicates, of the eight rural periodic day markets distinguished in the Akinyele ring, seven take place at eight-day intervals. These seven markets operate on successive days in such a way that each market takes place on a day on which it is the only one of the seven markets operating within the ring. After all markets have had their turn, there is one marketless day after which the process is repeated in the same order, one–seven. The eighth market, Ijaiye, occurs every four days and operates simultaneously with Akinyele on the first day and with Olorisaoko on the fifth day of each eight-day period. The Akinyele ring also operates in such a way that successive markets are not normally adjacent markets. The first market is at the centre at Akinyele, the second at the extreme southern edge of Ojo, and the third at the extreme northern edge at Iware; there are subsequent further movements back and forth until the seventh day, when the market near the centre at Onidundu takes place before the eighth or rest day.

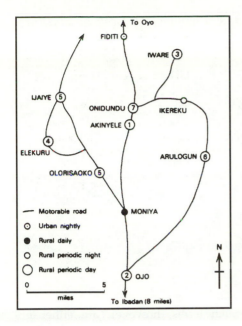
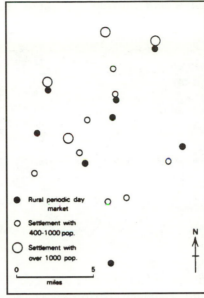

FIGURE 18.1 *Types of local markets north of Ibadan and the population of settlements and the location of day markets in the same area (after B.W. Hodder, 1961).*

In this way the timing of marketing activities is evened out over the whole ring so that no hamlet or other settlement is far from a market for more than three days. This integrated pattern and timing of markets in the Akinyele ring is thus most logical and convenient and is in some form or another characteristic of most parts of the Yoruba country. It is a wholly indigenous phenomenon, expressing an intelligent mutual self-interest among neighbouring village chiefs, or *bale* (Hodder, 1961: 152).

Attempts at interpretation via traditional central place theory have been criticized by Bromley *et al.* (1975). They argue for a more complex situation where the periodic market is in part the product of an economic system where making (producing) and selling have not been divorced, so that time is needed for 'manufacture' and part-time retailers demand periodic markets. Again, the organization of time is a controlling factor as in the Yoruba rings already discussed. The length of week, the days set aside by religion or for rest and the significance of astrology, all these determined the market timing. Finally, they argue that inertia often retains a system long after it has become economically unnecessary. The consequence of

these influences is that an analysis in purely central place terms is unlikely to be greatly productive.

18.3 INTRA-URBAN CHARACTERISTICS OF DEVELOPING COUNTRIES

Internal structure of towns

City plan is largely the product of the cultural and social characteristics of the group which lays it out (Carter, 1983: 114–29). Accordingly it is impossible to follow all the variants of plan produced in the developing world, or rather more appropriately, outside the European cultural sphere. It is worth pursuing the colonial theme introduced into this chapter, however, by considering the impact of English colonialism on the cities of India where, unlike America or Australia, movement was taking place into culturally alien areas with high densities of population and long-standing urban cultures. The two cultures of colonizer and colonized continued side by side, certainly interacting, but engendering a distinctive urban form made up

Table 18.2 *A typology of cities with differing ecological structures*

1. Pre-industrial cities influenced by colonialism.
2. Pre-industrial cities uninfluenced by colonialism, having indigenous invention of industrial technology.
3. Pre-industrial cities uninfluenced by colonialism *not* having indigenous invention of industrial technology.
4. Cities reflecting the 'pure' impact of industrial technology.

Source: After London and Flanagan, 1976.

Table 18.3 *Key terminology (scale or level) in the language of colonial urbanization*

Urban	Urban sector	Urban unit
Civil station	Bazaar	Barracks
Civil lines	Colony	Bungalow
Cantonment	Esplanade	Chummery
Lines	Mall	Club
Circuit house		Compound
Dak (bungalow)		Lodge
		Quarters
Residency		Rest house

Source: After King, 1976.

of the two elements. In writing on 'comparable urban ecology' London and Flanagan (1976) proposed a typology of cities as set out in Table 18.2.

Although the 'types' are based on ecological structure they can be as appropriately used for the examination of physical structure. In such a context no better example exists of the first type than that brought about by British colonial rule in India. That the interaction was not only one way can be seen in the many terms which found their way back to the colonizing country. In the physical or built structure of the town, what in the USA is called a 'ranch style house' still remains in Britain the term derived from colonial Asia, a bungalow. This whole problem of the colonial city in India has been the subject of an excellent and thorough study by A. D. King in his book entitled *Colonial urban development* (1976) and accordingly only a brief outline is presented here based largely on that source.

A most effective introduction to the colonial city is a consideration of the linguistic terms which were used to describe the various parts of the city. King sets these out in a table which is reproduced, in part, here (Table 18.3).

These words represent 'physical–spatial–social' elements in the urban system which are expressed in the metropolitan rather than in the indigenous language (i.e. English rather than Hindi). Such terms, however, are either not found in the system of urban nomenclature in the metropolitan society (e.g. civil lines, civil station, cantonment, chummery...) or, if they do exist, are used with different meanings to those prevailing in the ex-colonial society (e.g. compound, bungalow, bazaar).

(King, 1976: 71)

Perhaps the most common word which King does not mention, since it has no spatial connotation, though most definitely a social one, is the word 'pucka' which described a bungalow of masonry construction, brick or stone with tiled roof, and which brought a widely used adjective into English. The location of the berth on the ships taking the representatives of the colonial power to India also provided a key word. Port Out, Starboard Home gave not only a location away from the heat of the sun but in the mnemonic POSH, a lasting social categorization also. A review of these terms can be used to describe the major spatial elements of the colonial city, grouped as they were about the three elements critical in the morphology of Indian urbanism: the first was the garrison, the basic physical representation of military control; the second element was the civil administration developed by the colonizing power 'to run' the country; while the third was the native city

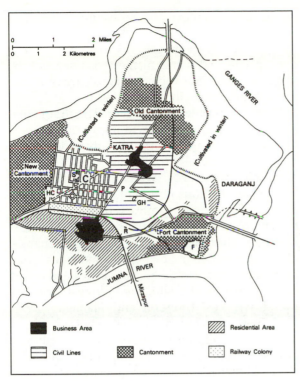

FIGURE 18.2 *Allahabad: the city's internal structure in a map of 1931 from the Imperial Gazeteer of India. C: Cannington (part of Civil lines); HC: High Court; GH: Government House; S: Secretariat; P: Alfred Park; R: City Railway Station. Reprinted (adapted) from the Geographical Review with the permission of the American Geographical Society.*

which predated the colonial grafting on of the newer parts.

Figure 18.2 shows how many of these elements characterized the city of Allahabad during the nineteenth and early twentieth centuries. The settlement was located at a confluence of the Ganges and the Jumna rivers and was a sacred site, the ancient city of Prayag. It was also supposedly at the confluence of the legendary Saraswati, the lost river of the Punjab (Spate and Ahmed, 1950). The Mogul emperor Akbar had built a fort at the angle where the two rivers met and this was taken over by the British to form the Fort Cantonment. It constituted, however, a very

limited area, far too small in extent for the effective deployment of the military arm of the Raj. The site of Allahabad was itself constricted, for it lay between the Ganges and the Jumna. The latter was bordered by steep bluffs while the Ganges was characterized by moving sandbanks liable to inundation. A more extensive old cantonment was developed within the confluence core, but eventually a still more extensive site was developed to the west called the New Cantonment. The demand for space, derived both from basic military needs and from the large compounds in which the bungalows were set, is well illustrated by the progressive development. Figure 18.2 clearly shows the distinctive elements. The New Cantonment was lavishly laid out in spatial terms. The core was dominated by the Brigade HQ which was surrounded by the European Infantry Lines, the Cavalry Lines and the Artillery Lines. Tucked away from these and well segregated to the northeast were the native Infantry Lines. Adjacent to them was the Bazaar. The burial ground was located well away from the living areas, from which it was separated by Macpherson Park.

The Civil Lines were situated to the east of the New Cantonment and were established on an extensive grid plan cut through by the main route which was related to the axis of the old Mogul city. Within the area were all the standard representatives of English culture in India: the Government Offices, the Law Courts, the Hotel, Hospital and the Allahabad Club. It was neatly segregated: to the north was the river Ganges while the New Cantonment was to the west: to the south were the railway lines and to the east Alfred Park provided a break. The native city was to the south, nearer the Jumna river and on the other side of the tracks. It was densely populated and unplanned maze of narrow streets in complete contrast to the Civil Lines. Not only the railway, but the Junction Station and the accommodation for the railway staff set out in severely rectangular fashion, separated Indian city from the European quarters.

Brush points not only to the ubiquity and the formal layout of the railway settlement, but also to its internal social distinctions:

The most distinctive features of the railway towns are the perfectly uniform grid of streets and the monotonous rows of brick dwellings, graded and rented strictly according to the wage scale, and occupational status of the employees. For the lowest-paid semi-skilled workers, who were from the start invariably Indians, a single room with attached kitchen and enclosed courtyard was considered sufficient. These dwelling units are built in contiguous rows of ten and twelve, or multiples thereof, allowing little or no open space between street and entrance. The quarters designed for the intermediate grades of personnel, often including Eurasians as well as Indians, have more rooms, some in a second storey, and are set back farther from the street. The two-storey bungalows built for the employees with the highest technical skills or managerial capacity and receiving the highest salaries, who at first were mainly British, stand amid spacious landscaped grounds with small row houses for the servants in the rear.

(Brush, 1962: 62–3)

In Allahabad the railway area was effectively deployed to block access to the north, there being no crossing for wheeled traffic for more than a mile. The railway thus completely shuts off the crowded and irregular Indian city from the broadly planned rectangular Cantonment and Civil Lines, a district of large bungalows and public offices in spacious compounds and gardens.

(Spate and Ahmed, 1950: 267).

There is no need to labour the point. Society in nineteenth-century India had a distinctive character and organization of its own. That organization with all the crudities and niceties of its ethnic and social distinctions was quite faithfully reflected in the forms of the town it developed.

Social areas of cities

In Chapter 12, dealing with the social areas of cities as part of residential patterning, some attention was directed to cities of the developing world in the context of establishing a dynamic model which could epitomize the mode of evolution of the urban mosaic. In this chapter it is appropriate that some further consideration be given towards the topic of social areas in the Third World cities in its own right.

There are problems, however. The first is that it would be inappropriate to imply that there is any such thing as uniformity within the greatly differing cultural traditions and politico-economic structures which make up that somewhat nebulous notion of a Third World, or even of developing countries. It is true that the evolutionary model noted above implies at least common characteristics of economic change. 'The accumulated evidence indicates that economic and technological factors are the dependent variables which tend to foster ecological differences' (London and Flanagan, 1976: 59). Even in relation to the penetration and impact of Western, industrial technology, however, too easy an assumption of a single path of change should be questioned until a good deal more is known from a wider range of examples. Nevertheless, a basic factor in all studies of the social areas of Third World Cities is the degree of modernization and the consequent existence of both 'traditional' and 'modern' areas.

The second problem is that data availability and economic development are highly correlated and as a result it is difficult to obtain adequate small area data to provide the input to detailed analyses of smaller towns in the least developed areas. Indeed, in such cases the very data used in Western studies could be completely inappropriate. It follows that factorial ecologies have been produced for those cities where 'development' is most advanced.

The most frequently quoted studies are those of Calcutta reported by Berry and Kasarda (1977) and of Cairo, Egypt, by Janet Abu-Lughod (1969). The latter piece of work, although now somewhat dated, presents a valuable insight into the social and physical structuring of the capital of a developing country and is worth a brief summary. Abu-Lughod used the same set of 13 variables (Table

Table 18.4 *Factor loadings, Cairo census tracts, 1960*

	Factor 1	Factor 2	Factor 3
Persons per room	−81	+01	+32
Persons per km²	−03	−10	+72
Sex ratio	−01	+97	−03
Fertility ratio	−89	−01	−00
Never married females	+95	+04	−03
Divorced females	+52	+03	+41
Never married males	+76	+49	+11
Handicapped	−40	+13	+32
Male literacy	+81	−07	−20
Female literacy	+92	+10	−19
Females employed	+85	+06	+15
Males employed	−31	−03	+18
Muslims (%)	−56	−06	+55

Source: After Abu-Lughod, 1969.
Note: Sex ratio is males per 100 females in age group 15–49. Fertility ratio is children under 5 per 100 women aged 15–49. The 'never married' rates largely reflect age of marriage. Values +/–20 are significant. Decimal points have been omitted.

18.4) for 216 census tracts for 1947 and 1960. A principal components analysis with varimax orthogonal rotation was employed and the loadings of the variables on three abstracted factors for 1960 are given in Table 18.4 and the average scores for the first factor for 13 communities grouped by type are given in Table 18.5. The 13 communities themselves are mapped in Fig. 18.3.

Factor 1 is identified as a 'Life Style' factor, contrasting those sections which were most modern, which can be read as European or cosmopolitan, with those which were predominantly traditional in their way of life. The high positive loading variables stress male and female literacy, female employment and, via the never married surrogate, later age of marriage. All these reflect the emancipation of women from the traditional roles and a modern–traditional polarization. The second factor is seen as identifying male dominance, for the sex ratio and never married males are the highest loading variables. This is brought about by the selective in-migration of unattached males. The third

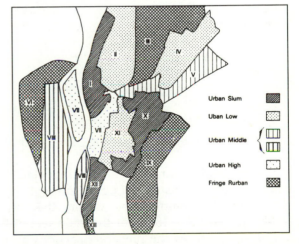

FIGURE 18.3 *Cairo, Egypt: Factorial ecology. The 13 sub-cities defined by Factor 1, Life Style. The northern and southern fringes of the city have been omitted (after Abu-Lughod, 1969).*

factor identifies social disorganization. There are clear contrasts with the factors that characterize the Western city, with the nature of regional and

Table 18.5 *The communities of Cairo 1960**

Community groupings by type	Average scores on Factor 1, 1960
Rural fringes	
North III	−1.56
West VI	−1.39
Cemetery IX	−1.54
South XIII	−1.84
Interior slums	
Bulaq I	−0.35
Medieval X	−0.37
Misr al Qadimah XII	−0.54
Urban working class, Low	
Zaytun IV	−0.08
Urban middle class	
Transition XI	+0.63
Shubra II	+0.95
Silver Coast VIII	+1.18
Upper middle or better	
Heliopolis V	+1.48
Gold Coast VII	+1.97

Source: After J. Abu-Lughod, 1969.
* For location of communities see Fig. 18.3.

intra-city migration appearing as critical controls.

It is too lengthy a task to follow the detailed spatial analysis which Abu-Lughod presents by identifying 13 communities within the city (Table 18.5 and Fig. 18.3) but the main features can be sketched. At the physical heart of Cairo is Community X which is virtually co-extensive with the medieval city. Through it runs the Qasabah along which all the market activities were located. It still retains that character, but away from it is 'a world never glimpsed by the tourist. It is a dark and dense residential *cum* commercial quarter, mud splattered and garbage strewn, penetrated by winding, narrow dirt paths, terminating here at the gate to a thirteenth century *khan*, terminating there at a cul-de-sac. . .' (Abu-Lughod, 1971: 190).

In complete contrast is Community VII, called The Gold Coast by the author, perhaps as a tribute to the role of Chicago in urban ecology for it must be emphasized that it is not a name used in Cairo. It is the Westernized modern central business area of the city, once the domain of foreign colonials but now the domain of the capital's élite. Between these two extremes lies Community XI which, forming a bridge between X and VII, a connection between the modernizing forces of the Golden Coast and the deterioration of the old city, is appropriately called 'Transition'. On the west bank of the river, that is the opposite flank of the Gold Coast, Abu-Lughod names Community VIII the Silver Coast since it is a somewhat down-market version of Community VII.

The old city, Community X, forms one of the city's slums, but the location of these poorest areas is neither characteristically 'inner city' nor peripheral, but rather related to the oldest sections. Community I, Bulaq and Community XII, Misr-al-Qadimah, were both port suburbs which have been engulfed by city growth, although extension along the river has given them a fringe character also. But the fringes are made up of distinctive communities, especially to the north where low and middle class residential areas penetrate sector-wise into the rural surrounds producing 'five sector cities'. The eastern fringe, however, made up of Community IX is dominated by 'the cemetery cities', and here is a section that can be compared to the squatter peripheral settlements classically associated with cities of the developing world.

From even this brief summary 'three types [of population] that co-exist within Cairo may be identified ... (1) the rural; (2) the traditional urban; and (3) the modern and industrial urban'. It is the differential intermingling of these that gives character to the constituent communities. In this context 'traditional' is defined as referring 'primarily to the persistence of economic activities, forms of social relationships, and systems of values which were once typical

Table 18.6 *Component structure for Accra*

Variables	Factors			
	I	II	III	IV
Local born	−77	−27	−17	30
Foreign Africans	39	76	−21	−10
Non-Africans	70	−37	08	−13
Child/women ratio	−77	10	04	17
Sex ratio	69	29	−18	−05
Educated adults	06	−82	06	−16
Unschooled children	10	85	11	−01
Working-age males	23	12	−53	−15
Employed males	11	11	10	−84
Unemployment	−10	23	05	78
Unemployed females	71	05	37	25
Density	−35	−06	−62	28
Distance	14	10	86	−07

Source: After Brand, 1972.

within the Cairo of a hundred years ago'. To a degree it can be linked to the notion of the lower of the two circuits of the urban economy in undeveloped countries proposed by Milton Santos in his book *Shared space* (Santos, 1979). The two circuits not only characterize the economy but must appear in the spatial structure of the city also, though never divorced into discrete areas for, as Santos avers, the space is shared.

A conclusion, generally in parallel with the above, emerged from a study of a city in a contrasted part of Africa. This was an analysis of the spatial organization of residential areas in Accra, Ghana, by R. R. Brand. Thirteen variables which could be derived from the 1960 census were assembled for 268 enumeration areas. Four components were abstracted from a principal components analysis (Table 18.6) explaining 28.5, 17.0, 12.2 and 9.8 per cent of the variance respectively.

Component 1 is characterized by Brand as bourgeois migrant communities given the high loadings of non-Africans. These sub-areas exhibited very low proportions of local-born persons, relatively few children per woman, and medium

to low gross population densities. They also had a large proportion of economically non-active females [and] large representations of non-African expatriates' (Brand, 1972: 291). The distribution pattern is clearly sectoral, with a wedge extending north-east from the city centre. There was one outlying area in Tesano (A on Fig. 18.4) 'a European enclave analogous to the wealthy commuter suburbs ringing many western cities'.

Component II shows high positive loading variables of foreign Africans and unschooled children, with a high negative loading on educated adults. This is interpreted as 'urban-villagers – the sub-culture of traditionalism among migrants'. Three areas can be identified on Fig. 18.4. The first two, Nima and New Town (B on Fig. 18.4) and Sabon Zongo (C) are areas of squatter dwellings and subdivided houses in a peripheral location, while Ussher Town and part of Jamestown (D on the map) are the city centre parallels. It is significant that Brand notes one sub-area, Tudu, as a popular destination node for long-distance migrants, especially Hansa and Yoruba traders, and Sabon Zongo as developed by Muslims. Component III is a

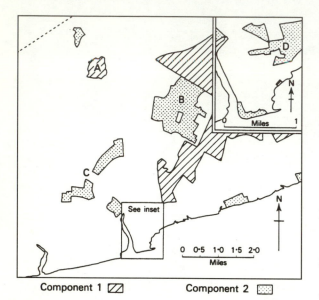

FIGURE 18.4 *Accra, Ghana. Factorial ecology. Areas with high scores on the first two components. Component 1: Bourgeois migrant communities (more than two standard deviations above the mean). Component 2: Urban villagers: the culture of traditionalism among migrants (more than one standard deviation above the mean) (after R.R. Brand, 1972).*

distance component contrasting the low density fringe with the centre, and IV an unemployment component.

Two immediate reflections arise from Brand's study. The first is the clear parallelism with Abu-Lughod's analysis of Cairo; the first and dominant contrast is between aspects of modernization and traditionalism. The second is the way in which ethnic associations, or tribal links, remain strongly reproduced in the urban situation. Urban ethnicity (Cohen, 1974) is itself the product of migration chains and a feature of the largest towns and cities. The characters of homogeneous settlements, smaller in size, and in areas little affected by Western notions, are very different. It is possible, therefore, to query many of the analyses of modernizing cities and using variables closely ranged as possible to those of Western factorial ecologies; it is little

wonder that results first contrast those parts which are amenable to such analyses with those that are less so.

At this point it is possible to revert to another contribution by Abu-Lughod where, in relation to North Africa, she identifies 'six fairly distinctive and co-existent urban arrangements which make up the ecological structure (Abu-Lughod, 1976: 202). These are:

1. **Medina core**. This is the medieval or traditional nucleus.
2. **Modern appendage**. This was either designed for and built by Europeans or constructed in European style.
3. **Rapidly proliferating uncontrolled settlements**.
4. **Peripheral suburbs**. These are built for an indigenous middle and upper class and expand in sector-like fashion between the uncontrolled settlements.
5. **Rural fringe**. This is made up of wedges of undeveloped rural land, in sharp contrast to the urban sectors.
6. **Transitional working class zones**. These are usually the intermediaries between 1 and 2.

This situation is illustrated in Abu-Lughod's study of Rabat in Morocco (Abu-Lughod, 1980). Rabat (this discussion will set aside the attached settlement of Sale to the east of the river, the Bou Regreg) was founded in the twelfth century but failed to develop and it was not until the early seventeenth century that the walled city was rebuilt by Muslims expelled from Spain. The two main elements of the layout were the castle, the Qasbah of the Udaya, and the Medina, the characteristic mercantile complex of narrow streets and alleys. In 1912 the French established a protectorate over Morocco and Rabat was made the administrative centre. The Medina was left as it was, surrounded by its defensive walls which were retained as a symbolic and legal barrier between the indigenous Moroccan population and the occupying French. As the British in India, so the French in Morocco built a new city, 'la ville nouvelle', to the south of the Medina (Fig. 18.5). It was composed of broad

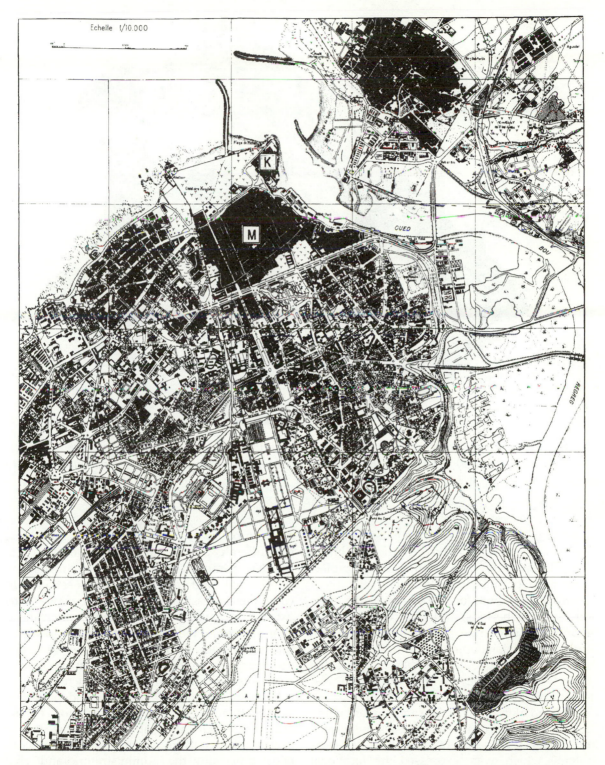

FIGURE 18.5 *Rabat-Sale, Morocco: A general map. M marks the location of the walled medina.*

☐	1.7500000 – 10.000000
▨	.5000000 – 1.740000
▨	–.1400000 – .490000
▨	–.3900000 – –.150000
▨	–.6400000 – –.400000
■	–10.0000000 – –.650000

FIGURE 18.6 *Rabat, Morocco. Factorial ecology. Factor 1 'Caste' (after Abu-Lughod, 1980). Source: J. L. Abu-Lughod, 1980: Rabat. Urban apartheid in Morocco. Princeton. NJ: Princeton University Press. Figure 13, p. 307.*

boulevards and large round points (*places*) in classic French planning style. All the new buildings, administrative and cultural, of the protectorate were located in this 'ville nouvelle'.

Growth in population, especially in the inter- and post-war periods, and particularly by that urbanward migration of population which characterized developing countries, led to an influx of rural people. Unable to enter the property market they settled on rough and unwanted ground along the cliffs of the Bou Regreg and along the barren sandy areas behind the coastline. Out of these processes there came into being what Abu-Lughod calls a 'caste system' which was expressed in locational terms. The old Medina still retained the consid-

erable mixture of populations of the traditional core of crafts and of minor trade. To the south, the new administrative city was dominated by the French and other European settlers. In complete contrast, and in the form of two arms extending along the river and the Mediterranean shore, were the bidonvilles and shanty towns of the immigrants. As in British India, so in Rabat, the sharp divisions brought about by colonial rule were replicated in the city structure.

With the coming of Moroccan independence in 1956 there was an obvious basis for change, but one which Abu-Lughod epitomizes in a chapter title as 'from caste to class'. A Moroccan élite, especially the administrators and civil servants who could afford the high rents and

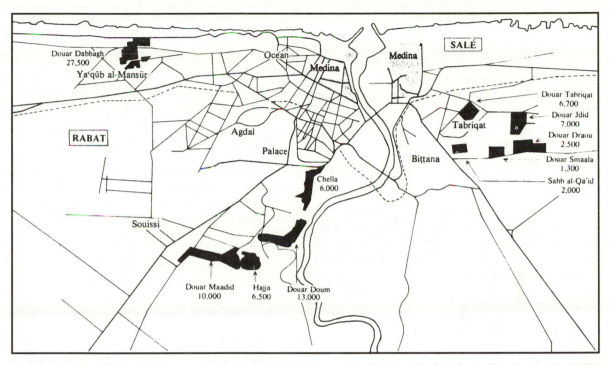

FIGURE 18.7 *Rabat, Morocco. The location of the major 'bidonvilles' (after Abu-Lughod, 1980). Source: As Fig. 18.6. Fig. 11, p. 284.*

building costs, took over the 'ville nouvelle'. Its population increased only slightly. In contrast, the poor peripheral areas experienced that rapid growth which, as has been seen (p. 21), so often goes along with political freedom. By 1970 there were some 77,000 living in bidonvilles. 'Throughout the metropolitan area segregation by economic class was supplanting the trimodal segregation by caste that had been the dominant pattern during the colonial era' (Abu-Lughod, 1980: 268–9).

All this is reflected in the factorial ecology of Rabat as set out by Abu-Lughod using some 27 variables derived from the 1971 census of Housing and Population. Only Factor 1 is considered here. It has high positive loadings on the percentage born abroad, the percentage literate in a language other than Arabic, the percentage of females over 15 who can read and write, the percentage of the male labour force in technical and managerial occupations and the percentage of females in employment.

The main negative loadings are on the percentage of Moroccan Muslims, the persons per room ratio and average number of persons per household. Abu-Lughod writes, 'factor 1 is clearly a socio-economic factor in which ethnicity and status are inextricably entwined' (Abu-Lughod, 1980: 299). Significant among the variables loading highly are both male and female literacy – 'the rate of female literacy indicates in particularly sensitive fashion not merely status but "degree of modernity". Literacy rates are "style of life" measures as much as they are indicators of economic power' (Abu-Lughod, 1980: 299).

The distribution of Factor 1 is shown in Fig. 18.6, while Fig. 18.7 outlines the general structure of the city. Clearly the old divisions still exist and are epitomized in Abu-Lughod's identification of five urban types.

1. Upper class areas of the central plateau section.

2. Lower and middle working class areas relatively small in extent and characterizing parts once occupied by less affluent foreigners at the edges of the upper class area. Ocean is an example (Fig. 18.7).
3. The Medina which has resisted degradation and is highly complex in character.
4. The rural fringe which is beginning to be influenced by urban extension.
5. Illegal or squatter settlement; the bidonvilles (see Fig. 18.7).

The Medina core still emerges as distinct with an intermediate position on Factor 1. The 'ville nouvelle' constitutes an upper class area to the south, while the two arms of the poorest illegal settlement are immediately apparent. Independence has not given rise to a more equal city; class has replaced caste.

At the beginning of this section the significance of economic and technological factors was noted and their influence in setting up convergent patterns suggested. Most Third World authors react fairly sharply against the notion that their cities are becoming more like those of the West, and still more consistently against any view that their cities are following the paths of Western cities at earlier periods. National and cultural pride are potent factors in such reactions: not for the first time in this book it has to be noted that the perception of the world is a relative matter. It would not be difficult to find parallels for the six 'urban arrangements' of Abu-Lughod (p. 348) in eighteenth- and nineteenth-century Britain, while Rabat shows clear evidence of parallelism with Western cities. Abu-Lughod even uses the word 'apartheid' and the combination of an élite occupying privileged and advantaged areas in contrast to the misery of the bidonvilles certainly recalls the city of apartheid (p. 276).

But most certainly since modern communications, both via the mass media and in the more direct physical sense of moving populations, are so much the greater, the problems in developing areas are greatly exacerbated. One of the greatest of these, certainly the one with the most extensive literature, is that of the uncontrolled peripheral settlement.

Spontaneous or squatter settlements

The huge extent of urbanward migration has already been identified. Within the cities themselves it produces immense social problems, the most immediate of which is inadequate housing. Partly as a result of the size of the problem, partly as a result of the limitations of national and municipal resources and partly as a result of social systems which inhibit the equitable division of resources, little shelter is provided through formal and planned channels. The result is that the immigrant population collects in extensive peripheral areas of settlement:

> Peripheral settlements account for at least one quarter, and often one third or even one half of the total urban populations in most primate cities. Thirty-five per cent of the people in Manila, thirty-three per cent in Calcutta, twenty-five per cent in Jakarta, forty-five per cent in Istanbul, twenty-seven per cent in Rio de Janeiro and forty-six per cent of the population in Mexico City live in slums and uncontrolled settlements.
>
> (Ulack, 1978: 535)

The names for this phenomenon are widely varied, both in the academic and planning literature and in the vernacular languages. Uncontrolled settlements, as in the quotation above, spontaneous settlements, autonomous settlements, illegal settlements, all of these have been used from time to time. Drakakis-Smith has argued for the use of the term squatter settlements or squatter housing, defining it as:

> any housing which contravenes existing legislation on the occupation of land or the construction of dwellings. The juridicial definition is preferred to currently fashionable neologisms such as 'spontaneous' or 'uncontrolled' urban settlements, because the illegal nature of such housing strongly affects all aspects of life in the squatter community, from the physical environment, through types of economic activity to relations with the urban authorities.
>
> (Drakakis-Smith, 1978)

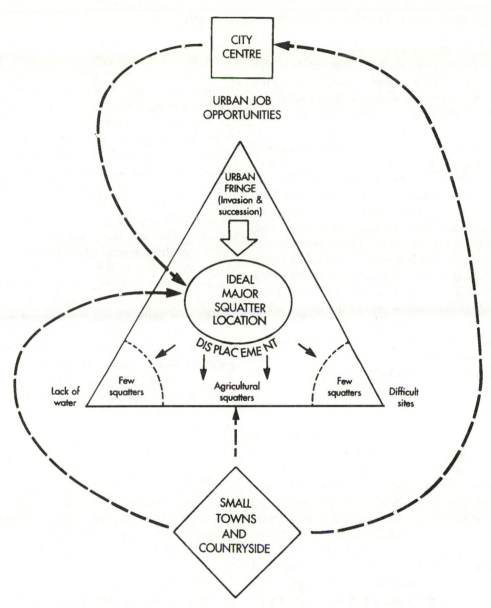

FIGURE 18.8 *Typical movements of low-income population in cities in developing countries (after D. J. Dwyer, 1975).*

In this he is undoubtedly correct since the distinction between these settlements and slums is that the residents of the latter, however dilapidated the physical conditions, own the legal title or formally rent the property. In the favellas, barrios, bidonvilles and bustees, perhaps the most widely used names for the squatter settlements, there is no legal title to the land occupied.

It has been suggested that urban poverty is simply rural poverty displaced and such a view seems appropriate in relation to the spontaneous settlements, for the move to the city, as has been indicated, is powered by rural change and rural poverty. Once arrived at the city the migrants are faced with a group of related problems. They have no skills to offer in a market which increasingly seeks skilled workers so that the best that can be hoped for is either sporadic labouring employment or petty trading of the most limited kind. Under these circumstances no bid can be made for housing and the only way is to construct shelter from available materials, galvanized sheeting, tin cans, odd pieces of wood, cardboard boxes, attempting gradually to improve the structure. It is in this fashion, and unprovided with water, sewerage or electricity that the spontaneous settlements have developed. Dwyer begins his book on *People and housing in Third World cities* with a section entitled 'A major urban form' (Dwyer, 1975: 34) and by so doing properly calls attention to the fact that these settlements do constitute one of the major urban features of the contemporary world.

Four further features of these settlements can be briefly considered: location, character, evolution and future development.

Location

Dwyer has constructed a diagram reproduced as Fig. 18.8 which shows typical movements of low-income population in the cities of developing countries (Dwyer, 1975: 34). Migration takes place both characteristically to the inner city areas of reception as well as to the periphery. But, as the central city expands, those least able to meet rising rents are pushed to the periphery. They are successfully and continually displaced into worsening environments in terms of water access and site conditions, such as steep slopes. At the same time the periphery becomes the prime point of entry for migrants (Conway, 1985). Inevitably a situation arises where the poorest have to pay high journey to work costs. Also, peripheral land, low in quality because of steepness of slope, is likely to be that which can be 'squatted on' most easily. The spontaneous settlements, therefore, take up the least desired locations, virtually when land values have fallen to zero, for the terrain can be used neither for agriculture nor for good quality housing.

Character

Much has been written on the degrading conditions of these settlements. Perhaps the most effective public statement was that of the São Paulo Justice and Peace Commission in their report entitled *São Paulo growth and poverty* (1978). That volume pointed out that São Paulo is the fastest growing city in the world having grown from a small town 100 years ago to 12 million at present, and if growth rates are maintained, 25 million by the year 2000. The result has been a failure to provide essential facilities. Only 40 per cent of the 8000 kilometres which make up the road network for local traffic in the Metropolitan Region are paved. Approximately 489,000 inhabitants live in houses without electricity. Only about 30 per cent of houses in the Metropolitan Region have drains and only 53 per cent piped water' (São Paulo Justice and Peace Commission, 1978: 31). In terms of particular areas within the municipality of São Paulo, conditions are considerably worse. In 1968 at Itaquera, 89 per cent of households lacked drains, 97 per cent had no piped water and 72 per cent had no refuse collection. With the growth of population the situation is deteriorating, not improving.

Evolution

L. J. Eyre in a study of the shanty towns of Montego Bay in Jamaica (Eyre, 1972) proposed stages of development for these forms of settlement. These he identified as follows:

1. The initial occupation stage
2. The transitional stage
3. The stage of attaining secure tenure
4. The stage of absorption.

This suggested evolution implies a gradual change from an initial state of insecurity and lack of organization, to a gradual achievement of a feeling of permanence. This is accompanied by the improvement of dwellings and the gradual establishment of small businesses so that eventually the settlement becomes part of the city. 'Other things being equal, the older settlements have the greatest potential to become viable city communities. More recently established settlements, if their residents are not evicted and if they are upwardly mobile in a socio-economic sense, also have the potential to become part of the urban system because the city is expanding its opportunities and amenities to these peripheral areas' (Ulack, 1978: 539). Ulack, in a study of Cagayan de Oro in the Philippines, confirmed three hypotheses regarding the change with time, that the age of settlement is the most critical variable in determining the role of squatter settlements; that the oldest settlements have the best locations, closest to employment and urban amenities; that the oldest settlements have the highest socio-economic status as measured by education, income levels, jobs and housing conditions. Because of this, he argues, 'squatters in the older settlements have the potential to contribute more to the city. The existence of small business, membership in association, better housing and employment in more skilled jobs are all indications that the older settlements are more highly integrated into the city' (Ulack, 1978: 548–9).

Future development

J. F. C. Turner, one of the earliest and most influential writers on squatter settlements, argued that, 'It is clear that ... the peripheral squatter settlements do, in fact, perform the principal functions demanded by their inhabitants. So, in spite of the many, and often severe drawbacks, they often act as forward moving vehicles of social and economic change' (Turner, 1969). This is the role of these settlements, similar to that of the so-called slums of hope, that is positive and can be built upon. The direct task of rehousing the populations of the squatter settlements by the municipalities themselves is one on such a scale that it cannot be the only, even the major policy. The pattern of change has to be encouraged. The first issue is that of property titles, for the inhabitants need the security of ownership in order to have the incentive for improvement. Most of the households are too poor to want standard credit

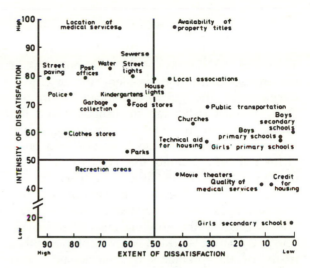

FIGURE 18.9 *Attitudes to public and private services in the Lima 'barricades' (after F. M. Andrews and Philips, 1970).*

arrangements for housing improvement, preferring to add to their dwelling as opportunity permits.

Figure 18.9 shows the attitudes to public and private services in the Lima barricadas after a survey by Andrews and Phillips (1970). The intensity of dissatisfaction registers the importance of property titles and, not unexpectedly given the environmental conditions, the demand for accessible medical services. The extent of dissatisfaction indicates the demand for standard urban services. It is the provision of these services to the squatter sites that has become the dominant policy in many areas, the so-called site and services approach. Given the basic facilities, the initiative of the populations itself is sufficient to engender the gradual process of upgrading. Certainly solutions on Western housing development concepts will be completely inadequate in the face of the scale of growth in developing countries. Notions of 'slum clearance' have no meaning whatsoever under the conditions which have been discussed. The management of the spontaneous settlements and their progressive improvement is one of the critical problems in practical urban research.

18.4 CONCLUSION

This short chapter has sought to set out some of the characteristics and problems of the city in developing and underdeveloped countries. Those characteristics are so distinctive and the problems are so great that they have engendered an extensive and growing literature. In a book written mainly around the Western city, one limited chapter can do little more than act as a token, an acknowledgement that the stresses and disadvantages that occur within Western cities are dwarfed by the growing crisis of urbanization in the Third World. It is pertinent to add that many of the problems of homelessness and shanty towns are seen by those on the radical left as the direct product of the penetration of developing countries by Western capitalism (Burgess, 1978, 1981). The causal chain linked to that penetration has been implicit throughout this chapter. The impact of capitalist modes of production undermines the stability of rural peripheries and engenders the mass migration to cities. Moreover, the control of production exercised from advanced countries leads to a switch from labour-intensive to capital-intensive methods of production which restricts the demand for labour so that employment chances for migrants are diminished. At the same time, these poorest countries become dependent on foreign capital for much of the essential investment and incur greater debt service payments. All this implies that the squatter settlement problem is the product of deeper lying economic mechanisms and site and services and self-help programmes are merely tinkering with the symptoms rather than curing the illness – indeed they can be regarded as supporting the system which created the housing problem in the first place. But it is difficult to envisage any major change in the established system and in that context amelioration of conditions is a necessary policy. Perhaps the greatest test of what can be done will come in South Africa and the extent to which the new democratic government can fulfil its promises to those now living in squatter camps or poor quality housing, and that at a time when increases in the black population are likely to be substantial.

TOWN PLAN AND TOWNSCAPE

It might seem somewhat illogical, even perverse, or at the worst an afterthought, that a chapter on town plan should appear at the end of this volume rather than at its beginning. For surely it can be argued, the plan must be considered as constituting the stage on which the action, or the processes with which so much of the book has been concerned, is played out. To change the metaphor, although to no great extent, in order to describe any game, be it football or tennis, cricket or baseball, boxing or golf, the first essential is to set out the dimensions and markings of the field, pitch, court, ring or course on which the game or contest is to take place. It would seem sensible that a similar procedure should mark the description and analysis of towns and town life. In so far as towns throughout history have been laid out with set street plans, there is something to be said for such a view. But, in contrast, town plan, especially as an integral part of the more complex notion of townscape, is the crucial element in the total visual image of the city. And hence its interpretation is the final step, the last stage in which the influence of all those processes which this book has considered are brought together. There emerges the end result, the physical structures which make up the city, and although the inhabitants live and have their being within those structures, it is that living and being which created them. There is, therefore, every justification for placing a consideration of town plan as a last chapter. Ideally, all would come together in an integral presentation of the impact of all the forces and processes considered in the various chapters in the creation of townscape. Reality is unfortunately far removed from that, but even so, the argument presented is a firm indicator that this chapter is in its proper place.

The contrast between the ideal concept of the townscape as the summation of urban processes and the reality that such a convincing interpretation was rarely presented, dominated studies of town plan at a time when during the 1950s and 1960s the great leap forward in other aspects of urban geography was taking place. Thus, for example, at The Lund Symposium on urban geography in 1960 Garrison represented a standard view when he maintained that on urban morphology 'the studies we are now making are no richer than the studies we were making twenty or thirty years ago' (Garrison, 1962: 463) and blamed the situation on the lack of development of general theory. The succeeding decades, therefore, have been taken up by the attempt to provide for studies of urban morphology the sort of stimulus which central place theory provided for the study of urban functions and city location (Whitehand, 1992b).

The central difficulty with which these attempts have had to contend is that, of necessity, there must be a strong historical element in any study of town plan or morphology, for the moment treating the two as synonymous. Although it can be argued cogently that analyses of the urban hierarchy, or of the distribution of land values or social class within the city, need an historical input in interpretation, they can be, and have been most frequently studied in purely contemporary terms. Certainly in early published work on the urban hierarchy there was very little, if any, interest in the past. In most aspects of urban geography, and almost of necessity in model building, the interaction of contemporary forces was dominant. Manifestly in urban morphology such an attitude was much more difficult, if not impossible and approaches suggesting measurement were summarily dismissed. The derivation of the crucial elements of plot, street and building must necessarily be referred to the past when they were created. There follows the close relationship between morphological studies and urban conservation. Thus although conservation can be applied to functional status and the retention of a role, as in the attempts, mostly abortive, to retain coal mines and the well-being of settlements which depend on them, and certainly to urban land-use in the perceived need to safeguard city centre shopping as against out-of-town developments, nevertheless, it is most frequently applied to specific buildings or built-up areas of cities where the direct relationship with preserving something of value from the past is most obvious.

Given the central role of inheritance from the past in urban form, there has followed a strong tendency for morphological and townscape studies to take the nature of narrative histories of town planning. Beginning with the urban forms of the centres of urban origins, successive chapters deal with the Greek city, the Roman city, the nature of medieval urbanism, the town layouts of the Renaissance and Baroque periods, the emergence and character of the industrial town and thence, via the Garden City movement to modern and contemporary trends. Books in this field and with such an organization have proliferated, The early standard was set by Lavedan's *Histoire de l'urbanisme*, the first edition of the volume simply called *Antiquité*, being published in 1926, with succeeding volumes appearing much later; *L'Urbanisme au moyen age* in 1974, *Renaissance et temps modernes* in 1959 and *Epoque Contemporaine* in 1952. The doyen of such work in English was the American Lewis Mumford; *The culture of cities* was published in 1938 and an enlarged and revised edition as *The city in history* in 1961. A somewhat different and primarily regional approach was Gutkind's attempt at a world survey in the eight volumes of his *International History of City Development* published between 1964 and 1972 (Gutkind, 1964, 1965, 1969, 1970, 1971, 1972a and b). The other characteristic form of publication has been the town plan atlas (Lobel, 1969; Andrews and Simms, 1988). Some works on the theme are manifestly textbooks (Hiorns, 1956; Burke, 1971; Morris, 1972) while others move towards the 'coffee table' category (Lloyd, 1984; Girouard, 1985). Apart from Lavedan's volumes, the examples cited have been in English, but virtually every language has its representatives. Thus in Italian Benevolo's *Storia della citta* (1975) is the obvious work to quote, though it has been translated into English as *The history of the city* (Benevolo, 1980) and followed by a more succinct study *The European city* (1993) which was published simultaneously in five languages. A recent French series is the *Histoire de la France urbaine* edited by Duby (1980–5).

It would be wrong to decry in any way the scholarship of these books, but they are very repetitive, often using the same examples. They are essentially narrative histories which do not meet the challenge of creating a theoretical framework. There have been three reactions to this impasse. The first differs little from traditional approaches but puts an emphasis on the cultural, social and political contexts of town founding rather than on historical periods. The second is greatly different in that it attempts to analyse more directly the structure of towns, devising a terminology to do so and seeking to demonstrate universal elements. The third, which can be closely associated with the second, and indeed is represented in the first, moves, like all the other areas of urban geography which this book has considered, to a behavioural level and seeks to identify and explore the role of the decision makers in the formation of the townscape.

19.2 TOWN PLAN – ISOLATION OF DOMINANT CREATIVE FORCES

One of the earliest attempts to isolate an array of general forces which were instrumental in creating a particular plan was that by Stanislawski (1946). In a paper which was primarily concerned with the evolution of the grid pattern, he derived certain general conditions which, he maintained, were necessary for such a plan to be adopted. That is, Stanislawski presented an argument that independent of particular historical circumstances, a grid will emerge given certain conditions. The five conditions he isolated and proposed were :

1. A new town, or new part of a town is in question.
2. There should be centralized control.
3. Very often the new town has colonial status.
4. There should be a measured disposition of available land.
5. There should be a knowledge of the grid.

Inspection of these will reveal that, excepting the last, they can be reduced to the single notion of the operation of centralized political control in a colonial or quasi-colonial situation. But it is possible to carry the argument further. In more general terms, uniformity of plan (though this may take a wide variety of forms other than the grid) reflects organized central control, so that contrasts in the uniformity of plans are a reflection of the degree of the concentration of power.

This concentration can be interpreted via the decision-making process: thus when decision making is fragmented, plans are incoherent; when it is concentrated, plans are coherent. The merchants of medieval London built their houses where they wished within the plots allocated to them and from that fragmentation an incoherent plan of streets and alleys resulted. The modern suburbs are laid out by a municipal authority and a coherent plan appears.

This particular line of argument may be contested, although it is more likely to be judged self-evident. But the crucial point is that it is a line independent of **particular** historical input defining a unique situation. It is the essence of this argument that the variables which shape plan can be isolated independently of any one historical phase. It is possible to propose a list which would include, for example, political, economic, cultural and religious factors. These do not work independently, however, and most plan elements are a product of two or more, or indeed of all of them. In the present context they may be considered separately.

Political factors in town form

Under the heading political influences are included those forces which are related to the concentration of power either in the hands of a single person or a group of people. It has already been maintained that the conditioning forces of the grid plan as envisaged by Stanislawski nearly all pertain to the concentration of power. The same situation is seen in that phase of town planning which is usually given the name Renaissance and Baroque. Mumford, outlining the controlling conditions of the new urban complex, writes, 'between the fifteenth and eighteenth centuries a new complex of cultural traits took shape in Europe. ... The new pattern of existence sprang out of a new economy, that of mercantile capitalism; a new political framework, mainly that of a centralized despotism or oligarchy, usually embodied in a national state; and a new ideological form that derived from mechanistic physics' (Mumford, 1961: 345). Here Mumford associates three of the variables that have been isolated – economic, political and cultural – and from them derives urban form. *The city in history* is perhaps the outstanding study which sees plans as the product of such associated forces, although their analysis is confined to the usual historical progression. But in a section, headed appropriately 'The ideology of power', Mumford derives the obvious parallel with a much earlier period, 'thus the baroque rulers reinstated all the institutions of the original urban implosion' (Mumford, 1961: 367). But the comparison is not completely true for although the processional way of Ur might have been a response to the same basic forces as the avenues of Versailles, the residential areas were very differently disposed. The order and harmony of mechanistic physics had produced an emphasis on the complete balance and control of the whole plan, but such was certainly not the case at Ur where the detail of much of the city seems to have been the consequence of innumerable individual decisions. But it follows that the city plan is the reflection of the concentration of power and of the cultural forces at work.

Versailles has already been noted; Richelieu (Fig. 19.1) provides a smaller but no less appropriate example and can indeed be regarded as the precursor of Versailles (Lavedan, 1959: 228–32). Cardinal Richelieu decided in 1625 to transform an old residence and convert it into a chateau fitting to his estate. The architect was Jacques Lemercier. The chateau was finished by 1635, but in 1633 royal permission had been obtained 'fair batir un bourg clos, avec pouvoir d'y etablir des marches et foires ayant pareils privileges que celles de la ville de Niort et de Fontenay-le-Comte' (Lavedan, 1959: 228). The whole layout displays the unity of design that comes from being built at one time and under the complete control of one hand. The aspects of design, which Lavedan in his *Histoire de l'urbanisme* carefully notes, are a consequence of that control. It is possible to discuss at length the aesthetic and planning principles of this case but these details, such as the way the main squares are disposed, are a consequence of the design ideas of the time.

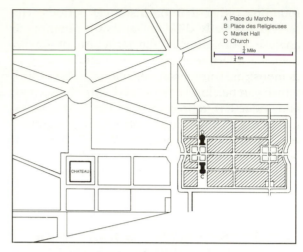

FIGURE 19.1   *Richelieu. This is a composite map which depicts in diagrammatic form the layout of the town and the château and its grounds.*

The same concentration of power, although in a somewhat different general context, led to the ground plan of Washington, appropriately a centre of federal government and the focal point of the political power of the United States. At the same time it had clear links with its European predecessors through L'Enfant as designer – who 'was the product of his age and the instrument through which certain principles of civic design that had been developed in Western Europe found expression on the Potomac river' (Reps, 1965: 252). The whole situation however, in which these principles could be put into effect, depended upon the unity of political control vested in the commissioners as the instrument of the president of the United States.

If, in contrast, one considers towns during the industrial period, two differing aspects appear. On the one hand the concentration of power led to the construction of company towns with very marked rectangular patterns. But at the same time the freedom of many places from any controlling national or municipal authority led to the conditions of near anarchy under which planless agglomerations of poor cottages came into being, creating the feature so often associated with industrialization (see p. 217–18). These conditions in turn stimulated a social reaction which itself generated carefully structured towns – giving what has been called 'the city as social conscience'

(Carter, 1983: 125).

Cultural factors in town form

To a large extent in the above discussions a second variable has been introduced but not identified and this is the one which was initially called 'cultural'. This is the system of social values which underlies town layout. Perhaps an appropriate example is Philadelphia (Fig. 19.2). 'The dominant force in the shaping of Philadelphia', writes A. N. B. Garvan, 'was the philosophy of William Penn' (Garvan, 1966: 197). This can be applied to the layout as well as to the general atmosphere of the early city. The charter which Penn obtained from Charles II in 1681 established him as governor and proprietor. Purchasers of a share in the province were also to be provided with a city plot within Philadelphia. The commissioners who accompanied the first settlers were given detailed instructions, dated 30 September 1681, for the laying out of the new town, and the regularity of a somewhat unusual town was clearly provided for by Penn in the instructions:

> Be sure to settle the figure of the town so that the streets hereafter may be uniform down to the water from the country bounds; let the place for the storehouse be on the middle of the key, which will yet serve for market and statehouse too. This may be ordered when I come, only let the houses built be in a line, or upon a line, or as much as may be ... Let every house be placed, if the person pleases, in the middle of its plot, as to the breadth way of it, that so there may be ground on each side for gardens or orchards, or fields, that it may be a green country town, which will never be burnt, and always be wholesome.
>
> (Reps, 1965: 160)

Garvan points out that this implies a very unorthodox city since Penn 'neglected the urban centre for his gentlemen's seats. He visualized, not a city but a residential district of regularly arranged parks with uniform streets down to the water from the country bounds' (Garvan, 1966: 190). The only commercial consideration is the 'place for storehouses'. The actual planning on the spot by Thomas Holme brought to bear some experience of town foundation in Ireland as well as much of the discussion of the reconstruction of London after the Great Fire, so that eventually 'little remained of Penn's original plan and instructions. Instead, an extremely interesting and complex plan evolved which owed little to the proprietor except the partial achievement of his intent. But nevertheless Holme firmly agreed with Penn's objectives to create a regional plan suitable for Quaker worship in which the proprietor's interests would be secure.' Garvan develops the contrast between Philadelphia and the New England settlements from which it had two major differences. The settlement at Philadelphia was widely scattered and settlers isolated and the whole of the three counties was covered more or less at once; there was no moving frontier. Only a population predominantly free of dependance, either upon a fixed ministry or on a place of worship, could make so rapid an advance. Friends free to worship with one another in their own houses found the scheme totally agreeable, while their contemporaries disputed in New England legislatures the precise location of Puritan meeting houses and adjacent home lots (Garvan, 1966: 196).

It is also instructive to observe the consequences for the plan once the concentration of decision making in Penn's hands was broken. The American Revolution destroyed the proprietary, which was transformed into a 'political administration for the benefit of democratic voters and settlers'. The result was the substitution of a varied haphazard 'control', in which individual speculation was a key factor. 'It would be at this point quite easy to condemn the *laissez-faire* city and point out that as an artefact of the mature philosophy of William Penn, Philadelphia's end was near' (Garvan, 1966: 198). The point here is neither to condemn nor praise nor even to evaluate plan in such terms at all but to demonstrate the way in which the variables that have been isolated act in fact as a complete

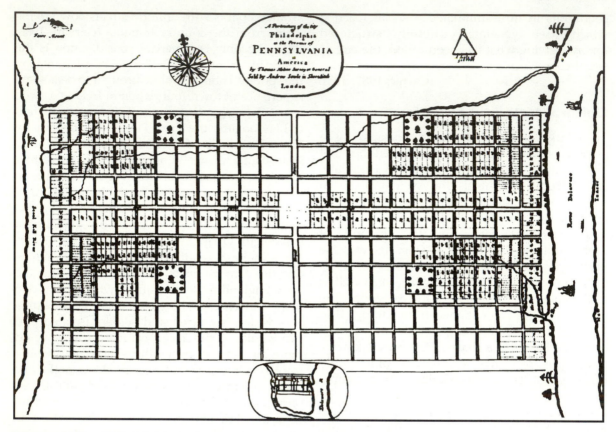

FIGURE 19.2   *Philadelphia in 1682.*

matrix. The single control of Penn represents the concentration of decision making already identified; the city he built was the creation of the cultural values by which he lived. As power was dispersed other values began to play their part, very different from the Quaker beliefs of Penn. The general layout of the whole area as well as part of the particular detail of the city of Philadelphia can, however, be fairly associated with the general cultural inheritance of the proprietor and his associates. In particular a contrast between the mode of settlement in the area and that of the Puritans of New England can be related to differences in the form of worship.

Religious factors in town form

This aspect, of religious influence on plan, is one which deserves more attention than it has

had. In extra-European areas cosmological forces have been particularly important and their significance within Europe has not been properly evaluated. Wheatley, in a critical review of Sjoberg's *Pre-industrial city*, argues that Sjoberg 'does less than justice to a principle related to that of symbolic centrality, and that is the construction of the city as an *imago mundi*, with the cosmogony as paradigmatic model' (Wheatley, 1963: 163). Wheatley further points out that southeast Asian capitals were often laid out as images of the universe and quotes Groslier's interpretation of the temple-city of Kambujadesa, 'un diagramme magique trace sur le parchemin de la plaine' or 'an expression in plastic terms of Khmer cosmological concepts' (Wheatley, 1963: 179). The use of cardinal axes has been extremely widespread and even in terms of *cardo* and *decumanus* must preserve a basically cosmological purpose.

Perhaps the most intensive study of the religious influence on town layout is that by Rykwert on *The idea of a town* with the subtitle of 'The anthropology of urban form in Rome, Italy and the ancient world'. He concludes, 'It is difficult to imagine a situation where the formal order of the universe could be reduced to a diagram of two intersecting coordinates in one plane. Yet this is exactly what did happen in antiquity: the Roman who walked along the *cardo* knew that his walk was the axis round which the sun turned, and that if he followed the *decumanus*, he was following the sun's course. The whole universe and its meaning could be spelt out of his civic institutions' (Rykwert, 1976: 202).

Again the well-known example of Angkor Thom demonstrates 'the minutiae of cosmic symbolism':

> The city was surrounded by a wall and moat forming a square almost two miles on each side, its sides being directed towards the four cardinal points. There are gates in the middle of each side and a fifth one on the East leading to the entrance of the royal palace. The towers above the gates are crowned with the same fourfold faces of Lokesvara as those of the central temple. Thus, that smaller world, the city of Angkor, and through its means the whole Khmer empire, were put under the protection of the 'Lord of the Universe'.
>
> (Heine-Geldern, 1958: 34)

In Burma, the ancient city of Srikshetra (Old Prome) was, according to tradition, built by the Gods themselves as an image of Indra's city, Sudarsana. The remains of the city in fact show a circular layout. In this context perhaps one of the most fascinating instances of cultural change is in the squaring of Circleville, a town situated in the state of Ohio in the United States (Reps, 1965: 484–90). 'The original settlement was established in 1810 after the establishment of Pickaway County. The site selected on the Scioto river was occupied by a large circular Indian earthwork and the new settlement was adapted

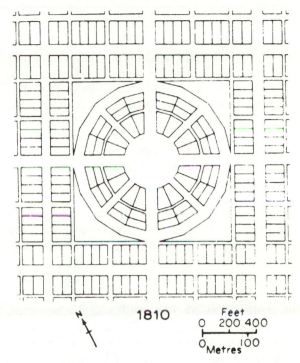

FIGURE 19.3   *Circleville, Ohio, in 1810 (after J. W. Reps, 1965).*

to the shape of the circular enclosure.' This produced Circleville (Fig. 19.3), a unique radial concentric scheme in the early town plans of America. Reps writes 'while one is inclined to give credit to an unknown tribal chief or priest, the honours must surely go to Driesbach [the founder] for realizing the potentialities of the site'.

Here then the magico-religious significance of an Indian shape had been transformed into the layout of an American town. But by the time that James Silk Buckingham visited the town in 1840 changes were in progress:

> So little veneration . . . have the Americans for ancient remains, and so entirely destitute do they appear to be, as a nation, of any antiquarian taste, that this interesting spot of Circleville is soon likely to lose all traces of its original peculiarities. The centre of the town contained, as its first building, an octagonal edifice . . . and

the streets beyond this were laid out in a circular shape. . . . But though the octagonal building remains, the circular streets are fast giving way, to make room for straight ones; and the central edifice itself is already destined to be removed, to give place to stores and dwellings; so that in half a century or less, there will be no vestige left of that peculiarity which gave the place its name, and which constituted the most perfect and therefore the most interesting work in ántiquity of its class in the country.

(Buckingham, 1842: 351)

Reps argues that the reasons for the change were that the lot shapes were awkward, that the central circle had soon degenerated and become unkempt and that the waste ground in the centre and at the angles where the circular section was joined to the surrounding grid could be profitably used. The result was the 'squaring of Circleville' between 1837 and 1849, as shown in Fig. 19.4. Although sound economic reasons can be advanced for this odd episode in the development of American town plans, nevertheless it is possible to speculate that at the root of the matter was the fact that radial-concentric plans were not consonant with the cultural background of the settlers. The grid was the form they knew and understood, and the symbolic preservation in the town form of the rites and usages of the Indians must have seemed particularly inappropriate. This seems to be at the basis of objections that the concentric scheme was a 'piece of childish sentimentalism' in the way it preserved the old tradition. It is certainly possible to regard the squaring of Circleville as an illustration of cultural forces at work and not merely as an exercise in economic convenience.

Economic factors in town form

The last phrase 'an economic exercise in convenience' introduces the problem of 'economic' influences on town plan. Perhaps the most distinctive work in this context is that of Ganshof (1943) in his study of the growth of towns between the Loire and the Rhine in the Middle Ages. His specific intention was to study

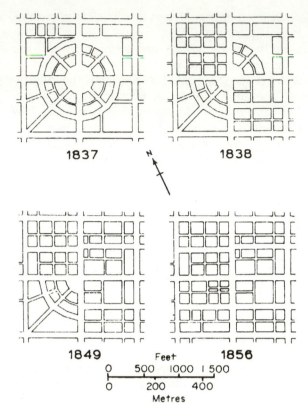

FIGURE 19.4 *The squaring of Circleville (after J. W. Reps, 1965).*

the physical extension of towns, 'nous nous sommes proposé de retracer le developpement des villes dans l'espace . . .' (Ganshof, 1943: 7). The key to the physical pattern is seen by Ganshof to reside in the 'pre-urban nucleus' which was a strong point, most often a Roman town or fort which survived the turbulent period of imperial collapse to provide a point of security in later times. 'Thanks to their fortifications, these *civitates* and *castra* played the role of pre-urban nuclei in the formation of medieval towns. In other words, it is around this pre-existing Roman element that the medieval town is formed' (Ganshof, 1943: 27). Other nuclei were found in the castles of the nobility, particularly in Flanders and Brabant, while a third type was found in the residences of the emerging ecclesiastic hierarchy.

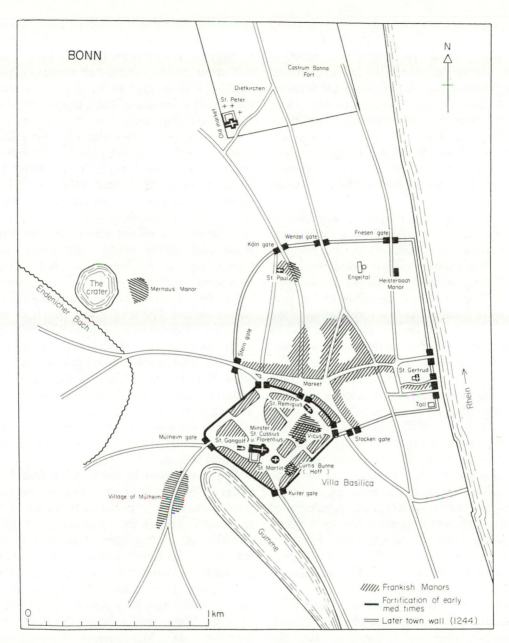

FIGURE 19.5   *Bonn: the historical topography (after K. Böhner, 1977). Source: K. Böhner 1977. Urban and rural settlement in the Frankish Kingdom. In M. W. Barley ed. European towns. Their archaeology and early history. London: Academic Press. Fig. 7, p. 197.*

All these had in common the offering of a degree of physical protection in a greatly disturbed world. But they were no more than *pre-urban* nuclei, the element generative of growth was the new merchant or commercial quarters which were added to them. Commercial activity revived after the disruptions of the 'Dark Ages' and as it did so, it needed security above all else. Revived commerce became attached to the earlier nuclei which could provide protection and added a new and growing quarter to the old settlements. These agglomerations or suburbs (sub urbium) had distinctive names, portus (poort), vicus (vik), or in France burgus (bourg). Thus there emerges the fundamental and characteristic difference which can be found in French towns between *cité* and *bourg*, between the old protective core and the new quarter of the merchants, the bourgeoisie.

Figure 19.5 indicates the basic structure of Bonn in West Germany. The Roman castrum on the Rhine bank was followed by the bishop's seat which was established some distance to the west with the basilica of St Cassius at the heart. Bonn had, therefore, two elements of the pre-urban triad of nuclei. The first fortification of the basilica was probably undertaken at the end of the eighth century. Subsequently, with commercial development, the market was established at a characteristic point in front of the gate of *Villa Basilicata*, the node about the basilica. Around this the new plan element collected and, as in most cases, this was walled in 1244. Most often the market was at a focus of tracks which, by use, became the main thoroughfares. Since the area was outside the close control of urban institutions, haphazard building led to plans which lacked regular form. At Bonn the regularity of the Roman fort contrasts with the loose block pattern of the later mercantile town. Here the argument reverts to degree of political and institutional control, bringing back into play a variable already discussed.

Ganshof's study is a clear indication of the way economic factors generate physical growth in particular locational situations. But the conclusions are directed towards general areas of growth rather than plan, and although there are implications for layout they cannot be made explicit in the context of his study. It is apparent, though, that many of these new areas had little regularity in their layout, for it was not until an integrated combination of the fortified strong point and the centre of economic development was created in the *bastide* that once

again the controlled chequer board layout was revived. This lapse of control is indeed mirrored in the way in which the later layout of Caerleon in Gwent, South Wales, had deviated from the original Roman grid as the imperial control fell away and a period of local usage modified the older plan. (Fig. 19.6)

It is perhaps in the context of circulation that economic influences on plan can best be made explicit. The basic function of the street system is to provide the most effective means of movement within the city. In nearly all cases, however, the present layout is inherited from a time when movement was on a very different scale and based on a completely different technology. It is from this that the traffic problems which now face all large cities result.

But these features are apparent at a very much earlier period. The simple way in which the grid of Greek colonial cities was laid down is related to the fact that no adaptation to site was necessary. At Miletus the simple plan outline does not show that the streets across the slope took in actuality the form of elongated staircases, an arrangement peculiar to a situation where wheeled vehicles were not dominant and much was carried by mules which could negotiate the stepped streets.

The distinction in medieval towns between street widths was often based on the function the streets were to perform. In larger towns, or in towns deliberately planned like the true bastides of southwestern France, the colonial towns of eastern Germany or Edward I's boroughs in Wales, it is generally not difficult to distinguish three functional types of streets, though frequently they are combined in varying degrees. Major traffic streets (Verkehrsstrassen, carrières at Montpazier) connecting the restricted points of exit from the walled town commonly had the greatest width. Residential streets (Wohnstrassen) carrying traffic to and from adjoining residential plots only, were often narrower. Occupation roads (Wirtschaftsstrassen) providing subsidiary access were the narrowest type (Conzen, 1960).

It follows from the close relationship between street layout and the functions which streets are

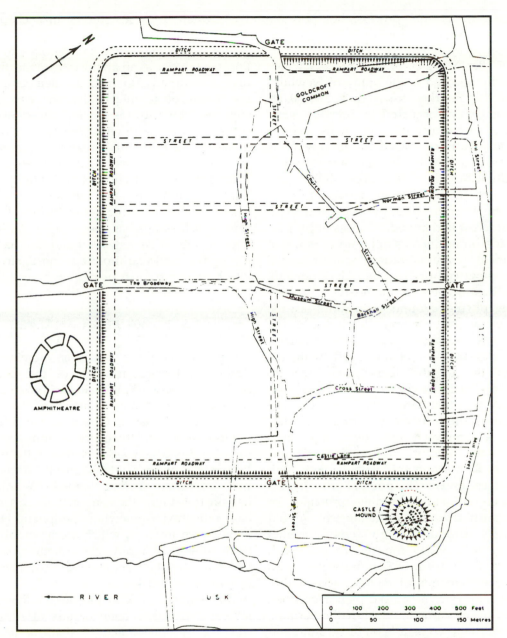

FIGURE 19.6 *Roman Isca and medieval Caerleon (after V. E. Nash-Williams, 1954).*

called upon to perform that one of the major themes of modern town planning should be 'traffic in towns'. The whole engineering and planning process is concerned with the analysis of traffic generation through origin and destination surveys and then the manipulation of the urban environment so that flows can be accommodated. This is not, therefore, a problem in street widening or reorientation, but it involves the whole city complex. The minimization of travel by the compacting of such land-use categories as need constant physical interaction is a vital part of this process, but also the complete or partial restructuring of the street pattern can be involved.

One of the instances of this process which can usefully be cited is the study of Newbury (Berkshire), a market town of some 30,000 population, which occurred in *Traffic in towns* (1963), a classic study of how towns were to cope with greatly increased traffic flows. The portion dealing with the town centre considers the relationship between accessibility and environment, an environment derived from the past and in which much of historical and architectural interest is retained. A series of proposals is made on an increasingly radical scale of development – minimal redevelopment, partial redevelopment comprehensive redevelopment. These scales involve a successively greater alteration of the ground plan until at the comprehensive stage wholesale clearance and rebuilding is envisaged. 'The northern part of Northbrook street is redeveloped with the whole of the ground level given over to parking and servicing with shopping and residential accommodation over . . .' The old layout under this scheme is virtually obliterated and a completely new plan comes into being in an attempt to combine 'highest standards of environment' with 'a very high level of accessibility'. The point at issue here is not concerned with planning principle or whether such drastic remodelling of an old town will produce high environmental standards, that is to say, the concern is not with the merits of any form of plan but solely with the fact that demands of modern accessibility can drastically reframe an old plan. Here then is another major variable which will determine the form of the town as it is on the ground and as it appears on the map.

This fairly extended review of the range of influences which can be discerned in town plan is intended to demonstrate that a generalized approach, independent of specific historical periods, is feasible through the isolation of the array of forces which underline the creation of town plan. This was the approach adopted in *An introduction to urban historical geography* (Carter, 1983) where separate studies in a chapter on town plan called 'In whose image is the city made?' are: Renaissance and Baroque towns in Europe; the city as an aristocratic estate; the grid plan and urban America; the democratic city; the towns of colonial India; the city of two cultures; the city as social conscience; the planned industrial town. Even so, it is difficult to isolate the individual creative influences. For example, it is not easy to distinguish between what can be called cultural and what political. The collapse of these two into a single complex leads directly back to a set of historical phases from which escape has been sought. A different approach must be generated and that has come through a concentration on those structures which make up town plan.

19.3 TOWN PLAN ANALYSIS

Town plan analysis has been dominated in Britain in the second half of the twentieth century by the work of M. R. G. Conzen and his successors who now constitute the Urban Morphology Research Group in the School of Geography at the University of Birmingham. In the USA it is closely associated with the work of Professor Conzen's son Michael Conzen. The very character of that limited concentration demonstrates how the concern with urban form has remained somewhat peripheral to the mainstream interests which are in social conditions rather than in physical structures, except in so far as a link between them can be determined (Chapters 14 and 15).

The seminal work in the field, although not his earliest, was Conzen's study of Alnwick – *Alnwick, Northumberland. A study in town-plan analysis* – published in 1960. It must be recognized that its inspiration was derived from an earlier German work (Whitehand, 1981) which Conzen brought into English language urban geography. Conzen's monograph on Alnwick generated an extended technical vocabulary. Thus Larkham and Jones's *A glossary of urban form* (1991) identifies some 69 of the defined terms as 'Conzenian'. There is, however, some

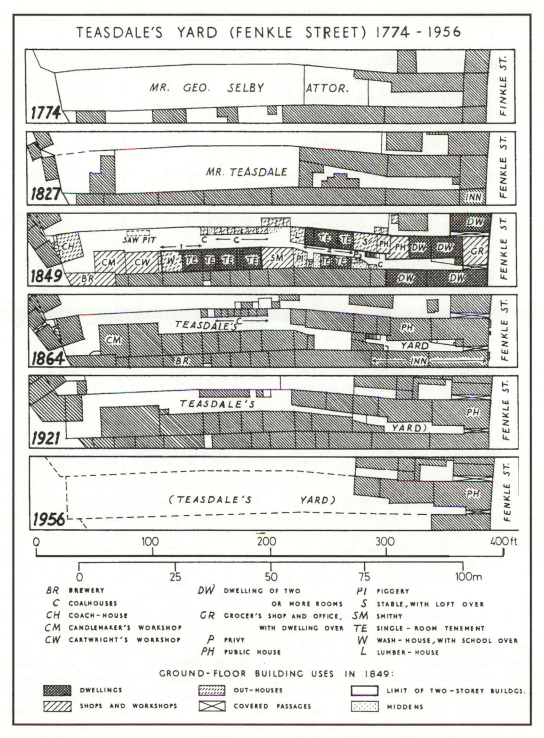

FIGURE 19.7 *The burgage cycle. Teasdale's Yard, Fenkle Street, Alnwick, 1774–1956 (after M. R. G. Conzen, 1960). Source: M. R. G. Conzen, 1960: Alnwick, Northumberland. A study in town-plan analysis. Transactions of the Institute of British Geographers Special Publication No. 27, p. 68, Fig. 14.*

confusion between those which Conzen invented and those which were already in currency which he employed. Thus 'kernel' and 'pre-urban nucleus' were widely used before Conzen's work appeared and with the same meaning that he gave them. Even so, there is little doubt that his work was the inspiration of nearly all the English language work on plan analysis which has subsequently appeared.

There are three basic units which make up any plan or layout. These are the plot, 'a parcel of land representing a land-use unit defined by boundaries on the ground'; the building, 'the abbreviated use in town-plan analysis for the block plan of a building'; and the street 'a town or village road that has more or less closed building development along its length'. The three are, therefore, the plan elements – 'the townplan may be divided into three constituent parts or elements: streets, and their arrangements into street systems; plots and their aggregation in street blocks; and buildings within those plots' (Larkham and Jones, 1991: 61).

From all the complexity of analyses generated by Conzen there has been but limited replication in subsequent studies. Perhaps the two concepts which have most clearly entered the literature are that of the burgage cycle and the fringe belt. The volume, *The built form of western cities*, edited by T. R. Slater (1990a) and presented to Professor Conzen on his eightieth birthday included detailed studies of towns by plot analysis, and on the fringe belt, to which a section is allocated.

The burgage cycle, illustrated in Fig. 19.7, epitomizes the way in which a single plot develops in terms of buildings placed upon it. The basic plan of the medieval town was the land allocated to the individual burgers, the burgage, 'defined in legal terms as a property unencumbered with manorial services which could be bought, sold or bequeathed freely without reference to any manorial authority' (Bond, 1990: 94). The various ways in which the development could take place are explicitly shown in Fig. 19.8 where an appropriate terminology is set out. It is possible that the plot is eventually built over by a combination of the plot dominant, plot accessories and tail end or derivative plots.

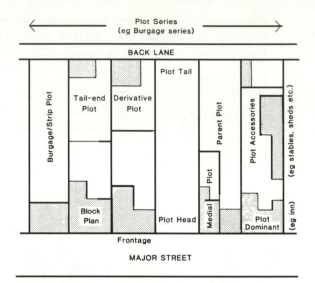

FIGURE 19.8 *Characteristic features in the development of a plot and a plot series (after T. R. Slater, 1991). Source: J. Larkham and A. N. Jones eds, 1991: A glossary of urban form. Urban Morphology Research Group, School of Geography, University of Birmingham. Fig. 22, p. 66.*

Given time, decay and congestion, complete clearance can initiate a new cycle of development. Plots, of course over time, could be divided or amalgamated so that a modified plot structure emerged. Metrological analysis (Slater, 1990a) can be used to 'generate a more detailed understanding of the ways in which surveyors laid out new towns or parts of towns' (Slater, 1990b: 71) and this type of study, of the basic unit in the formation of town plan, has most certainly provided the clearest insight into structure (Slater, 1981, 1988, 1990b). Moreover series of similar plots provide the bases for identifying the plan units of a town. The division of a town into plan units, or morphogenetic units where the term itself implies a time element, inevitably involves a study of historical development and the probability of a reversion to an historical phases approach. But in so far as the plan units are structural, or morphographic parts within the town, the basic identification is properly derived from form.

The fringe belt is a plan division on a larger scale. The basic tenet on which the concept is based is that urban extension is rarely continuous, but is rather cyclical with periods of rapid outward growth alternating with periods of standstill. The time of stability or litle movement is often defined by a fixation line which can be either a physical barrier such as a steep slope or river, or a human-made obstacle, such as a town wall. But the basic constraint on growth must come from a lack of economic dynamism, a downswing in the building cycle, a period of slump. From this type of situation two distinctive features follow.

The first is that there tend to be greatly contrasted patterns of land holding on either side of the fixation line. The most obvious situation is the long period of quiescence in urban development in the late medieval and early modern period which has already been identified in considering the growth of the city system. Many towns failed to expand beyond their medieval walls for a considerable time so that the extramural area became effectively a fringe belt. Under those circumstances there was a major contrast between the burgage plots within the walls and the open fields or common land which lay without. From the contrast a major difference in the grain of land patterning becomes apparent as the fringe belt begins to be developed.

The second feature which follows is probably the more important. During the period of stability the town still creates the need for extensive land uses on its margins. These fall into two categories. The first is the demand for large land-areas which are clearly not available within a fairly intensively built-up area. Such are cemeteries and hospitals, the latter especially at a time when those with infectious diseases need to be isolated. Later examples are water and sewerage works, cattle marts, public parks, playing fields and golf courses. Perhaps the earliest example would be the extensive defences with which early modern towns surrounded themselves. The second category consists of those uses, which could be construed as nuisances or pollutants and which were

therefore eventually dismissed from the city. Even in medieval times, although leather workers such as glovers carried on their trade within the walls, tanners and dyers were soon pushed out to locations where their offensive trades created fewer problems. In the eighteenth and nineteenth centuries heavy industry can be classified in this category and Burgess's third zone of workingmen's houses and heavy industry is a representation of its exclusion from the city centre, largely because of space demands rather than its offensive character.

It will be apparent that any one town will demonstrate not a single fringe belt around a primitive kernel but a succession of such belts related to phases of active growth. Dependent on the nature of that succession these are usually referred to as inner, middle and outer fringe belts.

Whitehand's interpretation of the fringe belts of Newcastle upon Tyne are shown in Fig. 19.9. In the late nineteenth century Whitehand notes a very clear fixation line to the east of the city in the gorge and valley of Jesmond Dene. The attraction of large residences to that location led to an internal differentiation of the belt with warehouses and industry developing at the Tyne confluence. Also during the Victorian period he adds into the picture the construction of various institutions, notably several isolation hospitals and a lunatic asylum, as well as cemeteries and waterworks.

> Where the geology permitted quarries and brickworks found similar peripheral locations. During the vigorous residential growth of the late Victorian period some of the less distant plots were swallowed up by the house builders, but the majority survived to form a discontinuous belt of varying width. This belt stretched northward from the riverside in the area of Elswick and Benwell and incorporated the open spaces of the Town Moor and Nun's Moor... . In the east the belt followed approximately the line of Jesmond Dene
>
> (Whitehand, 1967: 224)

There the distinct fixation line has already been noted.

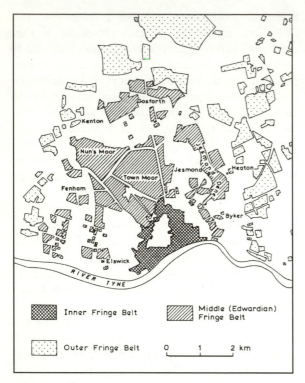

FIGURE 19.9 *The fringe belts of Newcastle upon Tyne (after J. W. R. Whitehand, 1967). Source: J. W. R. Whithand, 1967: Fringe belts: a neglected aspect of urban geography, Transactions of the Institute of British Geographers, 41, 225, Fig. 1.*

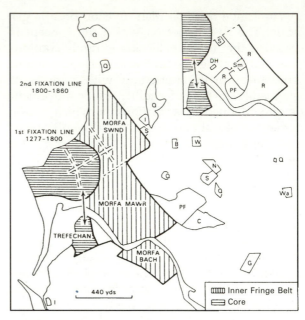

FIGURE 19.10 *Fixation lines and fringe belts in a small town: Aberystwyth in the nineteenth century. Inset: Morfa Mawr, fringe uses at the end of the century. The streets of the CBD in 1900 are indicated. I: Infirmary; S: School; W: Union Workhouse; Wa: Waterworks; G: Gasworks; PF: Playing Fields; N: Nursery; Q: Quarry; R: Railways; Sm: Smithfield; D: Drill Hall.*

It is useful to add two points relating to common nineteenth-century fringe-belt uses. Town Moor, of course, was part of the town's common land and these spaces were still available at the end of the nineteenth century when soccer was becoming organized into a national league. Characteristically, therefore, modern soccer stadiums occupy fringe-belt land. Turf Moor at Burnley is a clear example. Again, at Cardiff, the Arms Park and the Welsh national stadium, although on reclaimed river land, did form part of the inner fringe belt, as did the land already discussed, on which Cathays Park was developed. Because of its very late development, the inner fringe belt at Cardiff is a nineteenth-century feature. The second characteristic use is

derived from the fact that many universities were founded during the late nineteenth century and they, too, took over available land. Reference to Birmingham will show that Mason College, as the university then was, occupied what Broderwick described as 'the open fringe belt'.

The second example is the much smaller town of Aberystwyth in West Wales. The physical structure of Aberystwyth during the nineteenth century can be related to the existence of two fixation lines and two fringe belts (Fig. 19.10). The inner and first fixation line was that of the medieval town walls, together with the physical limits of the small extramural bridge-head settlement of Trefechan (literally 'little town'). Apart from Trefechan, no houses were built outside the walls until late in the 1790s, although minor encroachments had taken place during the eighteenth century. The medieval or inner first fixation line lasted therefore for some 500 years. To a large extent it remained inviolate because the town itself was but a skeleton of streets, a good deal of open land remained within the walls even in the second half of the nineteenth century. Beyond this fixation line lay the Inner Fringe Belt. It was composed of the extramural common lands of the borough, which were made up of three tracts of marshland which surrounded the small hill on which the town had been established in 1277. These were Morfa Swnd (Sand Marsh), Morfa Mawr (Great Marsh) and Morfa Bach (Little Marsh). Since there was ample intramural space, little specifically urban use was made of these lands, other than the location of the cattle market, the pound and the town gallows! Fringe uses, therefore, did not characterize them in any distinctive way, other than by the manner in which they contributed to those agricultural activities in which the burgesses themselves were involved. For small, remote towns in a period before effective transport, the provision of food can be regarded as a distinctive fringe use, although the conventional view of such uses is one which is dominated by the growth of nineteenth-century urban institutions. It was not until 1813 that these lands of the Inner Fringe Belt were formally divided and leased. In that year an entry in the Court Leet recorded, 'We, the jury, direct that part of the waste land called Morfa Swnd be mapped and divided into convenient spots for building.' The trigger for this decision was the demand for land brought about by the first growth phase as a resort town at the turn of the eighteenth century. Morfa Swnd was enclosed, divided and leased as was Morfa Mawr, but the latter remained in agricultural use.

With the extension of the town in this manner, a new and second fixation line was created. This was at the limit of borough common land, which had previously been classed as marshland, and which was clearly marked by the steep slopes of the Rheidol Valley sides both to north and south. Beyond this line, land was in private hands, mainly those of two prestigious local families, the Pryses of Gogerddan and the Powells of Nanteos. It is also interesting to observe that at two points a use, derived from the seaport role and demanding extensive linear land areas, marked the new fixation line, for ropewalks had been established both to the north of Morfa Swnd, where a lease had been granted in 1778, and to the south of Trefechan (1810). Land ownership, land tenure, a specific land-use and sharp breaks of slope all contributed to the emergence of this new fixation line, which was to last from the early nineteenth century, when the development of the Inner Fringe Belt first began by encroachment, until the 1870s when, after the coming of the railway in 1864, the second phase of population growth and physical extension pushed settlement beyond it and into the Middle Fringe Belt. This Middle Belt, beyond the second fixation line, was formed during the middle and later part of the nineteenth century when, related to a whole range of social legislation, towns were generating a variety of associated institutions. It had, therefore, to a much greater degree, the characteristic uses of a fringe belt. It was dominated by a series of quarries, worked into the valley side and developed as a source of stone for the phase of building after 1813. It also included the Union Workhouse and a militia barracks from the middle of the century and, by the end, an infirmary, gas works and the town cemetery, as well as schools and playing fields, including, characteristically, the town soccer ground (Fig. 19.10). After 1880 these were interspersed with the extending frontier of house building, as sites on the northern valley side were used. It is difficult to place Morfa Mawr in this context, Physically, it was part of the Inner Fringe Belt, but it remained undeveloped until the later part of the century when it was characterized by Middle Belt uses, including recreation grounds, railway yards, the smithfield, a drill hall and a school. To the south of the river the situation was simpler, for the very steep Rheidol slope meant that the first and second fixation lines were co-terminous; there was no distinction between an Inner and a Middle Fringe Belt.

By the end of the nineteenth century, therefore, the town was structured into three distinctive areas – Core, Inner Fringe Belt and Middle Fringe Belt. The physically structuring elements of the first fixation line, the town walls, and of the second, the ropewalks had been removed, but the influence of the two lines was clearly apparent. The physical character of the composing elements remains quite clear. The intricate block make-up of the core, the grid-like streets of the Inner Fringe Belt and the linear extensions through a much more open patterning of the Middle Belt are easily recognizable on a present-day map. It is perhaps worth adding that at Aberystwyth the university began not in the Inner Fringe Belt where it might have been expected but in the core itself by the takeover of an existing, or at least a half-finished building. But the constriction of space led to the College's purchase or lease of buildings in the Inner Fringe Belt until it could leapfrog out into preserved open land of the Middle Fringe Belt where it constitutes a characteristic institutional use. The National Library of Wales followed a similar path.

The main contribution of Whitehand has been to extend the fringe-belt concept which in its origins was related solely to urban morphology. He has argued that essential to its character is the replacement, in a phase where residential extension is not dominant, of house building by institutional uses. Once this is accepted then the tradition of the bid–rent curve is brought into consideration. That curve, which shows uses with less need for central locations being outbid in towards the city, is essentially static. But the existence of booms and slumps, of a building cycle, must mean it should be a dynamic feature since the willingness and ability to bid will vary over time. It is only because bids for residential development fall that a fixation line emerges with a bordering fringe belt where institutional uses demanding extensive land areas can effectively and successfully bid. In this way the fringe belt emerges not only as a morphological feature, but one closely related to the evolving land-use pattern of the city.

Whitehand has introduced a further feature into this situation. Each major phase of residential extension tends to bring with it distinctive building styles and housing fashions as well as more tangible improvements in domestic equipment. These are one of the generators of outward growth as they are demanded by the wealthier sections of the population at the leading edge of development. It follows that it is possible to identify innovations in house building and to integrate their diffusion and adoption into the process of fringe-belt creation, so that an holistic view of suburban character emerges at a much deeper level than the additive one with which the chapter began. The real difficulty is the practical one of achieving the integration of all the contributory elements to that character (Whitehead, 1972).

19.4 BEHAVIOURAL AND DECISION-MAKING APPROACHES

Perhaps this is not the most appropriate heading under which to present work which has largely been dominated by Whitehand. But it does epitomize the central concerns of a distinctive approach. It can be argued that the division of town plans usually made into 'planned' and 'unplanned' categories is pedagogically convenient but fundamentally false. All plans are the consequence of decisions of landowners (or squatters), developers and builders. All that varies is the scale of the decision-making authority which can cover a whole town or city, or be confined to a small plot. It is the product of all these decisions which constitutes the town plan. Plan, in brief, is the outcome of the perceptions, principles and policies of those individuals or groups who can effectively exercise the necessary power, within any constraints operated by the larger society of which they form part. The present author has demonstrated how one family in the mid-nineteenth century could initiate, lay out and effectively control the ground-plan and built form of a nineteenth-century seaside resort (Carter, 1970b). Urban historians, much more sensitive to the influence of individuals, and less concerned with theoretical viewpoints, have been much more successful in demonstrating how the owners of large urban estates were able to determine the character of their development (Cannadine, 1977, 1980).

If contemporary geography seeks to identify the roles of decision making through the operation of effective economic and social power, then it is proper that the geographer of an urban form, which is a relict from past periods, should also attempt to identify the perceptions, motivations and decisions which exercised the creators of plans. Perhaps the most useful exegesis has been made by George Gordon (1980) who sees plan as consisting of:

- **morphological elements**, e.g. plots, buildings, use, street plans, townscapes;
- **actors**, e.g. landowners, developers, builders, investors, financiers, architects, planners, lawyers, local authorities, national government, purchasers;
- **a stage**, e.g. sites, site characteristics, land value surfaces, accessibility surfaces, innovations, regulations, controls and constraints.

Whitehand's work, brought together in *The making of the urban landscape* (Whitehand, 1992a), has been dominated by the very detailed consideration of the second group identified by Gordon, the actors, seen as the creators of the morphological elements surrounded by the constraints of the stage, and including all the manipulations which modify the constraints.

It is not easy to present a brief summary of Whitehand's work but some flavour of it can be obtained by considering his analysis of residential areas where he has concentrated on the redevelopment process within existing housing areas. The infilling of the large gardens of houses built earlier in the century is a process which will be familiar to any student of townscape. It is typified in Fig. 19.11 where the large detached houses which were present in 1960 had by 1987 been replaced by estates of smaller dwellings (Whitehand, 1989, 1992a: 162). This shows the way in which a townscape is created but the process is complex with a variety of 'actors' involved. The actors are primarily the owners of the land, developers who seek to purchase and

exploit, and the local planning authority (LPA), itself consisting of the professional officers on the one hand and the planning committee of the elected members on the other. Whitehand seeks to explore the processes by which the type of transformation illustrated in Fig. 19.11 takes place. Thus, for example, he identifies a tendency for the LPA to reduce an initially proposed density of development, but eventually on resubmission to increase it. He writes:

> the tendency for density to be reduced between an initial application and a first approval, and for it to be increased between the first approval and the actual development may be seen in terms of the following sequence: first, there was an initial attempt to gain permission for as much development as possible on a site; secondly, in the face of LPA objections, often on the grounds of 'overdevelopment', a compromise permission was eventually arrived at, often following one or more rejections; thirdly, this permission was used, either by the same applicant or by another applicant to whom the plot had now been sold, to obtain permission for a more profitable, usually higher density, development. Consistent with this sequence, dwellings favoured for retention in planning officers' initial recommendations were frequently demolished in actual developments, and it was common for the building types constructed to be different from those originally approved.
>
> (Whitehand, 1992a: 163)

The basic character of the study is clear in this extract. Its essence lies in the attempt to analyse the backgrounds, motivations and actions of all those concerned in the creation of townscapes and how the inevitable conflicts between them are resolved. There remains a problem however. It is clearly far more difficult to carry out such studies in the past (Carter, 1970a) and almost impossible for the remote past. That is why detailed structural analysis, such as that by Slater (1988) referred to previously, is essential for from such detailed structural analysis at least some basis for the inference of motivation and action can be laid.

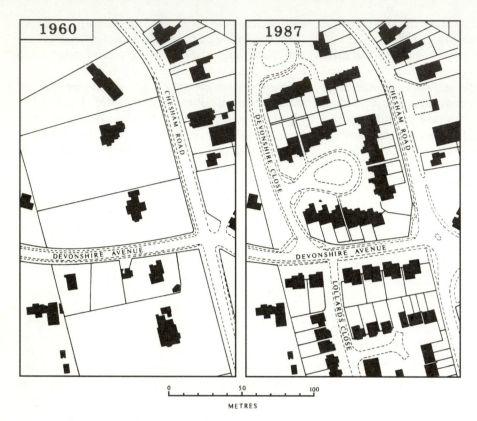

FIGURE 19.11 *The replacement of large houses in extensive grounds by estates of smaller dwellings: Chesham Road, Amersham, Bucks (after J. W. R. Whitehand, 1992a). Source: J. W. R. Whitehand, 1992a: The making of the urban landscape. Institute of British Geographers, Special Publication No. 16, p. 162, Figure 4.14.*

It is essential to add that Whitehand's investigations are much broader than the example given here and take in equivalent studies of the commercial core, especially the roles of developers and architects in its creation, and institutional and public areas. It certainly constitutes the most effective contribution to townscape study. But perhaps it is proper to note that it is no more than an intricate analysis of 'in whose image is the townscape made', a parallel in a democratic age to the scrutiny of the cultural inheritance and political ideology of a Renaissance prince as he built his ideal city.

19.5 CONSERVATION

Much of Conzen's work, as indeed that of his followers, was directed towards urban conservation issues. Whitehand begins his preface to an edition of the papers of Conzen with the sentence, 'During the last decade there has been an increasing concern in western countries for the visual aspects of the environment, particularly from the point of view of its amenity value' (Whitehand, 1981: v). Parts of towns certainly have a *genius loci*, a distinctive spirit of place, an atmosphere which epitomizes a past age or, indeed, a facet of contemporary living. That *genius loci* can be destroyed by unsympathetic development. The aesthetic pleasure of the visible scene in towns is not one widely treated by urban geographers. But new developments, both within old layouts and beyond them on greenfield sites, need scrutiny as contributions to townscape as well as economic stimuli or social benefits. But the call for some overarching theory against which measurement can take place is unlikely to be realized in a democracy where competing interests have to reach an acceptable compromise. Most of the great contributions to significant urban landscapes were made by those with dictatorial powers, but that has been largely the theme of this chapter. The stuff of town plan is lodged in the spirit of the age and it is in that context that study will be most fruitful.

19.6 CONCLUSION

At the outset of this chapter its placing at the end of the volume was discussed and justified. In reality it has pulled away from the holistic promise to become more a study of a single aspect. Even so, there is surely within it a glimpse, if no more, of an integrated urban geography where the end is the explanation of townscape, through function, size, internal ordering of uses into plots, buildings and streets. But perhaps that has little place in a post-modern era where the fissiparous fragments of urban study are pursued as ends in themselves.

CONCLUSION: URBANIZATION AND URBAN GEOGRAPHY

20.1 INTRODUCTION

The nature and consequences of urbanization are of prime significance to both the developed and less developed parts of the world. Indeed the whole future of the world is closely bound up with the changing character and role of the city. The urban geographer can only consider a small part of the totality which is expressed in the concept of urbanization. But of necessity, any geographer will be influenced by attitudes to the city which are common within the prevailing culture. At the same time urban geography itself changes over time as new trends and fashions manifest themselves, while the urban phenomenon itself is dynamic and constantly changing. In consequence there are three topics which this conclusion needs to consider. The first is the nature of attitudes to the city in western culture; the second is the changing nature of the city itself; the third is the impact of both of the above on urban geography itself.

20.2 ATTITUDES TO THE CITY

This book has been written from the viewpoint of the Western industrialized world. That is an inevitable consequence of the author's own background. In Western culture the attitude to the city has always been ambivalent; an ambivalence which it has grown more popular to identify and trace. More importantly, that ambivalence still remains.

It is evident that on the one side the city has been seen to use standard moral terms as the epitome of evil or, in the terminology of Wirth, anomic and alienating, disrupting the elemental cohesion of the social group and disintegrating the personality of the individual. The sources of this attitude lie deep in the cultural inheritance. Prime among them are the Christian religion and its Judaic antecedents. Christianity originated among a people of the desert fringes whose folk memory and preserved tradition in the Old Testament, were those of nomadic herdsmen. The city was regarded as destructive of the purity of religious belief and practice, a place where alien notions and false gods flourished. The cities of the plain were synonymous with luxury, looseness and idolatry: Sodom and Gomorrah perished to demonstrate the fate of the urban idea. Through Christanity these attitudes became embedded in the consciousness of the West, always below the surface in spite of a predominantly urban way of life.

In addition Europe, as it emerged as a meaningful concept after the collapse of the Roman Empire, was made up of a complex mosaic of small culture regions, most of which have retained their identity, and some of which still have distinct languages. Brittany and Wales, Catalonia and Galicia, Flanders and Fresia are examples. These culture regions, often governed in the form of principalities or duchies, were subject to a process of more or less arbitrary unification through battle or the marriage bed into the 'united' kingdoms of the later Middle Ages and of modern Europe. In that state-building operation the central and capital city played a leading role; indeed, the political unit was often forged about the city. Thus the name of the small area about Paris, the Ile de France, was extended to cover the whole country. By these means both symbolically and physically the city came to represent the cosmopolitan ideal, the destroyer of folk cultures and identities in the interests of state unity.

The industrial city of the late eighteenth and nineteenth centuries provided another impetus to anti-urban attitudes, not only by contributing further to the erosion of the older folk cultures, but by creating the most appalling physical conditions and the concomitant social degradation. Wirth, following the views of Tonnies and Durkheim, was able to adapt into a nascent urban sociology the inherited prejudices of the past.

In the present century the history of urbanism has provided enhanced stimuli for anti-urban attitudes. The obsolescence of the inner city has

contributed to the creation of slum and ghetto, the location of an impoverished underclass shot through with drug abuse and crime. The consequent escape from city centre problems by the process of suburbanization, together with increased car ownership, have brought immense pressures to bear on urban transport systems. Such are the pressures that urban motorways intended to solve circulation difficulties, destroy houses and add to accommodation problems. The disposal of waste of all kinds, from plastic containers to sewage, and the exhaust fumes from vehicles have created major crises of pollution. The attempt to escape from the noise and dirt of the city generates saturating demand for what open space is easily accessible; yesterday's wildernesses are tomorrow's urban parks, and, in consequence, access to national parks has to be rationed. Finally the vulnerability of the city has been revealed. It is exposed to any breakdown in its complex sustaining systems. It is riven with crime and its streets are unsafe. The frail and the elderly exist at night behind locked and bolted doors, afraid to walk the streets. The wealthy, where they still live in the city, employ security guards or elaborate alarm systems. The city is vulnerable to the urban guerilla and to the direct violence of a few determined and fanatical individuals.

From all these sources comes an attitude which decries the city, at least in its contemporary Western form. That attitude also has the core of traditional romanticism buried within it. It reverses Christian dogma for those on whom religion exerts no overt influence. Human beings are viewed not as being inherently wicked by virtue of original sin and in need of redemption, but rather as innocent beings corrupted by the vicious regime of the competitive, capitalist world. That corruption is most marked in the prime symbol of that world, the city, the least 'natural' feature of the earth's surface and representing 'the maximum possibility of humanizing the natural environment and naturalizing the human heritage' (Mumford, 1938: 6).

Such a romantic reaction perpetually rises to the surface in urban writing. It is nicely illustrated in Bunge's (1971) study of Fitzgerald (Detroit) in which although the author attempts to deny such a basis, it nevertheless appears, almost creating a latter day version of the *Lyrical Ballads*, complete with updated representations of Ruth and Michael. Even the idea that if man or woman is inevitably corrupted in the urban world then the nearest to innocence and virtue must be the child, is strongly pressed. Bunge writes, 'this is a book in defence of children. This book is designed to make us grow up, that is, to make us more childish' (Bunge, 1971: 242). This is a latter day version of Wordsworth :

Not in entire forgetfulness,
And not in utter nakedness,
But trailing clouds of glory do we come
From God who is our home:
Heaven lies about us in our infancy!
Shades of the prison-house begin to close
Upon the growing Boy.

Wordsworth's views are more tangibly anti-urban than Bunge's:

He in the dissolute city gave himself
To evil courses: ignominy and shame
Fell on him.

Even so the romantic viewpoint, most often through the notion of the geography of a revolution, has exerted an active influence on geographers.

On the other side of this ambivalence to the city is the view which maintains that the city is the greatest of all human creations, displaying in physical form all the achievements of the world's peoples – naturalizing the human heritage. In the city different cultures and traditions meet and mix and it is from that mixing and cross-fertilization that innovation is born. New ideas and new techniques are diffused from the city down the urban hierarchy and thence into the countryside. Philosophical revolution and technical achievement are not normally associated with the conservative and unchanging tradition of the countryside, but with the creative flux of the city. Even if

Christianity originated in a peasant community, it had to become Greek in philosophy before it could become universal and to be spread by Roman technology.

It is worthy of note that in attempting to produce a unifying theory of the town, Claval has stressed that the common ground between inter-urban and intra-urban aspects of geography lies in the maximization of social interaction (Claval, 1973). This is not far removed from Meier's communication theory of urban growth (Meier, 1962). Whenever the crux of urbanism is sought it tends to emerge as simply a way of optimizing intercourse between people; it stresses the role of the town as the meeting and mixing point out of which comes that emancipation from the restrictions of custom which is the core of the freedom which the city dweller gains. Only away from the close control of tradition can the freedom to experiment, or to be different, be realized: only in the anonymous crowds of the city is true liberation possible. That very anonymity which Wirth saw as destructive of the individual creates the conditions of liberation in which radical change is possible.

It must be added, however, that radical change is never acceptable to totalitarian regimes of the left or the right, or to those who are convinced they have absolute truth in their keeping. So the city has spawned a means of inhibiting the very freedom it generates. This is the secret or thought police, who replace in an urban situation the restraints operated by convention in rural and small-scale communities.

Finally, detachment from the confinement of the rural community transforms an immobile situation into one where mobility is an accepted part of life. 'How', a popular American song after the First World War asked, 'are you going to keep them down on the farm, now that they've seen Paree?' Along with physical mobility comes social mobility for in the city is opportunity. Each individual can stretch talents to their full extent. The constraints of the rural and the remote have always been recognized:

But knowledge to their eyes her ample page
Rich with the spoils of time did ne'er unroll;
Chill penury repressed their noble rage
And froze the genial current of the soul. . . .
Far from the madding crowd's ignoble strife
Their sober wishes never learn'd to stray;
Along the cool sequester'd vale of life
They kept the noiseless tenor of their way.

Truth or advantage does not always lie in one of the two polar extremes and discussions of urbanism have long been dominated by the notions of an ideal midway situation. Most explicit of all the attempts to derive such a median view was that by Ebenezer Howard in *Garden cities of tomorrow* where the two extremes were set out as magnets of attraction (Fig. 8.1) (Howard, 1965). Perhaps earlier views of the *citta ideale* saw the way through the creation of an apt physical environment, while a Marxist interpretation would presumably argue for a socialist economy and the elimination of competition as a means of producing the ideal city. At this point the problem becomes more clearly defined because presumably under a Marxist state, as in all ideal worlds, there could be no further development. Yet the city is a centre of change. The city is the place where people meet, where great hopes are generated but where many must be disappointed; where success is balanced by disaster, affluence with degradation. Only where people become units of production, all alike, will the static city emerge in an environment of intellectual atrophy and decay.

It is probable that most people's attitudes to the city will take in both polar extremes, now moving to the one, now to the other. The opinionated Samuel Johnson provides a characteristic example. On the one hand in *London* (1738) he wrote:

For who would leave, unbribed, Hibernia's strand
Or change the rocks of Scotland for the Strand? . . .
Here malice, rapine, accident conspire,
And now a rabble rages, now a fire;

Their ambush here relentless ruffians lay,
And here the fell attorney prowls for prey;
Here falling houses thunder on your head,
And here a female atheist talks you dead . . .
LONDON! the needy villain's general home,
The common shore of Paris and of Rome:
With eager thirst, by folly or by fate,
Sucks in the dregs of each corrupted state.

Two hundred and fifty years seem to have seen very little change! But on the other hand (Boswell's *Life of Johnson*) Johnson declares:

'Sir, if you wish to have a just notion of the magnitude of this city, you must not be satisfied with seeing its great streets and squares, but you must survey the innumerable little lanes and courts. It is not in the showy evolution of buildings, but in the multiplicity of human habitations which are crowded together, that the wonderful immensity of London consists.'
BOSWELL: 'The only disadvantage is the great distance at which people live from one another.'
JOHNSON: 'Yes, Sir; but that is occasioned by the largeness of it, which is the cause of all other advantages.' BOSWELL: 'Sometimes I have been in the humour of wishing to retire to a desert.'
JOHNSON: 'Sir, you have desert enough in Scotland.'

Or again: 'When a man is tired of London, he is tired of life; for there is in London all that life can afford.'

Perhaps at this point it is pertinent to repeat a comment made before in this book. The very wealthy solved the problem of town–country ambivalence by possessing both town houses and country estates. It is also relevant that for the less wealthy, suburbia was an attempt to combine both in one location.

This argument has veered away from urban geography as such. Even so attitudes to the city have shot through urban geography conditioning the way it is perceived and written. But the city has not remained a constant in spite of the relevance of Johnson's poem some 250 years on. It is to the changing nature of the city which attention must now be directed.

20.3 CHANGING NATURE OF THE CITY

In Chapter 2 of this book the problem of defining what is urban was reviewed and it was suggested that one difficulty was that both the reality and the concept of what is urban are subject to change. The publication in 1961 by Jean Gottmann of his book *Megalopolis*, subtitled 'The urbanized northeastern seaboard of the United States', was a clear indication that a new urban structure had been formed. Gottmann defined Megalopolis as 'an almost continuous system of deeply interwoven urban and suburban areas, with a total population of about 37 million people in 1960. . . .The cradle of a new order in the organization of inhabited space.' But the nature of that new order is not easy to pin down.

Peter Hall in his massive and masterly study of megalopolis England points out that 'for all his brilliant evocation of a megalopolis Gottmann never succeeds in defining it as a unique entity, and so never proves that it exists' (Hall, 1973: 47). Hall, after a careful and broadly based examination of evidence, accepted the reality of a megalopolis England but maintained that it was:

> a functional rather than a physical reality. Just like the comparable northeastern urban complex of the United States – Gottmann's megalopolis – it is a giant urban area only in the sense that here is a large tract of the earth's surface where the great majority of people depend on urban jobs and urban services; and where the impact of those jobs and services in terms of measurements like commuter zones, service areas and the exchange of goods and information, expands to involve each part of the area in a complex series of interactions with other parts. It is not, and does not conceivably seem likely to be, a giant urban area in the sense that the physical growth of its parts will gradually coalesce into a continuous sprawl from London to Birmingham and Manchester. That is a nightmare which has no foundation whatsoever in reality.
>
> (Hall, 1973: 320)

That would equally be true of the Great Lakes megalopolis, the Japanese megalopolis extending from Tokyo to Osaka, and the northwest European megalopolis extending generally about the Rhine from Randstat-Holland to Stuttgart.

Although simple physical sprawl is rejected as the basis of identity, nevertheless the complex interdependence of parts accepted, not only queries the definition of what is urban but, more significantly, challenges the idea that any meaningful distinction can be made between what is urban and what is rural. If that be the case then the sorts of study presented in this volume, all of which to a large degree assume discrete, free-standing urban entities, take on a somewhat dated appearance.

There is some overlap between the subject-matter of this book and Hall's study of urban England, but his main concern was with regional structure and much of his consideration was devoted to the broader problems of regional and economic planning. Even so, studies by such organizations as the Centre for Urban and Regional Studies have been of necessity quoted in this book. It is possible to maintain, as this book does, that there are specific urban problems demanding their own treatment, but the blurring of the urban–rural difference presents a real difficulty to urban geography as a systematic branch of the discipline of geography.

In spite of these difficulties, it would be wrong to imply that conventional urban geography, as it has been presented here, is now in some way outmoded. Although it is not as easy as it once was to build a study around the notion of the free-standing city clearly divorced from its rural surrounds, the basic ideas outlined in the book still remain central to the urban analyst. Perhaps the key issue is no more than one of scale. The studies of regional structure, already referred to in this section, attempt to resolve the mechanics of the massive agglomeration and its rural interstices. On the other hand, in studies of central places and of residential location it was seen to be necessary to move into micro-scale behavioural approaches. In spite of these extensions the basic pattern of inter- and intra-urban distributions present the same problems for analysis, and the changes in the nature of the urban phenomenon from discrete city to megalopolis, to regional city, and the development of behavioural research, reveal no more than a hierarchically scaled situation where the basic and essential middle ground is taken by this book.

It is probable that the real challenge to urban geography will come from a different but related source to which only one brief chapter was devoted. As was indicated there, most of the analyses presented in this book have assumed a *laissez-faire* situation, one of relatively untrammelled competition in a capitalist system. But such conditions appertain in reality in hardly any country in the world, for some form of control is exercised by local or central government, or by private corruption. The planning process determines land-uses, shapes the form of towns, and by the creation of new towns or regional policies, structures the city system. Planning permissions for out-of-town shopping centres or hypermarkets determine the nature of the retail system. In this way the study of urban geography necessarily becomes closely involved with the planning mechanisms of countries and the way they operate. This is certainly nothing new for the earliest towns were planned towns. Even so, a major difficulty for the urban geographer is a progressive involvement with the detail of planning legislation. Whitehand's studies of the development of suburban areas which were considered in Chapter 19 are admirable illustrations of that involvement. But planning itself does not take place in a vacuum and must be based on an initial examination of the patterns initiated by free operation under market conditions or a simulation of them. To that extent urban geography provides the basis from which planning moves forward.

Consideration of past changes in the nature of the city and the concept of 'moving forward' inevitably precipitate speculations as to its future.

20.4 THE CITY OF THE FUTURE

Discussion of the future has to be set against two quite opposite views which have been held by urban theorists as to what constitutes the urban ideal. These two views can be labelled the concentration or claustrophilic stance and the dispersal or agoraphilic stance.

Claustrophilia is the love of enclosed spaces. It is derived from the word for a monastic cloister and links the ideal structuring of a town with that of a monastery. Le Corbusier, one of the most revered names in the history of town planning in recent times, made the link between his ideal and the monastery quite explicit. His view was of a city which was concentrated, compact and lively throughout the 24 hours of the day. It was an attitude derived from the pre-industrial, free-standing town where lively communal traditions were expressed in colourful rituals. Phythian-Adams (1979) has traced the loss of public functions in Coventry during the sixteenth century. He observes 'the complete collapse of the ceremonial system as an ideal mirror of community. ... As outdoor ceremonies vanished and the cathedral was demolished, so too did the ritual centre of the city lose its significance' (Phythian-Adams, 1979: 275). As that occurred so the culture which emerged 'was no longer a public, visual affair concerning all who participated or cared to watch; but a private matter for the individual citizen in the furnishing or decorating of his house, his reading matter and his worship' (Phythian-Adams, 1979: 278). That crucial transition broke the ties of citizen to city, and transformed communal into private living thus creating the necessary conditions for suburbanization. Much of planning theory of the claustrophilic order has been directed towards re-creating the old communal order. The most notable was Corbusier's *Unités-d'Habitation*. They were to be large single blocks within which all the necessary urban functions were to be located – apartments, shops, banks, restaurants – so that a whole town was encompassed in one large building. To revert to the analogy introduced earlier, they were to be like monasteries where the cell provided maximum privacy but where living was communal and cooperative (Curtis, 1986).

Corbusier's views became widely influential but in practice the main derivative was the high-rise residential block the problems of which have already been discussed (Chapter 13, pp. 254–57). But even as these blocks are being destroyed, the idea still lives on. One of the recurrent themes in science-fiction is of a so-called plug-in city, made up of easily transported modules assembled either in space or on the surface of the moon. Space travellers always encounter compact cities, often with cell-like structuring, usually located in desolated surrounds. There is in all these one common factor. The external environment, be it space, or on the surface of a planet or indeed the sea bed, is totally hostile. Concentration and compactness are enforced by the hostility and insecurity without the walls, just as it was in the medieval city, or indeed, just as the monastery isolated itself from the evils of the secular world.

In complete contrast is the view of the future as inevitably dominated by dispersal because mobility, both physical and intellectual, has been made, or will be made absolute by the high advances in communication systems. The advantages of the city have been totally reversed. Once security lay within the walls and danger was shut out by town gates. Now with pollution, crime and terrorism, danger lies within the city, safety, peace and security are to be found in the countryside. All this is encapsulated in the counterurbanization process which has characterized every developed country. This has been allied to that anti-urbanism which was identified earlier in this chapter as characteristic of the Western inheritance

> proud and gay
> And gain-devoted cities; thither flow
> As to a common and most noiseome sewer
> The dregs and faeculence of every land. . . .
> God made the country, and man made the town.
> (William Cowper, 1785)

As with the ideal of the agglomerated city, so with the dispersed city, reality presents a

degenerate form. Surburbia, subtopia, as it was labelled in a diatribe widely quoted at the time by Ian Nairn (1955) in a book called *Outrage*, was to be regarded with contempt. The back of the dust-cover of the book reproduces a piece from the *Daily Mail*. 'If the suburban spread continues at its present speed we shall all be living in a fake-urban, fake-rural fake.' (See p. 254).

Here, too, in spite of the condemnation of sprawl, science-fiction has painted a future which is now nearly come about. Computers linked by modems and with the storage capacity of CD Rom disks can now bring the whole world into the remote home. Physical propinquity is no longer the iron control it once was and those crucial bonds which were part of the urban idea have been loosened if not broken. When the friction of distance is universally reduced to zero there will be no advantaged locations. Then there will be no towns.

If these two ideals – or indeed trends in modern urbanism for such they are – are set against contemporary reality, then both are seen to be in operation. All the attempts at central area and inner city regeneration, as well as gentrification, point towards concentration. They are manifest attempts to offset the dispersal of the last century. Moreover, they are strongly supported by 'green' interests for agglomeration cuts down on the demand for open land and minimizes the total journey to work and hence reduces pollution. On the other hand, those who vote with their wheels (Davie, 1972: 72) opt for the dispersed city where there is at least some escape from the claustrophobia of city living. If the wealthy, as has been noted, have both estate and town house, those less wealthy can have an imagined estate in surburbia or in the metropolitan village. Perhaps the tension between the two ideals, although in a trivial way, is epitomized in the controversy between city centre renewal and out-of-town centres. Inevitably the future will not wholly emphasize either view for outside of total dictatorship, democratic decisions are incoherent and messy. Thus a government in Britain which condemns out-of-town developments, sanctions the creation in Kent at Ebbsfleet of a new 'Euro City' described as 'a gleaming high-technology metropolis populated largely by middle class families wanting to escape the inner cities' (*The Sunday Times*, 21 August 1994: 1–9). The development is said to include Bluewater Park with some 2.3 million square metres of shops and leisure facilities, 40,000 new houses for some 200,000 people, 14 million square metres of office and light industrial land. All this will be linked to a science park and a new university campus. There will be complete cabling of all the houses to provide every possible electronic domestic facility. This, of course, although precipitated by the new channel tunnel, is no more nor less than the classic edge city of the Washington beltway; it is another West Edmonton Mall. It is the apotheosis of out-of-town development! But decisions are always made pragmatically with little relation to principle, still less to theory. But they do have consequences. The phrase above – middle class families wanting to escape the inner cities – has inevitable repercussions on social equality and social justice.

20.5 THE CITY AND SOCIAL JUSTICE

The worst scenario for the city is that where the affluent leave behind the decaying core and create a new, mobile suburban life, while that core is left to an underclass without hope and seeking solace not in rural spaciousness and greenery or edge city consumerism, but in the hallucinations of drugs. With such a future in view, geographers have turned more directly to concerns with social justice. Two quotations can introduce the problem. The first is appropriately from Chairman Mao, 'If you want knowledge you must take part in the practice of changing reality. If you want to know the taste of a pear you must change the pear by eating it yourself.' For a Western culture it is a great pity the Mao did not refer to an apple rather than a pear, for the follow through of the implications might have made this section unnecessary. The second extract is from David Harvey's book *Social justice*

and the city (1973), one that has been quoted before in this book: 'An urbanism founded upon exploitation is a legacy of history. A genuinely humanizing urbanism has yet to be brought into being. It remains for revolutionary theory to chart the path from an urbanism based on exploitation to an urbanism appropriate for the human species. And it remains for revolutionary practice to accomplish such a transformation.'

Throughout this book reference has occasionally been made to radical and Marxist alternatives to the presentation of what can be called traditional urban geography. The most widely disseminated source of such alternatives is the urban sociology of Manuel Castells (Castells, 1977). Stripped of its dense verbiage the crux of Castells's argument is akin to that of Harvey, that urbanism cannot be empirically examined and interpreted as a thing in itself for it is a manifestation, and only one, of the condition of advanced capitalist societies. 'An urban sociology founded on urbanism is an ideology of modernity ethnocentrically identified with the crystallization of the social forms of liberal capitalism' (Castells, 1978). His criticism of the Chicago ecologists stresses this view:

> The question is no longer to know how the social life of a neighbourhood is organized *vis-à-vis* the dominant culture, but how are to be determined the housing and infrastructure policies of the city from the relations of force between social groups structurally determined by their interests. New concepts and theories become necessary to consider problems of conflict and the decision-making processes which appear to determine the whole of urban organization.
>
> (Castells, 1978: 7)

The consideration of the housing market in Chapter 13 is presumably, if only superficially, a move in Castells's direction. He is even more dismissive of geographical empiricism which, he argues uses 'increasingly sophisticated statistical techniques but which [is] devoid of meaning in relation to theoretical categories and social issues' (Castells, 1978: 8). Indeed, the derivation of a theoretical perspective must be based on a rejection of the determinism of space, 'space in itself has no meaning because it is socially constructed' (Castells, 1978: 181), a phrase which could well be found in the writings of Vidal or Febvre fifty years ago! 'Urban social science must therefore move away from a spatial empiricism, precisely in order to carry out empirical research into the spatial expression of social relations' (Castells, 1978: 181).

Put in this way there is little new in Castells's attitude. The demand for enquiry into process rather than argument from pattern is one of the longest standing in geography.

The accepted wisdom at the present is that post-war urban geography made successive movement from a basis in logical positivism, to one in phenomenology, to one in Marxism. This smacks of notions of evolutionary progress. In many ways what has occurred is a pendulum swing between attitudes which can be subsumed under the all-embracing terms 'classical' and 'romantic'. The first assumes logic, order and control from a well-defined basis, a determinism whether it be of the environment, the geometry of space or the economic system. The second accepts mutability and variability, a disorder to be comprehended through culture and personality, through the spirit rather than the system. As the pendulum swings, so approaches change resulting in an eclecticism which is likely to characterize the work of many geographers, even as it will offend those who regard it as untenable.

But the crucial problem of an ever widening separation between an alienated urban underclass and the remainder will constitute a critical spatially based problem.

This book, in the light of the literature quoted above, and with overriding social issues in mind, appears as a manual, a sort of nuts and bolts study of the city from the point of view of a British geographer, or to use Harvey's description of such works, a 'partial analysis framed within the safety of a disciplinary womb'. The impatience, and occasionally the *naïvéte* of geographers who, inspired with a desire to eliminate the ills of the city, wish to establish overriding theory or radically change the socio-

economic system on which the contemporary city is based, may at times dispose them to have little patience with this constrained level of enquiry.

It might well be that the implicit philosophy of this book is a sort of nineteenth-century social Darwinianism: it accepts the basic drive of competition for scarce resources and of conflict over advantageous locations, however muted, modified and regulated they may be by the operation of constraints derived from community intrusion in formal political terms either at the local or national level. Competition, the survival of the ablest competitor and the demise of the weakest, is always the key whether it be for central place status among towns, for retail sites by entrepreneurs or for residential locations by families.

No doubt when the millenium arrives and such competition is done away with, some geographer will need to write a very different book; critical discussion of the existing order will then be inconceivable for sophistry decrees that of perfection there can be no criticism which, like other aspects of Dystopia, will have withered away. In passing it is interesting to recall that in More's Utopia all cities were identical and mirror images of each other for there was a complete separation and no competition. Even so, given something a little less than perfection and with the goals set by the community rather than by competing individuals and irresponsible interests, priorities will still have to be identified and there will still be competitors for the physically limited resource of urban land and the advantaged locations: even in the best of possible worlds, even with the information super highway in full operation, everything cannot be located on the same spot. And so inequality is written into the system in a more relevant form of environmental determinism.

20.6 CONCLUSION

To write a conclusion to a conclusion is somewhat indulgent. But if, as has been contended, academic fashion in geography swings between the classic, neat, all-encompassing model on the one hand, and the romantic view of a chaotic assemblage of incoherent forces on the other, then it is the latter which seems at present in vogue. It is called post-modernism. Fragmentation is the dominant characteristic of present geographic studies as innumerable aspects of city life are examined, many with little spatial relevance. Moreover, an emphasis on process has shifted geographers away from the spatial base to far wider enquiries. Indeed, it might well be contended that a specific and discrete urban geography was no more than a stage in a developmental sequence in geographical studies. As such, a past stage, it no longer has much relevance. But however wide apart the divergent fragments move, at some point it is essential to consider them in relation to a logical reference point – to the city in space and the city as space. It is to that end which this book has been written.

REFERENCES

Abercrombie, P. 1945: *Greater London Plan 1944*. London: HMSO.

Abu-Lughod, J. 1969: Testing the theory of social area analysis: the case of Cairo, Egypt. *American Sociological Review* 34, 198–211.

Abu-Lughod, J. 1971: *Cairo: 1001 years of the city victorious*. Princeton NJ: Princeton University Press.

Abu-Lughod, J. 1976: Developments in North African urbanism: the process of decolonization. In Berry, B.J.L. (ed.), *Urbanization and counter-urbanization*. Ch. 8, p. 202. Beverley Hills: Sage. Urban Affairs Ann. Review, vol. 11.

Abu-Lughod, J. 1980: *Rabat: urban apartheid in Morocco*. Princeton, NJ: Princeton University Press.

Alexandersson, G. 1956: *The industrial structure of American Cities*. London and Stockholm: Allen and Unwin. Almquist and Wissell.

Alonso, W. 1964a: *Location and land use*. Cambridge, Mass: Harvard University Press.

Alonso, W. 1964b: The historical and structural theories of urban form: their implications for urban renewal. *Land Economics* 40, 227–31.

Ambrose, P. and Colenutt, B. 1975: *The property machine*. Harmondsworth, Middx.: Penguin Books.

Andrews, F.M. and Phillips, G.W. 1970: The squatters of Lima: who they are and what they want. *Journal of Developing Areas* 4, 211–24.

Andrews, H.F. 1970: Working notes and bibliography on central place studies 1965–1969. University of Toronto: Dept. of Geography, Discussion Paper 8.

Andrews, J.H. and Simms, A. 1988: *The Irish historic towns atlas*. Dublin: Royal Irish Academy.

Applebaum, W. 1961: Teaching marketing geography by the case method. *Economic Geography* 37, 48.

Appleyard, D. 1973: Notes on perception and knowledge. In Downs, R.M. and Stea, D. (eds), *Image and environment*. London: Edward Arnold.

Archer, J.C. and White, E.R. 1985: A service classification of American metropolitan areas. *Urban Geography* 6, 122–51.

Ardrey, R. 1966: *The territorial imperative*. New York: Atheneum.

Armstrong-Wright, A. 1991: Urban transport: World Bank policy. In Herity M. (ed.), *Urban transport in developing countries*. London: PTRC Education and Research Services Limited, pp. 19–24.

Ashworth, W. 1954: *The genesis of modern British town planning*. London: Routledge and Kegan Paul.

Atkinson, R. and Moon, G. 1994: *Urban policy in Britain: the city, the state and the market*. London: Macmillan Press.

Auerbach, F. 1913: Das Gesetz der Bevolkerungskonzentration. *Petermanns Mitt*. 59, 74.

Aurousseau, M. 1921: The distribution of population: a constructive problem. *Geographical Review* 11, 563.

Aurousseau, M. 1924: Recent contributions to urban geography: a review. *Geographical Review* 14, 444.

Badcock, B. 1984: *Unfairly structured cities*. Oxford: Basil Blackwell.

Balchin, P.N. 1985: *Housing policy: an introduction*. London: Croom Helm.

Banham, R. 1971: *Los Angeles. The architecture of four ecologies*. London: Allen Lane, Penguin Press.

Bannister, D. 1962: Personal construct theory: a summary and experimental paradigm. *Acta Psychologica* 20, 104.

Bannon, M.J. 1983: Urbanization in Ireland: growth and regulation . In Blackwell, J. and Convery, F.J. (eds), *Promise and performance: Irish environmental policies analysed*. Dublin: Resource and Environmental Policy Centre, University College, Dublin.

Bannon, M.J., Bourne, L.S. and Sinclair, R. (eds), 1991: Urbanization and urban development. Recent trends in a global context. Dublin: Service Industries Research Centre, University College Dublin.

Barber, G. 1986: Aggregate characteristics of urban travel. In Hanson, S. (ed.), *The geography of urban transportation*. New York: The Guilford Press, Ch. 4, pp. 73–90.

Barley, M.W. (ed.) 1977: *European towns. Their archeology and early history*. London: Academic Press.

Barton, D. 1978: The creation of centrality. *Annals of the Association of American Geographers* 68(1), 44.

Baskin, C.W. 1957: A critique and translation of Walther Christaller's Die Zentralen Orte in Suddeutschland. PhD dissertation. Virginia: University of Virginia.

Baskin, C.W. 1966: *Central places in southern Germany*. Translated from Christaller, W., *Die zentralen orte in Suddeutschland*. Englewood Cliffs. NJ: Prentice-Hall.

Basset, K. and Short, J.R. 1980: *Housing and residential structure. Alternative approaches*. London: Routledge and Kegan Paul.

Bater, J.H. 1980: *The Soviet city. Ideal and reality*. London: Edward Arnold.

Beauregard, R.A. 1989a: *Atop the urban hierarchy*. Totowa, NJ: Rowman and Littlefield Publishers.

Beauregard, R.A. 1989b: Urban restructuring in comparative perpective. In Beauregard, R.A. (ed.), *Atop the urban hierarchy*. Totowa, NJ: Rowman and Littlefield, Publishers, Ch. 7, pp. 239–74.

Beavon, K.S.O. 1977: *Central place theory: a re-interpretation*. London: Longman.

Benevolo, L. 1975: *Storia della citta*. Rome: Editori Laterza.

Benevolo, L. 1980: *The history of the city*. Translated by G. Culverwell. Cambridge, Mass: MIT Press.

Benevolo, L. 1993: *The European city*, Translated by C. Ipsen. Oxford: Blackwell.

Berry, B.J.L. 1959: The spatial organization of business land uses. In Garrison, W. *et al.* (eds), *Studies of highway development and geographic change*. Seattle: University of Washington Press, Ch. 3, p. 62.

Berry, B.J.L. 1962: *The commercial structure of American cities; a review*. Chicago: Community Renewal Program.

Berry, B.J.L. 1967: *Geography of market centres and retail distribution*. Englewood Cliffs NJ: Prentice-Hall.

Berry, B.J.L. 1972: Latent structure of the American urban system. In Berry B. J. L. (ed.), *City classification handbook: methods and applications*, New York: John Wiley & Sons, Ch. 1, pp. 11–60.

Berry, B.J.L. 1976: The counterurbanisation process: urban America since 1970. In Berry, B.J.L. (ed), *Urbanization and Counterurbanization*. Beverley Hills, Sage Publishers.

Berry, B.J.L. and Barnum, H.G. 1962: Aggregate relations and elemental components of central place systems. *Journal of Regional Science* 4, 35.

Berry, B.J.L. and Garrison, W.L. 1958a: Alternate explanations of urban rank–size relationships. *Annals of the Association of American Geographers* 48, 83.

Berry, B.J.L. and Garrison, W.L. 1958b: Functional bases of the central place hierarchy. *Economic Geography* 34, 145.

Berry, B.J.L. and Kasarda, J.D. 1977: *Contemporary urban ecology*. New York: Macmillan.

Berry, B.J.L. and Pred, A. 1961: *Central place studies: a bibliography of theory and application. Supplement 1965*. University of Pennsylvania: Regional Science Research Institute, Bibliography Series 1.

Betjeman, J. 1958: *John Betjeman's Collected Poems*. London: John Murray.

Blakely, E. 1991: The new technology city: infrastructure for the future community in Brotchie, J. *et al*. (eds), *Cities of the 21st Century*. Cheshire: Longman.

Blanchard, R. 1911: *Grenoble, étude de geographie urbane*. Paris:

Boal, F.W. 1969: Territoriality on the Shankhill–Falls divide, Belfast. *Irish Geographer* 6, 33–41.

Boal, F.W. 1978: Territoriality on the Shankhill–Falls divide, Belfast: the perspective from 1976. In Lannegran, D.A. and Palm, R. (eds), *An invitation in geography*. New York: McGraw-Hill, pp. 56–77.

Boal, F. W. 1982: Segregation and mixing space: space and residence in Belfast. In Boal, F.W. and Douglas, J.N.H. (eds), *Integration and division: geographical perspectives on the Northern Ireland problem*. London: Academic Press.

Boal, F.W. and Poole, M.A. 1973: Religious residential segregation in Belfast in mid-1969: a multi-level analysis. In Clark, B.D. and Gleave, M.B. (eds), *Social patterns in cities*. Institute of British Geographers. Special Publication No. 5.

Boal, F.W. and Royle, S. 1986: Belfast: boom, blitz and bureaucracy. In Cherry, G. (ed.), *Regional cities in the U.K. 1890–1980*. London: Harper and Row.

Böhner, K. 1977: Urban and rural settlements in the Frankish kingdom. In Barley, M.W. (ed.), *European towns. Their archaeology and early history*. London: Academic Press, Ch. 11, pp. 185–202.

Bond, C.J. 1990: Central place and medieval new town: the origins of Thame, Oxfordshire. In Slater, T.R. (ed.), *The built form of western cities*, Leicester: Leicester University Press, Ch. 5, pp. 83–108.

Bourne, L.S. 1967: *Private redevelopment of the Central City*. Chicago: University of Chicago, Department of Geography, Research papers 112.

Bourne, L.S., Sinclair, R. and Dziewonsky, K. 1984: *Urbanization and settlement systems. International perspectives*. Oxford: Oxford University Press.

Bourne, L. S. *et al*. (eds) 1989: *The changing geography of urban systems. Perspectives on the developed and developing worlds*. Pamplona: Cervicio de Publicaciones de la Universidad de Navarra.

Boyer, R. and Savageau, D. 1989: *Places rated almanac. Your guide to finding the best places to live in America*. New York: Prentice-Hall Travel.

Bracey, H.E. 1953: Towns as rural service centres. *Transactions of the Institute of British Geographers* 19, 95.

Brand, R.R. 1972: The spatial organization of residential areas in Accra, Ghana, with particular reference to aspects of modernization. *Economic Geography* 48, 284–98.

Briggs, L. 1951: *The ancient Khmer empire*. Philadelphia: American Philosophical Society.

Bromley, R.J. 1974: The organization of Quito's urban markets: towards a reinterpretation of periodic central places. *Transactions of the Institute of British Geographers* 62, 45.

Bromley, R.J., Symanski, R. and Good, C.M. 1975: The rationale of periodic markets. *Annals of the Association of American Geographers* 65, 530–7.

Bromley, R. and Thomas, C. 1993: The retail revolution, the careless shopper and disadvantage. *Transactions of the Institute of British Geographers* NS18, 222–36.

Brooke, C.N.L. 1975: *London, 800–1216; the shaping of a city*. London: Secker and Warburg.

Brookfield, H.C. 1952: Worthing. *Town planning review*, 23, 145–62.

Brookfield, H.C. and Hart, D. 1971: *Melanesia: a geographical interpretation of an island world*. London: Methuen.

Brotche, J. (ed.) 1991: *Cities of the 21st century: new technologies and spatial systems*. Melbourne: Longman Cheshire.

Brown, L.A. and Moore, E.G. 1970: The intra-urban migration process, a perspective. *Geografiska Annaler*, Series B, 52, 1–13.

Brown, S. 1987: The complex model of city centre retailing: an historical application. *Transactions of the Institute of British Geographers*, New Series, 12, 4–18.

Brunet, R. 1989: *Les Villes Européenes*. Paris: La Documentation Française.

Brush, J.E. 1953: The hierarchy of central places in southwestern Wisconsin. *Geographical Review* XL, 380.

Brush, J.E. 1962: The cities of India. In Turner, R. (ed.), *India's urban future*. Berkeley: University of California Press.

Bryant, C.R., Russwurm, L.H. and McLellan, A.G. 1982: *The city's countryside. Land and its management in the rural–urban fringe*. London: Longman.

Buchanan, C. 1963: Traffic in towns. Report of the steering and working group appointed by the Minister of Transport. London: HMSO.

Buckingham, J.S. 1842: *The eastern and western states of America*. 3 vols. London: Fisher, Son and Company.

Bunge, W. 1962: *Theoretical geography*. Lund: Gleerup.

Bunge, W. 1971: *Fitzgerald: the geography of a revolution*. Cambridge, Mass.: Schenkman Pub. Co.

Burgess, E.W. 1925: The growth of the city: an introduction to a research project. In Park, R.E. and Burgess, E.W. (eds), *The city*. Chicago: Chicago University Press, Ch. 2, pp. 47–62.

Burgess, R. 1978: Petty commodity housing or dweller control? A critique of John Turner's views on housing policy. *World Development* 6, 1105–13.

Burgess, R. 1981: Ideology and urban residential theory in Latin America. In Herbert, D.T. and Johnston, R.J. (eds), *Geography and the urban environment* 4. Chichester: John Wiley, pp. 57–144.

Burke, G. 1971: *Towns in the making*. London: Edward Arnold.

Burnett, J. 1980: *A Social History of Housing, 1815–1870*. London: Methuen.

Burtenshaw, D., Bateman, M. and Ashworth, G.J. 1991: *The European City. A Western Perspective*. London: David Fulton.

Butler, E.W. and Chapin, F.S. 1969: *Moving behaviour and residential choice. A national survey*. Washington DC: US Highway Research Board.

Bylinsky, G. 1974: California's great breeding ground for industry. *Fortune 89*, 128–35.

Cannadine, D. 1977: Victorian cities: how different? *Social History 4*, 457–82.

Cannadine, D. 1980: *Lords and landlords: the aristocracy and the towns: 1774–1967*. Leicester: Leicester University Press.

Carol, H. 1960: The hierarchy of central functions within the city. *Annals of the Association of American Geographers* 50, 419.

Carter, H. 1965: *The towns of Wales; a study in urban geography*. Cardiff: University of Wales Press.

Carter, H. 1970a: A decision making approach to town plan analysis: a case study of Llandudno. In Carter, H. and Davies W.K.D. (eds), *Urban essays: studies in the geography of Wales*. London: Longman.

Carter, H. 1970b: *The growth of the Welsh city system*. Cardiff: University of Wales Press.

Carter, H. 1972: *The study of urban geography*. 1st edn. London: Edward Arnold.

Carter, H. 1980: Transformations in the spatial structure of Welsh towns in the nineteenth century. *Transactions of the Honourable Society of Cymrodorion*, 175–200.

Carter, H. 1983: *An introduction to urban historical geography*. London: Edward Arnold.

Carter, H. and Lewis, C.R. 1983: *Processes and patterns in nineteenth century cities*. Unit 15 in Open University course D301 Historical sources and the social scientist. Milton Keynes: Open University Press.

Carter. H. and Lewis, C.R. 1990: *An urban geography of England and Wales in the nineteenth century*. London: Edward Arnold.

Carter, H. and Rowley, G. 1966: The morphology of the central business district of Cardiff. *Transactions of the Institute of British Geographers* 38, 119–34.

Carter, H. and Wheatley, S. 1982: *Merthyr Tydfil in 1851. A study of the spatial structure of a Welsh industrial town*. University of Wales, Board of Celtic Studies, Social Science Monograph, No. 7. Cardiff: University of Wales Press.

Castells, M. 1967: Theory and ideology in urban sociology. In Pickvance, C.G. (ed.), *Urban sociology: critical essays*. London: Tavistock Publications.

Castells, M. 1977: *The urban question: a Marxist approach*. Translated Sheridan, A. London: Edward Arnold.

Castells, M. 1978: *City, class and power*. Translated Lebas, E. London and Basinstoke: Macmillan.

Castells, M. and Hall, P. 1994: *Technopoles of the World. The making of the twenty-first century industrial complexes*. London and New York: Routledge.

Census 1981: Office of populations census and surveys 1981: Preliminary report for towns. London: HMSO.

Census Users Guide 1990: Department of Commerce, Washington, DC.

Centre for Urban and Regional Development Studies (CURDS). 1983: *Functional regions; definitions, applications, advantages*. Newcastle upon Tyne: CURDS Factsheet 1.

Champion, A.G. (ed.), 1989: *Counterurbanization. The changing pace and nature of population deconcentration*. London: Edward Arnold.

Champion, T. and Green, A. 1987: The booming towns of Britain: the geography of economic performance in the 1980s. *Geography* 72(2), 97–108.

Champion, T. and Green, A. 1991: Britain's economic recovery and the north–south divide. *Geography* 76, 249–54.

Chapin, F.S. 1965: *Urban land use planning*. Urbana, Ill.: University of Illinois Press.

Charles, S. 1977: *Housing economics*. London: Macmillan.

Chase-Dunn, C.K. 1985: The system of world cities, 800–1975. In Timberlake, M. (ed.), *Urbanization in the world-economy*, Orlando, Fla.: Academic Press, Ch. 12, pp. 269–92.

Childe, V.G. 1950: The urban revolution. *Town Planning Review*, 21, 3–17.

Christaller, W. 1933: *Die zentraler orte in Suddeutschland*. Jena: Fischer.

Clark, T.A. 1979: *Blacks in suburbs. A national perspective*. New Brunswick, NJ: Rutgers University Center for Urban Policy Research.

Clark, W.A.V., Deurloo, M. and Dielman, F.M. 1984: Housing consumption and residential mobility. *Annals of the Association of American Geographers* 74, 29–43.

Clausen, J.A. 1986: *The life course: a sociological perspective*. Englewood Cliffs, NJ: Prentice-Hall.

Claval, P. 1973: Une théorie unitaire de la ville. *Canadian Geography* 17, 276–9.

Cohen, A. (ed.), 1974: *Urban ethnicity*. London: Tavistock Publications.

Cohen, M.A. 1974: *Urban policy and politial conflict in Africa: a study of the Ivory Coast*. Chicago: Chicago University Press.

Collison, P. 1963: *The Cutteslowe Walls. A study in social class*. London: Faber and Faber.

Community Development Project (CDP 1976: *Profits against houses. An alternative guide to housing finance*. London: CDP Information and Intelligence Unit.

Connell, J. 1973: Social networks in urban society. In Clark, B. and Gleave, M.B. (eds), *Social patterns in cities*. London: Institute of British Geographers. Special Publication No. 5, pp. 41–52.

Conway, D. 1985: Changing perspectives on squatter settlements, intra-urban mobility, and constraints on housing choice of the Third World urban poor. *Urban Geography* 6, 170–97.

Conzen, M.R.G. 1960: *Alnwick, Northumberland. A study in town-plan analysis. Transactions of the Institute of British Geographers*, 27.

Coombes, M.G. and Dixon, J.S. 1982: Functional regions for the population census of Great Britain. In Herbert, D.T. and Johnson, R.J. (eds), *Geography and the urban environment. Progress in research and applications*, vol. 5. Chichester: John Wiley.

Cooper, S. 1985: *Public housing and private property 1970–1984*. Aldershot: Gower Publishing.

Coupland, A. 1992: Every job in office jobs. In Thornley, A. (ed.) 1992: *The crisis of London*, London: Routledge.

Cowan, P., Daniel, F., Ireland, J., Jordan, C., Mercer, D. and Sears, A. 1969: *The office. A face of urban growth*. London: Heinemann Educational.

Cox, K.R. and Johnston, R.J. (eds) 1982: *Conflict, politics and the urban scene*. London: Longman.

Coxon, A.P.M. and Davies, P.M. 1986: *Images of social stratification: occupational structure and class*. London: Sage.

Cross, D.F.W. 1992: *Counterurbanization in England and Wales*. Aldershot: Avebury.

Crowe, P.R. 1938: On progress in geography. *Scottish Geographical Magazine* 54, 18.

Curtis, W.J.A. 1986: *Le Corbusier: ideas and forms*. Oxford: Phaedon.

Dacey, M.F. 1960: The spacing of river towns. *Annals of the Association of American Geographers* 50, 59.

Dacey, M.F. 1962: The analysis of central places and point patterns by nearest neighbour method. In Norborg, K. (ed.), *The IGU Symposium in Urban Geography Lund 1960*. Lund: Dept of Geography. Royal University of Lund, Sweden.

Dacey, M.F. 1964: Modified poisson probability law for point patterns more regular than random. *Annals of the Association of American Geographers* 54, 559.

Dacey, M.F. 1967: Some properties of order distance for random point distributions. *Geografiska Annaler* 49(b), 25.

Daniels, P.W. 1979a: *Movement in cities: spatial perspectives on urban transport and travel*. London: Methuen.

Daniels, P.W. (ed.) 1979b: *Spatial patterns of office growth and location*. Chichester: John Wiley.

Darden, J.T. 1973: *Afro-Americans in Pittsburg: the residential segregation of a people*. Lexington: D.C. Heath.

Darden, J.T. 1985: The housing situation of blacks in metropolitan areas of Michigan in the state of Black Michigan. East Lansing, Michigan State University, pp. 11–12.

Darden, J.T. 1987: Choosing neighbors and neighborhoods: the role of race in housing preference. In Tobin, G.A. (ed.), *Divided neighborhoods. Changing patterns of racial segregation*. Newbury Park, Calif.: Sage Publishers.

Davidoff, L. 1973: *The best circles: society, etiquette and the season*. London: Croom Helm.

Davie, M. 1972: *In the future now: a report from California*. London: Hamilton.

Davies, D.H. 1965: *Land use in Central Cape Town: a study in urban geography*. Cape Town: Longman, SA.

Davies, K. and Sparks, L. 1989: Superstore retailing in Great Britain 1960–1980: results from a new database. *Transactions of the Institute of British Geographers*. NS 14, 74–89.

Davies, R. 1981: The spatial formation of the South African city. *GeoJournal*. Supplementary Issue 2, 59–72.

Davies, R.L. 1972a: Structural models of retail distribution: analogies with settlement and land use theories. *Transactions of the Institute of British Geographers*, 57, 59–82.

Davies, R.L. 1972b: Structural models of retail distribution. *Transactions of the Institute of British Geographers* 57, 59–82.

Davies, R.L. 1972c: The retail pattern of the central area of Coventry. In *The retail structure of cities*. Urban Geography Study Group. Institute of British Geographers Occasional Publication No. 1. London.

Davies, R.L. 1973: The retail pattern of the central area of Coventry. Institute of British Geographers Occasional Publications No. 1, 1–42.

Davies, R.L. and Champion, A.G. (ed.) 1983: *The future for the city centre*. London: Academic Press for the Institute of British Geographers.

Davies, W.K.D. 1964: The hierarchy of commercial centres: a case study in South Wales. University of Wales. Unpublished PhD thesis.

Davies, W.K.D. 1967: Centrality and the central place hierarchy. *Urban Studies* 4, 61.

Davies, W.K.D. 1983: *Urban social structure. A multivariate-structural analysis of Cardiff and its region*. Cardiff: University of Wales Press.

Davies, W.K.D. 1984: *Factorial ecology*. Aldershot: Gower.

Davies, W.K.D. 1989: Towards a conceptual integration of the urban system literature. In Bourne, L.S. *et al.* (eds), *The changing geography of urban systems*. Pamplona: Cervicio de Publicaciones de la Universidad de Navarra, SA, Ch. 3, pp. 61–86.

Davies, W.K.D. 1991: Urban systems development policies: the Alberta experience. In Bannon, M.J., Bourne, L.S. and Sinclair, R. (eds), *Urbanization and urban development. Recent trends in a global context*. Dublin: Service Industries Research Centre, University College Dublin.

Davies, W.K.D. and Lewis, G. 1973: The urban dimensions of Leicester, England. In B. Clark and B. Gleave (eds), *Social patterns in cities*. London: Institute of British Geographers Special Publication No. 5, pp. 95–107.

Davies, W.K.D. and Lewis, G. 1974: The social patterning of a British city, Leicester, England. *Tidschrift voor Economische en Sociale Geografie* 65, 95–107

Davis, K. 1972: *World urbanization 1950–1970*, 2 vols. Berkeley. Calif.: University of California.

Davis, M, 1990: *City of quartz. Excavating the future of Los Angeles*. London and New York: Verso.

Dawson, J.A. 1979: *The marketing environment*. London: Croom Helm.

Dawson, J.A. (ed.), 1980: *Retail geography*. London. Croom Helm.

Dawson, J.A. 1983: *Shopping centre development*. London: Longman.

Dawson, J.A. 1988: Futures for the high street. *Geographical Journal* 154, 1–12.

Dawson, J.A. and Lord, J.D. (eds) 1985: *Shopping centre development: policies and prospects*. London: Croom Helm.

Dear, M. 1977: Psychiatric patients and the inner city. *Annals of the Association of American Geographers* 67(4), 588–94.

Dear, M. and Scott, A.J. 1981: *Urbanization and urban planning in capitalist societies*. London and New York: Methuen.

Denham, C. 1984: Urban Britain. *Population Trends*, pp. 11–18.

Department of the Environment 1973: *Using predictive models for structure plans*. London: HMSO.

Department of the Environment 1977: *Housing policy. A consultative document*. London: HMSO.

Department of the Environment 1987: *Housing, the Government's proposals*. London: HMSO.

Department of the Environment 1993: *Housing and construction statistics*. London: Department of the Environment.

Department of Transport 1993: *Transport statistics Great Britain*. London: HMSO.

de Vries, J. 1984: *European urbanization 1500–1800*. Cambridge, Mass.: Harvard University Press.

Dick, R.S. 1961: Variations in the occupational structure of central places of the Darling Downs, Queensland. *University of Queensland Papers* 1, 2.

Dickinson, B.B. 1901–2: The position of towns. *Geographical Teacher* 1, 97–108.

Dickinson, R, E. 1947: *City region and regionalism*. London: Kegan Paul, Trench, Trubner and Co.

Dickinson, R.E. 1951: *The west European city. A geographical interpretation*. London: Routledge and Kegan Paul.

Dickinson, R.E. 1964: *City and region. A geographical interpretation.* London: Rutledge and Kegan Paul.

Dingemans, D. 1979: Redlining and mortgage lending in Sacramento. *Annals of the Association of American Geographers* 69, 225–39.

Doherty, P. 1989: Ethnic segregation levels in the Belfast urban area. *Area* 21, 151–9.

Downs, R.M. 1970a: Geographic space perception: past approaches and future prospects. *Progress in Geography* 2, 65–108.

Downs, R.M. 1970b: The cognitive structure of an urban shopping centre. *Environment and Behaviour* 2(1), 13.

Drakakis-Smith, D.W. 1978: Shelter: overview. In Rimmer, P.J., Drakakis-Smith, D.W. and Mc Gee, T.G. (eds), *Food, shelter and transport in Southeast Asia and the Pacific.* Canberra: Australian National University, Department of Human Geography Publication No. 12.

Duby, G. 1980–5: *Histoire de la France urbaine* (5 vols). Paris: Seuil.

Duffy, F. and Henney, A. 1989: *The changing city.* London: Bulstrode Press.

Duncan, O.D. and Duncan, B. 1955: Residential distribution and occupational stratification. *American Journal of Sociology* 60, 493–503.

Duncan, O.D. and Reiss, A.J. 1956: *Social characteristics of urban and rural communities.* New York: John Wiley.

Dwyer, D.J. 1975: *People and housing in third world cities. Perspectives on the problems of spontaneous settlements.* London: Longman.

EPAD. ud: *La défense.* Paris: Info Défense.

Evans, D.J. 1984: The segregation of the New Commonwealth population in Wolverhampton Municipal Borough, 1961–1981. Newcastle under Lyme, North Staffs. Polytechnic, Department of Geography and Recreation Studies, North Staffs Polytechnic, *Occasional Papers* No. 3.

Eyre, L.J. 1972: The shanty towns of Montego Bay, Jamaica. *Geographical Review* 62, 394–413.

Feagin, J.R. and Beauregard, R.A. 1989: Houston: hyper development and the sunbelt. In Beauregard, R.A. (ed.), *Atop the urban hierarchy.* Totowa, NJ: Rowman and Littlefield Publishers.

Feldman, M.M.A. 1985: Biotechnology and local economic growth: the American pattern. In Hall, P. and Markusen, A. (eds), *Silicon landscapes.* Boston: George Allen and Unwin, Ch. 5, pp. 65–79.

Feldman, M.M.A. and O'Malley, E. 1982: *The biotechnology industry in California.* Sacramento: California Commission on Industrial Innovation.

Ferrer, M. and Precedo, A. 1984: The national settlement system in Spain. In Bourne L.S., Sinclair, R. and Dziewonski, K. (eds), *Urbanization and settlement systems. International perspectives.* Oxford: Oxford Unversity Press, Ch. 14, pp. 301–17.

Ferrer, M. *et al.* 1989: Recent evolution of the Spanish settlement system. In Bourne, L.S. *et al. The changing geography of urban systems,* Pamplona: Cervicio de Publicaciones de la Universidad de Navarra, Ch. 1, pp. 21–38.

Fielding, G.J. 1986: Transit in American Cities. In Hanson, S. (ed.) *The geography of urban transportation.* New York: Guilford Press.

Figueroa, O. and Henry, E. 1991: Analysis of the underground systems in Latin America. In Heraty, M. (ed.), *Urban transport in developing countries.* London: PTRC Education and Research Services, pp. 232–40.

Firey, W. 1945: Sentiment and symbolism as ecological variables. *American Sociological Review* 10, 140–8.

Firey, W. 1947: *Land use in central Boston.* Cambridge, Mass.: Harvard University Press.

Fishman, R. 1977: *Urban utopias in the twentieth century. Ebenezer Howard, Frank Lloyd Wright and Le Corbusier.* New York: Basic Books.

Fleure, H.J. 1920: Some types of cities in temperate Europe. *Geographical Review* 10, 357.

Fleure. H.J. 1924: Cities of the Po basin: an introductory study. *Geographical Review* 14, 345.

Form, W.H. 1954: The place of social structure in the determination of land use: some implications for a theory of urban ecology. *Social Forces* 32, 317.

Forman, S. and Riegelhaupt, J.F. 1970: Market place and marketing system: toward a theory of peasant economic integration. *Comparative Studies in Sociology and History* 12, 188.

Foster, R.S. and Berger, R.A. (eds) 1982: *Public–private partnership in American cities. Seven case studies*. Lexington, Mass.: D.C. Heath and Co., Lexington Books.

Francescato, D. and Mebane, W. 1973: How citizens view two great cities, Milan and Rome. In Downs, R.M. and Stea, D. (eds), *Image and environment*. London: Edward Arnold, pp. 131–47.

French, R.A. and Hamilton, F.E.I. (eds) 1979: *The socialist city. Spatial structure and urban policy*. Chichester: John Wiley.

Fuchs, R.J. and Demko, G.J. 1979: Geographic inequality under socialism. *Annals of the Association of American Geographers* 69, 304–18.

Ganshof, F.L. 1943: *Etude sur le developpement des villes entre Loire et Rhin au moyen age*. Paris: Presses Universitaires de France.

Garner, B.J. 1966: *The internal structure of retail nucleations*. Northwestern University Studies in Geography No. 12. Evanston. Ill.: Northwestern University.

Garner, B.J. 1970: Towards a better understanding of shopping patterns. In Osborne, R.H., Barnes, F.A. and Doornkamp, J.C. (eds), *Geographical essays in honour of K.C. Edwards*. Nottingham: Nottingham University Press.

Garrison, W. L. 1962: In Norborg, K. (ed.), *Proceedings of the IGU Symposium in Urban Geography, Lund 1960*. Discussion, Urban Morphology 463. Lund: C.W.K. Gleerup Publishers.

Garrison W.L. *et al.* 1959: Empirical verification of concepts of spatial structure. In Garrison W.L. *et al.* (eds), *Studies of highway development and geographic change*. Seattle: University of Washington.

Garvan, A.N.B. 1966: Proprietary Philadelphia as an artefact. In Handlin, O. and Burchard, J. (eds), *The historian and the city*. Cambridge, Mass.: MIT Press.

Gaskell, S.M. 1990: *Slums*. Leicester: Leicester University Press.

Geddes, P. 1949: *Cities in evolution*. Edited by The Outlook Tower Association and The Association for Planning and Regional Reconstruction. Edited reprint of the 1915 edition. London: Williams and Norgate.

Gibson, J. 1981: *The new housing subsidy system and its interaction with the block grant*. Birmingham: University of Birmingham Institute of Local Government Studies.

Giggs, J.A. 1970: Fringe expansion and suburbanization around Nottingham: a metropolitan area approach. *East Midlands Geographer*, **5**, 9–18.

Giggs, J.A. 1973: The distribution of schizo-phrenics in Nottingham. *Transactions of the Institute of British Geographers* 59, 55–76.

Giggs, J.A. 1983: Schizophrenia and ecological structure in Nottingham. In McGlashan, N.D. and Blunden, J.R. (eds), *Geographical Aspects of Health*. London: Academic Press, pp. 197–222.

Gilbert, A. 1982: Urban and regional systems: a suitable case for treatment. In Gilbert, A. and Gugler, J: *Cities, Poverty and Development*. Oxford: Oxford University Press. Chapter 8, 162–97.

Girouard, M. 1985: *Cities and people. A social and architectural history*. New Haven: Yale University Press.

Glade, W. 1969: *The Latin American economics: a study of their institutional evolution*. New York.

Glendinning, M. and Muthesius, S. 1994: *Tower blocks: modern public housing in England, Scotland, Wales and Northern Ireland*. Yale: Yale University Press.

Gober, P., Mc Hugh, K.E. and Reid, N. 1991: Phoenix in flux: household instability and neighbourhood change. *Annals of the Association of American Geographers*. 81, 80–8.

Goddard, J. 1968: Multivariate analysis of office location patterns in the city centre, a London example. *Regional Studies*, 2, 69–85.

Goddard, J. 1970: Functional regions within the city centre: a study by factor analysis of taxi flows in central London. *Transactions of the Institute of British Geographers* 49, 161–82.

Golledge, R.G. 1960: Sydney's metropolitan fringes: a study in urban–rural relations. *Australian Geographer* 7, 243–55.

Golledge, R.G. 1970: Some equilibrium models for consumer behaviour. *Economic Geography* 46, 417–25.

Golledge, R.G. 1993: Geography and the disabled: a survey with special reference to vision impaired and blind populations. *Transactions of the Institute of British Geographers*. New Series 18, 63–85.

Golledge, R.G., Rushton, G. and Clark, W.A.Y. 1966: Some spatial characteristics of Iowa's dispersed farm population and their implication for the grouping of central place functions. *Economic Geography* 42, 261.

Goodey, B. *et al.* 1971: *City-scene: an exploration into the image of central Birmingham*. University of Birmingham, Centre for Urban and Regional Studies. Research memorandum No. 10.

Gordon, G. 1980: *Urban morphology – structure and process*. University of Strathclyde, Dept. of Geography. Research Seminar Series No. 10.

Gottmann, J. 1961: *Megalopolis. The urbanized northeastern seaboard of the United States*. New York: Twentieth Century Fund.

Gradman, R. 1916: Schwabische städter. *Zeitschrift Ges. Erdkunde* Berlin:

Gray, F. 1976: Selection and allocation in council housing. *Transactions of the Institute of British Geographers*. New Series No. 1, 34–46.

Green, F.H.W. 1950: Urban hinterlands in England and Wales, *Geographical Journal* 116, 64–81.

Green, L.P. 1959: *Provincial metropolis*. London: Allen and Unwin.

Grimm, F. 1984: The settlement system of the German Democratic Republic: its structure and development. In Bourne L.S., Sinclaire R. and Dziewonski, K. (eds), *Urbanization and settlement systems. International perspectives*. Oxford: Oxford University Press.

Grodzins, M. 1958: *The metropolitan area as a racial problem*. Pittsburgh: University of Pittsburgh Press.

Gugler, J. (ed.). 1988: *The urbanization of the third world*. Oxford: Oxford University Press.

Gugler, J. and Flanagan, W.G. 1978: *Urbanization and social change in West Africa*. Cambridge: Cambridge University Press.

Guiliano, G. 1986: Land use impacts of transportation investments: highway and transit. In Hanson, S. (ed.) *The geography of urban transportation*. New York: The Guilford Press. Chap. 11, 247–279.

Gutkind, E.A. 1964–71: *International history of city development*. 1. *Central Europe (1964)*; 2. *The Alpine and Scandinavian Countries (1965)*; 3. *Southern Europe, Spain and Portugal (1969)*; 4. *Southern Europe, Italy and Greece (1969)*; 5. *Western Europe, France and Belgium (1970)*; 6. *Great Britain and the Netherlands (1971)*. New York: The Free Press of Glencoe. 7. East Central Europe. Poland, Czechoslovakia and Hungary (1972a); 8. Eastern Europe, Romania and the USSR (1972b).

Guttenberg, A.Z. 1960: Urban structure and urban growth. *Journal of the American Institute of Planners* 25, 104.

Guy, C.M. 1988: Information technology and retailing: the implications for analysis and forecasting. In Wrigley N. (ed.), *Store choice, store location and market analysis*. London and New York: Routledge.

Hadden, J.K. and Borgatta, E.F. 1965: *American cities; their social characteristics*. Chicago: Rand McNally and Co.

Hagerstrand, T. 1952: The propagation of innovation waves. Lund Studies. Geography, Series B. *Human Geography* 4, 16.

Hagerstrand, T. 1970: What about people in regional science? *Regional Science Association Papers* 24, 7–21.

Hagerstrand, T. 1975: Space, time and human condition. In Karlqvisi, A., Lundqvist, L. and Snickers, F, (eds), *Dynamic allocation of urban space*. Farnborough: Saxon House, pp. 3–14.

Haggett, P. and Chorley, R.J. 1967: *Models in geography*. London: Methuen.

Haggett, P. 1965: *Locational analysis in human geography*. London: Edward Arnold, Fig. 5.4, p. 119.

Haig, R.M. 1926: Toward an understanding of the metropolis. *Quarterly Journal of Economics* 40, 421.

Haig, R.M. 1928: *Regional survey of New York, vol. 1. Major economic factors in metropolitan growth and arrangement.* New York:

Hall, J.M. 1990: *Metropolis now.* London and its region: Queen Mary and Westfield College, University of London.

Hall, P. 1973: *The containment of urban England.* 2 vols. London: George Allen and Unwin.

Hall, P. 1977: *The world cities*, 2nd edn. London: Weidenfeld and Nicolson.

Hall, P. 1988: *Cities of tomorrow.* Oxford: Basil Blackwell.

Hall, P. 1991: Moving information: a tale of four technologies. In Brotche, J. *et al., Cities of the 21st century.* Melbourne: Longman Cheshire.

Hall, P. and Markusen, A. 1985: *Silicon landscapes.* Boston: George Allen and Unwin.

Hallett, G. (ed.) 1993: *The new housing shortage. Housing affordability in Europe and the USA.* London: Routledge.

Halvorson, P. 1970: Residential location and the journey to work in Charleston, West Virginia. Cincinnati: University of Cincinnati, Unpublished PhD.

Hamilton, F.E.I. 1967: Models of industrial location. In Chorley, R.J. and Haggett, P. (eds), *Models in geography.* London: Edward Arnold, Ch. 10, pp. 361–417.

Hamilton, F.E.I. 1979: Spatial structure in East European cities. In French, R.A. and Hamilton, F.E.I. (eds), *The socialist city.* Chichester: John Wiley.

Hamilton. H.C. and Falconer W. (ed) 1912: *The geography of Strabo.* London: G. Bell and Son Ltd.

Hamnett, C. 1984: Gentrification and residential location theory: a review and assessment. In Herbert, D.T. and Johnston, R.J. (eds), *Geography and the urban environment. Progress in research and applications.* vol. 6. London: John Wiley, pp. 283–319.

Hamnett, C. 1991: The blind men and the elephant: the explanation of gentrification. *Transactions of the Institute of British Geographers.* New Series 16, 173–89.

Hanson, S. 1986: Dimensions of the urban transportation problem. In Hanson, S. (ed.), *The geography of urban transportation.* New York: The Guilford Press.

Harloe, M. (ed.). 1977: *Captive cities. Studies in the political economy of cities and regions.* London: John Wiley.

Harris, C.D. 1943: A functional classification of cities in the United States. *Geographical Review* 33, 86.

Harris, C.D. and Ullman, E.L. 1945: The nature of cities. *Annals of the American Academy of Political Science* 242, 7–17.

Harrison, J. and Saare, P. 1971: Personal construct theory in the measurement of environmental images. *Environment and Behaviour* 3, 351.

Hartenstein, W. and Staack, G. 1967: Land use in the urban core. In Heinmeijer, W.F., van Hulton, M. and Reilingh, H.D. de Vreis (eds), *Urban core and inner city.* Leiden: E.J. Brill.

Hartshorne, R. 1959: *Perspective on the nature of geography.* London: Rand McNally.

Harvey, D. 1973: *Social justice and the city.* London: Edward Arnold.

Harvey, D. 1974: Class monopoly rent, finance capital and the urban revolution. *Regional Studies* 8, 239–55.

Harvey, D. 1978: The urban process under capitalism: a framework for analysis. *International Journal of Urban and Regional Research* 2, 101–31.

Harvey, D. 1985a: *Consciousness and the urban experience. Studies in the history and theory of capitalist urbanization.* Oxford: Basil Blackwell.

Harvey, D. 1985b: *The urbanization of capital.* Oxford: Basil Blackwell.

Harvey, D. 1989: *The condition of post modernity.* Oxford: Blackwell.

Hassert, K. 1907: *Die städte geographisch betrachtet.* Leipzig: B.G. Teubner.

Haughton, G. 1994: Birthpangs in Utopia: plans for a high-tech sustainable city in Australia. *Geography* 79, 42–52.

Heine-Geldern, R. 1958: Conceptions of state and kingship in southeast Asia. *Southeast Asia Program, Department of Asia Studies*. Data paper 18, Cornell University.

Heinemeijer, W.F. *et al.* (eds) 1967: *Urban core and inner city*. Leiden: E.J. Brill.

Henderson, J. and Karn, V. 1987: *Race, class and state housing: inequality and the allocation of public housing in Britain*. Aldershot: Gower Publishing.

Heraty, M. (ed.) 1991: *Urban transport in developing countries. Lessons in innovation*. London: PTRC Education and Research Services.

Herbert, D.T. 1972: *Urban geography. A social perspective*. Newton Abbot: David and Charles.

Herbert, D.T. 1976a: Social deviance in the city: a spatial perspective. In Herbert, D.T. and Johnston, R.J. (eds), *Social areas in the city*. vol. 2. Chichester: John Wiley, Ch. 3, pp. 89–122.

Herbert, D.T. 1976b: The study of delinquency areas: a social geographical approach. *Transactions of the Institute of British Geographers*. New Series 1, 472–92.

Herbert, D.T. 1977: Crime, delinquency and the urban environment. *Progress in Human Geography* 1, 208–39.

Herbert, D.T. 1979: Urban crime: a geographical perspective. In Herbert, D.T. and Smith, D.M. (eds), *Social problems and the city. Geographical Perspectives*. Oxford: Oxford University Press, Ch. 7, 117–38.

Herbert, D.T and Johnston, R.J. 1978: *Social areas in cities. Processes, patterns and problems*. Chichester: John Wiley.

Herington, J. 1984: *The outer city*. London: Harper and Row.

Hill, F. and Michelson, W. 1981: Towards a geography of urban children and youth. In Herbert, D.T. and Johnston, R.J. (eds), *Geography and the urban environment*. vol. IV. Chichester: John Wiley, Ch. 6, pp. 193–228.

Hiorns, F.R. 1956: *Town building in history*. London: Harrup.

HMSO 1987: Housing: The Government's Proposals. London: HMSO C214.

HMSO 1988: Action for cities. London: HMSO.

HMSO 1989: Progress in cities. London: HMSO.

HMSO 1990: Action for cities. London: HMSO.

HMSO 1991: Housing and construction statistics. London: HMSO.

Hodder, B.W. 1961: Rural periodic day markets in part of Yorubaland. *Transactions of the Institute of British Geographers* 29, 141–59.

Holcomb, B. 1986: Geography and urban women. *Urban Geography* 7, 448–56.

Holly, B.F. 1987: Regulation, competition and technology: the restructuring of the US commercial banking system. *Environment and Planning*, 19, 633–52.

Horvath, R.V. 1972: A definition of colonialism. *Current Anthropology* 13, 45–57.

Horwood, E.M. and Boyce, R.R. 1959: *Studies of the central business district and urban freeway development*. Seattle: University of Washington Press.

Hoselitz, B. 1955: Generative and parasitic cities. *Economic Development and Cultural Change* 3, 278.

Hoskins, W.G. 1935: *Industry trade and people in Exeter, 1688–1800*. Manchester: Manchester University Press.

Hourihan, K. 1991: Culture, politics and the recent urbanization of the republic of Ireland. In Bannon, M.J. *et al.* (eds), *Urbanization and urban development*. Dublin: Services Industries Research Centre, University College Dublin, Ch. 13, pp. 141–50.

Howard, E. 1965: *Garden cities of tomorrow*. Edited with a preface by F.J. Osborn. London: Faber.

Hoyt, H.H. 1939: *The structure and growth of residential neighborhoods in American cities*. Washington: Federal Housing Administration.

Huff, D.L. 1963: A probabilistic analysis of shopping centre trade areas. *Land Economics* 39, 81 .

Hurd, R.M. 1924: *Principles of city land values*. New York: The Record and Guide.

Isard, W. 1956: *Location and space economy*. New York: John Wiley.

Jackson, E.L. and Johnson, D.B. 1991: Geographic implications of mega-malls, with special reference to West Edmonton Mall. *The Canadian Geographer*, 35(3), 226.

Jackson, J.A. 1968: *Social stratification*. Cambridge: Cambridge University Press.

Jackson, R.T. 1987: Commuter mining and the Kidston gold mine: goodbye to mining towns? *Geography* 72(2), 162–4.

Jakle, J.A. and Wheeler, J.O. 1969: The changing residential structure of the Dutch population of Kalamazoo, Michigan. *Annals of the Association of American Geographers* 59, 441–60.

Jefferson, M. 1939: The law of the primate city. *Geography Review* 29, 227.

Jennings, J. 1971: Geographical implications of the municipal housing programme in England and Wales. *Urban Studies* 8, 121–38.

Johnson, D.B. 1991: Structural features of West Edmonton Mall. *The Canadian Geographer* 35(3), 249.

Johnston, R.J. 1977: Urban geography: city structures. *Progress in Human Geography* 1(i), 118–29 and 2(i), 148–52.

Johnston, R.J. 1984: *Residential segregation, the state and constitutional conflict in American urban areas*. London: Academic Press for the Institute of British Geographers.

Jones, E. 1960: *A social geography of Belfast*. London: Oxford University Press.

Jones, E. 1990: *Metropolis*. Oxford: Oxford University Press.

Jones, P.N. 1967: The segregation of immigrant communities in the city of Birmingham. Hull: University of Hull *Occasional Papers in Geography* No. 7.

Jones, P.N. 1970: Some aspects of the changing circumstances of coloured immigrants in Birmingham, 1961–66. *Transactions of the Institute of British Geographers* 50, 199–220.

Jones, P.N. 1976: Some aspects of the changing distribution of coloured immigrants in Birmingham, England. *Annals of the Association of American Geographers* 66, 89–103.

Jones, K.G. and Simmons, J.W. 1987: *Location, location, location: Analysing the Canadian retail environment*. Toronto: Nelson.

Kain, I K. 1987: Housing market discrimination and health suburbanization in the 1980s. In Fabin, G.A. (ed.). *Divided neighbourhoods*. Newbury Parts, CA: Sage Publications

Karn, V. 1993: Remodelling a HAT: including implementation of the Housing Action Trust legislation 1987–92. In Malpass, P. and Means, R. (eds), *Implementing housing policy*. Birmingham: Open University Press.

Keeble, D.E. 1967: Models of economic development. In Chorley, R.J. and Haggett, P. (eds), *Models in geography*. London: Methuen, Ch. 8, pp. 243–302.

Kellett, J.R. 1969: *The impact of railways on Victorian cities*. London: Routledge and Kegan Paul.

Kelly, G.A. 1955: *The psychology of personal constructs*. New York: Norton.

Kerner, O. 1968: *Report of the National Advisory Commission on Civil Disorders*. New York: Bantam Books.

King, A.D. 1976: *Colonial urban development. Culture, social power and environment*. London: Routledge and Kegan Paul.

King, L.J. 1961: A multivariate analysis of the spacing of urban settlements in the United States. *Annals of the Association of American Geographers* 51, 222.

King, L.J. 1962: A quantitative expression of the pattern of urban settlements in selected areas of the USA. *Tijdschrift voor Economische en Sociale Geografie* 50, 1.

King, L.J. 1984: *Central place theory*. Beverley Hills, CA: Sage Publications.

Kirby, D.A. 1992: Employment in retailing: unsociable hours and Sunday trading. Paper presented at the Institute of British Geographers' conference, Swansea.

Klein, H.-J. 1967: The delimitation of the town-centre in the image of its citizens. In Heinemeijer, W.F. *et al.* (eds), *Urban core and inner city*. Leiden: E.J. Brill.

Kneebone, D.C. 1991: Transport characteristics of Chinese cities. In Herity, M. (ed.), *Urban transport in developing countries*. London: PTRC Education and Research Services, pp. 150–4.

Knos, D. 1962: *Distribution of land values in Topeka.* Lawrence: University of Kansas Press.

Knox, P.L. 1991: The restless urban landscape, economic and socioculture change and the transformation of metropolitan Washington DC. *Annals of the Association of American Geographers* 81, 181–204.

Kochethov, A.A. and Listengurt, F.M. 1977: A strategy for the distribution of settlement in the USSR. Aims, problems and solutions. *Soviet Geography Review and Translation* 18, 660–74.

Kowinski,W.S. 1978: The malling of America. *New Times Magazine*, 1 May.

Kristol, I. 1972: An urban civilization without cities. *Washington Post*, 3 December.

Kutter, F. 1973: A model for individual travel behaviour. *Urban Studies* 10, 223–58.

Lakshamanan,T.R. and Hansen,W.G. 1965: A retail market potential model. *Journal of the American Institute of Planners* 31, 134.

Lambert, C. 1976: *Building societies, surveyors and the older areas of Birmingham.* Centre of Urban and Regional Studies, University of Birmingham. Working Paper No. 38, 30–5.

Lampard, E.E. 1955: The history of cities in economically advanced areas. *Economic Development and Cultural Change* 3, 92.

Lampard, E.E. 1968: The evolving system of cities in the United States. In Perloff, H.S. and Wingo, L. Jr. (eds), *Issues in urban economics*. Baltimore: Johns Hopkins Press, pp. 81–140.

Langton, J. 1975: Residential patterns in the pre-industrial cities; some case studies from seventeenth century Britain. *Transactions of the Institute of British Geographers* 65, 1–28.

Larkham, P.J. and Jones, A.M. 1991: *A glossary of urban form*. Institute of British Geographers. Historical Geography Research Series, 26.

Lavedan, P. 1952: *Histoire de l'urbanisme. Epoque contemporaine.* Paris: Henri Laurens.

Lavedan, P. 1959: *Histoire de l'urbanisme. Renaissance et temps modernes*. Paris: Henri Laurens.

Lavedan, P. 1926: *Histoire de l'urbanisme. Antiquité.* Paris: Henri Laurens. (Reprinted 1966).

Lavedan, P. and Hugueney, J. 1966: *Histoire de l'urbanisme. L'urbanisme au moyen age*. Geneva: Droz.

Lawless, P. and Brown, F. 1986: *Urban growth and change in Britain: an introduction.* London: Harper & Row.

Lawton, M.P. 1974: The human being and the institutional building. In Lang, J. *et al.* (eds), *Designing for human behavior. Architecture and the Behavioral Sciences*. Strondsburg, Penn.: University of Pennsylvania Press.

LeBeau, R. 1975: *La Suisse*. Paris: Musson.

Le Corbusier. 1947: *Concerning town planning.* Translated by C. Entwistle. London: Architectural Press.

Lee, R. 1992: London's docklands: the 'exceptional place'? An economic geography of inter-urban competition. In Ogden, P. (ed.), *London docklands: the challenge of development*. Cambridge: Cambridge University Press.

Lewis, C.R. 1970: The central place patterns of mid-Wales and the middle Welsh borderland. In Carter, H. and Davies, W.K.D. (eds), *Urban essays: studies in the geography of Wales*. London: Longman, Ch. 10, pp. 228–68.

Lewis, P. 1983: The galactic metropolis. In Platt, R.H. and Macinko, W. (eds), *Beyond the urban fringe*. Minneapolis: University of Minnesota Press, pp. 23–49.

Ley, D. 1986: Alternative explanations for inner city gentrification: a Canadian assessment. *Annals of the Association of American Geographers* 76, 521–35.

Ley, D. and Cybriwsky, R. 1974: Urban graffiti as territorial markers. *Annals of the Association of American Geographers* 64, 491–505.

Light, I. 1988: Los Angeles. In Doyan, M. and Kasarda, J. D. (eds), *The Metropolis Era. Vol. 2 Mega-cities.* Newbury Park, CA: Sage Publications, Chapter 2, 56–96.

Likert, R. 1932: A technique for the measurement of attitudes. *Archives of Psychology* 140, 299–336.

Lindau, L.A. 1991: Bus priority systems in Brazil: from theory to practice. In Herity, M. (ed), *Urban transport in developing countries.* London: PTRC Education and Research Services, pp. 209–17.

Linsky, A.S. 1965: Some generalizations concerning primate cities. *Annals of the Association of American Geographers* 55, 506.

Linton, D.L. 1958: Millionaire cities today and yesterday. *Geography* 43(4), 253–8.

Little, J., Peake, L. and Richardson, P. 1988: *Women in cities. Gender and the urban environment.* Basingstoke: Macmillan Educational.

Liu, B.-C. 1976: *Quality of life indicators in US metropolitan areas: a statistical analysis.* New York: Praeger.

Llewelyn-Davies, Weeks, Forester, Walker and Bor, 1976: *Inner area study, Birmingham. Housing policies for the inner city.* London: HMSO.

Lloyd, D.W. 1984: *The making of English towns. A vista of 2000 years.* London: Gollancz.

Lobel, M.D. 1969: *Historic towns. Maps and plans of towns and cities in the British Isles, with historical commentaries from earliest times to 1800.* Oxford: Lovell Johns-Cook, Hammond and Kell Organization.

Loewenstein, L.K. 1963: The location of urban land uses. *Land Economics* 39, 406–20.

London, B. and Flanagan, W.G. 1976: Comparative urban ecology: a summary of the field. In Walton. J. and Masotti, L.H. (eds), *The city in comparative perspective,* New York: Sage, Ch. 3, pp. 41–66.

Lord, J.D. 1984: Shifts on the wholesale trade status of US Metropolitan Areas. *Professional Geographer* 36, 49–64.

Lord, J.D. and Guy, C.M. 1991: Comparative retail structure of British and American cities: Cardiff (UK) and Charlotte (USA). *International Review of Retail, Distribution and Consumer Research* 1, 391–436.

Lorenz, K. 1961: *On aggression.* New York: Harcourt Brace and World.

Los Angeles Department of City Planning. 1971: *The visual environment of Los Angeles.* Los Angeles: Department of City Planning.

Lösch, A. 1954: *The economics of location.* Newhaven: Yale University Press.

Lukermann, F. 1966: Empirical expressions of nodality in a circulation manifold. *East Lakes Geographer* 2, 20.

Lyall, K. 1982: A bicycle built-for-two: public–private partnership in Baltimore. In Foster, R.S. and Berger, R.A. (eds), *Public–private partnership in American cities.* Lexington, Mass.: D.C. Heath and Co., Lexington Books, Ch. 2, pp. 17–58.

Lynch, K. 1960: *The image of the city.* Cambridge, Mass.: MIT Press.

Macdonald, R. 1968: *The instant energy.* London: Collins.

McDowell, L. 1985: Some gloomy thoughts from Britain: a response to Suzanne Mackenzie on developing gender sensitive theory. *Women and Environment* 7, 10–11.

Mackeith, M. 1985: The changing high street. *The Planner* 71, 9–12.

McNeil, F.B. 1970: *The psychoses.* London.

Malpass, P. and Means, R. (eds) 1993: *Implementing housing policy.* Birmingham: Open University Press.

Malpass, P. and Murie, A. 1990: *Housing policy and practice,* 2nd edn. Basingstoke: Macmillan.

Malpass, P. and Warburton, M. 1993: The new financial regime for local authority housing. In Malpass, P. and Means, R. (eds), *Implementing housing policy.* Birmingham: Open University Press, Ch. 7, pp. 91–105.

Marinelli, O. 1916: Dei tipi economici dei centri abitati a proposito di alcune citta italiane ed americane. *Rivista Geographia Italia* 23, 413.

Marshall, J.U. 1989: *The structure of urban systems.* Toronto: University of Toronto Press.

Martyn, W.F. 1793: *The geographical magazine: or new system of geography.* London: Harrison and Co.

Maruyama, M. 1963: The second cybernetics: deviation amplifying mutual causal processes. *General Systems* 8, 233.

Masser, I. and Gould, W.T.S. 1975: *Inter-regional migration in tropical Africa*. London: Institute of British Geographers. Special Publication No. 8.

Mattila, J.M. and Thompson, W.R. 1955: The measurement of the economic base of the metropolitan area. *Land Economics* 31, 215.

Mayer, H.M. 1954: Urban geography. In James P. and Jones C.F. (eds), *American geography; inventory and prospect*. Syracuse, NY: Syracuse University Press.

Mayer, H.M. 1967: A survey of urban geography. In Hauser, P.M. and Schnore, L.F. (eds), *The study of urbanization*. New York: John Wiley, Ch. 3, pp. 81–114.

Mayer, A.J. and Hoult, T.F. 1962: *Race and residence in Detroit*. Wayne State University, Institute for Regional and Urban Studies.

Medvedkov, Y.V. 1967a: The concept of entropy in settlement pattern analysis. *Papers of the Regional Science Association* 18, 165.

Medvedkov, Y.V. 1967b: The regular component of settlement patterns as shown on maps. *Soviet Geography* 8, 50.

Meier, R.L. 1962: *A communications theory of urban growth*. Cambridge, Mass.: MIT Press.

Mercer, C. 1975: *Living in cities. Psychology and the urban environment*. Harmondsworth: Penguin Books.

Merrett, S. 1979: *State housing in Britain*. London: Routledge and Kegan Paul.

Merrett, S. and Gray, F. 1982: *Owner-occupation in Britain*. London: Routledge and Kegan Paul.

Meuriot, P. 1911: De la mesure des agglomerations urbains. *Bulletin of the Institute of International Statistics* 19, 158.

Michelson, W. 1977: *Environmental choice, human behaviour and residential satisfaction*. New York: Oxford University Press.

Midwest Research Institute 1975: *The quality of life in the US 1970. Index rating as statistics*. Kansas City: University of Kansas.

Minford, P., Peel, M. and Ashton, P. 1987: *The housing morass. Regulation, immobility and underemployment*. London: Institute of Economic Affairs.

Mingione, E. 1981: *Social conflict and the city*. Oxford: Basil Blackwell.

Ministry of Housing and Local Government 1962: *Residential areas: higher densities*. London: Ministry of Housing and Local Government Planning Bulletin 2.

Ministry of Housing and Local Government 1970: *Families living at high density*. Design Bulletin 21, 33.

Mitchell, J.C. (ed.). 1969: *Social networks in urban situations*. Manchester: Manchester University Press.

Montague, A. 1968: *Man and aggression*. London: Oxford University Press.

Morgan, B.S. 1976: The basis of family status segregation: a case study in Exeter. *Transactions of the Institute of British Geographers*. New Series 1, 83–107.

Morgan, W.T.W. 1961: A functional approach to the study of office distribution. *Tijdschrift voor Economische en Sociale Geografie* 52, 207–10.

Moriarty, B.M. 1970: A test of alternative hypotheses of urban residential growth. *Proceedings of the Association of American Geographers* 2, 97–101.

Morrill, R. 1962: Simulation of central place patterns over time. In Norborg, K. (ed.), *IGU, Symposium in Urban Geography. Lund, 1960*. Lund: The Royal University of Lund, Sweden, Dept. of Geography, pp. 109–20.

Morrill, R. 1965a: *Migration and the spread and growth of urban settlement*. Lund: The Royal University of Lund, Sweden, Dept. of Geography.

Morrill, R. 1965b: The Negro ghetto: problems and alternatives. *Geographical Review* 55, 339–61.

Morris, A.E.J. 1972: *History of urban form. Prehistory to the Renaissance*. London: George Godwin.

Morris, D. 1967: *The naked ape*. New York: McGraw-Hill.

Morris, T. 1957: *The criminal area: a study in social ecology*. London: Routledge and Kegan Paul.

Moser, C.A. and Scott, W. 1961: *British towns. a statistical study of their social and economic differences*. Edinburgh and London: Oliver and Boyd.

Moss, M.L. 1991: The information city in the global economy. In Brotche, J. *et al*. (eds), *Cities of the 21st century*. Melbourne, Cheshire, Longman.

Muller, P.O. 1981: *Contemporary surburban America*. Englewood Cliffs, NJ: Prentice-Hall.

Mumford, L. 1938: *The culture of cities*. London: Secker and Warburg.

Mumford, L. 1961: *The city in history*. London: Secker and Warburg.

Munton, R. 1983: *London's green belt: containment in practice*. London: George Allen and Unwin.

Murdie, R.A. 1965: Cultural differences in consumer travel. *Economic Geography* 41, 211.

Murdie, R.A. 1969: *The factorial ecology of Toronto*. Chicago: Department of Geography, University of Chicago, Research Paper No. 116.

Murphy, R.E. 1972: *The central business district*. London: Longman.

Murphy, R.E. and Vance, J.E. 1954a: Delimiting the CBD. *Economic Geography* 30, 189–222.

Murphy, R.E. and Vance, J.E. 1954b: A comparative study of nine central business districts. *Economic Geography* 30, 301–36.

Murphy, R.E., Vance J.E. and Epstein, B.J. 1955: Internal structure of the CBD. *Economic Geography* 31, 21–46.

Nairn, I. 1955: *Outrage*. London: The Architectural Press.

Nash-Williams, V.E. 1954: *The Roman frontier in Wales*. Cardiff: University of Wales Press.

Nelson, H.J. 1955: A service classification of American cities. *Economic Geography* 31, 189.

Nelson, N. 1988: How women and men get by: the sexual division of labour in a Nairobi squatter settlement. In Gugler, J. (ed.), *The urbanization of the third world*. Oxford: Oxford University Press, Ch. 11, pp. 183–203.

Nelson, R.L. 1958: *The selection of retail locations*. New York: Dodge.

Nemeth, R.J. and Smith, D.A. 1985: The political economy of contrasting urban hierarchies in South Korea and the Philippines. In Timberlake, M. (ed.), *Urbanization in the world economy*, Orlando, Fla.: Academic Press, Ch. 9, pp. 183–206.

Newman, D. 1985: Integration and ethnic spatial concentration: the changing distribution of the Anglo-Jewish community. *Transactions of the Institute of British Geographers*. New Series 10, 360–76.

Newman, D. 1987: Data collection and the identification of Jewish ethnic community patterns: a reply. *Transactions of the Institute of British Geographers. New Series* 12, 113–15.

Newman, O. 1972: *Defensible space*. New York: Macmillan.

Nicolson, A. 1993: Doughnuts business boom leaves hole in town's heart. *The Sunday Telegraph*, 28 November, p. 22.

O'Brien, L. and Harris, F. 1991: *Retailing, shopping, society, space*. London: David Fulton.

Oehme, R. 1965: *Old European cities*. Hamburg: Johannes Asmus Verlag.

Office of Population Censuses and Surveys (OPCS) 1981: *Census 1981. Preliminary report for towns*. London: HMSO.

Office of Population Censuses and Surveys (OPCS) 1984: *Census 1981. Key statistics for urban areas*. London: HMSO.

Olsson, G. 1967: central place systems, spatial interaction and stochastic processes. *Regional Science Association Papers* 18, 13.

Osgood, C., Sucie, G. and Tannenbaum, P. 1957: *The measurement of meaning*. Urbana, Illinois: University of Illinois Press.

Pahl, R. 1965: *Urbs in rure. The metropolitan fringe in Hertfordshire*. London: London School of Economics, Geographical Papers 2.

Pahl, R. 1968: The rural–urban continium. In Pahl, R. (ed.) *Readings in urban sociology*. Oxford: Pergamon Press.

Pahl, R. 1970: *Whose city?* London: Longman.

Palm, R. 1976a: Real estate agents and geographical information. *Geographical Review* 66, 266–80.

Palm, R. 1976b: The role of real estate agents as information mediators in two American cities. *Geografiska Annaler, Series B* 58, 28–41.

Park, R.E. and Burgess, E.W. 1925: *The city*. Chicago: University of Chicago Press.

Pas, E.I. 1986: The urban transportation planning process. In Hanson, S. (ed.), *The geography of urban transportation*. New York: The Guilford Press, Ch. 3, pp. 49–70.

Paz, O. 1967: *The labyrinth of solitude: life and thought in Mexico*. Translated by Kemp, L. London: Allen Lane.

Pelliot, P. 1951: *Mémoires sur les coutumes du Camboge de Tcheon Ta-Kouan*. Paris.

Penny, N.J. and Broom, D. 1988: The Tesco approach to store location. In N. Wrigley (ed.), *Store choice, store location and market analysis*, London and New York: Routledge, Ch. 4, pp. 106, 119.

Petersen, G.L. 1967: A model of preference: quantitative analysis of the perception of the visual appearance of residential neighbourhoods. *Journal of the Regional Science Association* 7, 19–31.

Philbrick, A.K. 1957: Principles of areal functional organization in regional human geography. *Economic Geography* 33.

Phillips, J.C. and Veal, A.J. 1979: *Research on the urban fringe*. Birmingham, UK: Centre for Urban and Regional Studies.

Phythian-Adams, C. 1979: *Desolation of a city – Coventry and the urban crisis of the late Middle Ages*. Cambridge: Cambridge University Press.

Piche, D. 1981: The spontaneous geography of the urban child. In Herbert, D.T. and Johnson, R.J. (eds), *Geography and the urban environment*. vol. IV. Chichester: John Wiley, Ch. 7, 229–56.

Pinkerton, J. 1807: *Modern geography; a description of the empires, kingdoms, states and colonies; with the oceans, seas and islands: in all parts of the world*, 2 vols. London: T. Cadell and W. Davies.

Pinkney, A. 1984: *The myth of black progress*. Cambridge: Cambridge University Press.

Plane, D.L. 1981: The geography of urban commuting fields. *The Professional Geographer* 33, 182–8.

Popper, F.J. 1981: *The politics of land use reform*. Madison: University of Winsconsin Press.

Precedo Ledo, A. 1989: Tertiary structure of the Spanish system of cities. In Bourne, L.S. *et al.* (eds), *The changing geography of urban systems*, Pamplona: Cervicio de Publicaciones de la Universidad de Navarra, Ch. 2, pp. 39–60.

Pred, A.R. 1964: The intra-metropolitan location of American manufacturing. *Annals of the Association of American Geographers* 54, 165–80.

Pred, A.R. 1966: *The spatial dynamics of the US urban-industrial growth 1800–1914: interpretive and theoretical essays*. Cambridge, Mass.: M.I.T. Press.

Pred, A.R. 1973: Urbanization, domestic planning problems and Swedish geographic research. *Progress in Geography* 5, 37.

Pred, A.R. 1977: *City systems in advanced economies*. London: Hutchinson.

Preston, R.E. and Griffin, D.W. 1966: A restatement of the transition zone concept. *Annals of the Association of American Geographers* 56, 339–50.

Preston, R.E. and Mitchell, C.J.A. 1989: Notes on a combined economic base – central place theory and the study of urban systems. In Bourne, L.S. *et al.* (eds), *The changing geography of urban systems*. Pamplona: Cervicio de Publicaciones de la Universidad de Navarra, Ch. 4, pp. 87–102.

Preston, V. and Taylor, S.M. 1981: Personal construct theory and residential choice. *Annals of the Association of American Geographers* 71, 437–51.

Pringle, D.G. 1991: Urban growth and economic change in the Republic of Ireland, 1971–1981. In Bannon, M.J. *et al.* (eds), *Urbanization and urban development*. Dublin: Services Industries Research Centre, University College Dublin, Ch. 14, pp. 151–62.

Proudfoot, M.J. 1937a: City retail structure. *Economic Geography* 13, 425.

Proudfoot, M.J. 1937b: The outlying business centers of Chicago. *Journal of Land and Public Utility Economics* 13, 57–70.

Pullan, B. 1971: *The rich and poor in Renaissance Venice*. Oxford: Oxford University Press.

Pumain, D. and Saint-Julien, T. 1989: Recent changes in the French urban system. In Bourne, L.S. *et al.* (eds), *The changing geography of urban systems*. Pamplona: Cervicio de Publicaciones de la Universidad de Navarra, Ch. 13, pp. 239–50.

Punter, J. 1992: Classic carbuncles and mean streets: contemporary urban design and architecture in central London. In Thornley, A, (ed.), *The crisis of London*. London: Routledge. Ch. 6, pp. 69–89.

Purdom, C.B. 1925: *The building of satellite towns*. London: J.M. Dent.

Quinn, J.A. 1939: The nature of human ecology: re-examination and re-definition. *Social Forces* 18, 166.

Quinn, J.A. 1950: *Human ecology*. New York: Prentice-Hall.

Racine, J.-B. and Cunha, A. 1989: Invariance and structural changes in the Swiss urban system: the leading role of tertiary activities. In Bourne, L.S. *et al.* (eds), *The changing geography of urban systems*. Pamplona: Cervicio de Publicaciones de la Universidad de Navarra.

Rainwater, L. 1970: *Beyond ghetto walls: black family life in a federal slum.* Chicago: University of Chicago Press.

Rannells, J. 1956: *The core of the city: a pilot study of changing land uses in central business districts.* New York: Columbia University Press.

Ratcliff, R.V. 1949: *Urban land economics*. New York: McGraw-Hill.

Ravetz, A. 1980: *Remaking cities. Contradictions of the recent urban environment*. London: Croom Helm.

Report of the Select Committee on the Health of Towns. 1840: Volume IV. London: British Parliamentary Papers, XI.

Reps, J.W. 1965: *The making of urban America. A history of urban planning in the United States.* Princeton, NJ: Princeton University Press.

Reynolds, F. 1986: *The problem housing estate*. Aldershot: Gower Publishing.

Reynolds, J. 1992: International perspectives in shopping centre developments. Paper presented at the Institute of British Geographers' conference, Swansea.

Reynolds, S. 1977: *An introduction to the history of English medieval towns*. Oxford: Clarendon Press.

Roberts, B. 1978: *Cities of peasants*. London: Edward Arnold.

Robson, B.T. 1969: *Urban analysis. A study of city structure: with special reference to Sunderland.* Cambridge: Cambridge University Press.

Robson, B.T. 1973: *Urban growth: an approach.* London: Methuen.

Robson, B.T. 1975: *Urban social areas*. Oxford: Oxford University Press.

Rogerson, R.J. 1989: *Measuring quality of life: methodological issues and problems*. Glasgow: University of Glasgow Applied Population Research Unit.

Rogerson, R.J., Findlay, A.M. and Morris, A.S. 1988: The best cities to live in. *Town and Country Planning* 57, 270–3.

Rowley. G. 1967: The middle order towns of Wales. Cardiff: University of Wales: Unpublished PhD thesis.

Rowley. G. 1993: Prospects for the central business district. In Bromley, R.F. and Thomas, C.J. (eds), *Retail change. Contemporary issues*. London: UCL Press.

Rukavishnikov, V.O. 1978: Ethnospatial aspects of population distribution in cities of Tartaria. *Soviet Sociology* 8, 59–79.

Rushton, G. 1966: Spatial pattern of grocery purchases by the Iowa rural population. *University of Iowa Studies in Business Economics. New Series* 9.

Rushton, G. 1969: Analysis of spatial behaviour by revealed space preference. *Annals of the Association of American Geographers* 59, 391.

Rykwert, J. 1976: *The idea of a town. The anthropology of urban form in Rome, Italy and the ancient world*. London: Faber and Faber.

Sanders, E.P. 1993: *The historical figure of Jesus.* London: Allen Lane.

Santos, M. 1971: *Les villes du tiers monde.* Paris: Editions M.-Th. Genin.

Santos, M. 1979: *The shared space. The two circuits of the urban economy on underdeveloped countries.* Adapted for publication in English by Gerry, C. London and New York: Methuen.

São Paulo Justice and Peace Commission 1978: *São Paulo growth and poverty.* London: The Bowerdean Press/Catholic Institute for International Relations.

Saunders, P.R. 1977: *Housing tenure and class interest.* Brighton: University of Sussex Urban and Regional Studies, Working Paper 6.

Saxenian, A. 1985: The genesis of Silicon Valley. In Hall, P. and Markusen, A. (eds), *Silicon landscapes.* Boston: Allen and Unwin, Ch. 2, pp. 20–34.

Schaffer, R. and Smith, N. 1986: The gentrification of Harlem? *Annals of the Association of American Geographers* 76, 347–65.

Schiller, R. 1988: Retail decentralisation – property view. *Geography Journal* 154, 17–19.

Schnore, L.F. 1965: On the spatial structure of cities in the two Americas. In Hauser, P.M. and Schnore, L.F. (eds), *The study of urbanization.* New York: John Wiley, Ch. 10, pp. 347–98.

Schnore, L.S. and Knights, P.R. 1969: Residence and social structure: Boston in the ante-bellum period. In Thernstrom, S. and Sennett, R. (eds), *Nineteenth century cities; essays in the new urban history.* New Haven: Yale University Press.

Schuler, Y.H.J. 1979: A disaggregated store choice model of spatial decision making. *Professional Geographer* 31, 147–9.

Self, P. 1972: *New towns: the British experience.* London: Charles Knight and Co. for The Town and Country Planning Association.

Semple, R.K. and Golledge, R.G. 1970: An analysis of entropy changes in a settlement pattern over time. *Economic Geography* 46, 167.

Senior, D. 1969: *Memorandum of dissent. Royal Commission on Local Government in England 1966–1969,* vol. 2. London: HMSO.

Sewel, J., Twine, F. and Williams, N. 1984: The sale of council houses. Some empirical evidence. *Urban Studies* 21, 439–50.

Shaw, C.R. 1929: *Delinquent areas; A study of the geographical distribution of school truants, juvenile delinquents and adult offenders in Chicago.* Chicago: Chicago University Press.

Shepherd, P.M. and Rowley, G. 1978: The association of retail functions within cities. *Tijdschrift voor Economische en Sociale Geografie* 69.

Shevky, E. and Bell, W. 1955: *Social area analysis.* Stanford, Calif.: Stanford University Press.

Shevky, E. and Williams, M. 1949: *The social areas of Los Angeles.* Los Angeles: University of California Press.

Sibley, D. 1981: *Outsiders in urban society.* Oxford: Basil Blackwell.

Silk, J.A. 1971: *Search behaviour: general characterization and review of literature in the behavioural sciences.* Reading: University of Reading Geographical Papers 7.

Silk, J.A. 1972: *Comments on 'historical' factors influencing residential choice and associated research problems.* Aberystwyth, IBG Urban Study Group. Social patterns in cities.

Sim, D. 1982: *Change in the city centre.* Aldershot: Gower Publishing.

Sjoberg, G. 1965: *The pre-industrial city past and present.* New York: The Free Press (Collier-Macmillan).

Skellington, R. 1981: How blacks lose out in council housing. *New Society* 55, 187–9.

Skinner, G.W. 1964: Marketing and social structure in rural China. *Journal of Asian Studies* 24, 5–31.

Slater, T.S. 1976: Estate ownership and nineteenth century suburban development. In McWhirr, A. (ed.), *Archaeology and history of Cirencester.* Oxford: British Archaeological Reports, 30.

Slater, T.S. 1981: The analysis of burgage patterns in Medieval towns. *Area* 13, 211–16.

Slater, T.S. 1988: English Medieval town planning. In Deneche, D. and Shaw, G. (eds), *Urban Historical Geography.* Cambridge: Cambridge University Press, Ch. 8, pp. 93–108.

Slater, T.R. 1990a: *The built form of western cities*. Leicester: University of Leicester Press.

Slater, T.R. 1990b: English medieval towns with composite plans: evidence from the Midlands. In Slater T.R. (ed.), *The built form of western cities*. Leicester: University of Leicester Press, Ch. 4, pp. 60–82.

Smailes, A.E. 1944: The urban hierarchy of England and Wales. *Geography* 29, 41.

Smailes, A.E. 1946: The urban mesh of England and Wales. *Transactions of the Institute of British Geographers* 11, 87.

Smailes, A.E. 1947: The analysis and delimitation of urban fields. *Geography* 32, 151.

Smith, C.A. 1976: Regional economic systems: linking geographic models and socio-economic problems. In Smith, C.A. (ed.), *Regional analysis*, vol. 1. New York: Academic Press, Ch. 1, pp. 3–63.

Smith, C.A. 1985a: Theories and measures of urban primacy: a critique. In Timberlake, M. (ed.), *Urbanization in the world-economy*, Orlando, Flo.: Academic Press, Ch. 6, pp. 87–117.

Smith, C.A. 1985b: Class relations and urbanization in Guatemala: toward an alternative theory of urban primacy. In Timberlake, M. (ed.), *Urbanization in the world-economy*, Orlando, Flo.: Academic Press, Ch. 7, pp. 121–67.

Smith, D.M. 1974: *Crime rates as territorial indicators: the case of the United States*. London: Queen Mary College, Occasional Papers, 11.

Smith, D.M. 1989: *Urban inequality under socialism*. Cambridge: Cambridge University Press.

Smith, M.P. 1980: *The city and social theory*. Oxford: Basil Blackwell.

Smith, N. 1979: Toward a theory of gentrification: a back to the city movement by capital not people. *Journal of the American Planning Institute* 45, 538–48.

Smith, R.V., Flory, S.F., Bashshur, R.L. and Shannon, G.W. 1967: *Community interaction and racial integration in the Detroit area: an ecological analysis*. Grand Rapids: Eastern Michigan University.

Smolensky, E. and Ratajczak, D. 1965: The conception of cities. *Explorations in Entrepreneurial History, Second series* 2, 90–131.

Soja, E., Morales, R. and Wolff, G. 1983: Urban restructuring, an analysis of social and spatial change in Los Angeles. *Economic Geography* 59, 195–230.

Soja, E., Morales, R. and Wolff, G. 1989: Urban manufacturing: an analysis of social and spatial change in Los Angeles. In Beauregard, R.A. (ed.), *Atop the Urban Hierarchy*. Totowa, NJ: Roland and Litchfield, Ch. 3, 87–122.

Sombart, W. 1902: *Der moderne Kkapitalismus*. 3 vols. Leipzig: Duncker and Humbolt.

Spate, O.H.K. 1948: The growth of London AD 1660-1800. In Darby, H.C. (ed.), *An historical geography of England before 1800*. Cambridge: Cambridge University Press, Ch. XIV, pp. 529–48.

Spate, O.H.K. and Ahmad, E. 1950: Five cities of the Gangetic plain. *Geographical Review* 40, 260–78.

Stanislawski, D. 1946: The origin and spread of the grid pattern town. *Geographical Review* 36, 105–20.

Stewart, C.T. 1958: The size and spacing of cities. *Geographical Review* 48, 222.

Stewman, S. and Tarr, J.A. 1982: Four decades of public–private partnerships in Pittsburg. In Foster, R.S. and Berger, R.A. (eds), *Public–private partnerships in American cities*. Lexington: D.C. Heath and Co., Ch. 3, pp. 59–128.

Stone, P.A. 1959: The economics of housing and urban development. *Journal of the Royal Statistical Society. Series A (General)* 122, 426–36.

Sudjic, D. 1993: *The 100 mile city*. London: Flamingo/Harper Collins.

Suttles, G.D. 1968: *The social order of the slum, ethnicity and territory in the inner city*. Chicago: University of Chicago Press.

Suttles, G.D. 1972: *The social structure of communities*. Chicago: University of Chicago Press.

Talbot, P.A. 1926: *The Peoples of Southern Nigeria*. London.

Taylor, G. 1949: *Urban Geography*. London: Methuen.

Theodorson, G.A. 1961: *Studies in human ecology*. New York: Harper and Row.

The Sunday Times 1967: Why so few can have the houses they want. London: *The Sunday Times*, 21 May.

Thompson, I. 1990: The bottom of the heap. *Sunday Times Magazine*. London: *The Sunday Times*.

Thompson, K.J. 1981: *Farming in the fringe*. Cheltenham: Countryside Commission, CCP 142.

Thompson W.R. 1965: *A preface to urban economics*. Baltimore: John Hopkins Press.

Thompson, W.R. 1968: Internal and external factors in the development of urban economies. In Perloff, S. and Wingo, L. (eds), *Issues in urban economics*. Baltimore: Johns Hopkins Press, pp. 43–63.

Thorndike, E.L. 1939: *Your city*. New York: Harcourt, Brace and Co.

Thorndike, E.L. 1940: *144 smaller cities*. New York: Harcourt, Brace and Co.

Thorpe, D. 1988: The changing high street. *The Geographical Journal* 154(1); 19.

Thurstone, L. and Chave, E. 1929: *The measurement of attitude*. Chicago: University of Chicago Press.

Timberlake, M. 1985: *Urbanization in the world-economy*. Orlando, Fla.: Academic Press.

Timms, D.W.G. 1971: *The urban mosaic*. Cambridge: Cambridge University Press.

Tolley, R. (ed.) 1993: *The greening of urban transport. Planning for walking and cycling in western cities*. London: Belhaven Press.

Turner, J.F.C. 1969: Uncontrolled urban settlement: problems and policies. In Breese, G. (ed.), *The city in newly developing countries: readings on urbanism and urbanization*. Englewood Cliffs, NJ: Prentice-Hall.

Twine, F. and Williams, N. 1983: Social segregation in public sector housing: a case study. *Transactions of the Institute of British Geographers. New Series* 8, 253–66.

Ulack, R. 1978: The role of urban squatter settlements. *Annals of the Association of American Geographers* 68, 535.

Ullman, E.L. and Dacey, M.F. 1962: The minimum requirements approach to the urban economic base. In Norborg, K. (ed.), *Proceedings of the IGU Symposium in Urban Geography, Lund 1960*. Lund: Royal University of Lund, Sweden. Dept. of Geography, pp. 121–43.

United Nations 1955: *United Nations demographic yearbook 1952*. New York: United Nations.

United Nations 1969: *Growth of the world's urban and rural population 1920–2000*. New York: UN Dept. of Economic and Social Affairs, Population Studies 44.

United Nations 1990: *United Nations demographic yearbook 1988*. New York: United Nations.

USA Department of Commerce 1990: *Data Users Guide*. Washington DC: Bureau of Census.

Valentine, G. 1989: The geography of women's fear. *Area* 21, 385–90.

Vapnarsky, C.A. 1969: On rank-size distribution of cities: an ecological approach. *Economic Development and Cultural Change* 17, 584–95.

Varady, D.P., Mantel, S.J. Jr, Hinitz-Washofsky, C. and Halpern, H. 1981: Suburbanization and dispersion, a case study of Cincinnati's Jewish population. *Geographical Research Forum* 3, 5–15.

Vining, D.R. 1985: The growth of core regions in the Third World. *Scientific American* 252(4), 24–31.

von Thunen, J.H. 1826: *Der isolieter staat in beziehung auf landwirtschaft unt nationalokonomie*. Hamburg: Perthes.

Wade, R.C. 1964: *Slavery in the cities. The south, 1820–1860*. London: Oxford University Press.

Wallerstein, I. 1974: *The modern world system*, vol. 1. *Capitalist agriculture and the origins of the European world-economy in the sixteenth century*. New York: Academic Press.

Wallerstein, I. 1979: *The capitalist world economy*. Cambridge: Cambridge University Press.

Wallerstein, I. 1980: *The modern world-system*, vol. 2. *Mercantilism and the consolidation of the European world economy 1600–1750*. New York: Academic Press.

Walmsley, D.J. 1988: *Urban living. The individual in the city*. Harlow: Longman.

Walters, P.B. 1985: Systems of cities and urban primacy: problems of definition and measurement. In Timberlake, M. (ed.), *Urbanization in the world-economy*. Orlando, Fla.: Academic Press, Ch. 5, pp. 63–85.

Ward, D. 1966: The industrial revolution and the emergence of Boston's central business district. *Economic Geography* 42, 152–71.

Warner, S.B. Jr 1968: *The private city. Philadelphia in three periods of its growth*. Philadelphia: University of Pennsylvania Press.

Waterman, S. and Kosmin, B.A. 1987: Residential change in a middle-class suburban ethnic population: a comment. *Transactions of the Institute of British Geographers. New Series* 12, 107–12.

Wavell, S. 1993: Soho is reborn as the centre of the universe. *Sunday Times*, 15 August, 1–7.

Webb, J.W. 1959: Basic concepts in the analysis of small urban centres of Minnesota. *Annals of the Association of American Geographers* 49, 55.

Webber, M. 1964: The urban place and the non-place urban realm. In Webber, M.M. (ed.), *Explorations into urban structure*. Philadelphia: University of Pennsylvania Press.

Weclawowicz, G. 1979: The structure of socio-economic space in Warsaw in 1931 and 1970: a study in factorial ecology. In French, R.A. and Hamilton, F.E.I. (eds), *The socialist city*. Chichester: John Wiley.

Wehrwein, G.S. 1942: Rural–urban fringe. *Economic Geography* 18, 217–28.

Weimer, A.M. and Hoyt, H. 1939: *Principles of real estate*. New York: The Ronald Press.

Western, J. 1981: *Outcast Cape Town*. London: Allen and Unwin.

Wheatley, P. 1963: What the greatness of a city is said to be. *Pacific Viewpoint* 4, 163.

Wheeler, J.O. and Mitchelson, A.L. 1989: Information flows among major metropolitan areas in the United States. *Annals of the Association of American Geographers* 79, 523–43.

Whitehand, J.W.R. 1967: Fringe belts: a neglected aspect of urban geography. *Transactions of the Institute of British Geographers* 41, 223–33.

Whitehand, J.W.R. 1972: Urban rent theory, time series and morphogenesis: an example of eclecticism in geographical research. *Area* 4, 215–22.

Whitehand, J.W.R. 1981: *The urban landscape: historical development and management*. London: Academic Press. IBG Special Publications, 13.

Whitehand, J.W.R. 1989: *Residential development under restraint: a case study in London's rural–urban fringe*. Birmingham: University of Birmingham, School of Geography, Occasional Papers, 28.

Whitehand, J.W.R. 1992a: *The making of the urban landscape*. Oxford: Blackwell. IBG Special Publications, 26.

Whitehand, J.W.R. 1992b: Recent advances in urban morphology. *Urban Studies* 29, 619–36.

Whitelegg, J. 1985: *Urban transport*. Basingstoke: Macmillan Education.

Williams, F.B. 1922: *The law of city planning and zoning*. New York: Macmillan.

Williams, P. 1976: *The role of financial institutions and estate agents in the private housing market: a general introduction*. Birmingham: University of Birmingham Centre for Urban and Regional Studies. Working Paper No. 39.

Williams, P. 1978: Building societies and the inner city. *Transactions of the Institute of British Geographers. New Series* 3, 23–34.

Wilson, A.G. 1977: *Models of cities and regions*. New York: John Wiley.

Wilson, W.J. 1978: *The declining significance of race: blacks and changing American institutions*. Chicago: University of Chicago Press.

Wilson, W.J. 1985: The urban underclass in an advanced industrial society. In Peterson, P.E. (ed.), *The new urban reality*. Washington DC: The Brookings Institution.

Winchester, S. 1974: They've come a long way in St Louis. London: *Guardian*, 15 May, 16.

Wingo, L. 1961: *Transportation and urban land*. Washington:

Wingo, L. 1963: *Cities and space. The future use of urban land*. Baltimore: Johns Hopkins Press.

Wirth, L. 1928: *The ghetto*. Chicago: University of Chicago Press.

Wissink, G.A. 1962: *American cities in perspective; with special reference to their fringe areas*. Assen: Van Gorcen.

Wolf, E.P. 1963: The tipping point in racially changing neighbourhoods. *Journal of the American Institute of Town Planning* 29, 217–22.

Wolpert, J. 1964: The decision process in a spatial context. *Annals of the Society of American Geographers* 54, 558.

Wolpert, J. 1965: Behavioural aspects of the decision to migrate. *Papers Regional Science Association* 15, 159–69.

Wrigley, N. *et al.* 1985: The Cardiff consumer panel: methodological aspects of the conduct of a long-term panel survey. *Transactions of the Institute of Professional Geographers* NS10, 63–76.

Wrigley, N. 1988: *Store choice, store location and market analysis*. London and New York: Routledge.

Yantsky, O. 1986: Urbanization in the USSR theory, tendencies and policies. *International Journal of Urban and Regional Research* 10, 265–87.

Zipf, G.W. 1941: *National unity and disunity*. Bloomington. Ill.: Principia Press.

Zukin, S. 1988: *Loft living. Culture and capital in urban change*. London: Radius (Century Hutchinson).

INDEX